Edited by
Matthias Dehmer,
Frank Emmert-Streib
Armin Graber, and
Armindo Salvador

Applied Statistics
for Network Biology

Related Titles

Emmert-Streib, F., Dehmer, M. (eds.)

Medical Biostatistics for Complex Diseases

2010

ISBN: 978-3-527-32585-6

Dehmer, M., Emmert-Streib, F. (eds.)

Analysis of Complex Networks

From Biology to Linguistics

2009

ISBN: 978-3-527-32345-6

Emmert-Streib, F., Dehmer, M. (eds.)

Analysis of Microarray Data

A Network-Based Approach

2008

ISBN: 978-3-527-31822-3

Junker, B. H., Schreiber, F.

Analysis of Biological Networks

2008

ISBN: 978-0-470-04144-4

Stolovitzky, G., Califano, A. (eds.)

Reverse Engineering Biological Networks

Opportunities and Challenges in Computational Methods for Pathway Inference

2007

ISBN: 978-1-57331-689-7

Quantitative and Network Biology
Series Editors M. Dehmer and F. Emmert-Streib
Volume 1

Applied Statistics for Network Biology

Methods in Systems Biology

Edited by
Matthias Dehmer, Frank Emmert-Streib, Armin Graber,
and Armindo Salvador

The Editors

Matthias Dehmer
UMIT
Institute for Bioinformatics
and Translational Research
Eduard Wallnöfer Zentrum 1
6060 Hall, Tyrol
Austria

Frank Emmert-Streib
Queen's University Belfast
Center for Cancer Research and Cell Biology
97, Lisburn Road
Belfast BT9 7BL
United Kingdom

Armin Graber
UMIT
Institute for Bioinformatics
and Translational Research
Eduard Wallnöfer Zentrum 1
6060 Hall, Tyrol
Austria

and

Novartis Pharmaceuticals Corporation
Oncology Biomarkers and Imaging
One Health Plaza
East Hanover, NJ 07936
USA

Armindo Salvador
University of Coimbra
Center for Neuroscience and
Cell Biology, Department of Chemistry
3004-535 Coimbra
Portugal

Composition Thomson Digital, Noida, India
Printing and Binding betz-druck GmbH, Darmstadt
Cover Design Adam Design, Weinheim

Limit of Liability/Disclaimer of Warranty: While the publisher and author have used their best efforts in preparing this book, they make no representations or warranties with respect to the accuracy or completeness of the contents of this book and specifically disclaim any implied warranties of merchantability or fitness for a particular purpose. No warranty can be created or extended by sales representatives or written sales materials. The Advice and strategies contained herein may not be suitable for your situation. You should consult with a professional where appropriate. Neither the publisher nor authors shall be liable for any loss of profit or any other commercial damages, including but not limited to special, incidental, consequential, or other damages.

Library of Congress Card No.: applied for

British Library Cataloguing-in-Publication Data
A catalogue record for this book is available from the British Library.

Bibliographic information published by the Deutsche Nationalbibliothek
The Deutsche Nationalbibliothek lists this publication in the Deutsche Nationalbibliografie; detailed bibliographic data are available on the Internet at http://dnb.d-nb.de.

© 2011 Wiley-VCH Verlag & Co. KGaA, Boschstr. 12, 69469 Weinheim, Germany

Wiley-Blackwell is an imprint of John Wiley & Sons, formed by the merger of Wiley's global Scientific, Technical, and Medical business with Blackwell Publishing.

All rights reserved (including those of translation into other languages). No part of this book may be reproduced in any form – by photoprinting, microfilm, or any other means – nor transmitted or translated into a machine language without written permission from the publishers. Registered names, trademarks, etc. used in this book, even when not specifically marked as such, are not to be considered unprotected by law.

Printed in the Federal Republic of Germany
Printed on acid-free paper

ISBN: 978-3-527-32750-8

Contents

Preface *XVII*
List of Contributors *XIX*

Part One Modeling, Simulation, and Meaning of Gene Networks *1*

1 **Network Analysis to Interpret Complex Phenotypes** *3*
  Hong Yu, Jialiang Huang, Wei Zhang, and Jing-Dong J. Han
1.1 Introduction *3*
1.2 Identification of Important Genes based on Network Topologies *5*
1.2.1 Degree *5*
1.2.2 Betweenness *6*
1.2.3 Network Motifs *6*
1.2.4 Hierarchical Structure *7*
1.3 Inferring Information from Known Networks *8*
1.3.1 Understanding Biological Functions based on Network Modularity *8*
1.3.2 Inferring Functional Relationships and Novel Functional Genes Through Networks *8*
1.3.3 Unraveling Transcriptional Regulations from Expression Data through Transcriptional Networks *9*
1.3.4 Extracting the Pathway-Linked Regulators and Effectors based on Network Flows *10*
1.4 Conclusions *10*
  References *11*

2 **Stochastic Modeling of Gene Regulatory Networks** *13*
  Tianhai Tian
2.1 Introduction *13*
2.2 Discrete Stochastic Simulation Methods *14*
2.2.1 SSA *15*
2.2.2 Accelerating τ-Leap Methods *16*
2.2.3 Langevin Approach *19*

2.3	Discrete Stochastic Modeling	20
2.3.1	Stochastic Modeling Method	20
2.3.2	Toggle Switch with the SOS Pathway	22
2.3.3	Other Models for the Genetic Toggle Switch	24
2.4	Continuous Stochastic Modeling	26
2.4.1	Deterministic Models for the λ Phage Network	26
2.4.2	Stochastic Models for External Noise	28
2.4.3	Deterministic Models with Threshold Values	29
2.4.4	Stochastic Switching	30
2.5	Stochastic Models for Both Internal and External Noise	31
2.5.1	Stochastic Models for Microarray Gene Expression Data	33
2.6	Conclusions	34
	References	35
3	**Modeling Expression Quantitative Trait Loci in Multiple Populations**	**39**
	Ching-Lin Hsiao and Cathy S. J. Fann	
3.1	Introduction	39
3.1.1	Data Structure in eQTL Studies	39
3.1.2	Current eQTL Studies	40
3.1.2.1	eQTL Studies in a Single Human Population	40
3.1.2.2	eQTL Studies in Multiple Human Populations	43
3.1.3	An Illustrated Example	45
3.1.4	Notations	46
3.2	IGM Method	47
3.2.1	Modeling SNP–GE Association in a Single Population	47
3.2.2	Integrating Hypotheses to Identify Common eQTL	48
3.2.3	Applying the IGM Method to HapMap Data	48
3.2.3.1	Characterizing Putative eQTL Identified by the IGM	49
3.3	CTWM	50
3.3.1	Modeling SNP–GE Association in Pooled Data by CTWM	50
3.3.2	Applying CTWM to HapMap Data	52
3.3.2.1	Characterizing Putative eQTL Identified by CTWM	52
3.3.2.2	Justification of Model Assumptions	53
3.4	CTWM-GS Method	53
3.4.1	Solving Normal Equations in CTWM	54
3.4.2	Estimators of BD and GS	55
3.4.3	Testing BD and GS	56
3.4.4	Applying CTWM-GS to HapMap Data	56
3.4.4.1	Applying the GS to Population Studies	57
3.5	Discussion	60
	References	61

Part Two   Inference of Gene Networks   67

4	**Transcriptional Network Inference Based on Information Theory** 69
	Patrick E. Meyer, Catharina Olsen, and Gianluca Bontempi
4.1	Introduction 69
4.1.1	Notation 69
4.1.2	Formalization 71
4.1.3	Performance Measures in Undirected Network Inference 72
4.1.3.1	Precision–Recall (PR) Curves 73
4.1.3.2	F-Scores 74
4.1.4	Causal Subset Selection 74
4.2	Inference Based on Conditional Mutual Information 76
4.2.1	Constraint-Based Methods 77
4.2.2	Approximated Conditional Mutual Information 78
4.2.3	Variable Selection Algorithms 78
4.3	Inference Based on Pairwise Mutual Information 80
4.3.1	Relevance Network (RELNET) 80
4.3.2	Context Likelihood of Relatedness (CLR) 81
4.3.3	Chow–Liu Tree 81
4.3.4	Algorithm for the Reconstruction of Accurate Cellular Networks (ARACNE) 82
4.3.5	Minimum Redundancy Networks (MRNET) 83
4.4	Arc Orientation 84
4.4.1	Assessing Arc Orientation Methods 86
4.5	Conclusions 87
	References 87
5	**Elucidation of General and Condition-Dependent Gene Pathways Using Mixture Models and Bayesian Networks** 91
	Sandra Rodriguez-Zas and Younhee Ko
5.1	Introduction 91
5.2	Methodology 92
5.2.1	Network Learning Algorithms: Frequentist- and Bayesian MCMC-based algorithms 93
5.3	Applications 95
5.3.1	Elucidation of Gene Networks 95
5.3.2	Discovery of Condition-Dependent Gene Relationships 96
5.3.3	MCMC Mixture Bayesian Network 99
5.3.4	Computational Considerations 101
5.4	Conclusions 101
	References 102

6 Multiscale Network Reconstruction from Gene Expression Measurements: Correlations, Perturbations, and "A Priori Biological Knowledge" *105*

Daniel Remondini and Gastone Castellani

6.1 Introduction *105*
6.1.1 Complex Networks *105*
6.1.2 Gene Interaction Networks from Gene Expression Measurements *107*
6.2 "Perturbation Method" *108*
6.3 Network Reconstruction by the Correlation Method from Time-Series Gene Expression Data *109*
6.4 Network Reconstruction from Gene Expression Data by *A Priori* Biological Knowledge *110*
6.5 Examples and Methods of Correlation Network Analysis on Time-Series Data *112*
6.6 Examples and Methods for Pathway Network Analysis *117*
6.6.1 Gene Selection and Pathway Grouping *118*
6.6.2 Pathway Significance and Pathway Network *118*
6.6.3 Results *121*
6.7 Discussion *126*
6.8 Conclusions *127*
References *128*

7 Gene Regulatory Networks Inference: Combining a Genetic Programming and H_∞ Filtering Approach *133*

Lijun Qian, Haixin Wang, and Xiangfang Li

7.1 Introduction *133*
7.2 Background *134*
7.2.1 Noise in Gene Expression *134*
7.2.2 Modeling of Gene Regulatory Networks with Noise *136*
7.2.2.1 Boolean Networks Model with Noise *136*
7.2.2.2 Bayesian Networks Model with Noise *136*
7.2.2.3 Linear Additive Regulation Model with Noise *137*
7.2.2.4 Neural Networks Model with Noise *137*
7.2.3 Proposed Nonlinear ODE Model with Noise *138*
7.3 Methodology for Identification and Algorithm Description *139*
7.3.1 GP *140*
7.3.2 H_∞ Filter *142*
7.4 Simulation Evaluation *144*
7.4.1 Synthetic Data *144*
7.4.2 Microarray Data *145*
7.5 Conclusions *146*
References *150*

8	**Computational Reconstruction of Protein Interaction Networks** *155*
	Konrad Mönks, Irmgard Mühlberger, Andreas Bernthaler, Raul Fechete,
	Paul Perco, Rudolf Freund, Arno Lukas, and Bernd Mayer
8.1	Introduction *155*
8.1.1	Selecting Relevant Features from Omics Profiles *156*
8.1.2	Analyzing Omics Data on a Network Level *157*
8.2	Protein Interaction Networks *159*
8.2.1	Network Categories *159*
8.2.1.1	Metabolic Networks *159*
8.2.1.2	Paralog Networks *160*
8.2.1.3	Physical Interaction Networks *160*
8.2.2	Parameters for Protein Annotation *161*
8.2.2.1	Gene Expression Profiles *161*
8.2.2.2	Subcellular Location *161*
8.2.2.3	Gene Annotation *161*
8.2.2.4	Transcription Factors *162*
8.2.2.5	MicroRNA *162*
8.2.2.6	Pathways *162*
8.2.3	Data Preparation *163*
8.2.3.1	Integration of Data Sources *163*
8.2.3.2	Obtaining Edge Weights *164*
8.2.3.3	Data Completeness *166*
8.2.3.4	Data Normalization *166*
8.2.4	Deriving Models *166*
8.2.4.1	Basic Considerations *167*
8.2.4.2	Choosing an Algorithm *169*
8.2.5	Validation Procedures *169*
8.2.5.1	Model Performance Evaluation *169*
8.2.5.2	Network Structure Assessment *170*
8.3	Characterization of Computed Networks *171*
8.3.1	Evaluation of the Specific Protein–Protein Interactions *171*
8.3.2	Application of the Specific Protein–Protein Interactions *175*
8.4	Conclusions *177*
	References *178*

Part Three Analysis of Gene Networks *181*

9	**What if the Fit is Unfit? Criteria for Biological Systems Estimation Beyond Residual Errors** *183*
	Eberhard O. Voit
9.1	Introduction *183*
9.2	Model Design *184*
9.3	Concepts and Challenges of Parameter Estimation *187*
9.3.1	Typical Parameter Estimation Problems *190*
9.3.1.1	Data Fit is Unacceptable *190*

9.3.1.2	Differently Structured Candidate Models are Difficult to Compare *191*	
9.3.1.3	Fit is Acceptable, But... *192*	
9.3.1.4	Needed: A Better Fit! Or Not? *195*	
9.4	Conclusions *197*	
	References *198*	

10 Machine Learning Methods for Identifying Essential Genes and Proteins in Networks *201*
Kitiporn Plaimas and Rainer König

10.1	Introduction *201*	
10.2	Definitions and Constructions of the Network *202*	
10.3	Network Descriptors *203*	
10.3.1	Network Topological Features for Undirected Graphs *204*	
10.3.1.1	Connectivity *205*	
10.3.1.2	Clustering Coefficient *205*	
10.3.1.3	Centrality Measures *205*	
10.3.2	Network Topological Features for a Bipartite Graph of Metabolic Networks *206*	
10.3.2.1	Stoichiometric Properties *206*	
10.3.2.2	Chokepoints *207*	
10.3.2.3	Load Scores *207*	
10.3.2.4	Deviations *207*	
10.3.2.5	Damage in Global Networks *208*	
10.4	Machine Learning *208*	
10.5	Some Examples of Applications *210*	
10.5.1	Validating an Experimental Knock-Out Screen *210*	
10.5.2	Training with Data from One Organism to Predict Essential Genes for Another Organism *211*	
10.5.3	Further Reported Investigations *211*	
10.6	Conclusions *212*	
	References *213*	

11 Gene Coexpression Networks for the Analysis of DNA Microarray Data *215*
Matthew T. Weirauch

11.1	Introduction *215*	
11.2	Background *216*	
11.2.1	Gene Transcription *216*	
11.2.2	DNA Microarrays *217*	
11.2.3	Networks *218*	
11.3	Construction of GCNs *218*	
11.3.1	Data Format and Representation *219*	
11.3.2	Calculating Pairwise Gene Scores *219*	
11.3.2.1	Overview *219*	

11.3.2.2	Euclidean Distance *220*
11.3.2.3	Mutual Information *220*
11.3.2.4	Pearson's Correlation Coefficient *221*
11.3.2.5	Spearman's Rank Correlation Coefficient *221*
11.3.3	Choosing a Threshold *222*
11.3.3.1	Simple Thresholding Methods *222*
11.3.3.2	Fisher's Z-Transformation *222*
11.3.3.3	Permutation-Based Thresholds *223*
11.3.3.4	Other Methods *223*
11.4	Integration of GCNs with Other Data *224*
11.4.1	Integration of Multiple Expression Datasets *225*
11.4.1.1	Integrating Data within a Species *226*
11.4.1.2	Integrating Data across Species *226*
11.4.2	Integration of Heterogeneous Data Sources *227*
11.4.2.1	Union and Intersection-Based Methods *227*
11.4.2.2	Probabilistic Methods *228*
11.4.2.3	Integration of Expression Data with Specific Other Data Types *229*
11.5	Analysis of GCNs *230*
11.5.1	Network Hubs *231*
11.5.2	Network Motifs *232*
11.5.3	Gene Modules *232*
11.5.3.1	Gene Vector Clustering Methods for the Identification of Gene Modules *233*
11.5.3.2	Network-Based Methods for the Identification of Gene Modules *233*
11.6	GCNs for the Study of Cancer *237*
11.7	Conclusions *237*
	References *238*
12	**Correlation Network Analysis and Knowledge Integration** *251*
	Thomas N. Plasterer, Robert Stanley, and Erich Gombocz
12.1	Introduction *251*
12.2	Systems Biology Data Quandaries *252*
12.3	Semantic Web Approaches *252*
12.4	Correlation Network Analysis *253*
12.4.1	Selecting Nodes and Edges for Networks *255*
12.4.2	Distributions of Correlation Statistics *258*
12.5	Knowledge Annotation for Networks *259*
12.5.1	HRP and the Paired-Plaque Study *260*
12.5.2	Annotation with Public Sources and Ontologies *261*
12.5.3	Results and Benefits of the Approach *262*
12.5.3.1	Integral Informatics Approach *263*
12.6	Future Developments *274*
12.6.1	Improved Background Corrections *274*

12.6.2	Better Tools for Stratifying Key Observations 274
12.6.3	Integration of Specialized Content: Chemical Structure and Images 275
12.6.4	Expanded Sharing and Integration of Public Datasets 275
12.6.5	Improved Integration of Text and Structured Data 276
12.6.6	New Classes of Knowledge-Based Applications Such as Network Pattern Based Screening and Prediction 277
	References 278

13 Network Screening: A New Method to Identify Active Networks from an Ensemble of Known Networks 281
Shigeru Saito and Katsuhisa Horimoto

13.1	Introduction 281
13.2	Methods 282
13.2.1	Causal Graph 283
13.2.2	Quantification of Consistency Between a Model and Data 284
13.2.3	Statistical Methods for Evaluation 284
13.2.3.1	GEV Distribution 284
13.2.3.2	Nonparametric Statistic 285
13.2.4	Simulation Study 286
13.2.5	Network Screening 287
13.3	Example Applications 289
13.3.1	Evaluation of the *E. coli* SOS Network 289
13.3.2	Network Screening for *E. coli* Networks Under Anaerobic Conditions 290
13.3.3	Network Screening for ChIP networks on iPSCs 292
13.4	Discussion 295
	References 295

14 Community Detection in Biological Networks 299
Gautam S. Thakur

14.1	Introduction 299
14.2	Centrality Measures 300
14.2.1	Degree Centrality 300
14.2.2	Eigenvector Centrality 300
14.2.3	Betweenness Centrality 301
14.2.4	Closeness Centrality 301
14.3	Study of Complex Systems 302
14.4	Overview 302
14.5	Proposed Algorithm 304
14.5.1	Overview 304
14.5.2	Algorithm Description 305
14.5.3	Random Walk 306
14.5.4	Local Optimization 308
14.5.4.1	Vertex Ranking 308

14.5.4.2	Community Ranking	*310*
14.5.5	Mutual Exchange	*311*
14.5.5.1	Dynamic Optimization	*311*
14.6	Experiments	*312*
14.6.1	Benchmarking	*312*
14.6.1.1	Evaluation	*312*
14.6.2	Computer-Generated Experiments	*314*
14.6.3	Application to Real Networks	*317*
14.6.3.1	Zachary Karate Club	*318*
14.6.3.2	Neurotransmitter Receptor Complexes	*319*
14.6.4	Study of Wireless Mobile Users	*321*
14.7	Further Improvements	*323*
14.8	Conclusions	*324*
	References	*325*

15 On Some Inverse Problems in Generating Probabilistic Boolean Networks *329*

Xi Chen, Wai-Ki Ching, and Nam-Kiu Tsing

15.1	Introduction	*329*
15.2	Reviews on BNs and PBNs	*331*
15.2.1	BNs	*331*
15.2.2	PBNs	*332*
15.3	Construction of PBNs from a Prescribed Stationary Distribution	*333*
15.3.1	CG Method Approach	*334*
15.3.2	Estimation of $C_j^{(i)}$	*335*
15.3.3	Numerical Example	*336*
15.3.4	Computational Cost Analysis	*338*
15.4	Construction of PBNs from a Prescribed Transition Probability Matrix	*338*
15.4.1	Heuristic Algorithms	*339*
15.4.2	Numerical Demonstration	*340*
15.4.3	Computational Cost Analysis	*341*
15.4.4	Modifications of Algorithms 15.1 and 15.2	*341*
15.4.5	Numerical Example	*342*
15.5	Conclusions	*345*
	References	*346*

16 Boolean Analysis of Gene Expression Datasets *349*

Debashis Sahoo

16.1	Introduction	*349*
16.2	Boolean Analysis	*350*
16.3	Main Organization	*351*
16.4	StepMiner	*351*
16.5	StepMiner Algorithm	*352*

16.5.1	Fitting One- or Two-Step Functions	*352*
16.5.2	Selecting the Best Step Function	*353*
16.5.3	Degrees of Freedom	*354*
16.5.4	FDR	*355*
16.6	BooleanNet	*355*
16.6.1	Boolean Implications in Gene Expression Microarray Data	*357*
16.6.2	Biological Interpretations of Boolean Implications	*360*
16.6.3	Conserved Boolean Implications	*361*
16.6.4	Comparison against Correlation Network	*364*
16.6.5	Boolean Implication Networks are Not Scale-Free	*365*
16.6.6	Computational Efficiency of BooleanNet	*367*
16.7	BooleanNet Algorithm	*368*
16.7.1	Data Collection and Preprocessing	*368*
16.7.2	Discovery of Boolean Relationships	*368*
16.7.3	Computation of FDR	*371*
16.7.4	Correlation Network for Human CD Genes	*371*
16.7.5	Discovery of Conserved Boolean Relationships	*371*
16.7.6	Connected Component Analysis	*371*
16.8	Conclusions	*371*
	References	*373*

Part Four   Systems Approach to Diseases   *377*

17   **Representing Cancer Cell Trajectories in a Phase-Space Diagram: Switching Cellular States by Biological Phase Transitions**   *379*
Mariano Bizzarri and Alessandro Giuliani

17.1	Introduction	*379*
17.2	Beyond Reductionism	*380*
17.3	Cell Shape as a Diagram of Forces	*381*
17.4	Morphologic Phenotypes and Phase Transitions	*382*
17.5	Cancer as an Anomalous Attractor	*386*
17.6	Shapes as System Descriptors	*388*
17.7	Fractals of Living Organisms	*389*
17.8	Fractals and Cancer	*390*
17.9	Modifications in Cell Shape Precede Tumor Metabolome Reversion	*391*
17.10	Conclusions	*395*
	References	*396*

18   **Protein Network Analysis for Disease Gene Identification and Prioritization**   *405*
Jing Chen and Anil G. Jegga

18.1	Introduction	*405*
18.2	Protein Networks and Human Disease	*405*

18.2.1	Ranking Algorithms for Network Analyses	406
18.2.2	Prioritization Methods	407
18.2.3	Evaluation of Protein Interaction Network Topological Features	409
18.3	ToppGene Suite of Applications	409
18.4	Conclusions	410
	References	411

19 Pathways and Networks as Functional Descriptors for Human Disease and Drug Response Endpoints 415

Yuri Nikolsky, Marina Bessarabova, Eugene Kirillov, Zoltan Dezso, Weiwei Shi, and Tatiana Nikolskaya

19.1	Introduction	415
19.2	Gene Content Classifiers and Functional Classifiers	416
19.3	Biological Pathways and Networks Have Different Properties as Functional Descriptors	418
19.4	Applications of Pathways as Functional Classifiers	420
19.5	Single Pathway Learning for Identifying Functional Descriptor Pathways	425
19.6	Multiple-Path Learning (MPL) Algorithm for Pathway Descriptors	427
19.7	Applications of MPL-Deduced Pathway Descriptors	428
19.7.1	Cross-Tissue Prediction of Compound Toxicity Using Functional Descriptors	428
19.7.2	Cross-Platform Reproducibility Using Pathways (MPL Algorithms)	429
19.7.3	Predictive Classifiers for Breast Cancer Endpoints	430
19.7.4	Using Pathways as the Measure of Congruency Between Datasets	431
19.8	Combining Advantages of Pathways and Networks	432
19.9	Key Upstream and Downstream Interactions of Genetically Altered Genes and "Universal Cancer Genes"	435
19.10	Conclusions	437
	References	438

Index 443

Preface

For the field of systems biology to mature, novel statistical and computational analysis methods are needed to deal with the growing amount of high-throughput data from genomics and genetics experiments. This book presents such methods and applications to data from biological and biomedical problems. Nowadays, it is widely recognized that *networks* form a very fruitful representation for studying problems in systems biology [1, 2]. However, many traditional methods do not make explicit use of a network representation of the data. For this reason, the topics treated in this book explore statistical and computational data analysis aspects of *networks* in systems biology [3–6].

Biological phenotypes are mediated by very intricate networks of interactions among biological components. This book covers extensively what we view as two complementary but strongly interrelated challenges in network biology. The first lies in inferring networks from experimental observations of state variables of a system. Interactions among molecular components are traditionally characterized through equilibrium binding or kinetic experiments *in vitro* with dilute solutions of the purified components. However, such experiments are typically low throughput and unable to properly account for the conditions prevailing *in vivo*, where factors such as molecular crowding, spatial heterogeneity, and the presence of ligands might strongly modify the interactions of interest. The possibility of inferring network connectivity and even quantitative interaction parameters from observations of intact living systems is attracting considerable research interest as a way of escaping such shortcomings. The fact that biological networks are complex, that problems are often poorly constrained, and that data are often high dimensional and noisy makes this challenge daunting. The second and perhaps equally difficult challenge lies in deriving results that are both biologically relevant and reliable from incomplete and uncertain information about biological interaction networks. We hope that the contributions in the subsequent chapters will help the reader understand and meet these challenges.

This book is intended for researches and graduate and advanced undergraduate students in the interdisciplinary fields of computational biology, biostatistics, bioinformatics, and systems biology studying problems in biological and biomedical sciences. The book is organized in four main parts: Part One: Modeling, Simulation, and Meaning of Gene Networks; Part Two: Inference of Gene Networks; Part 3: Analysis of Gene Networks; and Part Four: Systems Approach to Diseases. Each part consists of chapters that emphasize the topic of the corresponding part, however, without being disconnected from the remainder of the book. Overall, to order the different parts we assumed an intuitive – problem-oriented – perspective moving from *Modeling, Simulation, and Meaning of Gene Networks* to *Inference of Gene Networks* and *Analysis of Gene Networks*. The last part presents biomedical applications of various methods in *Systems Approach to Diseases*.

Each chapter is comprehensively presented, accessible not only to researchers from this field but also to advanced undergraduate or graduate students. For this reason, each chapter not only presents technical results but also provides background knowledge necessary to understand the statistical method or the biological problem under consideration. This allows to use this book as a textbook for an interdisciplinary seminar for advanced students not only because of the comprehensiveness of the chapters but also because of its size allowing to fill a complete semester.

Many colleagues, whether consciously or unconsciously, have provided us with input, help, and support before and during the preparation of this book. In particular, we would like to thank Andreas Albrecht, Gökmen Altay, Subhash Basak, Danail Bonchev, Maria Duca, Dean Fennell, Galina Glazko, Martin Grabner, Beryl Graham, Peter Hamilton, Des Higgins, Puthen Jithesh, Patrick Johnston, Frank Kee, Terry Lappin, Kang Li, D. D. Lozovanu, Dennis McCance, James McCann, Alexander Mehler, Abbe Mowshowitz, Ken Mills, Arcady Mushegian, Katie Orr, Andrei Perjan, Bert Rima, Brigitte Senn-Kircher, Ricardo de Matos Simoes, Francesca Shearer, Fred Sobik, John Storey, Simon Tavaré, Shailesh Tripathi, Kurt Varmuza, Bruce Weir, Pat White, Kathleen Williamson, Shu-Dong Zhang, and Dongxiao Zhu and apologize to all who have not been named mistakenly. We would also like to thank our editors Andreas Sendtko and Gregor Cicchetti from Wiley-VCH who have been always available and helpful.

Finally, we hope that this book will help to spread out the enthusiasm and joy we have for this field and inspire people regarding their own practical or theoretical research problems.

March 2011
Belfast, Hall/Tyrol, and Coimbra

Matthias Dehmer,
Frank Emmert-Streib,
Armin Graber,
and Armindo Salvador

References

1 Barabasi, A.L. and Oltvai, Z.N. (2004) Network biology: understanding the cell's functional organization. *Nat. Rev. Genet.*, **5**, 101–113.

2 Emmert-Streib, F. and Glazko, G. (2011) Network biology: a direct approach to study biological function. *WIREs Syst. Biol. Med.*, in press.

3 Alon, U. (2006) *An Introduction to Systems Biology: Design Principles of Biological Circuits*, Chapman & Hall/CRC.

4 Bertalanffy, L. von. (1950) An outline of general systems theory. *Br. J. Philos. Sci.*, **1** (2)

5 Kitano, H. (ed.) (2001) *Foundations of Systems Biology*, MIT Press.

6 Palsson, B.O. (2006) *Systems Biology: Properties of Reconstructed Networks*, Cambridge University Press.

List of Contributors

Andreas Bernthaler
Vienna University of Technology
Institute of Computer Languages
Theory and Logics Group
Favoritenstrasse 9
1040 Vienna
Austria

Marina Bessarabova
Thomson Reuters
Healthcare & Life Sciences
169 Saxony Road
Encinitas, CA 92024
USA

Mariano Bizzarri
Sapienza University
Department of Experimental Medicine
Viale Regina Elena 324
00161 Rome
Italy

Gianluca Bontempi
Université Libre de Bruxelles
Computer Science Department
Machine Learning Group
Boulevard du Triomphe
1050 Brussels
Belgium

Gastone Castellani
Università di Bologna
Department of Physics
INFN Bologna Section and
Galvani Center for Biocomplexity
40127 Bologna
Italy

Jing Chen
University of Cincinnati
Department of Environmental Health
Cincinnati, OH 45229
USA

Xi Chen
The University of Hong Kong
Department of Mathematics
Pok Fu Lam Road
Hong Kong
China

Wai-Ki Ching
The University of Hong Kong
Department of Mathematics
Pok Fu Lam Road
Hong Kong
China

Zoltan Dezso
Thomson Reuters
Healthcare & Life Sciences
169 Saxony Road
Encinitas, CA 92024
USA

Cathy S. J. Fann
Academia Sinica
Institute of Biomedical Sciences
Academia Road, Nankang
115 Taipei
Taiwan

Raul Fechete
Emergentec Biodevelopment GmbH
Gersthofer Strasse 29-31
1180 Vienna
Austria

Rudolf Freund
Vienna University of Technology
Institute of Computer Languages
Theory and Logics Group
Favoritenstrasse 9
1040 Vienna
Austria

Alessandro Giuliani
Istituto Superiore di Sanità
Department of Environment
and Health
Viale Regina Elena 299
00161 Rome
Italy

Erich Gombocz
IO Informatics Inc.
2550 Ninth Street
Berkeley, CA 94710-2549
USA

Jing-Dong J. Han
Chinese Academy of Sciences
Institute of Genetics and
Developmental Biology
Center for Molecular Systems Biology
Key Laboratory of
Molecular Developmental Biology
Lincui East Road
100101 Beijing
China

and

Chinese Academy of Sciences–
Max Planck Partner Institute for
Computational Biology
Shanghai Institutes for
Biological Sciences
Chinese Academy of Sciences
320 Yue Yang Road
200031 Shanghai
China

Katsuhisa Horimoto
National Institute of Advanced
Industrial Science Technology
Computational Biology Research Center
2-4-7, Aomi, Koto-ku
135-0064 Tokyo
Japan

Ching-Lin Hsiao
Academia Sinica
Institute of Biomedical Sciences
Academia Road, Nankang
115 Taipei
Taiwan

Jialiang Huang
Chinese Academy of Sciences
Institute of Genetics and
Developmental Biology
Center for Molecular Systems Biology
Key Laboratory of
Molecular Developmental Biology
Lincui East Road
100101 Beijing
China

Anil G. Jegga
Cincinnati Children's Hospital
Medical Center
Division of Biomedical Informatics
Cincinnati, OH 45229
USA

and

University of Cincinnati
Department of Biomedical Engineering
Cincinnati, OH 45229
USA

and

University of Cincinnati
College of Medicine
Department of Pediatrics
Cincinnati, OH 45229
USA

Eugene Kirillov
Thomson Reuters
Healthcare & Life Sciences
169 Saxony Road
Encinitas, CA 92024
USA

Younhee Ko
University of Illinois at
Urbana-Champaign
Department of Animal Sciences
1207 W. Gregory Dr.
Urbana, IL 61801
USA

and

University of Illinois at
Urbana-Champaign
Institute for Genomic Biology
1205 W. Gregory Drive
Urbana, IL 61801
USA

Rainer König
University of Heidelberg
Institute of Pharmacy and Molecular
Biotechnology
Bioquant
Im Neuenheimer Feld 267
69120 Heidelberg
Germany

Xiangfang Li
Texas A&M University
Genomic Signal Processing Laboratory
TAMU 3128
College Station, TX 77843
USA

Arno Lukas
Emergentec Biodevelopment GmbH
Gersthofer Strasse 29-31
1180 Vienna
Austria

Bernd Mayer
Emergentec Biodevelopment GmbH
Gersthofer Strasse 29-31
1180 Vienna
Austria

Patrick E. Meyer
Université Libre de Bruxelles
Computer Science Department
Machine Learning Group
Boulevard du Triomphe
1050 Brussels
Belgium

Konrad Mönks
Vienna University of Technology
Institute of Computer Languages
Theory and Logics Group
Favoritenstrasse 9
1040 Vienna
Austria

and

Emergentec Biodevelopment GmbH
Gersthofer Strasse 29-31
1180 Vienna
Austria

Irmgard Mühlberger
Emergentec Biodevelopment GmbH
Gersthofer Strasse 29-31
1180 Vienna
Austria

Tatiana Nikolskaya
Thomson Reuters
Healthcare & Life Sciences
169 Saxony RD
Encinitas, CA 92024
USA

Yuri Nikolsky
Thomson Reuters
Healthcare & Life Sciences
169 Saxony Road
Encinitas, CA 92024
USA

Catharina Olsen
Université Libre de Bruxelles
Computer Science Department
Machine Learning Group
Boulevard du Triomphe
1050 Brussels
Belgium

Paul Perco
Emergentec Biodevelopment GmbH
Gersthofer Strasse 29-31
1180 Vienna
Austria

Kitiporn Plaimas
University of Heidelberg
Institute of Pharmacy and Molecular
Biotechnology
Bioquant
Im Neuenheimer Feld 267
69120 Heidelberg
Germany

Thomas N. Plasterer
Northeastern University
Department of Chemistry and
Chemical Biology
360 Huntington Ave.
Boston, MA 02115
USA

and

Pharmacogenetics Clinical Advisory
Board
2000 Commonwealth Avenue, Suite 200
Auburndale, MA 02466
USA

Lijun Qian
Texas A&M University System
Prairie View A&M University
Department of Electrical and
Computer Engineering
MS2520, POB 519
Prairie View, TX 77446
USA

Daniel Remondini
Università di Bologna
Department of Physics
INFN Bologna Section and
Galvani Center for Biocomplexity
40127 Bologna
Italy

Sandra Rodriguez-Zas
University of Illinois at
Urbana-Champaign
Department of Animal Sciences
1207 W. Gregory Drive
Urbana, IL 61801
USA

List of Contributors

Debashis Sahoo
Instructor of Pathology and Siebel
Fellow at Institute of Stem Cell Biology
and Regenerative Medicine
Lorry I. Lokey Stem Cell Research
Building
265 Campus Drive, Rm G3101B
Stanford, CA 94305
USA

Shigeru Saito
Infocom Corp.
Chem & Bio Informatics Department
Sumitomo Fudosan Harajuku Building
2-34-17, Jingumae, Shibuya-ku
150-0001 Tokyo
Japan

Robert Stanley
IO Informatics Inc.
2550 Ninth Street
Berkeley, CA 94710-2549
USA

Gautam S. Thakur
University of Florida
Department of Computer and
Information Science and Engineering
Science
PO Box 116120
Gainsville, FL 32611-6120
USA

Tianhai Tian
University of Glasgow
Department of Mathematics
University Gardens
Glasgow G12 8QW
UK

Nam-Kiu Tsing
The University of Hong Kong
Department of Mathematics
Pok Fu Lam Road
Hong Kong
China

Eberhard O. Voit
Georgia Tech and Emory University
The Wallace H. Coulter Department
of Biomedical Engineering
313 Ferst Drive
Atlanta, GA 30332
USA

Haixin Wang
Fort Valley State University
Department of Mathematics and
Computer Science
CTM 101A
Fort Valley, GA 31030
USA

Matthew Weirauch
University of Toronto
Banting and Best Department
of Medical Research and
Donnelly Centre for
Cellular and Biomolecular Research
160 College Street
Toronto, ON, M5S 3E1
Canada

Hong Yu
Chinese Academy of Sciences
Institute of Genetics and
Developmental Biology
Center for Molecular Systems Biology
Key Laboratory of
Molecular Developmental Biology
Lincui East Road
100101 Beijing
China

Wei Zhang
Chinese Academy of Sciences
Institute of Genetics and
Developmental Biology
Center for Molecular Systems Biology
Key Laboratory of
Molecular Developmental Biology
Lincui East Road
100101 Beijing
China

Part One
Modeling, Simulation, and Meaning of Gene Networks

1
Network Analysis to Interpret Complex Phenotypes
Hong Yu, Jialiang Huang, Wei Zhang, and Jing-Dong J. Han

1.1
Introduction

Gene network analysis is an important part of systems biology studies. Compared with traditional genotype/phenotype studies that focused on establishing the relationships between single genes and interested traits, network analysis give us a global view of how all the genes work together properly, which in turn leads to the correct biological functions [1].

Unlike the Mendelian "one gene–one phenotype" relationship, C.H. Waddington in 1957 came up with the "epigenetic landscape" to visually illustrate the multigene or network effects of genes on shaping the landscapes (various states) of cellular metabolism. Given our current knowledge, "cellular metabolism" in Waddington's landscapes model can be extended to "molecular networks," which turn steady states into network representations or snapshots. Such steady states and the transitions from one steady state to another have been computationally analyzed through simulated networks [2–4] and experimentally validated by checking gene expression profiles during proliferation/differentiation transitions, gene mutation perturbations, or environmental or physical stresses [5, 6]. The transition from one stable state to another is usually related to complex phenotypes, which could be both physiological and pathological, such as diabetes mellitus or cancerous proliferation (Figure 1.1) [7]. Gene function is not isolated, so we could not study their function separately. Not only the function of the individual gene products, but also their interaction with each other, which is increasingly more important to the success of higher organisms, determines the selective advantage of the genes and the networks they formed.

What can network analysis do? Here, we mainly talk about given a gene network, mostly validated by experiments, what information could be got from it? How could we understand the biological process with the help of a network? Basically, there are three aspects. The most traditional aspect is to identify the importance of each node in the network (e.g., which genes are more important or crucial, which genes are less

---

Applied Statistics for Network Biology: Methods in Systems Biology, First Edition.
Edited by M. Dehmer, F. Emmert-Streib, A. Graber, and A. Salvador.
© 2011 Wiley-VCH Verlag GmbH & Co. KGaA. Published 2011 by Wiley-VCH Verlag GmbH & Co. KGaA.

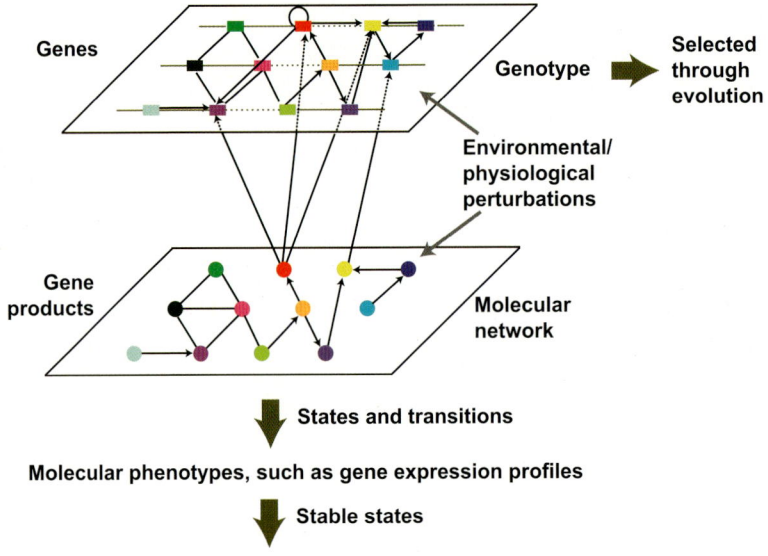

Figure 1.1 Complex phenotypes are determined by the steady state of the molecular network. A molecular network is encoded by the genetic network. The interplay of molecules in the network as well as their interactions with the environment and developmental cues determine the stable states of the network, which ultimately determines the phenotypes reflected by the system. (Adapted from [7].)

important or dispensable). Another aspect is to identify which genes are more functionally related through the whole network view, not only by measuring the direct connections, but also by considering the connections through the whole network. In this way, we could establish functional relationships between all the genes by protein–protein interaction networks or other kinds of experimentally validated networks. More recent studies have focused on identifying the paths or flows through the networks with known input and output genes. These methods could identify the unknown mediated genes and also identify which genes are more important in these processes. All these different aspects could serve well in understanding human diseases at different level and views. We will start by discussing these three aspects in detail, including some methods related to them, but not limited in pure network analysis in later sections.

Before we begin to talk about network analysis, we first explain several definitions that are very basic, but will be frequently mentioned in the following parts.

A network N consists of a set $V(N)$ of vertices (or nodes) together with a set $E(N)$ of edges (or links) that connect various pairs of vertices. Usually, nodes represent genes or proteins and edges represent interactions.

A network N is a weighted network if each of its edges has a number associated with it indicating the strength of the edge. Usually, the edge weights represent the confidences of interactions in biological experiments.

A network N is called a directed network if all of its edges are directed and a network N is called an undirected network if none of its edges is directed. Usually, signaling networks and transcriptional regulatory networks could be directed networks whose directions indicate signal transduction or transcriptional regulation.

For any network N and any particular vertex v in $V(N)$, the number of vertices v' in $V(N)$ that are directly linked to v is called the degree of v.

In particular, for any directed network N and any particular vertex v in $V(N)$, the number of vertices v' in $V(N)$ that are directly linked to v by an inward-pointing edge to v is called the in-degree of v and the number of vertices v' in $V(N)$ that are directly linked to v by an edge pointing outward from v is called the out-degree of v.

The minimum number of edges that must be traversed to travel from a vertex v to another vertex v' of a network N is called the shortest path length between v and v'. For any connected network N, the average shortest path length between any pair of vertices is called the network's "characteristic path length" (CPL).

1.2
Identification of Important Genes based on Network Topologies

Identification of important genes in biological processes is one of the most common and important aspects in all kinds of biology studies [8, 9]. The basic idea to achieve this goal in biological networks is to measure the influence or damage to the network by perturbing certain genes [10]. If removing a gene from a network leads to small changes or influences, this gene should be less important in maintaining the correct function of the biological network. In contrast, if it leads to the collapse or a large influence on the network, such as dividing the whole network into two subnetworks, this gene might play a crucial role in biological processes. This hypothesis has been increasingly supported by experimental data showing that genes with higher influences on the network were more lethal, more conserved through evolution, and basically more important in maintaining biological functions [11]. In order to evaluate genes' importance, several different measurements could be used due to different considerations.

1.2.1
Degree

The most intuitive consideration is that the more edges are removed, the more damage is taken by the network. Thus, the genes with high degrees, known as hubs in the network, should be more important. Evidence has shown that the perturbation of hubs leads to a more dramatic increase of CPL in a biological network than random perturbations [12]. Besides, other information could be further used, such as gene expression data, to find "date hubs" and "party hubs," which indicate different biological functions [12].

1.2.2
Betweenness

The centrality or connectivity of a network can be measured by the CPL. In biological networks, the CPL indicates the speed of signal transduction or the quickness of biological response. Thus, another consideration of a gene's importance is the CPL changes when perturbing it. These changes could be measured directly by recalculating the CPL when removing each gene from the network or indirectly using the betweenness of each gene. The betweenness of a vertex v is calculated as the number of shortest paths that pass through it divided by the number of all shortest paths. Compared with betweenness, recalculating the CPL is more accurate, but more time consuming. In fact, a very high correlation exists between the CPL recalculating results and the betweenness measurements, so basically measuring the betweenness of a gene is sufficient to see its influence on the CPL. We could also easily see that a gene with high betweenness is not necessarily a hub or has a very high degree, but in view of the whole gene set, betweenness does correlate with degree.

1.2.3
Network Motifs

Compared with the former two measurements, which could be applied to any kinds of networks, network motifs are usually employed in directed networks, such as transcriptional regulatory networks or transcription factor target networks. Network motifs could be regarded as the basic blocks to form the whole network [13], and they were shown to be important in maintaining robustness, perturbation buffering, quick responses, and accurate signal transductions [13–15]. Thus, the genes that take part in multiple network motifs should be more important and counting network motifs becomes one measurement for evaluating the importance of genes. Here, we introduce several commonly used network motifs (Scheme 1.1).

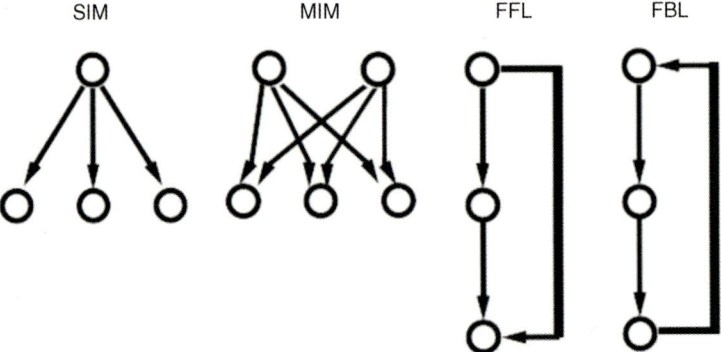

Scheme 1.1 Several commonly used network motifs.

- Single-input motifs (SIM): a group of nodes regulated by a single node without any other regulation.
- Multi-input motifs (MIM): a group of nodes regulate another group of nodes together.
- Feed-forward loops (FFL): a node regulates another and then these two nodes regulate a third one together.
- Feed-back loops (FBL; also known as a multicomponent loops (MCL): an upstream node is regulated by a downstream one.

In biological networks, genes in SIMs or MIMs usually determine the bottleneck of the network, which possibly indicates that the deletion or mutation of these genes is likely to cause lethal influences. FFLs and FBLs could enable precise control or quick response, which was precisely required in biological processes and responses. Network motifs are not limited to those mentioned above, but all the motifs that have been proved to have biological meanings. By searching for different kinds of network motifs, we could find important genes for certain functions that we are interested in.

1.2.4
Hierarchical Structure

In signal transduction networks or transcriptional regulatory networks, genes can be divided into several layers and the signals flow from top to bottom (with feedback allowed). This kind of structure is called a hierarchical structure. Apart from the degree and network motifs, genes on different layers or having different offspring nodes (regulated by this gene) could provide information on understanding biological processes [16].

These network topology-based analyses have been widely used in identifying important genes in multiple studies of different species. However, some other cautions should be announced in all of these measurements besides the fact that they are based on different considerations. First, it is hard to consider the combinatorial influence of the genes, such as when removing either one of two genes with very similar connections, the network will not be badly influenced because there is a backup gene, but when removing both of them, the whole network will collapse. Backup genes exist widely in real biological processes to ensure the robustness of organisms. Currently, it is possible to detect these combinatorial effects through applying newly developed IT methods, although calculations may be very time-consuming. Another problem is that the qualities of networks negatively influence the results, especially when the edges in the networks are biased. This does happen, especially in human studies. For instance, when using literature-supported protein–protein interactions (PPIs), the "hot" genes or interesting genes are much more intensively studied than the "cold" genes and they are more likely to be hubs, because most of their interactions are discovered, while for the "cold" genes, most of their interactions are unknown.

1.3
Inferring Information from Known Networks

1.3.1
Understanding Biological Functions based on Network Modularity

The existence of modular structures (clusters of tightly connected subnetworks) has been noticed in various biological networks. In biological networks, these modules often indicate particular biological functional processes [17, 18]. The modules can be identified by various algorithms, such as the Lin Log energy model (http://www.informatik.tu-cottbus.de/~an/GD/linlog.html), the MCODE algorithm (http://baderlab.org/Software/MCODE), and the Markov Clustering algorithm (http://www.micans.org/mcl/). Then, by examining the modules' enriched Gene Ontology (GO) terms, KEGG (Kyoto Encyclopedia of Genes and Genomes) pathways, and other functional annotations, we can discover their biological functions.

1.3.2
Inferring Functional Relationships and Novel Functional Genes Through Networks

In the past few years, more and more studies have focused on identifying functional relationships between genes. These studies came from the collaborations of human association studies and gene function prediction studies. These methods aim to identify unknown disease-related genes with a candidate list derived from association studies. Usually these methods include not only PPIs, but also many other kinds of information, which could be summarized into different kinds of edges. The basic idea is that genes sharing similar functions are usually highly connected in PPI networks. Thus, in order to identify novel disease-related genes from a candidate list, we just need to find the known genes with similar phenotypes in PPI networks.

Several studies analyzed Online Mendelian Inheritance in Man (OMIM) data using PPI and description similarity between genes and phenotypes, which is the result based on human association studies over recent decades [19, 20]. With the development of new technologies, more and more association studies have been finished on large populations and specific phenotypes at high coverage and high resolution levels. These genome-wide association studies (GWAS) provided opportunities for the application of all these methods. As the integration of different kinds of networks could be seen as a whole weighted network with different weights on different edges, we would mainly introduce one method with wide applications and a good computational performance, which is based on the random walk algorithm [21].

The random walk on graphs is defined as an iterative walker's transition from its current node to all its neighbors through all weighted edges starting at given source nodes, s. Each source node could take a different weight and basically the sum value could be normalized to 1, so this value could also be considered as the probability of the information transition through the whole network. Here, compared to the traditional random walk, it added another restart process that in every step, the signal restarts at node s with a probability r. It indicated that in every step of transition, only $(1-r)$ of total information is continuously transitioned, with r of total restart. The goal of this method is to add a continuous input and when the stable status is achieved, all the other nodes have a stable proportion of information to be output, the sum of which is r.

Formally, the random walk with restart is defined as:

$$P^{t+1} = (1-r)*W*P^t + r*P^0$$

where W is a matrix that is based only on the network topology; basically, it is the column-normalized adjacency matrix, each none zero value represents the weight of one edge in the network. P^t is a vector in which each element holds the probability of information on a node at step t. In this application, the initial probability vector P^0 was constructed as weighted probabilities where each probability represents the influence of a source gene on the disease we are interested in, with the sum of these probabilities equal to 1. When the difference between P^t and P^{t+1} is smaller than an arbitrarily given threshold, the steady-state P^N was obtained and considered as the result. Candidate disease-related genes are then ranked according to the values in P^N.

The performance of the random walk algorithm was shown to be better than the previous algorithms. Also, this algorithm is easily applied. One obvious benefit of this method is that P^N is additive, which makes this algorithm very convenient. Take one simple example, consider the steady state P^N of only one source node A or B to be $P^N(A)$ or $P^N(B)$. When we want to consider the combinatorial effect of A and B, we can apply the weighted probabilities of the two source nodes as a and $(1-a)$, and the steady state P^N of using both A and B as source nodes could be simply calculated as $P^N(AB) = a * P^N(A) + (1-a) * P^N(B)$. This formula could be extended to a set s of multiple source genes. Thus, basically, for a certain network, we do not have to recalculate P^N for each set of source genes. Instead, we could calculate each source gene individually and sum the weighted results. In this algorithm, different r indicates different affinity. High r indicates more influence of input genes and less transition in the network, while low r leads to more transition steps. Empirically, the stable result could be obtained within 30–50 steps considering different r and thresholds used, and the algorithm is not very time-consuming. Thus, it is possible to calculate P^N of each gene in a network.

As mentioned above in Section 1.2, all of these algorithms are negatively influenced by the quality of networks and those "hot" genes. We were very likely to be stuck in those "hot" genes if a biased network was used.

1.3.3
Unraveling Transcriptional Regulations from Expression Data through Transcriptional Networks

Transcription factors play a crucial regulatory role in various biological processes; however, they are unlikely to be detected from expression data due to their low, and often sparse, expression. To fill this gap, Reverter *et al.* proposed a regulatory impact factor (RIF) algorithm to identify critical transcription factors from gene expression data by integrating coexpression networks [22]. RIF analysis assigns a score to each transcription factor by considering both the correlation between the transcription factor and the differentially expressed genes and the expression level of the differentially expressed genes. In particular, for a given functional module, its potential regulators are scored by their absolute coexpression correlation averaged across all genes in the module [23].

1.3.4
Extracting the Pathway-Linked Regulators and Effectors based on Network Flows

Recently, high-throughput techniques have been widely used to detect the potential components of biological networks. So far, these high-throughput techniques cover two classes: (i) genetic screens including overexpression, deletion, or RNA interference library screens and (ii) mRNA profiling using microarray or RNA sequencing technology. By comparing the results of these two methods, Yeger-Lotem et al. found that genetic screens tend to identify regulators that are critical for the cell response, while the differentially expressed genes identified by mRNA profiling are likely their downstream effectors, whose changes indirectly reflect the genetic changes in the regulatory networks [24]. It is also true in diseases; using type II diabetes and hypertension as study cases [25], we found that the disease-causing genes, which have high probability to cause type II diabetes and hypertension phenotypes when perturbed, tend to be hubs in the interactome networks and enriched in signaling pathways, whereas the significantly differentially expressed genes identified by microarrays are mostly enriched in the metabolic pathways. The connection between these two gene sets is significantly tight.

To bridge the gap between the genetic screen data and the mRNA expression data using known molecular networks, Yeger-Lotem et al. developed an integrative approach called "Response Net" [24]. Briefly, Response Net is a flow optimization algorithm that redefines a crucial subnetwork that connects genetic hits (source) and differentially expressed genes (target) from a whole weight network, where each node or edge has been assigned a weight according to their biological importance or confidence. The cost of an edge is defined by the $-\log$ value of its weight. Thus, the goal of Response Net can be achieved by solving a linear programming optimization problem that minimizes the overall cost of the network when distributing the maximal flow from source to target. According to the solution, those edges with positive flow defined the predicted crucial subnetwork.

1.4
Conclusions

We have introduced basic methods and applications in network analysis to interpret complex phenotypes. Although these methods have many advantages, network biology still faces many challenges. Most of the methods rely on the quality of datasets, which determine the false-positives and limited coverage. Most edges in network maps are still lacking detailed attributes and directions. Post-transcriptional modifications are hardly monitored at a large scale. Tissue- and cell-type specificities are hard to consider. However, with the development of new technologies, such as high-throughput and single-cell dynamic measurement techniques, and with increasing accuracy and coverage of high-throughput technologies, the ever-accelerating data acquisition will raise further need for data integration and modeling at the network level. More and more methods have emerged, which provide important tools for network analysis. Mastering these methods is necessary, but far from sufficient for understanding biology. More important things to do are to ask the right questions, to choose proper network analysis tools, and to validate analysis results by solid experimentation. After all, network biology is biology and the fundamental goal is the same for network biology and molecular biology – to better understand basic biological processes and the mechanisms of human diseases.

References

1 Barabasi, A.L. and Oltvai, Z.N. (2004) Network biology: understanding the cell's functional organization. *Nat. Rev. Genet.*, **5**, 101–113.
2 Bergman, A. and Siegal, M.L. (2003) Evolutionary capacitance as a general feature of complex gene networks. *Nature*, **424**, 549–552.
3 Kauffman, S.A. (1969) Metabolic stability and epigenesis in randomly constructed genetic nets. *J. Theor. Biol.*, **22**, 437–467.
4 Li, F., Long, T., Lu, Y., Ouyang, Q., and Tang, C. (2004) The yeast cell-cycle network is robustly designed. *Proc. Natl. Acad. Sci. USA*, **101**, 4781–4786.
5 Chen, J.F., Mandel, E.M., Thomson, J.M., Wu, Q., Callis, T.E., Hammond, S.M., Conlon, F.L., and Wang, D.Z. (2006) The role of microRNA-1 and microRNA-133 in skeletal muscle proliferation and differentiation. *Nat. Genet.*, **38**, 228–233.
6 Huang, S., Eichler, G., Bar-Yam, Y., and Ingber, D.E. (2005) Cell fates as high-dimensional attractor states of a complex gene regulatory network. *Phys. Rev. Lett.*, **94**, 128701.
7 Han, J.D. (2008) Understanding biological functions through molecular networks. *Cell Res.*, **18**, 224–237.
8 Jeong, H., Mason, S.P., Barabasi, A.L., and Oltvai, Z.N. (2001) Lethality and centrality in protein networks. *Nature*, **411**, 41–42.
9 Tew, K.L., Li, X.L., and Tan, S.H. (2007) Functional centrality: detecting lethality of proteins in protein interaction networks. *Genome Inform.*, **19**, 166–177.
10 Albert, R., Jeong, H., and Barabasi, A.L. (2000) Error and attack tolerance of complex networks. *Nature*, **406**, 378–382.
11 He, X. and Zhang, J. (2006) Why do hubs tend to be essential in protein networks? *PLoS Genet.*, **2**, e88.
12 Han, J.D., Bertin, N., Hao, T., Goldberg, D.S., Berriz, G.F., Zhang, L.V., Dupuy, D., Walhout, A.J., Cusick, M.E., Roth, F.P. et al. (2004) Evidence for dynamically organized modularity in the yeast protein–protein interaction network. *Nature*, **430**, 88–93.
13 Milo, R., Shen-Orr, S., Itzkovitz, S., Kashtan, N., Chklovskii, D., and Alon, U. (2002) Network motifs: simple building blocks of complex networks. *Science*, **298**, 824–827.
14 Milo, R., Itzkovitz, S., Kashtan, N., Levitt, R., Shen-Orr, S., Ayzenshtat, I., Sheffer, M., and Alon, U. (2004) Superfamilies of evolved and designed networks. *Science*, **303**, 1538–1542.
15 Wuchty, S., Oltvai, Z.N., and Barabasi, A.L. (2003) Evolutionary conservation of motif constituents in the yeast protein interaction network. *Nat. Genet.*, **35**, 176–179.
16 Yu, H. and Gerstein, M. (2006) Genomic analysis of the hierarchical structure of regulatory networks. *Proc. Natl. Acad. Sci. USA*, **103**, 14724–14731.

17 Bader, G.D. and Hogue, C.W. (2003) An automated method for finding molecular complexes in large protein interaction networks. *BMC Bioinformatics*, **4**, 2.

18 Eisen, M.B., Spellman, P.T., Brown, P.O., and Botstein, D. (1998) Cluster analysis and display of genome-wide expression patterns. *Proc. Natl. Acad. Sci. USA*, **95**, 14863–14868.

19 Lage, K., Karlberg, E.O., Storling, Z.M., Olason, P.I., Pedersen, A.G., Rigina, O., Hinsby, A.M., Tumer, Z., Pociot, F., Tommerup, N. et al. (2007) A human phenome–interactome network of protein complexes implicated in genetic disorders. *Nat. Biotechnol.*, **25**, 309–316.

20 Wu, X., Jiang, R., Zhang, M.Q., and Li, S. (2008) Network-based global inference of human disease genes. *Mol. Syst. Biol.*, **4**, 189.

21 Kohler, S., Bauer, S., Horn, D., and Robinson, P.N. (2008) Walking the interactome for prioritization of candidate disease genes. *Am. J. Hum. Genet.*, **82**, 949–958.

22 Reverter, A., Hudson, N.J., Nagaraj, S.H., Perez-Enciso, M., and Dalrymple, B.P. (2010) Regulatory impact factors: unraveling the transcriptional regulation of complex traits from expression data. *Bioinformatics*, **26**, 896–904.

23 Hudson, N.J., Reverter, A., Wang, Y., Greenwood, P.L., and Dalrymple, B.P. (2009) Inferring the transcriptional landscape of bovine skeletal muscle by integrating co-expression networks. *PLoS ONE*, **4**, e7249.

24 Yeger-Lotem, E., Riva, L., Su, L.J., Gitler, A.D., Cashikar, A.G., King, O.D., Auluck, P.K., Geddie, M.L., Valastyan, J.S., Karger, D.R. et al. (2009) Bridging high-throughput genetic and transcriptional data reveals cellular responses to alpha-synuclein toxicity. *Nat. Genet.*, **41**, 316–323.

25 Yu, H., Huang, J., Qiao, N., Green, C.D., and Han, J.D. (2010) Evaluating diabetes and hypertension disease causality using mouse phenotypes. *BMC Syst. Biol.*, **4**, 97.

2
Stochastic Modeling of Gene Regulatory Networks
Tianhai Tian

2.1
Introduction

Recent studies through biological experiments have indicated that noise plays a very important role in determining the dynamic behavior of biological systems. Since the research work on stochastic modeling of the regulatory network of λ phage [1, 2], there have been an increasing number of studies in the last decade investigating the origins of noise in biological networks and its crucial role in determining the key properties of biological networks [3–5]. Experimental studies have demonstrated that noise in cellular processes may result from a small number of molecular species, intermittent gene activity, and fluctuations of experimental conditions [6–9]. Empirical discoveries have stimulated explosive research interests in developing stochastic models for a wide range of biological systems, including gene regulatory networks [10–12], cell signaling pathways [13–15], and metabolic pathways [16, 17].

It has been proposed that noise in the form of random fluctuations arises in biological networks in one of two ways: internal (intrinsic) noise or external (extrinsic) noise [18, 19]. The internal noise is mainly derived from the chance events of biochemical reactions in the system due to small copy numbers of certain key molecular species. External noise mainly refers to the environmental fluctuations or the noise propagation from the upstream biological pathways. In addition, there are two major types of response of biological systems to noise. In the first case, living systems are optimized to function in the presence of stochastic fluctuations, and biochemical networks must withstand considerable variations and random perturbations of biochemical parameters [20–22]. Such a property of biological systems is known as "robustness" [23, 24]. On the other hand, biological systems are also sensitive to environmental fluctuations and/or intrinsic noise in certain time periods. For example, noise in gene expression could lead to qualitative differences in a cell's phenotype if the expressed genes act as inputs to downstream regulatory thresholds [8, 25, 26].

Applied Statistics for Network Biology: Methods in Systems Biology, First Edition.
Edited by M. Dehmer, F. Emmert-Streib, A. Graber, and A. Salvador.
© 2011 Wiley-VCH Verlag GmbH & Co. KGaA. Published 2011 by Wiley-VCH Verlag GmbH & Co. KGaA.

One of the major challenges in systems biology is the development of quantitative mathematical models for studying regulatory mechanisms in complex biological systems [27]. Although deterministic models have been widely used for analyzing gene regulatory networks, cell signaling pathways, and metabolic systems [28, 29], a deterministic model can only describe the averaged behavior of a system based on large populations, but cannot realize fluctuations of the system behavior in different cells. Recently, there has been an accelerating interest in the investigation of the effect of noise in genetic regulation through stochastic modeling. Although stochastic models have been developed based on detailed knowledge of biochemical reactions, data availability and regulatory information usually cannot provide a comprehensive picture of biological regulations. In recent years, a number of approaches have been proposed to develop either continuous or discrete stochastic models for the study of noise in large-scale gene regulatory networks. These methods include stochastic Boolean models [30, 31], probabilistic hybrid approaches [32], stochastic Petri nets [33, 34], stochastic differential equations (SDEs) [35, 36], and multiscale (hybrid) models that include both stochastic and deterministic dynamics [37, 38].

Systems of ordinary differential equations (ODEs) have been widely used to model biological systems and there are a large number of well-developed deterministic models for a broad range of biological systems. An important question in stochastic modeling is how to develop stochastic models by introducing stochastic processes into deterministic models for the external and/or internal noise. This chapter will use a number of modeling approaches and biological systems to address this issue. The remaining part of this chapter is organized as follows. Section 2.2 discusses numerical methods for simulating chemical reaction systems. These methods are the theoretical basis for designing stochastic models in the following sections. A general modeling approach for developing discrete stochastic models is discussed in Section 2.3. Section 2.4 provides a number of techniques for designing continuous stochastic models by using SDEs.

2.2
Discrete Stochastic Simulation Methods

Since many cellular processes are governed by effects associated with small numbers of certain key molecules, the standard chemical framework described by systems of ODEs breaks down. The stochastic simulation algorithm (SSA) represents a discrete modeling approach and an essentially exact procedure for numerically simulating the time evolution of a well-stirred reaction system [39]. The advances in stochastic modeling of gene regulatory networks and cell signaling transduction pathways have stimulated growing research interests in the development of effective methods for simulating chemical reaction systems. These effective simulation methods in return provided innovative methodologies for designing stochastic models of biological systems.

2.2.1
SSA

It is assumed that a chemical reaction system is a well-stirred mixture at constant temperature in a fixed volume Ω. This mixture consists of $N \geq 1$ molecular species $\{S_1, \ldots, S_N\}$ that chemically interact through $M \geq 1$ reaction channels $\{R_1, \ldots, R_M\}$. The dynamic state of this system is denoted as $\mathbf{x}(t) \equiv (x_1(t), \ldots, x_N(t))^T$, where $x_i(t)$ is the molecular number of species S_i in the system at time t. For each reaction R_j ($j = 1, \ldots, M$), a propensity function $a_j(\mathbf{x})$ is defined for a given state $\mathbf{x}(t) = \mathbf{x}$ and the value of $a_j(x)dt$ represents the probability that one reaction R_j will fire somewhere inside Ω in the infinitesimal time interval $[t, t + dt)$. In addition, a state change vector $\mathbf{v}_j$ is defined to characterize reaction R_j. The element v_{ij} of $\mathbf{v}_j$ represents the change in the copy number of species S_i due to reaction R_j. The $N \times M$ matrix $\mathbf{v}$ with elements v_{ij} is called the stoichiometric matrix.

The SSA is a statistically exact procedure for generating the time and index of the next occurring reaction in accordance with the current values of the propensity functions. In each time step, two random numbers are generated to determine the time step and the index of the next reaction. There are several forms of this algorithm. The widely used direct method works as described in Method 2.1.

Method 2.1 Direct Method [39]

Step 1: Calculate the values of propensity functions $a_j(\mathbf{x})$ based on the system state x at time t and $a_0(\mathbf{x}) = \sum_{j=1}^{M} a_j(\mathbf{x})$.

Step 2: Generate a sample r_1 of the uniformly distributed random variable $U(0, 1)$ and determine the time of the next reaction:

$$\mu = \frac{1}{a_0(\mathbf{x})} \ln\left(\frac{1}{r_1}\right)$$

Step 3: Generate an independent sample r_2 of $U(0, 1)$ to determine the index k of the next reaction occurring in $[t, t + \mu)$:

$$\sum_{j=1}^{k-1} a_j(\mathbf{x}) < r_2 a_0(\mathbf{x}) \leq \sum_{j=1}^{k} a_j(\mathbf{x})$$

Step 4: Update the state of the system by:

$$\mathbf{x}(t + \mu) = \mathbf{x}(t) + \mathbf{v}_k \tag{2.1}$$

Step 5: Go to Step (1) if $t + \mu \leq T$, where T is the end time point. Otherwise, the system state $\mathbf{x}(T) = \mathbf{x}(t)$.

Another exact method is the first reaction method that uses M random numbers at each step to determine the possible reaction time of each reaction channel [40]. The reaction firing in the next step is that needing the smallest reaction time. Compared to the direct method, the first reaction method is not effective since it discards $M-1$ random numbers at each step. To improve the efficiency of the first reaction method, Gilson and Bruck [41] proposed the next reaction method by recycling the generated random numbers. The putative step size of a reaction channel is updated based on the step size of this channel at the previous step and values of the propensity function at these two steps. In addition, a so-called dependency graph was designed to reduce the computing time of propensity functions. Numerical results indicated that the next reaction method is effective for simulating systems with many species and reaction channels.

The SSA assumes that the next reaction will fire in the next reaction time interval $[t, t+\mu)$ with small values of μ. For systems including both fast and slow reactions, however, this assumption may not be valid if the slow reactions take a much longer time than the fast reactions. The large reaction time of slow reactions should be realized by time delay if we hope to put both fast and slow reactions in a system consistently and to study the impact of slow reactions on the system dynamics [42]. Recently, the delay SSA (delay stochastic simulation algorithm DSSA) was designed to simulate chemical reaction systems with time delays [43–45]. These methods have been used to validate stochastic models for biological systems with slow reactions [46, 47]. However, compared with the significant progress in designing simulation methods for biological systems without time delay [48, 49], only a few simulation methods have been designed to improve the efficiency of the DSSA [50, 51]. Similar to the effective methods for simulating biological systems without time delay, it is expected the progress in designing effective methods for simulating systems with time delay will also provide methodologies for modeling biological systems with time delay.

2.2.2
Accelerating τ-Leap Methods

Since the SSA can be very computationally inefficient, considerable attention has been paid recently to reducing the computational time for simulating stochastic chemical kinetics. Gillespie [52] proposed the τ-leap methods in order to improve the efficiency of the SSA while maintaining acceptable losses in accuracy. The key idea of the τ-leap methods is to take a larger time step and allow for more reactions to take place in that step. In the Poisson τ-leap method, the number of times that the reaction channel R_j will fire in the time interval $[t, t+\tau)$ is approximated by a Poisson random variable $\mathcal{P}(a_j(\mathbf{x})\tau)$ ($j = 1, \ldots, M$) based on the present state $\mathbf{x}(t)$ at time t [52]. Here, the leap size τ should satisfies the Leap Condition: a temporal leap by τ will result in a state change λ such that for every reaction channel R_j, $|a_j(\mathbf{x}+\lambda)-a_j(\mathbf{x})|$ is "effectively infinitesimal" [52]. This method is given in Method 2.2.

Method 2.2 Poisson τ-Leap Method [52].

Step 1: Calculate the values of propensity functions $a_j(x)$ based on the system state x at time t.

Step 2: Choose a value for the leap size τ that satisfies the Leap Condition.

Step 3: Generate a sample value of the Poisson random variable $\mathcal{P}(a_j(\mathbf{x})\tau)$ for each reaction channel ($j = 1, \ldots, M$).

Step 4: Perform the updates of the system by:

$$\mathbf{x}(t+\tau) = \mathbf{x}(t) + \sum_{j=1}^{M} v_j \mathcal{P}(a_j(\mathbf{x})\tau) \tag{2.2}$$

A major step of the Poisson τ-leap method is to choose an appropriate step size that satisfies the Leap condition. Gillespie first proposed a simple procedure to determine the leap size τ [52]. In this formula, the expected change of each propensity function $a_j(x)$ during $[t, t+\tau]$ should be bounded by $\varepsilon a_0(x)$ with a given error control parameter ε:

$$|a_j(\mathbf{x}+\lambda) - a_j(\mathbf{x})| \leq \varepsilon a_0(\mathbf{x}) \tag{2.3}$$

where λ is the expected net change in state in $[t, t+\tau)$, which can be calculated by $\lambda = \tau \sum_{j=1}^{M} v_j a_j(\mathbf{x})$. Later, more sophisticated methods have been proposed in order either to select the optimal leap size or to avoid the possible negative molecular numbers in simulation. For example, Gillespie and Petzold [53] proposed a method by considering both the mean and standard deviation of the expected change in the propensity functions. This method is an extension of the method (Equation 2.3) that only considered the mean of the expected change. It is worth noting that the leap size is a preselected deterministic value and is determined by the error control parameter ε. Like many other numerical methods, the leap size τ is related to the balance between computational efficiency and accuracy. In addition, our simulation results [54] indicated that the computing time for selecting the leap size is about a half of the total computing time when using the method of Gillespie and Petzold [53].

Since the samples of a Poisson random variable are unbounded, negative molecular numbers may be obtained if certain species have small molecular numbers and the propensity function involving that species has a large value. There are two ways of obtaining negative molecular numbers in stochastic simulations [55]. The first case is that the generated sample of reaction number is greater than one of the molecular numbers in that reaction channel. In the second case, a species involves a number of reaction channels and the total reaction number of these channels is greater than the copy number of that species, although the reaction number of each channel may be smaller than the molecular number.

For tackling the problem of negative numbers, binomial random variables were introduced to avoid the negative numbers of the first case by restricting the possible reaction numbers in the next time interval [55, 56]. In the binomial τ-leap method, the reaction number of channel R_j is defined by a sample value of the binomial random variable $\mathcal{B}(N_j, a_j(x)\tau/N_j)$ under the condition $0 \leq a_j(x)\tau/N_j \leq 1$. The maximal possible reaction number N_j has been defined for the widely used three types of elementary reactions. In addition, a sampling technique was designed for sampling the total reaction number of a group of reaction channels if a reactant species involves these reaction channels [55]. The binomial τ-leap method is given in Method 2.3.

Method 2.3 Binomial τ-Leap Method [55]

Step 0: Define the maximal possible reaction number N_j for each reaction channel. If a species involves two or more reaction channels $\{R_{j1}, \ldots, R_{jk}\}$, define a maximal possible total reaction number N_{jk} for these reaction channels.

Step 1: Calculate the values of propensity functions $a_j(x)$ based on the system state x at time t.

Step 2: Use a method to determine the value of leap size τ. Check the step size conditions $0 \leq a_j(x)\tau/N_j \leq 1$ of the binomial random variables. If necessary, reduce the step size τ to satisfy these conditions.

Step 3: Generate a sample value $\mathcal{B}_j$ of the binomial random variable $\mathcal{B}(N_j, a_j(x)\tau/N_j)$ for reaction channels in which species involve one single reaction. When a species involves two or more reaction channels, generate a total reaction number $\mathcal{B}\left(N_{jk}, \sum_{l=1}^{k} a_{jl}(x)\tau/N_{jk}\right)$ for these reaction channels and then generate the reaction number $\mathcal{B}_j$ of each reaction channel in this group.

Step 4: Perform the updates of the system by:

$$\mathbf{x}(t+\tau) = \mathbf{x}(t) + \sum_{j=1}^{M} v_j \mathcal{B}_j \quad (2.4)$$

In the τ-leap methods, it is assumed that, during a preselected time step τ, the number of fires of each reaction channel is a sample of a random variable. Another major type of leap method is the k_α-leap method [52]. In an implementation of the k_α-leap method, which is the so-called R-leap method [57], it was proposed to select a predefined number of firings L that may span several reaction channels. Then the leap step τ_L of these L reactions is a sample of the Γ random variable $\Gamma(L, 1/a_0(\mathbf{x}))$. Over the time interval $[t, t+\tau_L]$, the number of firings of each reaction channel K_j (satisfying $\sum_{j=1}^{M} K_j = L$) follows the correlated binomial distributions. A number of techniques have been proposed in the R-leap method to determine the total reaction number L and to sample the firing number K_j of each reaction channel [57]. A similar approach, which is called the K-leap method, was also proposed to achieve the computing efficiency over the exact SSA [58].

2.2.3
Langevin Approach

When the molecular numbers x_i ($i = 1, \ldots, N$) in a chemical reaction system are quite large, the value of $a_j(\mathbf{x})\tau$ in the Poisson τ-leap method may be large for an appropriately selected step size τ. In this case, the Poisson random variable $\mathcal{P}(a_j(\mathbf{x})\tau)$ can be approximated by a normal random variable with the same mean and variance, given by [59]:

$$\mathcal{P}(a_j(\mathbf{x})\tau) \approx N(a_j(\mathbf{x})\tau, a_j(\mathbf{x})\tau) = a_j(\mathbf{x})\tau + \sqrt{a_j(\mathbf{x})\tau}N(0,1)$$

Then the Poisson τ-leap method (2.2) can be approximated by the following formula with normal random variables:

$$\mathbf{x}(t+\tau) = \mathbf{x}(t) + \sum_{j=1}^{M} \mathbf{v}_j a_j(\mathbf{x})\tau + \sum_{j=1}^{M} \mathbf{v}_j \sqrt{a_j(\mathbf{x})\tau} N_j(0,1) \quad (2.5)$$

where the M normal variables $N_j(0,1)$ are all statistically independent. The above scheme is the explicit Euler method [60] for solving the chemical Langevin equation:

$$d\mathbf{x}(t) = \sum_{j=1}^{M} \mathbf{v}_j a_j(\mathbf{x})dt + \sum_{j=1}^{M} \mathbf{v}_j \sqrt{a_j(\mathbf{x})} dW_j(t)$$

where $W_j(t)$ is the Wiener process. If the molecular numbers in the system are very large, the value of $a_j(\mathbf{x})\tau$ may be still very large for a given step size τ. In this case, compared with the drift term $a_j(\mathbf{x})\tau$, the diffusion term $\sqrt{a_j(\mathbf{x})\tau}N_j(0,1)$ in (2.5) is neglectable. Finally, we obtained the explicit Euler method for solving the chemical rate equation:

$$\frac{d\mathbf{x}(t)}{dt} = \sum_{j=1}^{M} \mathbf{v}_j a_j(\mathbf{x})$$

The chemical Langevin equation links three types of important modeling regimes, namely the discrete stochastic models simulated by the SSA or Poisson τ-leap method, continuous stochastic models in terms of SDEs, and continuous deterministic systems of ODEs. In addition, the Langevin approach provides a method to describe internal noise of chemical reactions in the continuous SDE framework. When a reaction system has relatively large molecular numbers, the SDE models can be used to describe the system dynamics more efficiently than the discrete stochastic models. The chemical Langevin equation is also the theoretical basis of the multiscale simulation methods [61, 62]. Based on the molecular numbers and values of propensity functions, chemical reactions can be partitioned into a few reaction subsets at different time steps and then different simulation methods can be employed to simulate different subsets of chemical reactions. For example, Burrage et al. [63] proposed an adaptive approach to divide a reaction system into slow, intermediate, and fast reaction subsets, and used the SSA, Poisson τ-leap method, and SDEs to simulate the reactions in different subsets. Different partitioning techniques and different simulation methods have led to a number of effective methods and software for simulating chemical reaction systems [64–66].

2.3
Discrete Stochastic Modeling

Due to the lack of detailed knowledge of biochemical reactions, kinetic rates, and molecular numbers, stochastic models based on elementary chemical reactions may not always be the practical method to study chemical reaction systems. This section discusses a general approach to develop stochastic models based on widely used deterministic ODE models [67]. Instead of studying noise from detailed information of biochemical reactions, stochastic models will be developed by using macroscopic variables at some intermediate levels. Based on the stochastic simulation methods discussed in the previous section, the key idea of this method is to use Poisson random variables to represent chance events in protein synthesis, degradation, molecular diffusion and other biological processes. This technique is also consistent with other stochastic modeling approaches where Poisson random variables have been used for realizing the chance events in transcription and translation [68].

2.3.1
Stochastic Modeling Method

We first use a simple system to illustrate the relationship between a stochastic model, simulated by the Poisson τ-leap method, with the corresponding deterministic ODE model simulated by the Euler method. This system includes two reactions:

$$R_1: S_1 + S_2 \xrightarrow{c_1} S_3 \qquad (2.6)$$
$$R_2: S_3 \xrightarrow{c_2} S_4$$

By using the Poisson τ-leap method, the number of S_3 molecules within the time interval $[t, t+\tau)$ is updated by

$$x_3(t+\tau) = x_3(t) + \mathcal{P}[c_1 x_1(t) x_2(t)\tau] - \mathcal{P}[c_2 x_3(t)\tau]$$

By assuming the independence of x_1 and x_2, the mean $\bar{x}_i (= E(x_i))$ of molecular numbers in the above Poisson τ-leap method can be obtained by:

$$\bar{x}_3(t+\tau) = \bar{x}_3(t) + \tau[c_1 \bar{x}_1(t) \bar{x}_2(t) - c_2 \bar{x}_3(t)]$$

which is the Euler method for solving the ODE with respect to $\bar{x}_3$, given by:

$$\frac{d\bar{x}_3}{dt} = c_1 \bar{x}_1 \bar{x}_2 - c_2 \bar{x}_3$$

The above ODE is the chemical kinetic rate equation of species x_3 in the reaction system (2.6).

A further example is the enzymatic reaction:

$$E + S \underset{k_{-1}}{\overset{k_1}{\rightleftharpoons}} ES \xrightarrow{k_2} \text{Product} + E \tag{2.7}$$

with enzyme E and substrate S. The quasi-steady state assumption can be applied to approximate the concentration of the enzyme–substrate complex ES under certain conditions. Then the three reactions in (2.7) can be simplified into one single reaction:

$$S \xrightarrow{c} \text{Product} \tag{2.8}$$

where $c = V_{\max}/(K_m + s)$, $V_{\max} = k_2(E + ES)$, and $K_m = (k_{-1} + k_2)/k_1$. The SSA was used to simulate this simplified reaction [62]. When applying the Poisson τ-leap method to the one single reaction (2.8), the system is updated by:

$$s(t + \tau) = s(t) - \mathcal{P}\left[\frac{V_{\max}}{K_m + s(t)} s(t)\tau\right] \tag{2.9}$$

If the dynamics of the enzymatic reaction (2.7) is described by an equation with the Michaelis–Menten function:

$$\frac{d\bar{s}(t)}{dt} = -\frac{V_{\max}}{K_m + \bar{s}(t)} \bar{s}(t) \tag{2.10}$$

the mean $\bar{s}(t)$ of the substrate number obtained by the Poisson τ-leap method in (2.9) is well approximated by the Euler method for solving the differential equation (2.10).

These two examples indicate that we can construct discrete stochastic models for biochemical reaction systems from well-defined deterministic ODE models. Each item in the deterministic model for a biological process can be replaced by a Poisson random variable. Consider a system containing N species and described by the following deterministic model

$$\frac{d\bar{x}_i}{dt} = f_i(\bar{x}_1, \ldots, \bar{x}_N) - g_i(\bar{x}_1, \ldots, \bar{x}_N), \quad i = 1, \ldots, N$$

where $f_i(\bar{x}_1, \ldots, \bar{x}_N)$ and $g_i(\bar{x}_1, \ldots, \bar{x}_N)$ represent the increase and decrease processes in the value $\bar{x}_i$ of species S_i, respectively. Here, $\bar{x}_i$ normally represents the concentration of species S_i, whereas in stochastic models we use x_i to represent the molecular number of species S_i. It is assumed that the increase and decrease of the molecular number x_i in a time interval $[t, t + \tau)$ are samples of the Poisson random variables with mean $f_i(x_1, \ldots, x_N)\tau$ and $g_i(x_1, \ldots, x_N)\tau$, respectively, and the system is updated by:

$$x_i(t + \tau) = x_i(t) + \mathcal{P}[f_i(x_1, \ldots, x_N)\tau] - \mathcal{P}[g_i(x_1, \ldots, x_N)\tau]$$

Note that Poisson random variables in the above model can be approximated by binomial random variables in order to avoid the possible negative molecular numbers in stochastic simulations and to improve the computational efficiency [55].

If the increase process $f_i(\mathbf{x}) = f_i(x_1, \ldots, x_N)$ contains a number of macroscopic reactions, so that $f_i(x)$ can be written as:

$$f_i(\mathbf{x}) = f_{i1}(\mathbf{x}) + \cdots + f_{ik}(\mathbf{x})$$

where $f_{ij}(\mathbf{x})$ represents a process in which species S_i is involved. Then the Poisson random variable $\mathcal{P}[f_i(\mathbf{x})\tau]$ can be replaced by:

$$\mathcal{P}[f_{i1}(\mathbf{x})\tau] + \cdots + \mathcal{P}[f_{ik}(\mathbf{x})\tau]$$

This replacement is valid because the sum of two Poisson random variables $\mathcal{P}(\lambda_1)$ and $\mathcal{P}(\lambda_2)$ is also Poisson $\mathcal{P}(\lambda_1 + \lambda_2)$. Similar considerations can be applied to the decrease process $g_i(x_1, \ldots, x_N)$.

Although this modeling approach is based on the existing Poisson τ-leap method, the new modeling insight is that we do not have to go back to detailed first-principle biochemical reactions to develop stochastic models. Rather, we can take existing robust ODE models that may encapsulate detailed chemical kinetics by various Hill functions and quasi steady-state assumptions and apply the Poisson τ-leap method in the way described. This approach gives a general methodology for introducing intrinsic noise into robust deterministic models in a very simple manner.

2.3.2
Toggle Switch with the SOS Pathway

The genetic toggle switch, which is the first engineered switching network implemented on plasmids in *Escherichia coli* [26, 69] and in mammalian cells [70], is a robust bistable system comprised of two genes and regulated by a double-negative feedback loop. In Figure 2.1, this system consists of two genes, *lacI* and λ *cI*, that encode the transcriptional regulator proteins, LacR and λ CI, respectively [26, 69]. When *E. coli* cells are exposed to various concentrations of mitomycin C (MMC), the application of MMC causes DNA damage that leads to the activation of protein RecA. Activated RecA cleaves the λ CI repressor protein, resulting in the increase of the expression of gene *lacI*. Changes in gene expression levels will create an environment for cells to transfer from one steady state with high *λcI* expression level to the other steady state with high *lacI* expression level. In the absence of MMC, all cells exhibited little or no Green Fluorescent Protein (GFP) expression (low LacR expression). Nearly all of the cells expressed GFP (high LacR expression) after treatment with 500 ng/ml MMC. Bimodal population distributions were observed at

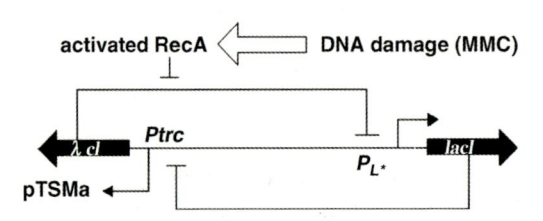

Figure 2.1 Genetic toggle switch interfaced with the SOS signaling pathway.

2.3 Discrete Stochastic Modeling

intermediate MMC concentrations for the cell numbers with different GFP expression levels [69].

A deterministic model was proposed for studying the bistability properties of the genetic toggle switch [26, 69]. Although this deterministic model can realize two distinct steady states and genetic switching, it cannot realize experimental results with different genetic switching in different cells under the same experimental conditions. Using this deterministic model and the modeling approach discussed in Section 2.3.1, a stochastic model was developed to realize experimental results with bimodal population distributions with regard to the expression levels of LacR:

$$u(t+\tau) = u(t) + \mathcal{P}\left[\varepsilon\left(\alpha_1 + \frac{\beta_1 K_1^3}{K_1^3 + v(t)^3}\right)\tau\right] - \mathcal{P}\left[\left(d_1 + \frac{\gamma s}{1+s}\right)u(t)\tau\right]$$

$$v(t+\tau) = v(t) + \mathcal{P}\left[\varepsilon\left(\alpha_2 + \frac{\beta_2 K_2^3}{K_2^3 + u(t)^3}\right)\tau\right] - \mathcal{P}[d_2 v(t)\tau]$$

(2.11)

where u and v are molecular numbers of λ CI and LacR, respectively, and s represents the effect of MMC on the degradation of λ CI. Detailed information of the model parameters can be found in Tian and Burrage [67].

As cells were exposed to various concentrations of MMC for 15 h [26], the degradation rate of λ CI is $d_1(=1)$ when $t \leq 60$ or $t \geq 960$, but it is $d_1 + \gamma s/(1+s)$ in $t \in [60, 960]$. When $60 \leq t \leq 960$, this large degradation rate in deterministic simulations will shift the system from the steady state with high λ CI expression level to an intermediate state and genetic switching will happen only if the concentration of λ CI is below a threshold value. Simulations showed that there was no switching for $s \leq 2.0$, but switching could occur for $s > 2.0$ (Figure 2.2a, $s = 2.12$). However, the situation with stochastic simulations is entirely different. Figure 2.2 gives two simulations for an unsuccessful transition in Figure 2.2(b) and a successful transition in Figure 2.2(c) with the same degradation parameter $s = 1.7$. In both simulations, the decrease of λ CI shifts the system from the steady state with high λ CI expression level to an intermediate unstable state. Intrinsic noise may also switch the system from this intermediate unstable state to the other steady state with high LacR expression level. If the transition between the steady states does not happen during $t \in [60, 960]$, the system will bounce back to the initial steady state. Figure 2.2(d) reports the percentages of switched cells based on different values of s. These percentages can be approximated by a Hill function with a Hill coefficient 4 that best fits the simulated percentages.

Figure 2.3 gives four bimodal distributions for the number of cells with different LacR molecular numbers when the degradation parameter s is 1.3, 1.7, 1.75, and 2, respectively. These simulated bimodal distributions are compared with experimental results that are derived from Figure 3(b) in [26]. Qualitative comparisons in Figure 2.3 indicate that simulation results are consistent with experimental results in terms of the percentages of switched cells.

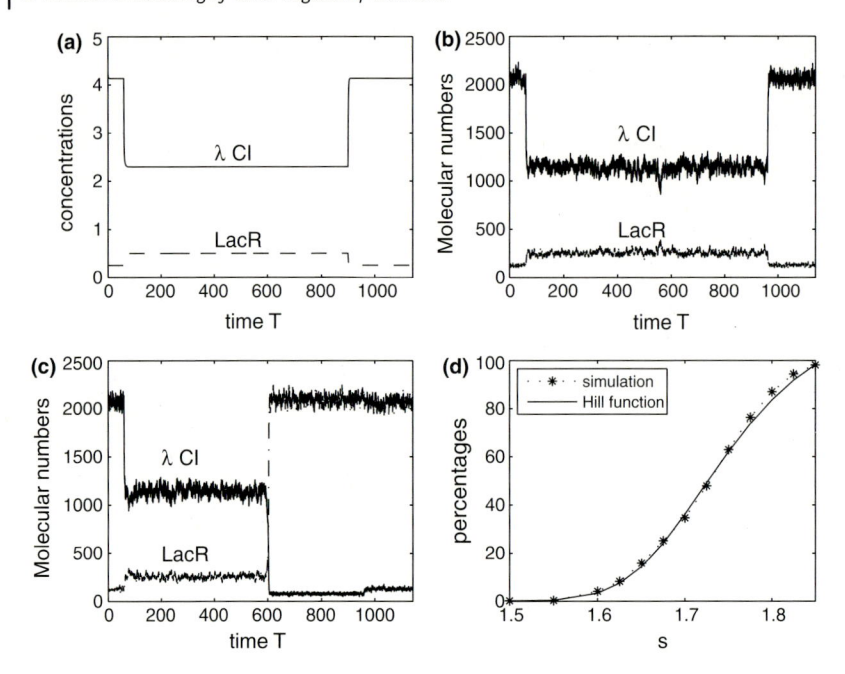

Figure 2.2 Simulations of the genetic toggle switch interfaced with the SOS signaling pathway. (a) Deterministic simulation of successful switching ($s = 2.12$). (b) Stochastic simulation of unsuccessful switching based on $s = 1.7$. (c) Stochastic simulation of successful switching also based on $s = 1.7$. (d) Percentages of switched cells in stochastic simulations based on different degradation parameter s and percentages obtained by a Hill function $p(s) = 1.2364*(s-s_0)4/(0.254 + (s-s_0)4)*100\%$, where 1.2364 is used to match the simulated percentage when $s = 1.85$.

2.3.3
Other Models for the Genetic Toggle Switch

Based on the deterministic model of the genetic toggle switch [69], a stochastic model including four macroscopic reactions was developed to study the sensitivity of discrete stochastic systems [71]. Each reaction in the system uses a nonlinear rate expression and represents either the production or removal of a molecule in the system. These four reactions are:

$$\emptyset \xrightarrow{a_1} \text{LacI}, \quad \text{LacI} \xrightarrow{a_2} \emptyset, \quad \emptyset \xrightarrow{a_3} \text{cIts}, \quad \text{cIts} \xrightarrow{a_4} \emptyset$$

where the a_i are propensity functions that come directly from the deterministic model [69] and are normalized to the system volume Ω:

$$a_1 = \frac{\alpha_1 \Omega}{1 + ([\text{cIts}_d/\Omega])^\beta}, \quad a_2 = \text{LacI}_d, \quad a_3 = \frac{\alpha_2 \Omega}{1 + ([\text{LacI}_d^*])^\gamma}, \quad a_4 = \text{cIts}_d$$

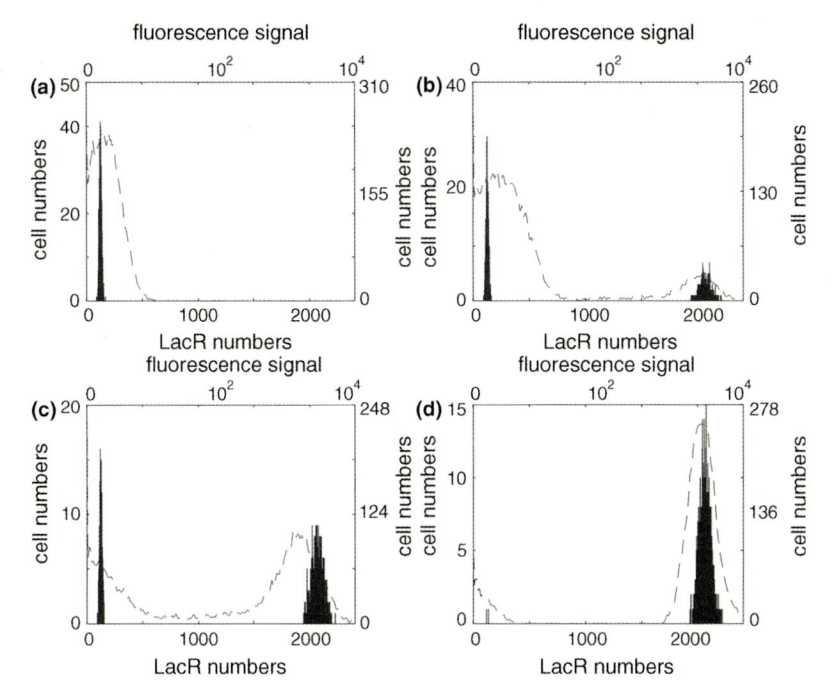

Figure 2.3 Comparison of simulation results with experimental observations for the genetic toggle switch interfaced with the SOS signaling pathway. Numbers of cells with different LacR molecular numbers are based on 1000 simulations and experimental observations in fluorescence signaling are derived from Figure 3(b) in [26] using the top and right labels. (a) $s = 1.3$: no cell has high LacR expression level and no MMC was applied in experiments. (b) $s = 1.7$: 35.9% of cells have high LacR expression levels and 1 ng/ml MMC was applied. (c) $s = 1.75$: 67.1% of cells have high LacR expression level and 10 ng/ml MMC was applied. (d) $s = 2.0$: all cells have high LacR expression levels and 500 ng/ml MMC was applied.

Then the SSA was used to simulate the dynamics of this stochastic system. This stochastic modeling approach is consistent with the method discussed in Section 2.3.1 in terms of using the macroscopic variables at certain intermediate levels. A similar modeling approach was used in an earlier work to study the circadian oscillations of the PER–TIM regulatory network in Drosophila. To assess the role of fluctuations in mRNA and protein copy numbers on circadian rhythmicity, two stochastic models were designed from a deterministic model [72]. Among them, a stochastic model was developed by decomposing the deterministic model into a system with 30 elementary reactions [73] and the other stochastic model used 10 macroscopic reactions with nonlinear rate expressions that were directly derived from the deterministic model [74]. A comparison of these two stochastic models suggested that these two approaches were equivalent and corroborated the predictions of the deterministic model [74, 75].

In addition, SDEs have been used to study the effect of external noise on the dynamics of the genetic toggle switch [76]. Similar to the stochastic models using stochastic degradation rates [35, 77], the proposed stochastic model is given by:

$$\frac{dx}{dt} = \frac{\alpha_1}{1+y^n} - (d_1 + \xi_1(t))x + \gamma_1$$

$$\frac{dy}{dt} = \frac{\alpha_2}{1+x^n} - (d_2 + \xi_2(t))x + \gamma_2$$

where the $\xi_i(t)$ are random terms with zero mean $\langle\xi_i(t)\rangle = 0$. In order to encapsulate rapid random fluctuations, it was assumed that the autocorrelation is δ-correlated (i.e., $\langle\xi_i(t)\xi_j(t')\rangle = D\delta_{i,j}(t-t')$). Simulation results indicated that the multiplicative noise in degradation rates could induce successive switching between two steady states for a moderate noise intensity D. However, low noise only enabled transitions from the upper state to the lower state, but the random kicks were not sufficient enough to let the system climb the steep barrier from the lower state back to the higher state [76].

A stochastic model was also developed to study the genetic toggle switch interfaced with a quorum sensing signaling pathway [67]. This stochastic model was developed from a deterministic ODE model by using the method in Section 2.3.1 and it successfully realized the bimodal distributions for the numbers of cells with different molecular numbers of LacR proteins expressed from the genetic toggle switch with different cell population densities. The simulation results are consistent with experimental results in terms of the percentages of switched cells [67].

2.4
Continuous Stochastic Modeling

This section uses the λ phage – a virus that infects the bacterium *E. coli* – as the test system to discuss a number of modeling approaches for gene regulatory networks in the framework of continuous differential equations. The λ phage has been studied extensively from both experimental and modeling perspective, because it is one of the simplest developmental switches [78, 79]. The biochemical reactions involved in the induction from lysogen to lysis are sufficiently well understood, which led to a number of well-developed deterministic differential equation models [80–85]. In the field of genetic network analysis, the λ regulatory network has often been used as a testing ground for modeling methodology [10, 86]. Specifically, Arkin *et al.* used the λ phage network to carry out the pioneering research on stochastic modeling of regulatory networks inside the cell [1, 2, 87].

2.4.1
Deterministic Models for the λ Phage Network

The λ phage is called a temperate phage because it can be in the form of either lysis or lysogen. In the lysogenic form, the virus will replicate passively whenever the

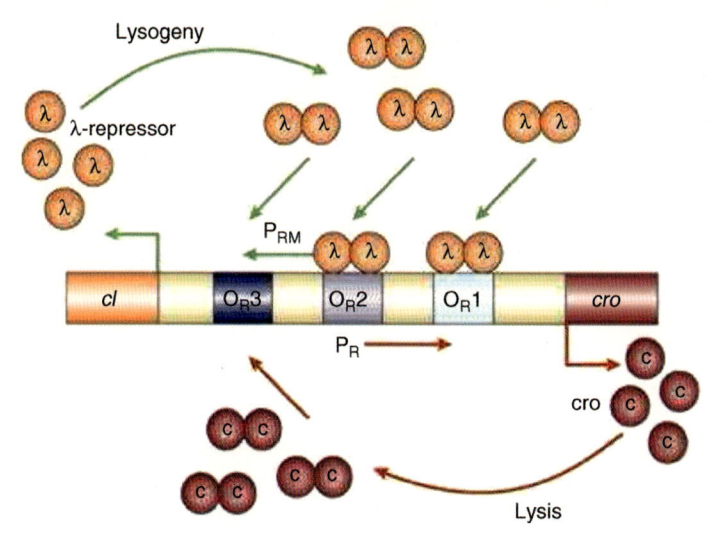

Figure 2.4 Right operator region (O_R) of λ phage. The three operators O_R1, O_R2, and O_R3 bind two proteins: λ repressor (λ) and cro (c). (Reproduced with permission from [10], © 2001 Macmillan Magazines.)

host bacterium replicates. Under the right conditions, a lysogen can be induced from the lysogenic pathway to the lysis pathway. Figure 2.4 shows that the right operator region O_R consists of three binding sites. The dimeric forms of repressor and cro bind to these binding sites to regulate the transcription of genes *cI* and *cro* [78, 79, 83]. Biochemical reactions in this system are classified into fast reactions and slow reactions. Fast reactions include two monomer–dimer reactions and binding reactions of ligands to the promoter sites, and they are assumed to be in an equilibrium state. Slow reactions represent transcription and degradation. A complete list of the possible binding configurations and the corresponding free energy of these configurations can be found in [85, 88]. Kinetic models with ODEs have been developed for describing the principle of the λ right operator control system [81–85]:

$$\frac{dx}{dt} = S_x P_x(x, y) - k_{dx} x$$
$$\frac{dy}{dt} = S_y P_y(x, y) - k_{dy} y \quad (2.12)$$

with:

$$P_x(x, y) = \alpha(f_{31} + f_{32} + f_{33}) + \sum_{m=34}^{40} f_m$$
$$P_y(x, y) = f_{28} + f_{29} + f_{30} + f_{40} \quad (2.13)$$

$$f_s = \frac{K_s(K_r x^2)^{i(s)}(K_c y^2)^{j(s)}[\text{RNAP}]^{k(s)}}{\sum\limits_{m=1}^{40} K_m(K_r x^2)^{i(m)}(K_c y^2)^{j(m)}[\text{RNAP}]^{k(m)}}$$

where x and y are the concentrations of repressor and cro, respectively, S_x and S_y are the synthesis rates of repressor and cro, respectively, k_{dx} and k_{dy} are the degradation rates for repressor and cro, respectively, f_s is defined by the sth configuration, $K_s = \exp(\Delta G_s/RT)$ is the equilibrium constant with the free energy ΔG_s, and $\alpha = 11$ is the indication of positive feedback of the repressor. More detailed information about the kinetic rates and equilibrium constant can be found in [77].

This deterministic model was used to determine the roles of several key molecular processes in the system that had been characterized by studies *in vitro*. For example, computer simulations demonstrated that the cooperative binding is the major mechanism to guarantee the stability of the lysogenic state against minor changes in CI repressor level inside the cell [85]. However, this kinetic model of λ phage exhibits no bistability property with experimentally estimated values. The possible reason may be due to some important aspects of the switching regulatory mechanisms in the standard "word model" that were omitted in the mathematical model [10, 84].

2.4.2
Stochastic Models for External Noise

SDEs have been used to study the influence of environmental fluctuations on the dynamics of the lysis–lysogeny pathway [35]. Rather than studying the pathway with all the 40 binding configurations, the proposed system included only a number of major binding configurations and nonspecific binding was ignored. Bifurcation analysis demonstrated that the proposed model for a system with three binding sites (O_R1, O_R2, and O_R3) had a much larger bistable region than a mutant system with only two binding sites (O_R2 and O_R3) [35].

Based on the model for the system with three binding sites, external noise was introduced into the deterministic model to study the influence of environmental fluctuations on the system bistability. Additive noise resource was first considered for the effect of randomly perturbations on the basal production rate of the repressor. The stochastic model with additive noise was given by:

$$\frac{dx}{dt} = \frac{\alpha(2x^2 + 50x^4)}{25 + 29x^2 + 52x^4 + 4x^6} - \gamma x + 1 + \xi(t)$$

where x is the dimensionless concentration of repressor, α is a measure of the degree in which the transcription rate is increased above the basal rate by repressor binding, γ is proportional to the relative strengths of the degradation and basal rates, and $\xi(t)$ is a rapidly fluctuating random term with zero mean ($\langle\xi(t)\rangle = 0$). In addition, a stochastic model with multiplicative noise was introduced to investigate the effects of a noise source that alters the transcription rate. The parameter α was allowed to vary stochastically, namely $\alpha \to \alpha + \xi(t)$. In this manner, the stochastic model was introduced as:

$$\frac{dx}{dt} = \frac{\alpha(2x^2 + 50x^4)}{25 + 29x^2 + 52x^4 + 4x^6} - \gamma x + 1 + \frac{(2x^2 + 50x^4)}{25 + 29x^2 + 52x^4 + 4x^6} \xi(t)$$

Hasty et al. [35] then showed how additive and multiplicative external noise could be used to regulate expression. In the additive case, simulations demonstrated that low noise enabled only transitions from the upper state to the lower state, because random kicks were not sufficient to let the system climb the steep barrier from the lower state, whereas high noise induced transitions between both of the states. Similar results about the genetic switching have also been observed in the study of the stochastic simulations of the genetic toggle switch system [76]. In the multiplicative case, it was shown that small deviations in the transcription rate could lead to large fluctuations in the production of protein and these fluctuations could be used to amplify protein production significantly. In addition, similar observations of the genetic switching have been observed in a more detailed model that includes three binding sites and major binding configurations of both repressor and cro [89].

2.4.3
Deterministic Models with Threshold Values

Using model (2.12) and parameter values determined by experiments, numerical simulations indicated that bistability did not exist in this system with the given parameter values [77]. The major reason of the nonexistence of bistability is the assumption of continuity for protein concentrations, which is valid if the molecular numbers may be large. Since the concentrations of CI and cro are low, functions are needed to indicate binding strengths and synthesis rates for different possibilities of biochemical reactions at different developmental stages.

Based on the different developmental stages of the lysogenic pathway, it was assumed that the synthesis rate $\bar{S}_x$ of repressor reaches the maximal constant value S_x if the repressor concentration exceeds a certain value x_1^*. However, when repressor concentration is below x_1^*, it was supposed that the synthesis rate is a linear function with value 0 at $x = 0$ and value S_x at $x = x_1^*$. This consideration leads to the following synthesis rate:

$$\bar{S}_x = \begin{cases} S_x, & x \geq x_1^* \\ S_x x / x_1^*, & x < x_1^* \end{cases}$$

The next function is associated with binding reactions to the operator site $O_R 3$. When the CI concentration is under a certain value, denoted by $x < x_2^*$, the binding reaction number of CI dimer to $O_R 3$ is likely to be zero. In this case we mathematically treated these relatively small probabilities to be zero since the reaction seldom occurs at that state condition. When the CI concentration is above the value x_2^*, the binding strength of CI dimer to $O_R 3$ will increase gradually in accordance with the increase of the repressor concentration. With the assumed threshold value x_2^*, the binding strength for a number of configurations is of the form:

$$\bar{K}_i = \frac{K_i}{k_x}(x-x_2^*)^+ = \begin{cases} \frac{K_i}{k_x}(x-x_2^*), & x > x_2^* \\ 0, & x \leq x_2^* \end{cases}$$

Based on the discussion above, the mathematical model of the λ phage system with threshold values is given by:

$$\frac{dx}{dt} = \bar{S}_x \bar{P}_x(x,y) - k_{dx} x$$

$$\frac{dy}{dt} = \bar{S}_y \bar{P}_y(x,y) - k_{dy} y$$

Here, functions $\bar{P}_x(x,y)$ and $\bar{P}_y(x,y)$ are defined by (2.13), but the binding probabilities of a number of configurations are given by:

$$\bar{f}_s = \frac{\bar{K}_s (K_r x^2)^{i(s)} (K_c y^2)^{j(s)} [RNAP]^{k(s)}}{\sum_{m=1}^{40} \bar{K}_m (K_r x^2)^{i(m)} (K_c y^2)^{j(m)} [RNAP]^{k(m)}}$$

This deterministic model successfully realized the bistability property of the λ phage system [77]. Numerical simulations showed that genetic switching could be realized by using a large degradation rate. This work represents an innovative technique to realize chemical reactions involving species with small copy numbers in the continuous deterministic regime, although more sophisticated techniques and/or mathematical functions are required to realize such chemical events with very small probabilities.

2.4.4
Stochastic Switching

A stochastic model was developed to realize the genetic switching in induction under the influence of intrinsic noise [77]. Since cleavage of the repressor is dominant in the process of induction, a stochastic degradation rate of the repressor was used to realize intrinsic noise in the system. As with the τ-leap method, the number of repressors degraded in the time period $[t, t+\tau)$ is a Poisson random variable $\mathcal{P}(k_{dx} x \tau)$, where x is the present number of the repressor. In order to be consistent with the continuous synthesis process, this Poisson random variable was approximated by a Gaussian random variable:

$$\mathcal{P}(k_{dx} x \tau) \approx N(k_{dx} x \tau, k_{dx} x \tau) = k_{dx} x \tau + \sqrt{k_{dx} x \tau} N(0,1)$$

This approximation is valid if the mean of the Poisson random variable is relatively large. This consideration formed the following stochastic equation model for describing the system dynamics in induction, given by:

$$dx = \bar{S}_x(\bar{P}_x(x,y) - k_{dx}x)dt - \sqrt{k_{dx}x}\,dW(t)$$

$$dy = S_y(\bar{P}_y(x,y) - k_{dy}y)dt \tag{2.14}$$

Figure 2.5 shows that the occurrence of switching for the stochastic model is determined by the intrinsic noise in the system. With the same control parameters including the repressor degradation rate, the λ phage system may switch from the lysogenic state to the lysis state (Figure 2.5a) or stay in the lysogenic state (Figure 2.5b). In addition, the stochastic model (2.14) was used to predict the proportions of the induced λ phage with different strengths of the inducer. Numerical simulations suggested that it was appropriate to use a linear function to represent the relationship between the degradation rate of repressor (k_{dx}) and strength of UV light [77].

2.5
Stochastic Models for Both Internal and External Noise

So far we have discussed stochastic models that were designed to study either external noise or internal noise. Computer simulations were used to illustrate the key functions of noise in determining the evolutionary pathways and genetic switching. Another type of research for noise is to derive the analytical expressions of the mean and variance of system dynamics via a simple mathematical model. Thattai and van Oudenaarden [68] used the expression of a single gene to show that the noise is essentially determined at the translational level. In addition, it has shown that a negative feedback loop could decrease the system noise. Swain et al. [90] decomposed the total variation in the level of expression into intrinsic and extrinsic components. Analytic expressions for intrinsic noise have been derived for a more detailed model that accounted for several steps in transcriptional initiation, as well as replication, cell growth, and division. Paulsson et al. [91, 92] studied the intrinsic and extrinsic noise

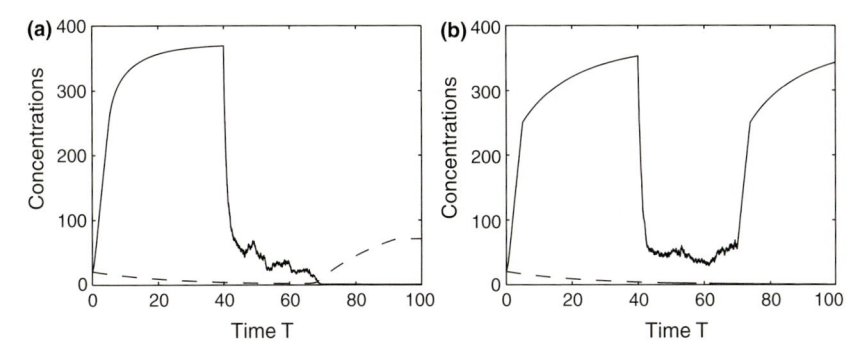

Figure 2.5 Simulations and statistical properties of the stochastic model (2.14). (a) Simulation of successful switching with $k_{dx} = 1.0$ at 40–70 min. (b) Simulation of unsuccessful switching with the same condition in (a). The occurrence of switching is determined by intrinsic noise in the system.

by assuming that transcription and translation are burst distributed random processes. These burst expression processes have been observed recently in biological experiments in single cells [93, 94].

Although it has been widely recognized that noise can be classified into internal noise and external noise, only a few research works have investigated the interacting effects of external and internal noise. One of the early studies of this question was based on a simple network including two species (e.g., mRNAs and proteins) [19]. The probability of having n_1 and n_2 molecules per cell of species X_1 and X_2 are determined by a birth-and-death Markov process:

$$n_1 \xrightarrow{R_1^{\pm}(n_1)} n_1 \pm 1 \quad \text{and} \quad n_2 \xrightarrow{R_2^{\pm}(n_1, n_2)} n_2 \pm 1$$

The stationary fluctuations around a stable fixed point follow:

$$\frac{\sigma_2^2}{\langle n_2 \rangle^2} \approx \underbrace{\frac{1}{\langle n_2 \rangle H_{22}}}_{\text{Intrinsic } n_2 \text{ noise}} + \underbrace{\frac{\sigma_1^2}{\langle n_1 \rangle^2}}_{n_1 \text{ noise}} \underbrace{\frac{H_{21}^2}{H_{22}^2}}_{\text{Susceptibility}} \underbrace{\frac{H_{22}/\tau_2}{H_{11}/\tau_1 + H_{22}/\tau_2}}_{\text{Time-averaging}}$$

$$\underbrace{}_{\text{Extrinsic } n_2 \text{ noise}}$$

where σ_i are standard deviations, $\langle n_i \rangle$ are averages, and τ_i are average lifetimes. The above results indicate that intrinsic noise depends on the averaged molecular number and also suggested that systematic adjustments (rate H_{22}/τ_2) quench spontaneous fluctuations. However, extrinsic noise instead depends on the magnitude of n_1 fluctuations and this formula shows how strongly the external (n_1) noise influences the system (n_2) noise [19].

Recently, Lei [95] used the transcriptional system of a single gene to derive the analytic expression about the interacting effects of external and internal noise. In this network, a gene moves between a repressed state and an activated state with rates λ_1^+ and λ_1^-, respectively. The activated gene synthesis mRNA at rate λ_2 and mRNA produces proteins at rate λ_3. The degradation rates of mRNA and protein are δ_2 and δ_3, respectively. This system with both external noise and internal noise was modeled by the chemical Langevin equations:

$$\frac{dX_1}{dt} = \lambda_1^+(n-X_1) - \lambda_1^- X_1 + \sqrt{\lambda_1^+(n-X_1)}\xi_1(t) - \sqrt{\lambda_1^- X_1}\xi_2(t)$$
$$+ f_{\lambda_1^+}(n-X_1)\mu_{\lambda_1^+} - f_{\lambda_1^-} X_1 \mu_{\lambda_1^-}$$

$$\frac{dX_2}{dt} = \lambda_2 X_1 - \delta_2 X_2 + \sqrt{\lambda_2 X_1}\xi_3(t) - \sqrt{\delta_2 X_2}\xi_4(t) \quad (2.15)$$
$$f_{\lambda_2} X_1 \mu_{\lambda_2}(t) - f_{\delta_2} X_2 \mu_{\delta_2}(t)$$

$$\frac{dX_3}{dt} = \lambda_3 X_2 - \delta_3 X_3 + \sqrt{\lambda_3 X_2}\xi_5(t) - \sqrt{\delta_3 X_3}\xi_6(t)$$
$$f_{\lambda_3} X_2 \mu_{\lambda_3}(t) - f_{\delta_3} X_3 \mu_{\delta_3}(t)$$

where X_1, X_2, and X_3 are the amounts of activated genes, mRNAs, and proteins, respectively, and n is a fixed gene copy number. Here fs are constants representing the standard deviations of noise perturbation. The, ξs and μs are independent white noises standing for intrinsic and extrinsic noises, respectively. In this stochastic model, the Itô interpretation was applied to the intrinsic noise, which is consistent with other studies using SDE models. However, an interesting result is that Stratonovich interpretation was applied to the extrinsic noise [95].

The analytical solution of the model (2.15) shows that the averages of the molecule numbers depend on the extrinsic noise strengths. Specifically, fluctuations in λ_1^-, δ_2, or δ_3 may increase the mean protein number, while fluctuation in λ_1^+ may decrease the mean protein number [95]. Mathematically, this effect originated from the Stratonovich interpretation for the extrinsic noises, which is consistent with the recent discoveries from stochastic simulation [96]. In addition, analytic results suggest that extrinsic noise affects the total fluctuations in protein numbers through multiple ways, including the extrinsic fluctuation, the correlation with intrinsic noises, the modification of the time averaging of transcription and translation, and the amplification of the overall fluctuations by the impact factors [95].

2.5.1
Stochastic Models for Microarray Gene Expression Data

The raw data from microarray assays often exhibit large, unexplained fluctuations. Microarray measurements are subject to many sources of variation that can be attributed to biological and technical causes [97]. In recent years, extensive research has been carried out on investigating the error models for describing the distributions of measurements of microarray experiments [98, 99]. In addition, stochastic models have been designed to investigate the impact of noise in microarray data on the inference of gene networks [100, 101]. Most recently, SDE models haven been developed to establish the relationship between the noise strength of stochastic models and parameters of the Rosetta error model for describing the distribution of the microarray measurements [102]. In the Rosetta error model [99], it was proposed that the intensity dependent error can be decomposed into:

$$\varepsilon = \varepsilon_{add} + \varepsilon_{Poisson} + \varepsilon_{frac}$$

where the additive error ε_{add} includes the background noise and cross-hybridization noise, the Poisson error $\varepsilon_{Poisson}$ represents the randomness of the hybridization binding process, and the multiplicative error ε_{frac} includes the variation of array spotting or probe synthesis. Therefore, noise in microarray gene expression datasets includes both external and internal noise. Following the notations introduced in [99], the variance of the modeled error is represented by:

$$\sigma^2 = \sigma_{add}^2 + \text{POISSON} \times I + \text{FRACTION}^2 \times I^2 \qquad (2.16)$$

where I represents the hybridization intensity measurement of a specific gene. The values of POISSON and FRACTION have been estimated from datasets obtained by some commonly used microarray technologies [99].

The target genes of the tumor suppressor gene p53 were used as the test problem. Since p53 is in the form of a tetramer as a transcriptional factor [103], a nonlinear model of a Hill function with coefficient 4 was used to model the dynamics of gene transcription [104]. Stochastic models:

$$dI = \left[a + b\frac{u^4}{K^4 + u^4} - kI\right]dt + \xi(t)$$

were developed by introducing additive noise ($\xi(t) = \mu dW(t)$), Langevin noise ($\xi(t) = +\mu\sqrt{kI}dW(t)$), or multiplicative noise ($\xi(t) = \mu kI dW(t)$) into the deterministic models [102]. Here, $W(t)$ is the Wiener process. Numerical simulations demonstrated that the simulated variance from models with different types of stochastic degradation process can be represented by a monomial in terms of the hybridization intensity measurement with different orders. Since the Rosetta error model (2.16) is a second-order polynomial, stochastic models with multiple stochastic processes were proposed to realize the noise predicted by the Rosetta error model. However, only the Langevin and multiplicative noise were studied by stochastic models since no quantitative information was available for additive noise in the Rosetta error model [99]. The following SDE model was proposed to generate simulations whose variance matches the POISSON and FRACTION noise in the Rosetta error model:

$$dI = \left[a + b\frac{u^4}{K^4 + u^4} - kI\right]dt + \mu_1\sqrt{kI}dW_1(t) + \mu_2 kI dW_2(t)$$

where $u(t)$ is the activity of transcription factor p53 at time t. W_1 and W_2 are two independent Wiener processes. Simulations were used to establish the relationship between the model parameters μ_i and the coefficients POISSON and FRACTION in the Rosetta error model [102]. In addition, this work established a general method to develop stochastic models from experimental information.

2.6
Conclusions

This chapter discusses a number of approaches for developing stochastic models of chemical reaction systems. We concentrate on the methods for introducing stochastic processes into the widely used deterministic ODE models with macromolecular variables. The Langevin approach and recently developed effective stochastic simulation methods provide innovative methodologies to represent the internal noise in either discrete or continuous mathematical models. When the molecular numbers in the system are relatively large, the Langevin approach uses SDE models to characterize the internal noise in the continuous regime in an effective way. Specifically, this chapter discusses an approach to use Poisson random variables that link the stochastic simulation of biochemical reaction systems to the Euler method for solving ODEs via the mean. On the other hand, the effects of external noise in reaction systems are widely studied by the addition of simple scaled noise processes such as additive noise or multiplicative noise based on Wiener processes.

Finally, an important question in mathematical modeling is how to represent chemical reactions that occur at low frequencies in a continuous modeling framework for biological systems with species of very low copy numbers. This chapter discusses a method to use functions instead of rate constants to represent binding strengths and synthesis rates in biochemical reactions with low molecule concentrations. The threshold values were used for indicating the effectiveness of positive and negative feedback regulatory mechanisms.

Although SDE models have been widely used to describe extrinsic noise, it is still a challenge to determine the noise strength in SDE models based on experimental information. A widely employed technique is to use the noise strengths as adjustable parameters to realize important system properties such as the genetic switching of bistable regulatory networks. A recent study using the p53 gene network showed that, for microarray gene expression data, the noise strength in stochastic models should be determined by the variance of microarray expression profiles [102]. Important properties of biological systems should be realized by both regulatory mechanisms and properly defined noise that reflects environmental fluctuations. It is expected that this important feature of stochastic modeling should be applied to other complex systems in which noise plays a key role in determining the system dynamics.

References

1 Ackers, G.K., Johnson, A.D., and Shea, M.A. (1982) *Proc. Natl. Acad. Sci. USA*, **79**, 1129.
2 Alon, U., Surette, M.G., Barkai, N., and Leibler, S. (1999) *Nature*, **397**, 168.
3 Arkin, A., Ross, J., and McAdams, H.H. (1998) *Genetics*, **149**, 1633.
4 Auger, A., Chatelain, P., and Koumoutsakos, P. (2006) *J. Chem. Phys.*, **125**, 084103.
5 Barkai, N. and Leibler, S. (1997) *Nature*, **387**, 913.
6 Barrio, M., Burrage, K., Leier, A., and Tian, T. (2006) *PLoS Comput. Biol.*, **2**, e117.
7 Bayati, B., Chatelain, P., and Koumoutsakos, P. (2009) *J. Comput. Phys.*, **228**, 5908.
8 Becskei, A. and Serrano, L. (2000) *Nature*, **405**, 590.
9 Blake, W.J., Kaern, M., Cantor, C.R., and Collins, J.J. (2003) *Nature*, **422**, 633.
10 Bratsun, D., Volfson, D., Tsimring, L.S., and Hasty, J. (2005) *Proc. Natl. Acad. Sci. USA*, **102**, 14593.
11 Burrage, K., Tian, T., and Burrage, P.M. (2004) *Prog. Biophys. Mol. Biol.*, **85**, 217.
12 Cai, X. (2007) *J. Chem. Phys.*, **126**, 124108.
13 Cai, X. and Xu, Z. (2007) *J. Chem. Phys.*, **126**, 074102.
14 Chaouiya, C. (2007) *Brief. Bioinform.*, **8**, 210.
15 Chatterjee, A., Vlachos, D.G., and Katsoulakis, M.A. (2005) *J. Chem. Phys.*, **122**, 024112s.
16 Chen, K.C., Wang, T.Y., Tseng, H.H., Huang, C.Y., and Kao, C.Y. (2005) *Bioinformatics*, **21**, 2883.
17 Cho, H. and Lee, J.K. (2004) *Bioinformatics*, **20**, 2016–2025.
18 Climescu-Haulica, A. and Quirk, M.D. (2007) *BMC Bioinformatics*, **8** (Suppl. 5), S4.
19 Darling, P.J., Holt, J.M., and Ackers, G.K. (2000) *J. Mol. Biol.*, **302**, 625.
20 de Jong, H. (2002) *J. Comput. Biol.*, **9**, 67–103.
21 Dodd, I.B., Perkins, A.J., Tsemitsidis, D., and Egan, J.B. (2001) *Genes Dev.*, **15**, 3013.
22 Elf, J., Li, G.W., and Xie, X.S. (2007) *Science*, **316**, 1191.
23 Elowitz, M.B., Levine, A.J., Siggia, E.D., and Swain, P.S. (2002) *Science*, **297**, 1183.

24 Gardner, T.S., Cantor, C.R., and Collins, J.J. (2000) *Nature*, **403**, 339.
25 Gibson, M.A. and Bruck, J. (2000) *J. Phys. Chem.*, **104**, 1876.
26 Gillespie, D.T. (1976) *J. Comput. Phys.*, **22**, 403.
27 Gillespie, D.T. (1977) *J. Phys. Chem.*, **81**, 2340.
28 Gillespie, D.T. (2000) *J. Chem. Phys.*, **113**, 297.
29 Gillespie, D.T. (2001) *J. Chem. Phys.*, **115**, 1716.
30 Gillespie, D.T. (2007) *Annu. Rev. Phys. Chem.*, **58**, 35.
31 Gillespie, D.T. and Petzold, L.R. (2003) *J. Chem. Phys.*, **119**, 8229.
32 Goldbeter, A. (1995) *Proc. R. Soc. London B*, **261**, 319.
33 Golding, I., Paulsson, J., Zawilski, S.M., and Cox, E.C. (2005) *Cell*, **123**, 1025.
34 Gonze, D., Halloy, J., and Goldbeter, A. (2002) *Proc. Natl. Acad. Sci. USA*, **99**, 673.
35 Gonze, D., Halloy, J., and Goldbeter, A. (2002) *J. Biol. Phys.*, **28**, 637.
36 Gonze, D., Halloy, J., Leloup, J.C., and Goldbeter, A. (2003) *C. R. Biol.*, **326**, 189.
37 Goss, P.J. and Peccoud, J. (1998) *Proc. Natl. Acad. Sci. USA*, **95**, 6750.
38 Gunawan, R., Gao, Y., Petzold, L., and Doyle, F.J. III (2005) *Biophys. J.*, **88**, 2530.
39 Harris, L.A. and Clancy, P.A. (2006) *J. Chem. Phys.*, **125**, 144107.
40 Haseltine, E.L. and Rawlings, J.B. (2002) *J. Chem. Phys.*, **117**, 6959.
41 Hasty, J., McMillen, D., Isaacs, F., and Collins, J.J. (2001) *Nat. Rev. Genet.*, **2**, 268.
42 Hasty, J., Pradines, J., Dolnik, M., and Collins, J.J. (2000) *Proc. Natl. Acad. Sci. USA*, **97**, 2075.
43 Hasty, J. and Collins, J.J. (2002) *Nat. Genet.*, **31**, 13.
44 Hasty, J., Issacs, F., Dolnik, M., McMillen, D., and Collins, J.J. (2001) *Chaos*, **11**, 207.
45 Hume, D.A. (2000) *Blood*, **96**, 2323.
46 Jõers, A., Jaks, V., Kase, J., and Maimets, T. (2004) *Oncogene*, **23**, 6175.
47 Johnson, A.D., Poteete, A.R., Lauer, G., Sauer, R.T., Ackers, G.K., and Ptashne, M. (1981) *Nature*, **294**, 217.
48 Kaern, M., Elston, T.C., Blake, W.J., and Collins, J.J. (2005) *Nat. Rev. Genet.*, **6**, 451.
49 Kar, S., Baumann, W.T., Paul, M.R., and Tyson, J.J. (2009) *Proc. Natl. Acad. Sci. USA*, **106**, 6471.
50 Kaznessis, Y.N. (2006) *Chem. Eng. Sci.*, **61**, 940.
51 Kitano, H. (2002) *Nature*, **420**, 206.
52 Kitano, H. (2004) *Nat. Rev. Genet.*, **5**, 826.
53 Kobayashi, H., Kaern, M., Araki, M., Chung, K., Gardner, T.S., Cantor, C.R., and Collins, J.J. (2004) *Proc. Natl. Acad. Sci. USA*, **101**, 8414.
54 Kramer, B.P., Viretta, A.U., El Baba, M.D., Aubel, D., Weber, W., and Fussenegger, M. (2004) *Nat. Genet.*, **22**, 867.
55 Lapidus, S., Han, B., and Wang, J. (2008) *Proc. Natl. Acad. Sci. USA*, **105**, 6039.
56 Levine, E. and Hwa, T. (2007) *Proc. Natl. Acad. Sci. USA*, **104**, 9224.
57 Lei, J. (2009) *J. Theor. Biol.*, **256**, 485.
58 Leier, A., Marquez-Lago, T.T., and Burrage, K. (2008) *J. Chem. Phys.*, **128**, 205107.
59 Llaneras, F. and Picó, J. (2008) *J. Biosci. Bioeng.*, **105**, 1.
60 Ma, L., Wagner, J., Rice, J.J., Hu, W., Levine, A.J., and Stolovitzky, G.A. (2005) *Proc. Natl. Acad. Sci. USA*, **102**, 14266.
61 Mao, L. and Resat, H. (2004) *Bioinformatics*, **20**, 2258.
62 Maurer, R., Meyer, B.J., and Ptashne, M. (1980) *J. Mol. Biol.*, **139**, 147.
63 McAdams, H.H. and Arkin, A. (1997) *Proc. Natl. Acad. Sci. USA*, **94**, 814.
64 McAdams, H.H. and Arkin, A. (1999) *Trends Genet.*, **15**, 65.
65 Meyer, B.J., Maurer, R., and Ptashne, M. (1980) *J. Mol. Biol.*, **139**, 163.
66 Monk, N.A.M. (2003) *Curr. Biol.*, **33**, 1409.
67 Nykter, M., Price, N.D., Aldana, M., Ramsey, S.A., Kauffman, S.A., Hood, L.E., Yli-Harja, O., and Shmulevich, I. (2008) *Proc. Natl. Acad. Sci. USA*, **105**, 1897.
68 Ozbudak, E.M., Thattai, M., Lim, H.N., Shraiman, B.I., and van Oudenaarden, A. (2004) *Nature*, **427**, 737.
69 Pahle, J. (2009) *Brief Bioinform.*, **10**, 53.
70 Paulsson, J. (2004) *Nature*, **427**, 415.

71 Paulsson, J. (2005) *Phys. Life Rev.*, **2**, 157.
72 Paulsson, J., Berg, O.G., and Ehrenberg, M. (2000) *Proc. Natl. Acad. Sci. USA*, **97**, 7148.
73 Pedraza, J. and Paulsson, J. (2008) *Science*, **319**, 339.
74 Ptashne, M., Jeffrey, A., Johnson, A.D., Maurer, R., Meyer, B.J., Pabo, C.O., Roberts, T.M., and Sauer, R.T. (1980) *Cell*, **19**, 1.
75 Ptashne, M. (1992) *A Genetic Switch: Phage Lambda and Higher Organisms*, Cell Press and Blackwell Scientific, Cambridge, MA.
76 Puchalka, J. and Kierzek, A.M. (2004) *Biophys. J.*, **86**, 1357.
77 Rao, C. and Arkin, A. (2003) *J. Chem. Phys.*, **118**, 4999.
78 Rao, C.V., Wolf, D.W., and Arkin, A.P. (2002) *Nature*, **420**, 231.
79 Raser, J.M. and O'Shea, E.K. (2004) *Science*, **304**, 1811.
80 Reinitz, J. and Vaisnys, J.R. (1990) *J. Theor. Biol.*, **145**, 295.
81 Rosenfeld, N., Young, J.W., Alon, U., Swain, P.S., and Elowitz, M.B. (2005) *Science*, **307**, 1962.
82 Roussel, M.R. and Zhu, R. (2006) *Phys. Biol.*, **3**, 274.
83 Salis, H., Sotiropoulus, V., and Kaznessis, Y.N. (2006) *BMC Bioinformatics*, **7**, 93.
84 Schlicht, R. and Winkler, G. (2008) *J. Math. Biol.*, **57**, 613.
85 Shahrezaei, V., Ollivier, J., and Swain, P.S. (2008) *Mol. Syst. Biol.*, **4**, 1.
86 Shea, M.A. and Ackers, G.K. (1985) *J. Mol. Biol.*, **181**, 211.
87 Shmulevich, I., Dougherty, E.R., Kim, S., and Zhang, W. (2002) *Bioinformatics*, **18**, 261.
88 Smolen, P., Baxter, D.A., and Byrne, J.H. (2000) *Neuron*, **26**, 567.
89 Stelling, J., Sauer, U., Szallasi, Z., Doyle, F.J. III, and Doyle, J. (2004) *Cell*, **118**, 675.
90 Swain, P.S., Elowitz, M.B., and Siggia, E.D. (2002) *Proc. Natl. Acad. Sci. USA*, **99**, 12795.
91 Thattai, M. and van Oudenaarden, A. (2001) *Proc. Natl. Acad. Sci. USA*, **98**, 8614.
92 Tian, T. (2010) *BioSystems*, 99 192.
93 Tian, T. and Burrage, K. (2001) *Appl. Numer. Math.*, **38**, 167.
94 Tian, T. and Burrage, K. (2004) *J. Chem. Phys.*, **121**, 10356.
95 Tian, T. and Burrage, K. (2004) *J. Theor. Biol.*, **227**, 229.
96 Tian, T. and Burrage, K. (2006) *Proc. Natl. Acad. Sci. USA*, **103**, 8372.
97 Tian, T. and Burrage, K. (2006) *ANZIAM J.*, **48**, C1022.
98 Tian, T., Burrage, K., Burrage, P.M., and Carletti, M. (2007) *J. Comput. Appl. Maths.*, **205**, 697.
99 Tian, T., Harding, A., Inder, K., Plowman, S., Parton, R.G., and Hancock, J.F. (2007) *Nat. Cell Biol.*, **9**, 905.
100 Tu, Y., Stolovitzky, G., and Klein, U. (2002) *Proc. Natl. Acad. Sci. (USA)*, **99**, 14031.
101 Wang, J. and Tian, T. (2010) *BMC Bioinformatics*, **11**, 36.
102 Wang, J., Zhang, J., Yuan, Z., and Zhou, T. (2007) *BMC Syst. Biol.*, **1**, 50.
103 Weng, L., Dai, H., Zhan, Y., He, Y., Stepaniants, S.B., and Bassett, D.E. (2006) *Bioinformatics*, **22**, 1111.
104 Wilkinson, D.J. (2009) *Nat. Rev. Genet.*, **10**, 122.
105 Zak, D.E., Gonye, G.E., Schwaber, J.S., and Doyle, F.J. III (2003) *Genome Res.*, **13**, 2396.

3
Modeling Expression Quantitative Trait Loci in Multiple Populations

Ching-Lin Hsiao and Cathy S. J. Fann

3.1
Introduction

Due to the rapid advancement of genotyping technology, there are a few well-designed, affordable commercial chips loaded with about 1 million single nucleotide polymorphisms (SNPs) that cover the whole genome, such as Affymetrix 6.0 and Illumina 1 M (Affymetrix, Santa Clara, CA and Illumina, San Diego, CA). In the past few years, these newly developed microarray chips have stimulated an enormous output of large-scale genome-wide association studies (GWAS) that have attempted to dissect complex traits such as type II diabetes and cardiovascular diseases. SNPs are probably the most commonly used genetic markers for studying genetic susceptibility of complex diseases; however, only a few studies have provided insight into the functional variant(s) or mechanism(s) underlying these diseases. SNPs that are located in coding regions and result in amino acid changes in the protein products of genes belong to the most identifiable category of functional SNPs, the most common of which are the nonsynonymous SNPs. However, the vast majority of significant signals underlying complex traits are found outside the coding region of the genes. Therefore, a survey of expression quantitative trait loci (eQTL) allows researchers to assess the mechanisms underlying these diseases using gene expression (GE) levels as an intermediate molecular phenotype [1, 2].

3.1.1
Data Structure in eQTL Studies

Data collection for eQTL studies is two-dimensional. One dimension examines GE levels, which are believed to contribute to phenotypic differences between individuals [3]. GE data obtained from microarray experiments allow scientists to monitor the expression levels of thousands of genes simultaneously. The primary purpose of GE analysis is to characterize the numerical difference between expression profiles

Applied Statistics for Network Biology: Methods in Systems Biology, First Edition.
Edited by M. Dehmer, F. Emmert-Streib, A. Graber, and A. Salvador.
© 2011 Wiley-VCH Verlag GmbH & Co. KGaA. Published 2011 by Wiley-VCH Verlag GmbH & Co. KGaA.

from two distinct cell samples, which may reveal clinical subtypes or otherwise clarify biological pathways involving certain genes [4]. Several previous studies have shown that transcript levels for many genes are indeed heritable. In humans, the levels of 16 678 transcripts have been found to be significantly heritable at a false discovery rate (FDR) of 0.05 and a median heritability of 22.5% [5], in which the heritability is defined as the proportion of phenotypic variance attributable to genetic variance in a population. Similarly, in mice, the median heritability was shown to be 14–17% for each tissue (fat, kidney, adrenal, and left ventricle) considered in the study [6]. With the advent of microarrays, it is possible to view the transcriptome as a suite of quantitative traits.

The second dimension of data collection for eQTL studies examines SNP genotypes in which variation in a given population is correlated with disease; through various mechanisms, these SNPs alter the amount of mRNA produced [7]. When individuals are subjected to both DNA sequence polymorphism array genotyping and microarray-based GE profiling, genome-wide joint analysis for identification of eQTL becomes feasible [8] (Figure 3.1). A central dogma of molecular genetics is that DNA is transcribed into RNA, which subsequently is translated into protein; alterations in proteins can influence human diseases. Therefore, in eQTL studies, the GE levels have been regarded as multiple quantitative traits such that traditional quantitative trait locusQTL analysis can be applied in eQTL identification [9].

3.1.2
Current eQTL Studies

The relative impact of SNPs on GE phenotypes has been studied in several species, including plants [10], yeast [11, 12], mice [13, 14], rats [15], and humans. With the availability of millions of genotypes determined for major world populations, the International HapMap project allows investigators to uncover the effects of DNA sequence variants on mRNA expression levels in humans. Expression data from previous studies in humans are derived from the Epstein–Barr Virus-transformed lymphoblastoid cell line (LCL) and are publicly available in the Gene Expression Omnibus (GEO) database. LCLs were used because they are created as a renewable resource of nucleic acids for the HapMap project. Table 3.1 lists the GE data that have been recently published in the GEO database. In eQTL studies, regulatory polymorphisms are often considered to be in *cis* or in *trans*, on the basis of their physical distance from the regulated gene. In eQTL studies on humans, *cis* is typically considered to comprise the segment of the chromosome spanning from 100 kb upstream to 100 kb downstream of the gene sequence of interest and *trans* is defined as the region outside of *cis*. The mapping efforts have led to the initial characterization of genetic regulation in *cis*, probably because of variations in the gene's own regulatory regions.

3.1.2.1 eQTL Studies in a Single Human Population
An early study measured GE levels in LCLs from members of 14 Center d'Etude du Polymorphisme Humain (CEPH) families and used their corresponding genotype

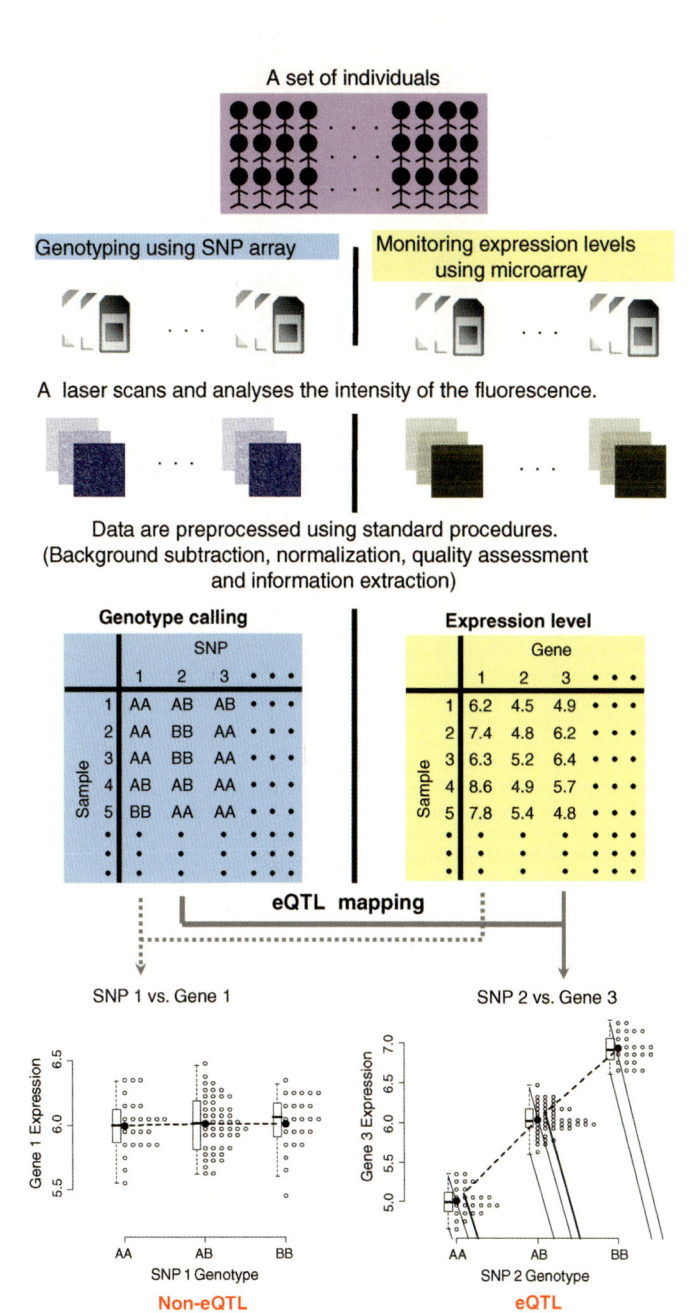

Figure 3.1 Microarray eQTL mapping pipeline.

Table 3.1 Publicly available GE data on the HapMap cell lines.

GEO ID	Technology platform	Samples	Number of targets	Public date	Reference
GSE2552	Affymetrix Human HG-Focus Target Array	57 CEPH	~8500	27-Oct-05	Cheung et al. [17]
GSE3612	Custom Illumina bead array	60 CEU	~630	1-Dec-05	Stranger et al. [18]
GSE5859	Affymetrix Human HG-Focus Target Array	60 CEU, 41 JPT, 41 CHB	~8500	7-Jan-07	Spielman et al. [20]
GSE6536	Sentrix Human-6 Expression BeadChip	60 CEU, 60YRI, 45 CHB, 45 JPT	~47 000	9-Feb-07	Stranger et al. [22]
GSE7236	Affymetrix Human HG-Focus Target Array	8 CEU, 8 YRI	~8500	14-Mar-07	Storey et al. [19]
GSE7851	Affymetrix Human Exon 1.0 ST Array [transcript version]	87 CEU, 89 YRI	~18 000	8-May-08	Zhang et al. [21]; Zhang et al. [30]; Huang et al. [31]
GSE9372	Affymetrix Human Exon 1.0 ST Array [transcript version]	57 CEU	~18 000	13-Jan-08	Kwan et al. [32]
GSE9703	Affymetrix Human Exon 1.0 ST Array [exon version]	87 CEU, 89 YRI	~1.4 million	23-Mar-09	Zhang et al. [21]; Zhang et al. [30]
GSE10824	Affymetrix Human HG-Focus Target Array	60 CEU, 60 YRI, 82 African-Americans	~8500	5-Dec-08	Price et al. [33]

data available in the public domain to perform genome-wide linkage analysis to identify eQTL [16]. A greater number of significant *trans* eQTL than *cis* eQTL were identified and some of the *trans* eQTL aggregated in genomic hotspots, which are defined as short genomic regions that exhibit higher levels of recombination than their surrounding sequences. To narrow down the candidate regulatory regions identified by the linkage study, a follow-up study was performed providing an association analysis with dense sets of SNP markers (more than 770 000 markers) on 57 unrelated CEPH individuals from the HapMap project. Replication of the linkage data was obtained for only about 20% of the genes; however, for the strongest 27 *cis* signals identified by linkage, the association analysis yielded confirmable evidence for about 50% of the genes [17]. In contrast, another study employed the densely genotyped (more than 1 000 000 SNPs) HapMap panel of 60 unrelated Utah pedigrees of individuals from the CEPH (CEU) to perform an eQTL analysis on 630 genes. This study applied three different methods of multiple test correction (including Bonferrone, FDR, and permutations), and showed that the signal proximal to the genes of interest is more abundant and stable than distal and *trans* signals across different methods of multiple-comparison correction [18].

3.1.2.2 eQTL Studies in Multiple Human Populations

The HapMap project has provided a significant resource for population differences in human genetics. The patterns of linkage disequilibrium in the human genome for three major human populations are freely available for research purposes. However, the study of population differences with respect to GE has only recently begun. Storey *et al.* showed that the proportion of GE variation due to interindividual variation is 85%, with the remaining 15% explained by differences between CEU and Yoruba in Ibadan, Nigeria (YRI) populations, which is nearly identical to the levels of population structure observed in extant patterns of human genetic variation [19]. Among the regions where population variations are found, the inflammatory pathways are strongly enriched for genes differentially expressed between the CEU and YRI samples, such as cytokines, and the chemokine receptors CCL5, CCR2, CCR7, and CXCR3. These receptors are relevant to numerous cardiovascular, infectious, and immune-related diseases. Between European-derived (CEU) and Asian-derived (Han Chinese in Beijing (CHB) and Japanese in Tokyo (JPT)) populations, about 25% of expression phenotypes of all tested genes have significant differences and identify elements that regulate transcription in human genomes [20]. These results also show that specific genetic variation among populations contributes appreciably to differences in GE phenotype. Taking the dependence structure between genes into account, 410 (4.5%) of the 9156 transcript clusters show significantly different expression between CEU and YRI trios [21]. These differentially expressed genes are not over-represented in any particular chromosome, but are related to certain biological processes.

For correlation between SNPs and GE levels, Spielman *et al.* identified 104 and 89 GE traits that show significant association with at least one SNP marker in the CEU and CHB + JPT samples, respectively. Common eQTL are identified when the SNP–GE association is significant in both populations; only 11 GE traits have common eQTL in these populations [20]. Stranger *et al.* showed that significant associations are detected with at least one SNP from 323, 348, 370, and 411 genes of the CEU, CHB, JPT, and YRI populations, respectively. These comprise a total of 888 nonredundant genes of which 331 (37%) are replicated at the same significance level in at least one other population and, of those, only 67 (8%) are significant in all four populations. In addition, for SNP–GE associations shared in populations, the direction of the allelic effect is generally the same across populations [22]. Identifying SNPs associated with the same expression trait in populations may reduce false-positive findings in the array-based eQTL mapping.

Although studies of HapMap populations have identified numerous eQTL, only a fraction of these eQTL are reproducible across populations. Combining samples across populations may increase sample size, and enhance both genetic dissimilarity and the range of GE variation, thereby increasing the statistical significance of identified eQTL [23–25]. Despite its advantage, combining various populations in an eQTL study is not straightforward due to the fact that allele frequency diversity and population-level expression differences are present in populations [26, 27]. To increase detection, Stranger *et al.* combined the data among populations and used conditional permutations to assess the significance of the SNP–GE associations [26]. Although conditional permutation can be used to manage inflation of the association p-value that is generated from population-level difference, SNP–GE associations in the combined dataset require appropriate adjustments for possible dissimilar population structure in the model. Veyrieras *et al.* combined samples from several populations using the normal quantile transformation method to avoid spurious eQTL arising from differences in population structure [28]. Quantile transformation to a standard normal distribution is a useful strategy and has been used with much success for normalizing GE data derived from different populations. Unfortunately, this technique may alter SNP–GE associations and result in misleading conclusions in eQTL studies when combining multiple populations, because expression values in each population are forced to have the same distribution and the difference in minor allele frequency (MAF) among populations has not been taken into consideration explicitly [29] (Figure 3.2).

Several potential approaches can be used for meta-analysis, which synthesizes and integrates individuals, test statistics, p-values, or statistical significances across a number of independent studies to estimate the associations. In this chapter, we provide an Independent Group Model (IGM) method [20], which is a notion of directly combining statistical significances derived from each population to identify common eQTL across populations. In addition, a constrained two-way model (CTWM) [29] is provided to identify eQTL in data that might contain genetic diversity among populations by combining individuals. Furthermore, the CTWM is extended to form a CTWM-genetic score constrained two-way model-genetic score (CTWM-GS) method to identify putative functional SNPs for which the difference of allelic frequency of a particular SNP induces a GE difference between populations. We focus on the genetic genomic data derived from two populations of unrelated individuals, and apply the methods to reanalyze data from unrelated Caucasian and Asian individuals available from HapMap.

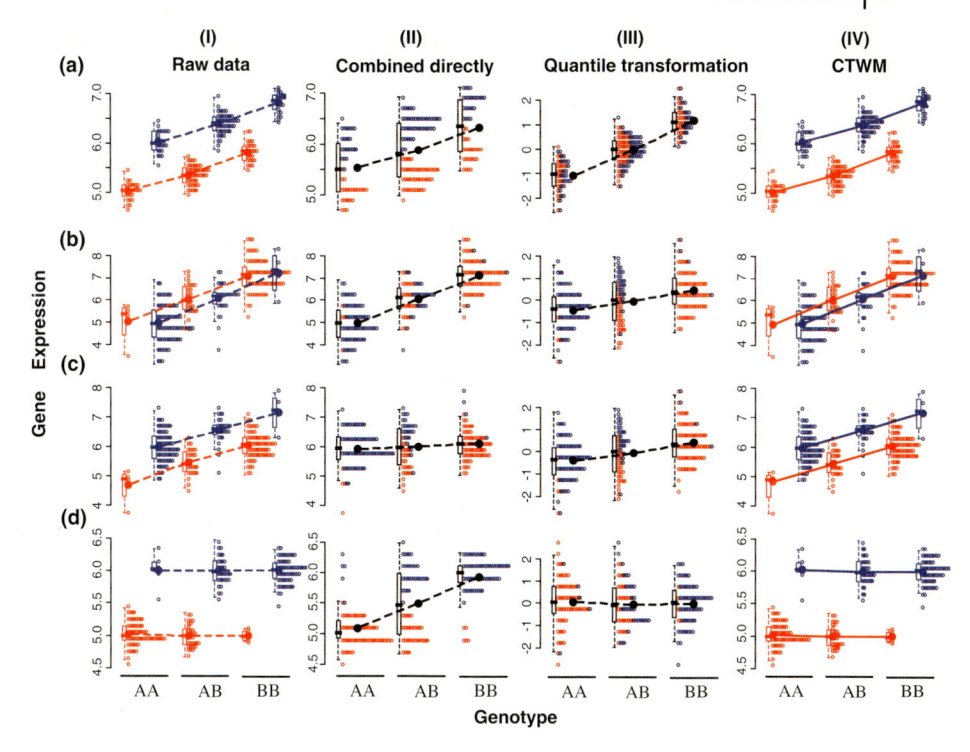

Figure 3.2 Effects of allele frequency and population-level expression differences on eQTL study with combined data. This graph depicts effects of allele frequency and population-level expression on eQTL mapping (as shown in I, raw data, red and blue colors represent two distinct populations) analyzed by three different methods (II, combined data directly; III, quantile transformation; IV, CTWM) in four different eQTL scenarios: (a) an eQTL with only population-level expression difference, (b) an eQTL with only allele frequency difference, (c) an eQTL with both population-level expression and allele frequency differences, and (d) a non-eQTL but with both population-level expression and allele frequency differences. For raw data (I), red and blue dots are the mean GE levels of the individuals in two populations for each genotype. These mean values are equal to those predicted expression levels using IGM method. For the remaining II–IV, the predicted GE level for each genotype is represented by dot. In (a), for (II), eQTL remained but with large variance; for (III), eQTL remained and with very small bias; for (IV), eQTL remained and also with very small bias. In (b), for (II), eQTL remained with very small bias; for (III), eQTL effect diluted; for (IV), eQTL remained with very small bias. In (c), for (II), eQTL effect diminished; for (III), eQTL effect diluted; for (IV), eQTL remained with very small bias. In (d), for (II), a spurious eQTL appeared; for (III), non-eQTL remained; similarly for (IV) non-eQTL remained.

3.1.3
An Illustrated Example

We reanalyzed the GE dataset from GEO (GSE6536) using 60 unrelated CEU and 90 CHB + JPT (Asian) individuals. The GE levels in this dataset were generated using

Illumina's commercial whole-genome expression array, the Sentrix Human-6 Expression BeadChip, which contains 47 296 unique 50-base oligonucleotides in total. Among these probes, 19 730 probes (41%) are targeted as Reference Sequence (RefSeq) transcripts [34] and the remaining 27 566 probes are for other putative transcripts, including known alternative splice regions, predicted by Gnomon and Unigene clusters. Recent studies have demonstrated that probe sequences including SNPs can influence the hybridization on microarrays and cause false eQTL identification [35]. The rationale for this effect is that the mRNA with identical sequence to the designed probe will hybridize better than nonidentical mRNA. To avoid false-positives of such eQTL, the subset of 14 456 expression traits excluding probes that have polymorphisms in the probe-target sequence – as reported in Stranger et al.'s study [26] – was thus used in the analysis. Based on physical location information, 10 197 well-annotated probes, mapped to 9333 nonredundant gene symbols, were used in this chapter.

The corresponding Affymetrix 500K expression data were downloaded from the Affymetrix website (www.affymetrix.com/estore) and SNP genotypes of each individual were called based on the Affymetrix BRLMM algorithm. This chip includes 489 922 SNPs from across the autosomal genome, and captures 65 and 66% of CEU and CHB + JPT common genetic variation (SNPs with MAF above 0.05), respectively [36]. To simplify the process in modeling eQTL, the missing genotypes were first imputed to form a complete SNP genotype dataset. The established imputation method "beagle" was used to impute the missing genotypes [37]. Beagle is a haplotype clustering-based algorithm that uses the localized haplotype cluster model to cluster haplotypes across SNPs and then defines a hidden Markov model to find the most likely haplotype pairs based on the individual's known genotypes. This program achieves nearly the same imputation accuracy, but runs faster than other methods when hundreds of samples and thousands of SNPs are analyzed. Using the beagle program, 15.8 and 16.5% of the missing genotypes were imputed as heterozygous for the CEU and Asian populations, respectively.

These data formed a genetic genomic dataset comprising 10 197 transcripts and 489 922 SNPs in the autosomal genome for each individual of the CEU and Asian populations. For each expression trait, SNP–GE associations were examined, provided that both the SNP and the gene were on the same chromosome. Taking multiple comparison correction into account, a conservative threshold with $p < 10^{-6}$ was used to identify significant SNP–GE associations for each trait.

3.1.4
Notations

We illustrate the models using a conventional gene-based mapping technique where an expression trait and a biallelic SNP are used to build a model. Consider n unrelated individuals observed from two different populations, $i = 0, 1$. Without loss of generality, we assumed that three genotypes are observed for a biallelic SNP in each population and the GE value for the kth replication (i.e., individual) on the jth genotype in the ith population is denoted by y_{ijk} (where $i = 0, 1$; $j = 0, 1, 2$, representing the number of minor alleles carried; $k=1, \ldots, n_{ij}$; $\sum_{j=0,1,2} n_{ij} = n_{i\bullet}$). A number of genetic models are applicable for modeling QTL, including additive, dominant, and recessive modes of inheritance. The codominant genetic model, in which three genotype effects are estimated, has good overall performance for identification of any simple inheritance patterns in QTL mapping when the inheritance pattern is unknown [38]. Throughout this chapter, the codominant genetic model was assumed and the GE values, y_{ijk} (for $k=1, \ldots, n_{ij}$), were assumed to be independently distributed as $N(\mu_{ij}, \sigma^2)$. In addition, we assigned a common genotype c (where $c = argmax_{j=0,1,2} \prod_{i=0,1} P_{ij}$ and P_{ij} is the proportion of individuals carrying genotype j in the ith population) between populations.

3.2
IGM Method

Owing to possible heterogeneity between populations, independent analysis of each population across a set of population-mixed genetic genomic data is a direct way to avoid spurious significant signals arising from any differences in population structure. The IGM, as implied by the name, is an independent approach in which GE levels are regressed on SNP genotypes independently in each population. Consequently, two hypothesis-testing procedures are carried out simultaneously in a dataset containing two populations where testing results are then summarized to identify common eQTL.

3.2.1
Modeling SNP–GE Association in a Single Population

The cell mean model has been used extensively for unbalanced data in which the number of observations varies from one genotype to another. Based on the cell means, within the population 0, the $n_{0\bullet}$ signals in a given gene are decomposed into:

$$y_{0jk} = \mu_{0j} + e_{0jk}$$

where μ_{0j} is the cell mean and e_{0jk} is assumed to be independently distributed as $N(0, \sigma^2)$. If the common genotype c is assumed to be 0 between populations, this model can be rewritten using another parameterization as:

$$y_{0jk} = \mu_{00} + \tau_{0j} + e_{0jk}$$

where μ_{00} is the expression baseline in the population 0; $\tau_{0j} = \mu_{0j} - \mu_{00}$ when $j \neq 0$, which represents the GE difference between individuals with genotype j and genotype 0. It is straightforward to observe maximum likelihood estimators through a normal equation, and the estimators of μ_{00} and τ_{0j} are $\bar{y}_{00\bullet}$ and $(\bar{y}_{0j\bullet} - \bar{y}_{00\bullet})$, respectively, where $\bar{y}_{0j\bullet} = \sum_{k=1}^{n_{0j}} y_{0jk}/n_{0j}$. Under this alternative parameterization model, the SNP–GE association in population 0 can be tested via the null hypothesis H_0: $\tau_{01} = \tau_{02} = 0$ with the F-statistic calculated from ANOVA as:

$$F = \frac{SSB/q}{SSE/\nu}$$

where SSB and SSE represent the between sum of squares $\sum_{j=0}^{2} n_{0j} (\bar{y}_{0j\bullet} - \bar{y}_{0\bullet\bullet})^2$ with $\bar{y}_{0\bullet\bullet} = \sum_{j=0}^{2} \sum_{k=1}^{n_{0j}} y_{0jk}/n_{0\bullet}$ and error sum of squares $\sum_{j=0}^{2} \sum_{k=1}^{n_{0j}} (y_{0jk} - \bar{y}_{0j\bullet})^2$, respectively; q and ν are the degrees of freedom for SSB and SSE, respectively, where for the case of the three genotypes observed $q = 3 - 1 = 2$ and $\nu = (n_{0\bullet} - 3)$. If the null hypothesis is true, F is distributed as $F(q, \nu)$ and the null hypothesis is rejected when $F > F(q, \nu, \alpha)$, in which $F(q, \nu, \alpha)$ is the upper α percentage point of the central F-distribution and α is the desired significance level of the test. This model can be directly used to identify eQTL when the data are derived from only one population.

3.2.2
Integrating Hypotheses to Identify Common eQTL

Using the IGM method, the SNP–GE associations for populations 0 and 1 are independently formed by the previous model with population index $i = 0$ and 1. The SNP–GE associations are therefore tested by the null hypotheses H_0: $\tau_{01} = \tau_{02} = 0$ and H_0: $\tau_{11} = \tau_{12} = 0$, respectively. Using this modeling method, two one-way ANOVAs are performed and the parameters are estimated independently in each population. Therefore, it is free of model assumptions, and allows for heterogeneity of SNP effects (τ_{ij}) between populations if there is an interactive effect of SNP and populations at the GE level. Many studies employ this approach to identify significant eQTL common to two populations [20, 22, 39, 40]. For instance, a common eQTL is identified if the two hypotheses H_0: $\tau_{01} = \tau_{02} = 0$ and H_0: $\tau_{11} = \tau_{12} = 0$ are simultaneously rejected.

3.2.3
Applying the IGM Method to HapMap Data

Using a pair of HapMap populations, 349 of 10 197 (3.4%) expression traits were mapped to at least one significant eQTL in both CEU and Asian populations by the IGM; however, after integrating statistical significances, only 77 of the 349 had common eQTL in both populations. We divided these eQTL into common and population-specific eQTL sets. For expression traits having more than one significant eQTL, the SNP with the smallest association p-value was selected as the putative eQTL. For instance, in the set of common eQTL, the SNP that had the minimum geometric mean of the p-values among SNPs with significance in CEU and Asian populations was chosen as the putative common eQTL. In the set population-specific eQTL, the SNP with the smallest p-value in the CEU and Asian population was picked as the CEU- and Asian-specific putative eQTL, respectively. Consequently, 77 unique

Table 3.2 Summary of common and population-specific eQTL identified by the IGM method.

eQTL set	Population	Number	Number of *cis*	Interaction test	Mean of MAF
Common	CEU	77	61 (79%)	11 (14%)	0.26
	Asian	77	61 (79%)	11 (14%)	0.25
Population specific	CEU	272	7 (3%)	209 (77%)	0.13
	Asian	272	20 (7%)	201 (74%)	0.10

SNP–GE pairs were shared between CEU and Asian populations; 272 unique SNP–GE pairs were population-specific for each of the CEU and Asian populations (Table 3.2).

3.2.3.1 Characterizing Putative eQTL Identified by the IGM

Investigating the putative eQTL map positions with respect to the two sets, 77 common eQTL were generally identified as *cis* eQTL; in contrast, most population-specific eQTL were *trans* eQTL. In detail, seven and 20 out of the 272 population-specific eQTL were considered *cis* eQTL for CEU and Asian populations, respectively, and 61 out of 77 common eQTL were in the *cis* (Table 3.2).

An issue relevant to eQTL cross-population analysis is whether the SNP–GE associations that are not shared between populations are attributed to factors of allele frequency or SNP effect differences between populations. We compared the MAFs of the putative eQTL in the two sets. Means of the MAFs were 0.26 and 0.25 for CEU and Asian populations in the common eQTL set, whereas those in the population-specific set were 0.13 and 0.10 for CEU and Asian populations, respectively (Table 3.2). Using four HapMap populations, Stranger *et al.* reported that many genes show *cis* associations in at least two populations and the heterozygosity differences between populations in shared eQTL are lower than those in population-specific eQTL [20]. We dissected the MAF of the putative eQTL and showed that the population-specific eQTL usually had smaller allele frequencies than those of common eQTL. A number of genetic genomic studies have shown that *cis* eQTL are more reproducible and are usually mapped with higher statistical significance than those in *trans* eQTL [5, 41] due to high heritability [6] and direct impact on the regulatory sequences [23, 42]. As the IGM is only able to identify common *cis* eQTL, rare minor alleles may partially account for the incapability of the IGM to identify common *trans* eQTL. That is, because the samples are separated into two independent studies, common *trans* eQTL will then be difficult to detect and reproduce in two populations with small sample size and/or small allele frequency.

When examining the effects of population–genotype interactions on GE by both the full and reduced two-way ANOVA models (the full model contains a population–genotype interaction term, whereas the reduced model does not), the interaction effect in the expression trait with population-specific eQTL was larger than that in the set with common eQTL. For example, among 272 population-specific putative eQTL, the number of eQTL with significant interaction effects was 209 and 201 for CEU and

Asian populations, respectively, and only 11 of the 77 common eQTL showed significant signals at a threshold of FDR $q < 0.05$ (Table 3.2). In other words, the SNP effects on GE levels were similar in populations for most common eQTL, but were dissimilar for population-specific eQTL. With a small sample size, estimating the SNP effect on GE is more likely to be perturbed by a higher degree of unbalanced genotypic data because the variation of the estimates of the SNP effect is larger than that with bigger MAF under the same SNP effect size. It is possible that population-specific eQTL have smaller allele frequencies, which subsequently contribute to population–genotype interaction effects.

The method of integrating statistical significances is unsatisfactory for identifying common eQTL and the power of this method is poor, even when the magnitude of the SNP effect is large [29]. In contrast, the proper combination of different ethnic datasets to increase the sample size may overcome statistical barriers inherent in unbalanced genotype data and result in association mapping that is more precise for identification of eQTL.

3.3
CTWM

In contrast to IGM, which integrates statistical significances, the CTWM combines two population data prior to statistical testing, and may enrich data variation and prove more effective for detection of common eQTL. Olkin and Sampson showed that meta-analysis performed using the best linear unbiased estimators computed from each study and using the pooled individuals modeled as the two-way model without interaction are equivalent [43]. To simplify interpretation, we employed a CTWM, in which different populations and different genotypes represent the two-way variables, to jointly assess SNP–GE associations in individuals from two populations. This model assumes that the mechanisms of SNP regulation of GE are similar between populations and the assumption is empirically supported by the results observed from the IGM. The model also allows for heterogeneous nongenetic effects on expression between populations to avoid deceptive SNP–GE associations generated as a result of GE differences in the population level. The word "constrained" refers to the constraint that no interaction exists between the two variables in the CTWM. In other words, the SNP effect sizes on GE in populations would be regarded as identical in the CTWM.

3.3.1
Modeling SNP–GE Association in Pooled Data by CTWM

As the results from the IGM were to be compared with those obtained from combined data, we considered $c = 0$ as the baseline genotype and individuals with the common genotype ($j = 0$) in each population were assigned a different GE mean value, μ_{i0} (for $i = 0, 1$), to allow for heterogeneity of the GE baseline. We assumed homogeneity of SNP regulatory mechanisms across populations by defining E

$(y_{ijk} - y_{i0k}) = \tau_{\bullet j}$ for all $j = 1, 2$ and $i = 0, 1$. Consequently, the n (where $n = n_{0\bullet} + n_{1\bullet}$) signals can be expressed as:

$$y_{ijk} = \mu_{i0} + \tau_{\bullet j} + e_{ijk}$$

where e_{ijk}s are assumed to be independently distributed as $N(0, \sigma^2)$. Thus, without displaying the redundant rows, the regression equation is:

$$E(Y) = D\theta \Rightarrow E\begin{pmatrix} Y_{00} \\ Y_{01} \\ Y_{02} \\ Y_{10} \\ Y_{11} \\ Y_{12} \end{pmatrix} = \begin{pmatrix} 1 & 0 & 0 & 0 \\ 1 & 0 & 1 & 0 \\ 1 & 0 & 0 & 1 \\ 0 & 1 & 0 & 0 \\ 0 & 1 & 1 & 0 \\ 0 & 1 & 0 & 1 \end{pmatrix} \times \begin{pmatrix} \mu_{00} \\ \mu_{10} \\ \tau_{\bullet 1} \\ \tau_{\bullet 2} \end{pmatrix}$$

The estimators of the parameter $\theta' = (\mu_{00}, \mu_{10}, \tau_{\bullet 1}, \tau_{\bullet 2})$ can be obtained through the normal equation $(X'X)\theta = X'Y$, in which the design matrix X is generated from D as:

$$X = \begin{pmatrix} \mathbf{1}_{n_{00}} & \mathbf{0}_{n_{00}} & \mathbf{0}_{n_{00}} & \mathbf{0}_{n_{00}} \\ \mathbf{1}_{n_{01}} & \mathbf{0}_{n_{01}} & \mathbf{1}_{n_{01}} & \mathbf{0}_{n_{01}} \\ \mathbf{1}_{n_{02}} & \mathbf{0}_{n_{02}} & \mathbf{0}_{n_{02}} & \mathbf{1}_{n_{02}} \\ \mathbf{0}_{n_{10}} & \mathbf{1}_{n_{10}} & \mathbf{0}_{n_{10}} & \mathbf{0}_{n_{10}} \\ \mathbf{0}_{n_{11}} & \mathbf{1}_{n_{11}} & \mathbf{1}_{n_{11}} & \mathbf{0}_{n_{11}} \\ \mathbf{0}_{n_{12}} & \mathbf{1}_{n_{12}} & \mathbf{0}_{n_{12}} & \mathbf{1}_{n_{12}} \end{pmatrix}$$

where $\mathbf{0}_{n_{ij}}$ and $\mathbf{1}_{n_{ij}}$ are vectors with length n_{ij}, and all of whose components are 0 and 1, respectively. As the two independent unbalanced one-way models in the IGM are combined into a CTWM, an analysis of SNP–GE association can be performed by testing the null hypothesis $H_0: \tau_{\bullet 1} = \tau_{\bullet 2} = 0$ using the partial F-statistic:

$$F = \frac{\left(B\hat{\theta}\right)'\left[B(X'X)^{-1}B'\right]^{-1}B\hat{\theta}/q}{SSE/v}$$

where B is in the matrix form $\begin{bmatrix} 0 & 0 & 1 & 0 \\ 0 & 0 & 0 & 1 \end{bmatrix}$, $q = 2$ is the rank of the matrix B, SSE represents the error sum of squares $\sum_i \sum_j \sum_k (y_{ijk} - \hat{y}_{ijk})^2$, with the prediction value $\hat{y}_{ijk}$ calculated from the model, and v represents the degree of freedom for the SSE (i.e., $v = n - 4$, in which 4 is the rank of design matrix X). The F-statistic is distributed as $F(q, v)$ if the null hypothesis is true. The design matrix can be directly used with the hypothesis testing formula because matrix $(X'X)$ is nonsingular.

Using simulation studies, it has been shown that CTWM outperforms the IGM method no matter what methods are used to integrate statistical significances (such as union–intersection or intersection–union tests) between two populations [29]. In addition, the CTWM provides unbiased estimates of the SNP effects on GE levels ($\tau_{\bullet j}$) and is not influenced by differences of allele frequency or expression baseline (μ_{i0}) between populations [29] (Figure 3.2).

3.3.2
Applying CTWM to HapMap Data

Using the same dataset analyzed in the IGM, 1839 of 10 197 (18%) expression traits had at least one SNP with a statistically significant SNP–GE association identified using CTWM. Notably, all 77 putative eQTL identified as common eQTL by the IGM were also statistically significant using CTWM with *p*-values in a range from 10^{-10} to 10^{-100}. In other words, the set of 77 expression traits with at least one eQTL detected by the IGM was a subset of those detected by CTWM. These results suggest that CTWM, as it polls individuals from two populations, has greater power to identify common eQTL than if individuals are analyzed separately in each population.

3.3.2.1 Characterizing Putative eQTL Identified by CTWM

To characterize the physical position of the putative eQTL identified by CTWM, the relative distances of the eQTL to their target transcripts were investigated. Among the 1839 putative eQTL, 438 (24%) were identified as *cis* eQTL, comprised of 169 and 269 eQTL residing within and in the vicinity of 100 kb of the target genes, respectively, and 1401 (76%) were identified as *trans* eQTL. In addition, the averages of the $-\log_{10}$ association *p*-values were 12.3 and 9.7 for the *cis* and *trans* putative eQTL, respectively, and the difference between these two means was significant ($p = 1.5 \times 10^{-6}$ by Welch's *t*-test). This suggests that eQTL near target genes have smaller SNP–GE association *p*-values than those further away from genes, as identified by CTWM. This finding is consistent with previous eQTL studies [23, 42]; however, the majority of the putative eQTL are distant from the gene sequences. This may have occurred because gene correlations are inherent in the same biochemical pathway [44]. For instance, if a group of genes are highly correlated and an eQTL is identified for one of them, then other expression traits in this group probably map to the same eQTL. Some other possible explanations include genes that are tightly regulated by hotspots [45] or a distant eQTL located in the gene sequence of a transcription factor that modulates the transcript level for the target genes. Increased effectiveness of CTWM to identify eQTL in distant regions may help recognize human eQTL hotspots where a single polymorphism impacts the majority of expression traits, which will improve our understanding of how gene networks are organized in an organism and clarify their potential impacts on disease phenotype.

Examination of the MAF for eQTL with respect to chromosomal position showed that the MAFs for *cis* and *trans* putative eQTL in the combined population were 0.28 and 0.1, respectively. The difference of these two frequencies was significant ($p < 1 \times 10^{-16}$ by Welch's *t*-test) and the finding coincided with that revealed by the IGM method. Diversified factors underlying the complexity of gene regulation may contribute to weak associations in distant eQTL, such as polygenic regulation, environmental input, and possibly interactions among loci [46, 47]. Therefore, differences of MAF only provide another possible interpretation for distant eQTL, as a low MAF might result in a weak association because small sample sizes and slight fluctuations in allele frequencies may affect the detection of significant SNP–GE associations by a conservative threshold simply due to sampling variance.

3.3.2.2 Justification of Model Assumptions

As in many other methods, CTWM also assumes that SNP effects are similar between populations in the combined data. This assumption presumably can largely reduce the complexity of the model and the number of parameters needed. The increased power derived from pooling individuals for identifying eQTL is evidence that eQTL have similar effects in different populations. After comparing the associations found among all populations, Veyrieras et al. suggested that significant SNP–GE associations are usually shared in different populations [28]. However, whether the magnitude of the SNP effect on GE is similar for any particular SNP–GE association in different populations remains unclear. We used full- and reduced-model strategies to justify the assumption that the magnitude of the SNP effect is homogeneous by examining the genotype–ethnic interaction term in the model. This analysis was performed using all 29 420 eQTL identified by CTWM rather than using those 1839 putative eQTL selected to partially follow the assumption of similar SNP effects between populations and indicated that 2968 (10%) eQTL had significant signals ($q < 0.05$) for interaction terms. In contrast, when we investigated the heterozygous nongenetic SNP effect assumption using full- and reduced-model strategies (the full model was fitted to different expression baselines, whereas the reduced model was fitted to an identical baseline between populations) for each of the 1839 SNP–GE pairs, there were 1298 (70.6%) traits with $q < 0.05$. These results emphasize the nature of putative eQTL identified by the model. That is, population-level differences in GE substantially affect eQTL identification and many more eQTL can be identified by assuming a simpler model without population–genotype interactions.

For genome-wide identification of eQTL, researchers often employ the additive genetic model as a basic inheritance pattern to assess the association between SNP genotype and GE. The additive model assumes an increasing trend in GE amount for genotypic groups 1 and 2 relative to 0, where the effect of genotype 1 is approximately half that of genotype 2. Using the full- and reduced-model strategies again (the full model using a codominant assumption and the reduced model using an additive assumption), 874 (48%) of 1839 putative eQTL were significantly associated with the codominant assumption ($q < 0.05$). Among the 874 eQTL, only 49 were in local regions. In other words, the additive genetic model assumption is applicable to most *cis* eQTL and the codominant genetic model assumption seems to be more suitable for *trans* eQTL.

3.4 CTWM-GS Method

In genetic genomic studies that incorporate data based on two phenotypes, GE, and population, investigators aim to identify not only the association between SNPs and GE, but also the relationship between SNPs and phenotypes. In general, investigators aim to identify SNPs that are associated with particular GE profiles and to discern whether the difference of allelic frequency of a particular SNP induces a GE difference between populations, termed expression single nucleotide polymorphismeSNP [48]. Using the CTWM described above, the GE differences between populations can be directly partitioned into two parts and independently tested. The first part represents baseline expression differences, which can be affected by nongenetic factors such as different environmental conditions across populations, and is termed the baseline difference (BD). The second part represents quantification of GE differences between populations resulting from genotype frequency differences (genetic) and is termed the genetic score (GS). This allows us to examine whether differentially expressed genes between populations are caused by genetic markers or nongenetic factors.

3.4.1
Solving Normal Equations in CTWM

Although it is possible to solve the normal equation numerically, the method is unable to display the invariant properties of the estimators. To this end, we solved the normal equation by maximum likelihood estimation and expressed the estimators $\hat{\theta}$ using the cell mean. Intuitively, the symmetrical matrix can be expressed as:

$$X'X = \begin{pmatrix} n_{0\bullet} & 0 & n_{01} & n_{02} \\ 0 & n_{1\bullet} & n_{11} & n_{12} \\ n_{01} & n_{11} & A_1 & 0 \\ n_{02} & n_{12} & 0 & A_2 \end{pmatrix}$$

where $A_j = n_{0j} + n_{1j}$. Using the blockwise inversion approach [49] we have:

$$(X'X)^{-1} = \frac{1}{cd} \times$$

$$\begin{pmatrix} n_{10}A_1A_2 + M_1A_2 + M_2A_1 & M_1A_2 + M_2A_1 & -n_{01}(n_{10}+n_{11})A_2 - M_2A_1 & -n_{02}(n_{10}+n_{12})A_1 - M_1A_2 \\ M_1A_2 + M_2A_1 & n_{00}A_1A_2 + M_1A_2 + M_2A_1 & -n_{11}(n_{00}+n_{01})A_2 - M_2A_1 & -n_{12}(n_{00}+n_{02})A_1 - M_1A_2 \\ -n_{01}(n_{10}+n_{11})A_2 - M_2A_1 & -n_{11}(n_{00}+n_{01})A_2 - M_2A_1 & n_{0\bullet}n_{12}(n_{10}+n_{11}) + n_{1\bullet}n_{02}(n_{00}+n_{01}) & n_{0\bullet}n_{11}n_{12} + n_{1\bullet}n_{01}n_{02} \\ -n_{02}(n_{10}+n_{12})A_1 - M_1A_2 & -n_{12}(n_{00}+n_{02})A_1 - M_1A_2 & n_{0\bullet}n_{11}n_{12} + n_{1\bullet}n_{01}n_{02} & n_{0\bullet}n_{11}(n_{10}+n_{12}) + n_{1\bullet}n_{01}(n_{00}+n_{02}) \end{pmatrix}$$

where $M_j = n_{0j} \times n_{1j}$ and $cd = \sum_{j=0}^{2}\left(M_j \prod_{t \neq j} A_t\right)$. The estimators $\hat{\theta}$ can be calculated by:

$$(X'X)^{-1}X'Y$$

where:

$$X'Y = \begin{pmatrix} y_{0\bullet} \\ y_{1\bullet} \\ y_{\bullet 1} \\ y_{\bullet 2} \end{pmatrix}$$

Hence, the estimators $\hat{\theta}$ can be represented using the cell mean as:

$$\hat{\mu}_{i0} = \left(\bar{y}_{i0}M_0\prod_{t=\{1,2\}}A_t + \sum_{t=\{1,2\}}M_tA_r\left(n_{i0}\bar{y}_{i0} + n_{s0}\bar{y}_{s0} + n_{s0}(\bar{y}_{it}-\bar{y}_{st})\right)\right)/cd, \text{ for } i = 0, 1$$

$$\hat{\tau}_{\bullet j} = \left(\begin{array}{c}\sum_{i=\{0,1\}}\left((\bar{y}_{ij}-\bar{y}_{i0})A_rn_{i0}n_{ij}(n_{s0}+n_{sj})\right) + \\ M_r\left(A_0\sum_{i=\{0,1\}}n_{ij}\bar{y}_{ij} - A_j\sum_{i=\{0,1\}}n_{i0}\bar{y}_{i0} + (n_{00}n_{ij}-n_{10}n_{0j})(\bar{y}_{0r}-\bar{y}_{1r})\right)\end{array}\right)\Bigg/cd, \text{ for } j = 1, 2$$

where $s = \{0,1\}\setminus\{i\}$, which means s is an element in $\{0,1\}$ but not in $\{i\}$, and $r = \{0,1,2\}\setminus\{j, t\}$, which means r is an element in $\{0,1,2\}$ but not in $\{j, t\}$.

3.4.2
Estimators of BD and GS

BD, by the previous definition, is the GE baseline difference between populations that can be parameterized as $(\mu_{00} - \mu_{10})$. Similarly, GS, which is the GE difference resulting from the genotype frequency differences between two populations, can be represented as $(P_{01} - P_{11})\tau_{\bullet 1} + (P_{02} - P_{12})\tau_{\bullet 2}$. Consequently, the estimators of GS and BD can be expressed as:

$$\widehat{BD} = \hat{\mu}_{00} - \hat{\mu}_{10} = \sum_{j=0}^{2}\left(M_j\left(\bar{y}_{0j} - \bar{y}_{1j}\right)\prod_{t \neq j} A_t\right) \Big/ cd$$

$$\widehat{GS} = (P_{01} - P_{11})\hat{\tau}_{\bullet 1} + (P_{02} - P_{12})\hat{\tau}_{\bullet 2}$$

$$= \sum_{j=0}^{2}\left((P_{0j} - P_{1j})\left(\left(M_j \sum_i \bar{y}_{ij} \prod_{t \neq j} n_{it}\right) - \left(\sum_{t \neq j} M_t \sum_i \bar{y}_{it} n_{ij} n_{sr}\right)\right)\right) \Big/ cd$$

In addition, the estimators of BD and GS can be represented using the designation $y_{ij\bullet}$ (for $i = 1, 2$ and $j = 1, 2, 3$), where $y_{ij\bullet}$ is the sum of GE values in population i and genotype j, $y_{ij\bullet} = \sum_{k=1}^{n_{ij}} y_{ijk}$:

$$\widehat{BD} = \sum_{i=0}^{1}(-1)^i \sum_{j=0}^{2}\left(y_{ij\bullet} n_{sj} \prod_{t \neq j} A_t\right) \Big/ cd$$

$$\widehat{GS} = \sum_{i=0}^{1}(-1)^i \sum_{j=0}^{2} y_{ij\bullet}\left((p_{ij} - p_{sj})\sum_{t \neq j} M_t A_r - \sum_{t \neq j}(p_{it} - p_{st}) n_{it} n_{sj} A_r\right) \Big/ cd$$

Therefore, it can be shown that the quantities of BD and GS can be represented as two nonoverlapping scores, and separated from the arithmetic mean difference between populations as:

$$(\bar{y}_{0\bullet\bullet} - \bar{y}_{1\bullet\bullet}) = (n_0)^{-1}\sum_{j=0}^{2} y_{0j\bullet} - (n_1)^{-1}\sum_{j=0}^{2} y_{1j\bullet}$$

$$= (\hat{\mu}_{00} - \hat{\mu}_{10}) + ((P_{01} - P_{11})\hat{\tau}_{\bullet 1} + (P_{02} - P_{12})\hat{\tau}_{\bullet 2}) = \widehat{BD} + \widehat{GS}$$

This method has the potential to distinguish population-level and genetic-level differences between populations. On the contrary, GS in the IGM method is represented as:

$$GS = \sum_{i=0,1}(-1)^i \sum_{j=1,2} P_{ij}\tau_{ij}$$

Applying the constraint $\Sigma_{j=0,1,2}P_{ij}=1$ for each i, the estimator of GS is:

$$\widehat{GS} = (\bar{y}_{0\bullet\bullet} - \bar{y}_{1\bullet\bullet}) - (\bar{y}_{00\bullet} - \bar{y}_{10\bullet})$$

This quantity cannot be tested directly because the second term in the estimator indicates that GS varies according to the reference genotype chosen.

3.4.3
Testing BD and GS

Under the CTWM, the estimators of BD and GS indicate that the estimates are invariant with respect to the baseline genotype chosen, as they are free of index j. Since we observed a unique solution for both BD and GS, these estimates can be tested using the null hypotheses H_0: $\mu_{00} = \mu_{10}$ and H_0: $(P_{01} - P_{11})\tau_{\bullet 1} + (P_{02} - P_{12})\tau_{\bullet 2} = 0$ via the partial F-statistic described in the CTWM method with $\mathbf{B} = (1, -1, 0, 0)$ when testing BD and $B = (0, 0, (P_{01} - P_{11}), (P_{02} - P_{12}))$ when testing GS, respectively. Hence, the test statistic F_{GS} for GS can be written as:

$$F_{GS} = \frac{\left(b'\hat{\theta}\right)'\left[b'(X'X)^{-1}b\right]^{-1}b'\hat{\theta}}{SSE/\nu} = \frac{\nu \times \left(b'\hat{\theta}\right)^2}{SSE \times b'(X'X)^{-1}b}$$

where $b' = (0, 0, (P_{01} - P_{11}), (P_{02} - P_{12}))$. We reject the null hypothesis if $F_{GS} \geq F_{(\alpha, 1, \nu)}$. Since F_{GS} has 1 and ν degrees of freedom F-distribution, we can equivalently use the t-distribution to test F_{GS} by the rejection criterion $(F_{GS})^{1/2} \geq t_{\alpha/2, \nu}$. If the null hypothesis for GS is rejected, then the SNP–GE association will be claimed as an eSNP. The test on GS has low power when the two populations have similar genotype frequencies and the power is increased as a function of the allele frequency differences between populations [29]. This may help prioritize case-control data to identify SNPs whose allele frequencies are different and are relative to the expression of a particular gene.

3.4.4
Applying CTWM-GS to HapMap Data

We have demonstrated the applicability of CTWM for combined populations. Thus, under CTWM, GS was substituted for SNP–GE associations to incorporate SNP diversity information into the analysis. Using the dataset of 10 197 transcripts, testing GS generated 1689 (16.5%) expression traits that included at least one SNP with a significant GS. Of these 1689 putative eSNPs, 914 (54%) were included in the set of 1839 putative eQTL identified by CTWM. A high number of intersecting eQTL was expected because both methods explore the random aspect of the SNP effect. To avoid unnecessary cancellation, we calculated the proportion of genetic factors by |GS| / (|BD| + |GS|). If the denominator was considered the total expression bias for a particular SNP-GS pair, most eSNPs had less than a 40% target GE bias between the

CEU and Asian cohorts. Furthermore, when examining the association between these proportions and the MAF differences using a locally weighted scatterplot smoothing method [50], the SNPs with higher proportions are those with greater MAF differences [29]. Although the CTWM-GS method still requires the CTWM to test the associations, it provides immediate insight into SNP effects on the variation of GE among populations.

3.4.4.1 Applying the GS to Population Studies

In microarray GWAS, it is more straightforward to map regulatory variation on GE rather than on complex clinical phenotypes because the direct effects of changes in nucleotides on molecular phenotypes are likely stronger than those on complex clinical phenotypes, which may be influenced by multiple genes and environmental interactions [51, 52]. The ultimate goal of microarray studies, which generally involve comparisons of two conditions within a single population or comparisons of two clinically distinct populations, is to generate a list of gene variants for further investigation. To achieve this objective, 20 SNP–GE pairs represented by 20 unique gene symbols and 19 unique SNPs were selected using a GS threshold of 0.5 ($|GS| \geq 0.5$) among the 1689 putative eSNPs. The GS combines two effects of a particular SNP–GE combination – the genotype frequency differences between populations and the differences in the GE levels directed by genotypes. Thus, a smaller GS p-value does not necessarily imply a greater allele frequency difference or a substantial SNP effect. To prioritize candidate eSNPs, it is more advantageous to rank the GS among a significant eSNP set rather than to rank their p-values directly. Most of the 20 pairs had significant SNP–GE association in at least one population tested using the IGM and the nonsignificant associations likely resulted from a low MAF (Table 3.3). For instance, the SNPs rs1419772, rs2337387, and rs604127 had MAFs ranging from 0.01 to 0.06, and their corresponding SNP–GE association p-values ranged from 0.07 to 0.7 in the Asian cohort.

Using these 20 putative eSNPs, the individuals in the two populations (CEU and Asian) could be clearly distinguished by a simple hierarchical clustering on genotypic data (Figure 3.3). In clustering, individuals homozygous or heterozygous for the upregulated allele or homozygous for the downregulated allele were assigned values of 2 or 1 or 0, respectively, for each eSNP, and the Spearman rank correlation coefficient was used to measure the similarity between individuals. The cluster divided eSNPs into two groups, in which group 1 comprised seven eSNPs from alleles associated with upregulated expression mostly found in the CEU cohort in contrast to the Asian cohort. Group 2 consisted of 13 eSNPs from alleles associated with upregulated expression mostly found in the Asian cohort compared to the CEU cohort. When the cluster order was fixed, the corresponding expression pattern was visualized using an expression heatmap. To avoid bias from nongenetic factor differences, we adjusted the population-level expression by adding the estimated BD score to each CEU individual for each expression trait. Similar to genotype clustering, the expression pattern was generally divided into two groups corresponding to regulatory SNPs and the majority of the individuals with genotype assignments of 2 or 1 had higher GE levels. These results indicate that GS not only filtered

Table 3.3 Summary of eSNPs.

Symbol	eSNP	Allele	MAF		CTWM-GS				IGM		Overall mean		p-value
			CEU	Asian	BD	p	GS	p	CEU	Asian	CEU	Asian	Diff.	
UTS2	rs161822	A/G	0.02	0.53	−0.62	5.5e-05	−1.38	2.3e-25	1.6e-09	3.2e-17	6.54	8.54	−2	1.3e-28
FLJ20444	rs41373149	G/C	0.01	0.29	0.26	3.7e-03	−0.72	1.0e-35	1.7e-03	7.0e-31	6.52	6.98	−0.46	1.9e-04
CHI3L2	rs11102223	T/G	0.51	0.26	−0.2	2.7e-01	−0.69	1.3e-18	1.1e-06	2.2e-17	10.63	11.53	−0.89	4.6e-04
CADM1	rs613699	G/T	0.41	0.00	1.39	1.6e-12	−0.64	9.2e-07	3.6e-04	NA	7.6	6.86	0.74	6.2e-06
TSPAN32	rs756920	C/G	0.06	0.72	0.01	9.3e-01	−0.6	1.5e-10	1.3e-05	1.5e-07	6.39	6.98	−0.59	1.6e-17
GPER	rs1419772	G/A	0.43	0.02	−0.02	9.0e-01	−0.6	1.3e-08	1.2e-06	1.1e-01	9.5	10.12	−0.62	1.9e-05
C1orf115	rs425437	T/C	0.06	0.32	−0.21	1.3e-01	−0.57	1.5e-16	8.6e-05	5.5e-13	7.07	7.85	−0.78	1.6e-06
C8orf13	rs998683	T/C	0.73	0.33	0.37	1.2e-03	−0.53	6.7e-15	4.4e-06	1.7e-08	7.86	8.02	−0.17	1.5e-01
KIAA0748	rs7962801	T/C	0.89	0.13	0.02	7.8e-01	−0.53	6.6e-14	4.3e-08	5.8e-07	6.87	7.38	−0.51	4.3e-22
LOC389493	rs11238381	C/G	0.18	0.55	−0.44	1.1e-05	−0.52	9.7e-22	1.3e-11	9.3e-12	6.71	7.66	−0.96	6.6e-14
SYNGR1	rs5757611	T/C	0.24	0.57	−0.05	6.9e-01	−0.51	1.0e-08	1.3e-03	1.0e-05	7.03	7.59	−0.56	8.4e-06
MOSC2	rs425437	T/C	0.06	0.32	−0.13	3.8e-01	−0.5	1.2e-11	5.6e-04	9.8e-11	7.69	8.32	−0.63	3.0e-04
IGK	rs604127	T/C	0.19	0.06	−3.39	5.3e-18	0.51	5.9e-07	1.5e-06	7.0e-01	8.03	10.92	−2.89	7.2e-12
C16orf75	rs4451969	T/C	0.72	0.22	0	9.7e-01	0.52	2.7e-18	2.0e-10	1.5e-07	9.94	9.42	0.52	1.0e-07
IGHV6-1	rs17710010	A/G	0.23	0.00	−1.47	1.9e-11	0.58	8.8e-07	3.0e-04	NA	8.84	9.73	−0.89	4.7e-05
ANKDD1A	rs871447	C/T	0.42	0.03	0.32	1.4e-07	0.67	7.9e-42	9.9e-19	3.8e-18	8.38	7.4	0.99	1.8e-15
NAPRT1	rs10112966	T/C	0.24	0.57	0.14	1.7e-01	0.75	1.1e-34	2.4e-12	2.4e-22	9.87	8.97	0.9	4.6e-09
HLA-DRB5	rs9271850	A/G	0.18	0.34	0.31	1.7e-01	0.85	5.1e-31	1.6e-08	4.3e-19	13.13	11.96	1.16	2.0e-04
IGHA2	rs10483288	A/C	0.22	0.00	−4.27	1.8e-19	1.37	3.5e-09	1.2e-05	NA	10.87	13.77	−2.9	1.3e-09
C3orf14	rs2337387	T/C	0.59	0.01	−1.71	8.5e-08	1.59	1.5e-08	5.4e-05	7.9e-02	8.7	8.82	−0.12	5.3e-01

IGM: p-values of the SNP–GE associations in the CEU and Asian populations, respectively, tested by the IGM method. Overall mean: arithmetic mean of GE values in the CEU and Asian populations, respectively; the difference of these two means is shown under the column "Diff." and the p-value of the difference was calculated by the Welch's two-sample t-test.

3.4 *CTWM-GS Method* | **59**

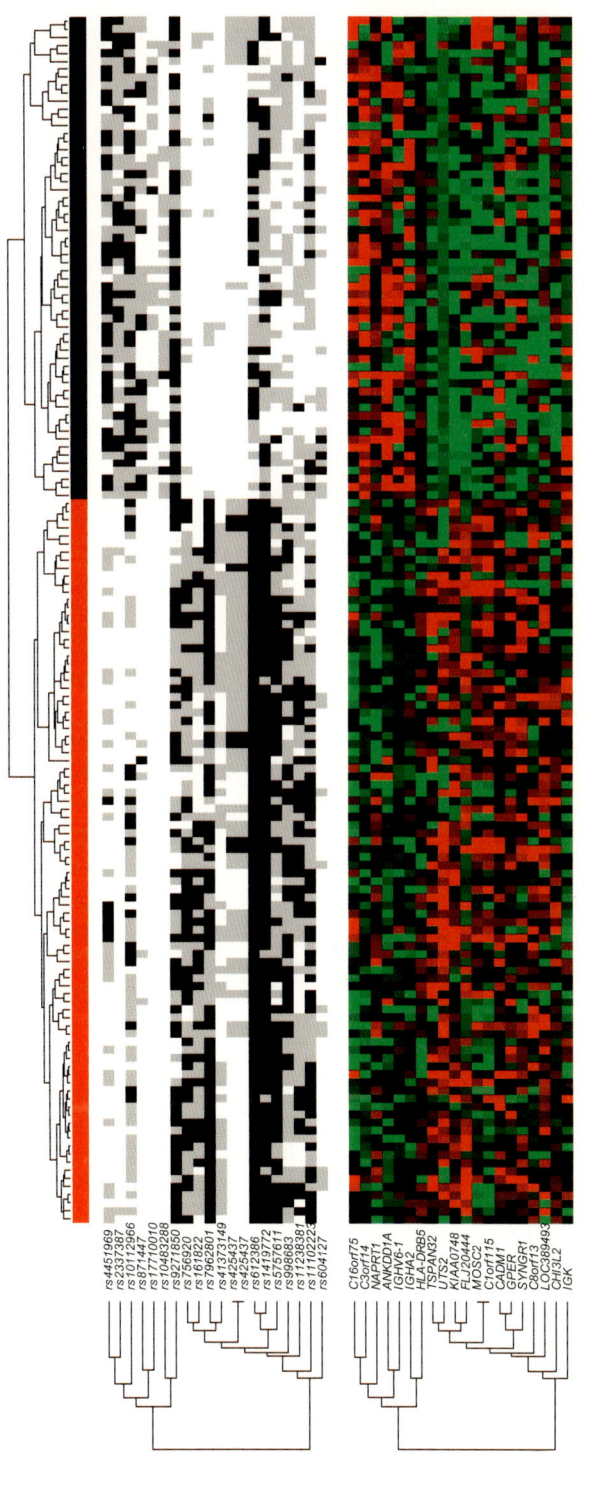

Figure 3.3 Heatmap of hierarchical clustering. The vertical hierarchical cluster (top) indicates that the CEU (black) and Asian (red) populations can be separated by clustering their genotypes as shown in the horizontal hierarchical tree. In the upper heatmap, black shading represents individuals homozygous for the upregulated allele and gray shading represents heterozygous individuals. White shading indicates individuals homozygous for the downregulated allele. The lower heatmap represents expression of the genes corresponding to the eSNPs in the upper heatmap; brightness of red is proportional to the degree of expression above the mean and brightness of green is proportional to the degree of expression below the mean.

functional SNPs, but also improved identification of those SNPs with causal potential.

3.5
Discussion

Several recent investigations have surveyed eQTL using human LCLs derived from healthy individuals in single or multiple ethnic populations to generate global regulatory networks in humans [53]. The efforts of current eQTL studies provide great insight into the biology of gene regulation and a shortcut to understanding disease biology by combining them with results from GWAS. The SCAN (SNP and CNV Annotation) database is a large-scale database of genetics and genomics data linked to a web interface, and it contains a category of SNPs that are classified according to their effects on expression levels [54]. SCAN associates over 13 000 transcript clusters with over 2 million common SNPs in the set of HapMap trios of CEU and YRI using family-based quantitative transmission disequilibrium test analysis. In addition, the mRNA-by-SNP Browser database of eQTL derived from asthma studies provides a general tool to investigate whether SNPs associated with any disease alter transcription of genes in *cis* or *trans* [23]. These resources illustrate that the genetic data included in analyses of differential GE is more powerful than using GE alone to identify candidate disease genes.

This chapter elucidates the genetics of GE in human populations, and reconstructs gene networks by integrating GE and genetic data. We investigated the statistical issues associated with common eQTL identification using data combined from different populations. We have demonstrated the impact of MAFs on identification of common eQTL using either separate or combined population data. As differentially expressed genes and SNPs with divergent allele frequencies are common among ethnic populations [55–57], combining genetic genomic datasets across populations likely improves identification of common eQTL. In addition to CTWM, other statistical approaches such as quantile transformation [28] are available for pooling individuals across studies to estimate the associations between genotype and GE. Combining genetic genomic data from different populations on transformed GE levels poses a number of statistical difficulties. These difficulties arise from the fact that the constituent datasets usually have different genetic structures in MAFs. As a consequence, directly combining transformed GE data does not explicitly take into account the diversity of MAFs among populations and may result in misleading statistical conclusions in eQTL studies [29]. Instead of transforming expression data, it has been shown that application of CTWM to model common eQTL directly onto the original expression data corrects the expression bias arising from genetic diversity among populations [29].

Combining a pair of HapMap populations using CTWM, 1839 expression traits with 29 420 common eQTL were identified with statistical significance and these traits included all of the 77 traits identified by the IGM. These results suggest that the CTWM is more powerful than the IGM for the identification of common eQTL and empirically supports the assumption of homogeneous SNP effects between populations for common eQTL. Meta-analysis detects modest associations and has been used extensively in GWAS [58]. As various genetic genomic datasets have recently become available, these results may form a basis for further meta-analyses of the datasets, in which different datasets are to be combined to increase the statistical power for identifying common eQTL with similar effects.

We have demonstrated that genetic effects on GE between populations can be calculated using the GS from data with three dimensions: GE level, SNP genotype, and ethnicity. It is possible to compare allelic frequencies of SNPs between populations and then test the association of these SNPs with GE data. However, the association between GE and populations cannot be directly implied given associations between populations and SNPs that are correlated with GE [48]. For example, the transcript *C8orf13* was associated with the putative eQTL "rs998683" ($p < 5 \times 10^{-14}$ tested by CTWM) and showed high discrepancy in allelic frequencies ($p < 10^{-9}$ tested by χ^2) between CEU and Asian cohorts. However, the overall expression level of that transcript was nondifferentially expressed between the two cohorts (Table 3.3). This phenomenon agrees with a previous genetic genomic study of childhood asthma, in which the expression level of *ORMDL3* was strongly associated with the disease-associated marker "rs7216389" [59]. However, the overall GE difference between nonasthmatic and asthmatic cohorts was not significant. The advantage of using CTWM-GS is that GE differences between populations can be partitioned into two parts – genetic differences (i.e., GS) and nongenetic differences (i.e., BD) – and both are tested independently.

Populations differ in the prevalence of many complex genetic diseases and few studies searching for susceptible genes underlying these diseases have been reproduced across populations. This may result from heterogeneity in disease etiology, ambiguity in definition and/or classification of phenotype, or the use of different genetic markers. One fundamental, yet less studied, confounding factor is the systemic variations in genomic structure for different populations. This presents a golden opportunity to study the ethnic differences in genomic structure and their impact on corresponding GE in the context of human disease gene mapping. The methods introduced in this chapter can also be used to examine genetic genomic data containing both case and control individuals. Such studies may provide new insights into disease etiology by identifying potential eSNPs with allele frequency differences or that are associated with different GE levels in the populations.

References

1 Nicolae, D.L., Gamazon, E., Zhang, W., Duan, S., Dolan, M.E., and Cox, N.J. (2010) Trait-associated SNPs are more likely to be eQTLs: annotation to enhance discovery from GWAS. *PLoS Genet.*, **6**, e1000888.

2 Nica, A.C. and Dermitzakis, E.T. (2008) Using gene expression to investigate the genetic basis of complex disorders. *Hum. Mol. Genet.*, **17**, R129–R134.

3 Cheung, V.G., Conlin, L.K., Weber, T.M., Arcaro, M., Jen, K.Y., Morley, M., and Spielman, R.S. (2003) Natural variation in

human gene expression assessed in lymphoblastoid cells. *Nat. Genet.*, **33**, 422–425.

4 van't Veer, L.J., Dai, H., van de Vijver, M.J., He, Y.D., Hart, A.A., Mao, M., Peterse, H.L., van der Kooy, K., Marton, M.J., Witteveen, A.T., Schreiber, G.J., Kerkhoven, R.M., Roberts, C., Linsley, P.S., Bernards, R., and Friend, S.H. (2002) G profiling predicts clinical outcome of breast cancer. *Nature*, **415**, 530–536.

5 Göring, H.H., Curran, J.E., Johnson, M.P., Dyer, T.D., Charlesworth, J., Cole, S.A., Jowett, J.B., Abraham, L.J., Rainwater, D.L., Comuzzie, A.G., Mahaney, M.C., Almasy, L., MacCluer, J.W., Kissebah, A.H., Collier, G.R., Moses, E.K., and Blangero, J. (2007) Discovery of expression QTLs using large-scale transcriptional profiling in human lymphocytes. *Nat. Genet.*, **39**, 1208–1216.

6 Petretto, E., Mangion, J., Dickens, N.J., Cook, S.A., Kumaran, M.K., Lu, H., Fischer, J., Maatz, H., Kren, V., Pravenec, M., Hubner, N., and Aitman, T.J. (2006) Heritability and tissue specificity of expression quantitative trait loci. *PLoS Genet.*, **2**, e172.

7 Chorley, B.N., Wang, X., Campbell, M.R., Pittman, G.S., Noureddine, M.A., and Bell, D.A. (2008) Discovery and verification of functional single nucleotide polymorphisms in regulatory genomic regions: current and developing technologies. *Mutat. Res.*, **659**, 147–157.

8 Jansen, R.C. and Nap, J.P. (2004) Regulating gene expression: surprises still in store. *Trends Genet.*, **20**, 223–225.

9 Franke, L. and Jansen, R.C. (2009) eQTL analysis in humans. *Methods Mol. Biol.*, **573**, 311–328.

10 Keurentjes, J.J., Fu, J., Terpstra, I.R., Garcia, J.M., van den Ackerveken, G., Snoek, L.B., Peeters, A.J., Vreugdenhil, D., Koornneef, M., and Jansen, R.C. (2007) Regulatory network construction in *Arabidopsis* by using genome-wide gene expression quantitative trait loci. *Proc. Natl. Acad. Sci. USA*, **103**, 1708–1713.

11 Storey, J.D., Akey, J.M., and Kruglyak, L. (2005) Multiple locus linkage analysis of genomewide expression in yeast. *PLoS Biol.*, **3**, e267.

12 Brem, R.B. and Kruglyak, L. (2005) The landscape of genetic complexity across 5,700 gene expression traits in yeast. *Proc. Natl. Acad. Sci. USA*, **102**, 1572–1577.

13 Schadt, E.E., Monks, S.A., Drake, T.A., Lusis, A.J., Che, N., Colinayo, V., Ruff, T.G., Milligan, S.B., Lamb, J.R., Cavet, G., Linsley, P.S., Mao, M., Stoughton, R.B., and Friend, S.H. (2003) Genetics of gene expression surveyed in maize, mouse and man. *Nature*, **422**, 297–302.

14 Hovatta, I., Zapala, M.A., Broide, R.S., Schadt, E.E., Libiger, O., Schork, N.J., Lockhart, D.J., and Barlow, C. (2007) DNA variation and brain region-specific expression profiles exhibit different relationships between inbred mouse strains: implications for eQTL mapping studies. *Genome Biol.*, **8**, R25.

15 Hubner, N., Wallace, C.A., Zimdahl, H., Petretto, E., Schulz, H., Maciver, F., Mueller, M., Hummel, O., Monti, J., Zidek, V., Musilova, A., Kren, V., Causton, H., Game, L., Born, G., Schmidt, S., Muller, A., Cook, S.A., Kurtz, T.W., Whittaker, J., Pravenec, M., and Aitman, T.J. (2005) Integrated transcriptional profiling and linkage analysis for identification of genes underlying disease. *Nat. Genet.*, **37**, 243–253.

16 Morley, M., Molony, C.M., Weber, T.M., Devlin, J.L., Ewens, K.G., Spielman, R.S., and Cheung, V.G. (2004) Genetic analysis of genome-wide variation in human gene expression. *Nature*, **430**, 743–747.

17 Cheung, V.G., Spielman, R.S., Ewens, K.G., Weber, T.M., Morley, M., and Burdick, J.T. (2005) Mapping determinants of human gene expression by regional and genome-wide association. *Nature*, **437**, 1365–1369.

18 Stranger, B.E., Forrest, M.S., Clark, A.G., Minichiello, M.J., Deutsch, S., Lyle, R., Hunt, S., Kahl, B., Antonarakis, S.E., Tavare, S., Deloukas, P., and Dermitzakis,

E.T. (2005) Genome-wide associations of gene expression variation in humans. *PLoS Genet.*, **1**, e78.
19 Storey, J.D., Madeoy, J., Strout, J.L., Wurfel, M., Ronald, J., and Aker, J.M. (2007) Gene expression variation within and among human populations. *Am. J. Hum. Genet.*, **80**, 502–509.
20 Spielman, R.S., Bastone, L.A., Burdick, J.T., Morley, M., Ewens, W.J., and Cheung, V.G. (2007) Common genetic variants account for differences in gene expression among ethnic groups. *Nat. Genet.*, **39**, 226–231.
21 Zhang, W., Duan, S., Kistner, E.O., Bleibel, W.K., Huang, R.S., Clark, T.A., Chen, T.X., Schweitzer, A.C., Blume, J.E., Cox, N.J., and Dolan, M.E. (2008) Evaluation of genetic variation contributing to differences in gene expression between populations. *Am. J. Hum. Genet.*, **82**, 631–640.
22 Stranger, B.E., Forrest, M.S., Dunning, M., Ingle, C.E., Beazley, C., Thorne, N., Redon, R., Bird, C.P., de Grassi, A., Lee, C., Tyler-Smith, C., Carter, N., Scherer, S.W., Tavare, S., Deloukas, P., Hurles, M.E., and Dermitzakis, E.T. (2007) Relative impact of nucleotide and copy number variation on gene expression phenotypes. *Science*, **315**, 848–853.
23 Dixon, A.L., Liang, L., Moffatt, M.F., Chen, W., Heath, S., Wong, K.C., Taylor, J., Burnett, E., Gut, I., Farrall, M., Lathrop, G.M., Abecasis, G.R., and Cookson, W.O. (2007) A genome-wide association study of global gene expression. *Nat. Genet.*, **39**, 1202–1207.
24 Rosa, G.J., de Leon, N., and Rosa, A.J. (2006) Review of microarray experimental design strategies for genetical genomics studies. *Physiol. Genomics*, **28**, 15–23.
25 Schliekelman, P. (2008) Statistical power of expression quantitative trait loci for mapping of complex trait loci in natural populations. *Genetics*, **178**, 2201–2216.
26 Stranger, B.E., Nica, A.C., Forrest, M.S., Dimas, A., Bird, C.P., Beazley, C., Ingle, C.E., Dunning, M., Flicek, P., Koller, D., Montgomery, S., Tavare, S., Deloukas, P., and Dermitzakis, E.T. (2007) Population genomics of human gene expression. *Nat. Genet.*, **39**, 1217–1222.
27 Idaghdour, Y., Storey, J.D., Jadallah, S.J., and Gibson, G. (2008) A genome-wide gene expression signature of environmental geography in leukocytes of Moroccan Amazighs. *PLoS Genet.*, **4**, e1000052.
28 Veyrieras, J.B., Kudaravalli, S., Kim, S.Y., Dermitzakis, E.T., Gilad, Y., Stephens, M., and Pritchard, J.K. (2008) High-resolution mapping of expression-QTLs yields insight into human gene regulation. *PLoS Genet.*, **4**, e1000214.
29 Hsiao, C.L., Lian, IeB., Hsieh, A.R., and Fann, C.S. (2010) Modeling expression quantitative trait loci in data combining ethnic populations. *BMC Bioinformatics*, **11**, 111.
30 Zhang, W. and Dolan, M.E. (2009) Use of cell lines in the investigation of pharmacogenetic loci. *Curr. Pharm. Des.*, **15**, 3782–3795.
31 Huang, R.S., Duan, S., Shukla, S.J., Kistner, E.O., Clark, T.A., Chen, T.X., Schweitzer, A.C., Blume, J.E., and Dolan, M.E. (2007) Identification of genetic variants contributing to cisplatin-induced cytotoxicity by use of a genomewide approach. *Am. J. Hum. Genet.*, **81**, 427–437.
32 Kwan, T., Benovoy, D., Dias, C., Gurd, S., Provencher, C., Beaulieu, P., Hudson, T.J., Sladek, R., and Majewski, J. (2008) Genome-wide analysis of transcript isoform variation in humans. *Nat. Genet.*, **40**, 225–231.
33 Price, A.L., Patterson, N., Hancks, D.C., Myers, S., Reich, D., Cheung, V.G., and Spielman, R.S. (2008) Effects of *cis* and *trans* genetic ancestry on gene expression in African Americans. *PLoS Genet.*, **4**, e1000294.
34 Pruitt, K.D., Tatusova, T., and Maglott, D.R. (2005) NCBI Reference Sequence (RefSeq): a curated non-redundant sequence database of genomes, transcripts and proteins. *Nucleic Acids Res.*, **33**, D501–D504.

35 Chen, L., Page, G.P., Mehta, T., Feng, R., and Cui, X. (2009) Single nucleotide polymorphisms affect both cis- and trans-eQTLs. *Genomics*, **93**, 501–508.

36 Li, M., Li, C., and Guan, W. (2008) Evaluation of coverage variation of SNP chips for genome-wide association studies. *Eur. J. Hum. Genet.*, **16**, 635–643.

37 Browning, B.L. and Browning, S.R. (2009) A unified approach to genotype imputation and haplotype-phase inference for large data sets of trios and unrelated individuals. *Am. J. Hum. Genet*, **84**, 210–223.

38 Lettre, G., Lange, C., and Hirschhorn, J.N. (2007) Genetic model testing and statistical power in population-based association studies of quantitative traits. *Genet. Epidemiol.*, **31**, 358–362.

39 Huang, R.S., Duan, S., Kistner, E.O., Zhang, W., Bleibel, W.K., Cox, N.J., and Dolan, M.E. (2008) Identification of genetic variants and gene expression relationships associated with pharmacogenes in humans. *Pharmacogenet. Genomics*, **18**, 545–549.

40 Duan, S., Huang, R.S., Zhang, W., Bleibel, W.K., Roe, C.A., Clark, T.A., Chen, T.X., Schweitzer, A.C., Blume, J.E., Cox, N.J., and Dolan, M.E. (2008) Genetic architecture of transcript-level variation in humans. *Am. J. Hum. Genet.*, **82**, 1101–1113.

41 Heap, G.A., Trynka, G., Jansen, R.C., Bruinenberg, M., Swertz, M.A., Dinesen, L.C., Hunt, K.A., Wijmenga, C., Vanheel, D.A., and Franke, L. (2009) Complex nature of SNP genotype effects on gene expression in primary human leucocytes. *BMC Med. Genomics*, **2**, 1.

42 Myers, A.J., Gibbs, J.R., Webster, J.A., Rohrer, K., Zhao, A., Marlowe, L., Kaleem, M., Leung, D., Bryden, L., Nath, P., Zismann, V.L., Joshipura, K., Huentelman, M.J., Hu-Lince, D., Coon, K.D., Craig, D.W., Pearson, J.V., Holmans, P., Heward, C.B., Reiman, E.M., Stephan, D., and Hardy, J. (2007) A survey of genetic human cortical gene expression. *Nat. Genet.*, **39**, 1494–1499.

43 Olkin, I. and Sampson, A. (1998) Comparison of meta-analysis versus analysis of variance of individual patient data. *Biometrics*, **54**, 317–322.

44 Wu, C., Delano, D.L., Mitro, N., Su, S.V., Janes, J., McClurg, P., Batalov, S., Welch, G.L., Zhang, J., Orth, A.P., Walker, J.R., Glynne, R.J., Cooke, M.P., Takahashi, J.S., Shimomura, K., Kohsaka, A., Bass, J., Saez, E., Wiltshire, T., and Su, A.I. (2008) Gene set enrichment in eQTL data identifies novel annotations and pathway regulators. *PLoS Genet.*, **4**, e100007.

45 Breitling, R., Li, Y., Tesson, B.M., Fu, J., Wu, C., Wiltshire, T., Gerrits, A., Bystrykh, L.V., de Haan, G., Su, A.I., and Jansen, R.C. (2008) Genetical genomics: spotlight on QTL hotspots. *PLoS Genet.*, **4**, e1000232.

46 Li, J. and Burmeister, M. (2005) Genetical genomics: combining genetics with gene expression analysis. *Hum. Mol. Genet.*, **2**, R163–R169.

47 Rockman, M.V. and Kruglyak, L. (2006) Genetics of global gene expression. *Nat. Rev. Genet.*, **7**, 862–872.

48 Cookson, W., Liang, L., Abecasis, G., Moffatt, M., and Lathrop, M. (2009) Mapping complex disease traits with global gene expression. *Nat. Rev. Genet.*, **10**, 184–194.

49 Horn, R.A. and Johnson, C.R. (1985) BT Matrix Analysis, Cambridge University Press, Cambridge.

50 Cleveland, W.S. and Devlin, S.J. (1979) Robust locally weighted regression and smoothing scatterplots. *J. Am. Stat. Ass.*, **74**, 829–836.

51 Gibson, G. and Weir, B. (2005) The quantitative genetics of transcription. *Trends Genet.*, **21**, 616–623.

52 Degnan, J.H., Lasky-Su, J., Raby, B.A., Xu, M., Molony, C., Schadt, E.E., and Lange, C. (2008) Genomics and genome-wide association studies: an integrative approach to expression QTL mapping. *Genomics*, **92**, 129–133.

53 Zhang, W., Ratain, M.J., and Dolan, M.E. (2008) The HapMap resource is providing new insights into ourselves and its application to pharmacogenomics. *Bioinform. Biol. Insights.*, **2**, 15–23.

54 Gamazon, E.R., Zhang, W., Konkashbaev, A., Duan, S., Kistner, E.O., Nicolae, D.L.,

Dolan, M.E., and Cox, N.J. (2010) SCAN: SNP and copy number annotation. *Bioinformatics*, **26**, 259–262.

55 Gilad, Y., Rifkin, S.A., and Pritchard, J.K. (2008) Revealing the architecture of gene regulation: the promise of eQTL studies. *Trends Genet.*, **24**, 408–415.

56 International HapMap Consortium (2005) A haplotype map of the human genome. *Nature*, **437**, 1299–1320.

57 International HapMap Consortium (2007) A second generation human haplotype map of over 3.1 million SNPs. *Nature*, **449**, 851–861.

58 Zintzaras, E. and Lau, J. (2008) Trends in meta-analysis of genetic association studies. *J. Hum. Genet.*, **53**, 1–9.

59 Moffatt, M.F., Kabesch, M., Liang, L., Dixon, A.L., Strachan, D., Heath, S., Depner, M., von Berg, A., Bufe, A., Rietschel, E., Heinzmann, A., Simma, B., Frischer, T., Willis-Owen, S.A., Wong, K.C., Illig, T., Vogelberg, C., Weiland, S.K., von Mutius, E., Abecasis, G.R., Farrall, M., Gut, I.G., Lathrop, G.M., and Cookson, W.O. (2007) Genetic variants regulating *ORMDL3* expression contribute to the risk of childhood asthma. *Nature*, **448**, 470–473.

Part Two
Inference of Gene Networks

4
Transcriptional Network Inference Based on Information Theory
Patrick E. Meyer, Catharina Olsen, and Gianluca Bontempi

4.1
Introduction

The reverse engineering of transcriptional regulatory networks from expression data is known to be a very challenging task because of the large amount of noise intrinsic to microarray technology, the high dimensionality, and the combinatorial nature of the problem [1]. Furthermore, a gene-to-gene network inferred on the basis of transcriptional measurements returns only an approximation of a complete biochemical regulatory network since many physical connections between macromolecules might be hidden by short-cuts.

In spite of these evident limitations the bioinformatics community has made important advances in this domain over the last few years using network inference methods like Boolean networks, Bayesian networks, and Association networks [2].

This chapter focuses on information-theoretic approaches that typically rely on the estimation of mutual information and conditional mutual information from data in order to measure the statistical dependence between gene expression. The adoption of mutual information in network inference can be traced back to Chow and Liu's tree algorithm [3]. Nowadays, two main categories of information-theoretic network inference methods hold the attention of the bioinformatics community: (i) methods based on conditional mutual information that are able to infer a larger set of relationships between genes, but at the price of a higher algorithmic complexity [4–6], and (ii) methods based on pairwise mutual information that infer undirected networks up to thousands of genes thanks to their low algorithmic complexity [7–10]. The strengths and weaknesses of these information-theoretic methods are depicted in this chapter.

4.1.1
Notation

Networks, like signal transduction networks or transcriptional regulatory networks, play a central role in biological systems. In order to get insights about their structure

---

Applied Statistics for Network Biology: Methods in Systems Biology, First Edition.
Edited by M. Dehmer, F. Emmert-Streib, A. Graber, and A. Salvador.
© 2011 Wiley-VCH Verlag GmbH & Co. KGaA. Published 2011 by Wiley-VCH Verlag GmbH & Co. KGaA.

and functionality from experimental data (e.g., microarray data), various methods of network inference have been proposed in the literature.

Network inference consists of representing the dependencies between the variables of a dataset by a graph [11]. The semantics of an arc in the graph may differ from one inference method to another. When network inference is used to reconstruct a transcriptional regulatory network from microarray data [1], arcs usually represent a regulator/regulated gene interaction where the genes are represented by nodes in the graph. This chapter focuses on a specific family of network inference methods that rely on the estimation of mutual information and conditional mutual information from data in order to measure the statistical dependence between gene expression.

Before introducing the basic notion of information theory, in order to help the understanding of the reader, we introduce the notation used throughout the chapter to denote gene expressions (i.e., the variables of our model) and to identify subsets of genes.

- $A = \{1, \ldots, n\}$: index set
- $X_A = (X_1, X_2, \ldots, X_n)$: a set of n random variables
- $X_k \in X_A$: the kth variable of the set
- $\mathcal{X}_k$: the domain of values of the kth variable
- $X_K \subset X_A$: a subset of variables
- $X_{-k} = X_A / X_k$: subset of X_A with the variable X_k set aside
- X_{-K}: subset of X_A with the subset of variables X_K set aside
- $X_{i,j} = \{X_i, X_j\}$: two variables of the set X_A
- $X_{-(i,j)}$: set of variables X_A without X_i and X_j
- $\mathcal{D}$: dataset

The concepts presented in this chapter are mainly based on two conventional information-theoretic measures.

Definition 4.1 [12]

Let X_i and X_j be two (discrete) random variables, $x_i \in \mathcal{X}_i$ and $x_j \in \mathcal{X}_j$ their respective realizations, and $p(x_i, x_j)$ their joint probability distribution. The *mutual information* between two random variables X_i and X_j is defined as:

$$I(X_i; X_j) = \sum_{x_i \in \mathcal{X}_i} \sum_{x_j \in \mathcal{X}_j} p(x_i, x_j) \log\left(\frac{p(x_i, x_j)}{p(x_i)p(x_j)}\right)$$

The main characteristics of mutual information are:

- It is a symmetric and non-negative quantity.
- It measures the divergence between the joint and the product of the two marginal distributions.
- $I(X_i; X_j)$ is maximal if X_i (X_j) is perfectly predictable once X_j (X_i) is known (e.g., deterministic dependence).

- $I(X_i; X_j) = 0$ if X_i (X_j) returns no information about X_j (X_i), that is X_i and X_j are independent.

Definition 4.2 [12]

Let X_i and X_j be two (discrete) random variables and $x_i \in \mathcal{X}_i$ and $x_j \in \mathcal{X}_j$ their respective realizations. Once a third random variable X_k is given, the *conditional mutual information* between X_i and X_j knowing X_k is:

$$I(X_i; X_j | X_k) = I(X_{i,k}; X_j) - I(X_k; X_j)$$

The main characteristics of the conditional mutual information are:

- It measures the amount of information that X_j (X_i) provides about X_i (X_j), once the value of X_k is known.
- $I(X_i; X_j | X_k) \geq 0$ with equality iff X_i and X_j are conditionally independent given X_k.

4.1.2
Formalization

Let us consider a transcriptional network composed of n genes where $X_j, j = 1, \ldots, n$ is the random variable representing the expression of the jth gene and $X_A = (X_1, X_2, \ldots, X_n)$ is the set of variables representing the expression of all genes. Let $X_{S_j} \subset X_{-j}$ denote the (potentially empty) subset of genes (whose products are) regulating the expression X_j of the jth gene.

This induces a network T of relationships between the n variables that can be represented either by a directed graph or equivalently by a square matrix where the i,j element is:

$$t_{ij} = \begin{cases} 1 & \text{if } X_i \in X_{S_j} \\ 0 & \text{else} \end{cases} \quad (4.1)$$

Note that the adopted convention for the directionality of directed edges in a graph is $X_i \rightarrow X_j \Leftrightarrow t_{ij} = 1$.

Once a set $\mathcal{D}$ of measurements of the expressions of the n genes is collected, the objective of a network inference algorithm $\mathcal{N}$ is to estimate a network $\hat{T}$ from $\mathcal{D}$:

$$\hat{T} = \mathcal{N}(\mathcal{D}) \quad (4.2)$$

that is as "close" (according to a given performance measure Φ) as possible to the unknown (or partially known) true network T:

$$\hat{T}^{max} = \arg\max_{\hat{T}} \Phi(\hat{T}, T) \quad (4.3)$$

Qualitatively, $\hat{T}$ is inferred from a dataset $\mathcal{D}$ that is generated according to the probability distribution $p(X)$, which is itself governed by the underlying network T of genetic interactions.

Typically the network inference task is decomposed into two steps: the first step consists of inferring from data the undirected network (also known as the skeleton) and the second one consists of orienting the arcs of the undirected network.

4.1.3
Performance Measures in Undirected Network Inference

An undirected network inference problem can be seen as a binary decision problem where the algorithm $\mathcal{N}$ plays the role of a classifier. Each pair of nodes $\hat{t}_{ij}$ is thus assigned either a positive label (1, i.e., an edge) or a null one (0, i.e., no edge). A reference network T is then required to assess the performances of the inferred network $\hat{T}$. Two strategies are usually adopted to define a reference network:

i) The dataset is synthetically generated by simulating a known reference network.
ii) The inference concerns data where several variable interactions are known (i.e., genetic interactions that have already been discovered by researchers). This list of known interactions can then be used as a reference network in order to discover new potential interactions.

A positive label (an edge) predicted by the algorithm is considered as a true-positive (*tp*) or as a false-positive (*fp*) depending on the presence or not of the corresponding edge in the reference network. Analogously, a null label is considered as a true-negative (*tn*) or a false-negative (*fn*) depending on whether the corresponding edge is absent or not in the underlying true network T, respectively (Table 4.1). Most of the inference methods discussed in this chapter return a weighted adjacency matrix $W = (w_{ij})_{i,j \in A}$ representing the network. Hence, a threshold value θ is used to remove the arcs of the network that have a too low score [2, 8, 13]:

$$\hat{t}_{ij} = \begin{cases} 1 & \text{if } w_{ij} \geq \theta \\ 0 & \text{otherwise} \end{cases} \tag{4.4}$$

For each threshold value θ, a different inferred network $\hat{T}(\theta, W)$ based on the weighted adjacency matrix W can be computed. As a result, a specific confusion matrix (Table 4.1) is obtained for each θ.

The two following sections introduce two accuracy measures Φ to assess the performance of a network inference technique.

Table 4.1 Confusion matrix.

	True edge (1)	No edge (0)
Inferred edge (1)	$tp = (\hat{t}_{ij} = t_{ij} = 1)$	$fp = (\hat{t}_{ij} = 1 \neq t_{ij} = 0)$
Deleted edge (0)	$fn = (\hat{t}_{ij} = 0 \neq t_{ij} = 1)$	$tn = (\hat{t}_{ij} = t_{ij} = 0)$

4.1.3.1 Precision–Recall (PR) Curves

Precision and recall are two useful measures, borrowed from the information retrieval literature, to assess the quality of an inferred network.

The *precision* quantity is given by:

$$pre = \frac{tp}{tp+fp} = \frac{tp}{\#(\hat{t}_{ij} = 1)} \quad (4.5)$$

and measures the fraction of real edges among the ones classified as positive.

The *recall* quantity (also called the true-positive rate (*tpr*)) is given by:

$$rec = tpr = \frac{tp}{tp+fn} = \frac{tp}{\#(t_{ij} = 1)} \quad (4.6)$$

and denotes the fraction of real edges that are correctly inferred. The objective of an inference method is to maximize both precision and recall. These quantities depend on the chosen threshold θ.

The precision–recall PR curve is a diagram that plots precision (*pre*) against recall (*rec*) for different values of the threshold [14] (see Figure 4.1 for a PR curve). In a network inference setting, this diagram illustrates the tradeoff between returning a small amount of arcs (low recall due to high threshold) with high confidence (high precision) and returning many arcs (high recall) with low confidence (low precision).

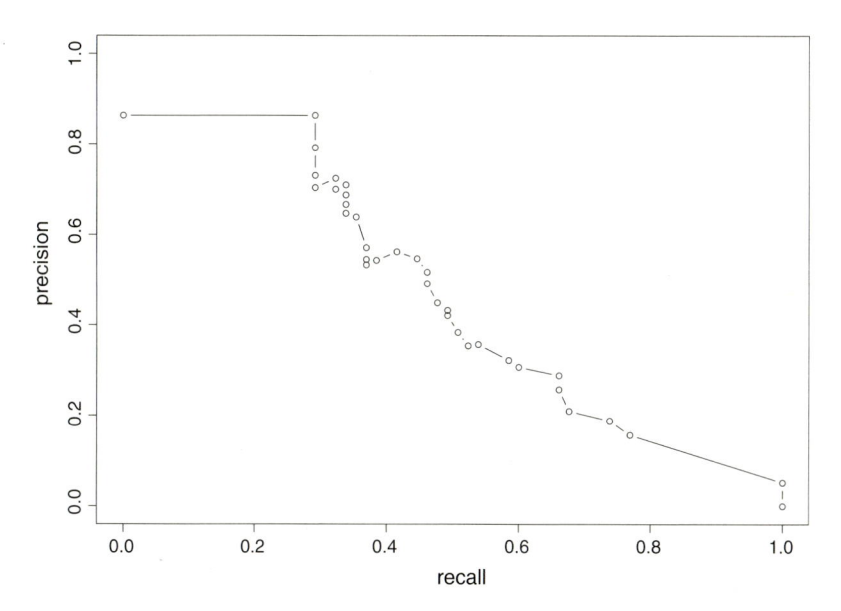

Figure 4.1 Example of PR curves (generated with the R package minet [15]).

4.1.3.2 F-Scores

The multicriteria nature of the inference problem can also be addressed by defining univariate measures that weight precision and recall. An example is given by the area under the PR curve [16].

Another well-known measure is the F-score quantity [14], which is a weighted harmonic average of precision and recall:

$$F_\beta(\hat{T}(\theta, W), T) = \frac{(1+\beta^2)(pre)(rec)}{\beta^2 pre + rec} \qquad (4.7)$$

where β is a non-negative real parameter weighting the importance of recall versus that of precision. The F-score takes value in the interval $[0, 1]$.

Three common values for the parameter β are:

- $\beta = 1$: assigning equal weight to precision and recall.
- $\beta = 2$: (the F_2-measure), which weights recall twice as much as precision.
- $\beta = 0.5$: (the $F_{0.5}$-measure), which weights precision twice as much as recall.

Note that in transcriptional network inference, precision is often valued higher than recall since the experimental cost to check for possible interactions is high.

A compact representation of the PR diagram can be returned by the maximum and/or the average (avg) F_β-score:

$$F_\beta^{max}(W, T) = \max_\theta(F_\beta(\hat{T}(\theta, W), T))$$
$$F_\beta^{avg}(W, T) = avg_\theta(F_\beta(\hat{T}(\theta, W), T)) \qquad (4.8)$$

where θ is the threshold parameter.

4.1.4
Causal Subset Selection

The inference of a network from data requires for each gene the identification of its own family of regulators. This problem is extremely complex since, given $n-1$ variables in X_{-j}, there is an exponential number 2^{n-1} of candidate subsets X_{S_j}.

An approach to solve this problem is given by *causal Bayesian networks* that provide a theoretical framework to identifying a causal subset X_{S_j} of a variable X_j.

Definition 4.3

X_i is a *cause* of X_j, denoted by $X_i \to X_j$, if there exists a value $x_i \in \mathcal{X}_i$ such that setting $X_i = x_i$ leads to a change in the probability distribution of X_j [17].

The definition of causality states that a causal relation between two variables creates a stochastic dependency between the probability distributions of causes and effects. Thus, two causally linked variables are not independent and therefore the mutual information is larger than zero:

$$X_i \leftrightarrow X_j \Rightarrow I(X_i; X_j) > 0 \qquad (4.9)$$

where $X_i \leftrightarrow X_j$ denotes an *undirected causal link* (i.e., $X_i \rightarrow X_j$ or $X_i \leftarrow X_j$). Unfortunately, since mutual information, unlike causality, is a symmetric measure, it is not possible to derive the direction of an edge.

However, the bivariate dependency stated in (4.9) is not always true. Two known examples are the cancellation of two causal pathways [17] and the XOR problem [18]. Since this occurs only under very special conditions, in the following we will keep on assuming that *causality implies stochastic dependency.*

However, the converse is not true – dependency does not imply causality:

$$I(X_i; X_j) > 0 \not\Rightarrow X_i \leftrightarrow X_j \tag{4.10}$$

One counterargument to the idea of *dependency implying causality* relies on the *common-cause effect*. That is, a dependency between two variable X_i and X_j can be created by a common cause (i.e., $X_i \leftarrow X_k \rightarrow X_j$). These two variables can be dependent, but manipulating one of them does not influence the other.

An illustration of this case is the well-known saying "curing the symptoms does not cure the disease."

In this context, the definition of direct causality provides a solution to the problem of deriving causal dependency from stochastic dependency.

Definition 4.4

X_i is a *direct cause* of X_j if X_i is a cause of X_j and there is no other variable X_k such that once the value of X_k is known, a manipulation of X_i no longer changes the probability distribution of X_j [17].

This definition states that if there are no sets of variables that cancel the dependency between two other variables, then one of the variables is a direct cause of the other.

In other words, it can be stated that if *two variables are dependent in every context*, then these variables have a causal relationship.

Once we make an additional assumption (known as *causal sufficiency condition*) the definition of direct cause provides the following implication in information-theoretic terms [4]:

$$\forall X_K \subseteq X_{-(i,j)} : I(X_i; X_j | X_K) > 0 \Rightarrow X_i \leftrightarrow X_j \tag{4.11}$$

The causal sufficiency condition requires that all variables that are causes to at least two effects (two variables in the dataset) be present in the set of measured variables. Indeed, if there is a common cause to two observable effects, the two effects are dependent in every context except when conditioning on the common cause. If the common cause is hidden then (4.11) can lead to false conclusions about the causal relationships between the variables.

However, it should be noted that the causal sufficiency condition does not concern intermediate unidentified variables along the *causal direction* $X_i \rightarrow X_k \rightarrow X_j$, as illustrated by the following example from [19].

Example 4.1

If A is the event of striking a match and C is the event of the match catching on fire and no other events are considered, then A causes C. However, if we add B, which stands for the sulfur on the match tip that achieved sufficient heat to combine with oxygen, then A no longer causes C directly. Rather A causes B and B causes C (i.e., $A \to B \to C$).

There is a fundamental difference between having hidden variables along each edge in the causal direction and hidden variables that are common causes of several effects. Along a causal direction it is intuitive to accept that the causality between a grandparent and a grandson is preserved once the parent is removed. This corresponds to make an assumption of *causal transitivity*.

$$\left. \begin{array}{c} X_i \to X_k \\ X_k \to X_j \end{array} \right\} \Rightarrow X_i \to X_j$$

However, if a common parent is missed, and a link between a variable and its sibling is added, then such link is no longer causal since acting on the sibling does not change the distribution of the variable.

Hence, the causal sufficiency condition can be rephrased by assuming that *there is no hidden common cause (to at least two effects)* in the set of considered variables:

$$\forall (X_i, X_j) \in X, \ (X_h \notin X) : X_i \leftarrow X_h \to X_j \tag{4.12}$$

The notions of information theory and causality introduced so far will be used in the rest of the chapter to present two main network inference approaches that hold the attention of the bioinformatics community: (i) methods based on conditional mutual information that are able to infer a larger set of relationships between genes, but at the price of a higher algorithmic complexity, and (ii) methods based on bivariate mutual information that infer undirected networks up to thousands of genes thanks to their low algorithmic complexity.

4.2
Inference Based on Conditional Mutual Information

Once causal sufficiency is assumed, the notion of conditional information allows the definition of a simple algorithm to infer an undirected network from observed data. The algorithm consists in setting an undirected causal link between all couples (X_i, X_j) such that:

$$\forall X_K \subseteq X_{-(i,j)} : I(X_i; X_j | X_K) > 0 \Rightarrow X_i \leftrightarrow X_j$$

In other terms this means to remove a link between the couples (X_i, X_j) if at least a conditioning set X_K such that $I(X_i, X_j | X_K) = 0$ exists. This is the strategy adopted in constraint-based Bayesian network algorithms, which are briefly discussed in the following section.

4.2.1
Constraint-Based Methods

Algorithm 4.1 Pseudo-Code of IC Algorithm

Start from an empty graph
foreach *pair of variables* (X_i, X_j) **do**
    if *there exists no subset X_K in $X_{-(i,j)}$ such that $I(X_i; X_j|X_K) = 0$* **then**
        | set an edge connecting (X_i, X_j)
    end
end

Algorithm 4.2 Pseudo-Code of SGS Algorithm

Start from the complete (fully connected) graph
foreach *pair of variables* (X_i, X_j) **do**
    if *there exists a subset X_K in $X_{-(i,j)}$ such that $I(X_i; X_j|X_K) = 0$* **then**
        | remove the edge connecting (X_i, X_j)
    end
end

IC (Algorithm 4.1 [18]) and SGS (Algorithm 4.2 [19]) are two state-of-the-art constraint-based algorithms that infer the network by carrying out a set of conditional independence tests. Note that IC proceeds in a forward manner by starting with an empty graph while SGS proceeds in a backward manner by removing progressively edges from a fully connected graph. However, in both algorithms, the IF instruction in the third line requires a computationally expensive procedure related to the search in the space of conditioning sets $X_K \subseteq X_{-(i,j)}$. This means that, given n variables, there are 2^{n-2} potential subsets for each couple (X_i, X_j). In order to address this issue, the PC algorithm has been proposed by [18] to speed up the SGS algorithm by replacing the IF test with the pseudo-code detailed in Algorithm 4.3. The rationale of the PC algorithm is to make conditional independence tests by using growing conditioning sets.

In spite of this improvement, these algorithms are not affordable in problems where the number n of variables is in the order of several thousands, such as in transcriptional network inference from microarray data.

Algorithm 4.3 Pseudo-Code of the Subset Search Procedure in PC Algorithm

foreach *subset size going from $|K| = 0$ to $|K| = n - 2$* **do**
    foreach *subset of variables X_K of size $|K|$* **do**
        if $I(X_i; X_j|X_K) = 0$ **then**
            | remove the edge connecting (X_i, X_j)
            | quit the two **for** loops
        end
    end
end

4.2.2
Approximated Conditional Mutual Information

In [6], the authors propose an information-theoretic translation of the PC algorithm that uses:

- A single conditioning variable X_k instead of the set of variables X_K, replacing $I(X_i; X_j | X_K)$ by $I(X_i; X_j | X_k)$.
- Two thresholds θ_1 and θ_2 representing estimation biases in the independence tests.

Algorithm 4.4 Pseudo-Code of [6]

Start from an empty graph
foreach *pair of variables* (X_i, X_j) **do**
    if $I(X_i; X_j) > \theta_1$ **then**
        if *there exists no variable X_k in $X_{-(i,j)}$ such that* $I(X_i; X_j | X_k) \leq \theta_2$ **then**
            set an edge connecting (X_i, X_j)
        end
    end
end
with θ_1 and θ_2 thresholds (parameters) of the method

Both modifications of the PC algorithm render this method adapted to real microarray datasets.

4.2.3
Variable Selection Algorithms

An alternative to the use of conditional independence tests is the use of variable selection methods for network inference. The idea is to perform network inference by repeating a variable selection step in the space X_{-j} for each variable $X_j \in X$.

A generic information-theoretic objective of variable selection can be formulated as: [20]:

> Given an output variable X_j and $n-1$ input variables X_{-j}, find the smallest subset $X_S \subseteq X_{-j}$ that maximizes the mutual information $I(X_S; X_j)$ between inputs and output.

Note that maximizing the (mutual) information of a subset of variables is equivalent to reducing the uncertainty (entropy) of the target variable.

Using variable selection strategies for network inference has many practical and theoretical advantages [21, 22]. For instance:

i) Some variable selection algorithms, like filters, can deal with thousands of variables in a reasonable amount of time. This makes inference scalable to large networks.

ii) Variable selection algorithms may be easily executed in parallel, since each of the n subset selection tasks is independent.
iii) Variable selection algorithms can use *a priori* knowledge. For example, knowing the list of regulator genes of an organism can improve the selection speed and the inference quality by limiting the search space of the selection step to a smaller list of genes.

Another advantage of a variable selection approach is that the subset that maximizes mutual information with the target contains all direct (causal) interactions [22].

The disadvantage is that this subset can also contain noncausal variables [22]. This results from the *explaining-away effect*.

Definition 4.9 [The explaining-away effect] [17]

Once the value of a common effect is given, it creates a dependency between its causes because each cause explains away the occurrence of the effect, thereby making the other cause less likely.

This is a common mechanism used by medical doctors when doing their diagnoses.

Example 4.2

Some cancer can cause headache, but a lot of more probable diseases, such as a cold, can also cause headache. Once a doctor has evidence that headaches are caused by a cold, the doctor stops searching for a cancer although having cold and having a cancer are two independent events (see Figure 4.2).

As a result, a variable selection algorithm should select the variables *cold* and *headache* since they reduce the uncertainty on the target variable *cancer*, while *cold* in this case is not a causal variable. Indeed, acting on *cold* does not modify the probability distribution of *cancer*.

In information-theoretic terms, the explaining-away effect can be expressed by a conditional mutual information higher than the mutual information [4, 5, 23]:

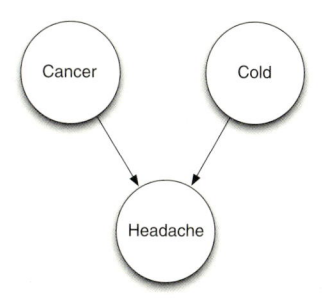

Figure 4.2 Illustration of the explaining-away effect.

$$X_i \rightarrow X_k \leftarrow X_j \Rightarrow I(X_i; X_j | X_k) > I(X_i; X_j) \tag{4.13}$$

This effect is also known as negative interaction [24], complementarity [20], or synergy [25].

In order to avoid this problem, one should determine which of the selected variables are indirect links and eliminate them from the selection. This can be done by modifying properly the PC algorithm (Algorithm 4.3) and using it to explore, instead of the whole search space $X_{-(i,j)}$, the selected subset X_S [26]. It follows that the procedure is still exponential, but in the size $|S|$ of the subset selected.

4.3
Inference Based on Pairwise Mutual Information

This section presents a set of algorithms for inferring networks from observed data relying only on the computation of pairwise mutual information. In these methods, a link between two nodes is set if its corresponding score (based on pairwise mutual information) is higher than the chosen threshold.

All these methods require the computation of the mutual information matrix $\text{MIM} = (\text{mim}_{ij})_{i,j \in A}$, a square matrix whose ijth element is given by,

$$\text{mim}_{ij} = I(X_i; X_j) \tag{4.14}$$

This is the mutual information between X_i and X_j, where X_i and X_j are random variables denoting the expression level of the ith/jth gene in a transcriptional regulatory network inference.

These methods have major advantages that enable them to deal with microarray data:

- An affordable computational complexity. This results from the fact that only $\frac{n}{2}(n-1)$ computations of mutual information, based on bivariate probability distributions, are required to obtain the mutual information matrix MIM [9].
- They do not require a large amount of samples, since only bivariate distributions are to be estimated. Hence, even basic entropy estimators perform well with these methods [27]. Most of these methods can be tested using the Bioconductor minet package [15].

4.3.1
Relevance Network (RELNET)

The relevance network approach [7] has been introduced for gene clustering, and successfully applied to infer relationships between RNA expression and chemotherapeutic susceptibility [13]. This method infers a network in which a pair of genes $\{X_i, X_j\}$ is linked by an edge if the mutual information $I(X_i; X_j)$ is larger than a given threshold θ. The complexity of the method is $O(n^2)$ since all pairwise interactions are considered.

This method relies on the assumption *causality implies dependency* (Section 4.1.4): $X_i \leftrightarrow X_j \Rightarrow I(X_i; X_j) > 0$. However, it does not eliminate all indirect interactions between genes. For example, if gene X_1 regulates both gene X_2 and gene X_3, this would cause a high mutual information between the pairs $\{X_1, X_2\}$, $\{X_1, X_3\}$, and $\{X_2, X_3\}$. As a consequence, the algorithm will set an edge between X_2 and X_3 although these two genes interact only through gene X_1.

Note that if one considers correlation instead of mutual information this approach boils down to building a correlation network [28].

4.3.2
Context Likelihood of Relatedness (CLR)

The CLR algorithm [8] is an extension of the RELNET algorithm. This algorithm derives a score from the empirical distribution of the mutual information for each pair of genes. In particular, instead of considering the information $I(X_i; X_j)$ between genes X_i and X_j, it takes into account the score $w_{ij} = \sqrt{z_i^2 + z_j^2}$ where:

$$z_i = \max\left(0, \frac{I(X_i; X_j) - \mu_i}{\sigma_i}\right) \quad (4.15)$$

where μ_i and σ_i are, respectively, the mean and standard deviation of the empirical distribution of the mutual information values $I(X_i, X_k)$, $X_k \in X_{-i}$. The pseudo-code of CLR is given in Algorithm 4.5. The CLR algorithm was successfully applied to decipher the *E. coli* transcriptional regulatory network [8]. CLR has a complexity in $O(n^2)$ once the MIM is computed.

Algorithm 4.5 Pseudo-Code of the Normal Version of CLR Algorithm

Input: $I(X_i; X_j)$, $\forall i, j \in A = \{1, 2, ..., n\}$
Output: the weighted adjacency matrix W (having elements w_{ij})
foreach *input X_i in the input space X* **do**
$\quad \mu_i \leftarrow \text{mean}(I(X_i; X_j), j \in \{1, 2, ..., n\})$
$\quad \sigma_i \leftarrow \text{variance}(I(X_i; X_j), j \in \{1, 2, ..., n\})$
end
foreach *pair of variables $X_{i,j}$ in the input space X* **do**
$\quad w_{ij} \leftarrow \max\left(0, \frac{1}{\sqrt{2}} * \left\{\frac{I(X_i; X_j) - \mu_i}{\sigma_i} + \frac{I(X_i; X_j) - \mu_j}{\sigma_j}\right\}\right)$
end

4.3.3
Chow–Liu Tree

The Chow and Liu approach consists of finding the maximum spanning tree on the complete graph whose edge weights are the mutual information between two nodes [3].

In graph theory, a tree is a graph in which any two vertexes are connected by exactly one path. A spanning tree is a tree that connects all the vertexes of the graph. The maximum spanning tree is the spanning tree whose sum of edge weights is greater than or equal to that of every other spanning tree.

A maximum spanning tree can be computed in $O(n^2 \log n)$ using, for example, Kruskal's algorithm [29]. The drawback of this method lies in the fact that the resulting network has typically a low number of edges. Also, precision and recall cannot be studied as a function of a parameter.

4.3.4
Algorithm for the Reconstruction of Accurate Cellular Networks (ARACNE)

ARACNE [9] is based on the Data Processing Inequality [30]. If gene X_i interacts with gene X_j through gene X_k, then $I(X_i; X_j) \leq \min(I(X_i; X_k), I(X_j; X_k))$. ARACNE begins by assigning to each pair of nodes a weight equal to their mutual information. Then, as in RELNET, all edges for which $I(X_i; X_j) < \theta$ are removed, with θ a given threshold. Eventually, the weakest edge of each triplet is interpreted as an indirect interaction and is removed (see pseudo-code Algorithm 4.6).

An extension of ARACNE removes the weakest edge only if the difference between the two lowest weights lies above a threshold η. Hence, increasing θ lowers the number of inferred edges, whereas the opposite happens when increasing η.

If the network is a tree including only pairwise interactions, the method guarantees the reconstruction of the original network, once it is provided with the exact MIM [9]. ARACNE's complexity is $O(n^3)$ since the algorithm considers all triplets of genes. In [9] method has been able to recover components of the transcription network in mammalian cells and has outperformed Bayesian inference networks on several inference tasks [9]. The Ch... a subnetwork of the network reconstructed by the AR...

Algor... ACNE Algorithm

Input: th... $?, ..., n\}$
Output: t... (having w_{ij} as elements)
foreach pai... X **do**
    foreach ... (i,j) **do**
        if $I(X_i...$
        | w_{ij} ...
        else
        | $w_{ij} \leftarrow I(X_i; X_j)$
        if $(I(X_i; X_j) < I(X_i; X_k)$ **and** $I(X_i; X_j) < I(X_j; X_k))$ **then**
        | $w_{ij} \leftarrow 0$
        end
    end
end

4.3.5
Minimum Redundancy Networks (MRNET)

MRNET [10] is a method based on the Maximum Relevance Minimum Redundancy (MRMR) variable selection procedure [31]. In other words, MRNET can be situated at the intersection of network inference methods based on pairwise mutual information (Section 4.3) and variable selection strategies (Section 4.2.3). Let X_j be the output variable, the MRMR methods ranks the set X_{-j} of inputs according to a score that is the difference between the mutual information with the output variable X_j (the relevance) and the average mutual information with the previously ranked variables (the redundancy) X_S. The rationale is that direct interactions should be well ranked, whereas indirect interactions (i.e., the ones with redundant information with the direct ones) should be badly ranked by the method. The MRMR method relies on a greedy search that starts by selecting the variable X_k providing the highest mutual information to the target X_j. The second selected variable X_i will be the one with a high information $I(X_i; X_j)$ to the target and at the same time a low information $I(X_i; X_k)$ to the previously selected variable. In the following steps, given a set X_S of selected variables, the criterion updates X_S by choosing the variable which maximizes the MRMR score:

$$X_i^{MRMR} = \arg\max_{X_i \in X_{-(i,j)}} \left(I(X_i; X_j) - \frac{1}{|S|} \sum_{k \in S} I(X_i; X_k) \right) \qquad (4.16)$$

At each step of the algorithm, the selected variable is expected to allow an efficient tradeoff between relevance and redundancy. The network inference approach MRNET consists in repeating this selection procedure for each target gene $X_j \in X$. For each pair $\{X_i, X_j\}$, MRMR returns two (not necessarily equal) scores s_i and s_j according to (4.16). The score of the pair $\{X_i, X_j\}$ is then computed by taking the maximum between s_i and s_j. A specific network can then be inferred by deleting all the edges whose score lies below a given threshold θ (as in RELNET, CLR, and ARACNE). Thus, the algorithm infers an edge between X_i and X_j either when X_i is a well-ranked predictor of X_j ($s_i > \theta$) or when X_j is a well-ranked predictor of X_i ($s_j > \theta$).

An effective implementation of the greedy search based on a similarity matrix is given in [32]. This implementation demands an $O(f \times n)$ complexity for selecting f variables. It follows that MRNET has an $O(f \times n^2)$ complexity since the variable selection step is repeated for each of the n genes. In other terms, the complexity ranges between $O(n^2)$ and $O(n^3)$ according to the value of f. In practice, the selection of variables is stopped when the average redundancy term $\frac{1}{|S|} \sum_{k \in S} I(X_i; X_k)$ exceeds the relevance term $I(X_i; X_j)$.

Although MRNET is based on a variable selection strategy, it does not suffer from the explaining-away effect as most variable selection methods do (see Section 4.2.3): $X_i \to X_k \leftarrow X_j \Rightarrow I(X_i; X_j | X_k) > I(X_i; X_j)$. Indeed, the MRMR criterion only relies on pairwise interactions, hence it does not measure the increase in information due to conditioning $I(X_i; X_j | X_k)$. Instead, it will consider the score $s_i = I(X_i; X_j) - I(X_i; X_k)$, where $I(X_i; X_j) = 0$ since X_i and X_j are independent, and $I(X_i; X_k) > 0$ since X_i is a cause of X_k (compare Example 4.3). This score s_i is negative and X_i will be badly ranked by MRMR. The same will happen for ranking X_j as a predictor of X_i.

Several papers [10, 27, 33, 34] have experimentally shown the accuracy of the MRNET method both in simulated and real data network inference tasks.

Algorithm 4.7 Detailed Pseudo-Code of the MRNET Algorithm (given the MIM)

Input: a matrix of weights MIM (with elements $I(X_i; X_j)$) of size n
Output: the weighted adjacency matrix W
Initialize the weighted adjacency matrix W to $n \times n$ zeros
foreach variable $j \in A = \{1, 2, ..., n\}$ **do**
　　Initialize search space: $R \leftarrow A \setminus \{j\}$
　　Initialize selected variable: $S \leftarrow \emptyset$
　　Initialize *relevance* vector: $relevance_j \leftarrow I(X_i; X_j), i \in R$
　　Initialize *redundancy* vector: $redundancy_j \leftarrow 0, i \in R$
　　while $w_{kj} > 0$ **do**
　　　　Select best variable: $k \leftarrow \arg\max_{i \in R}(relevance_i - redundancy_i/|S|)$
　　　　Update subset : $S \leftarrow \{S, k\}$
　　　　Update matrix: $w_{kj} \leftarrow relevance_k - redundancy_k/|S|$
　　　　Update search space : $R \leftarrow R \setminus k$
　　　　Update *redundancy* vector:
　　　　$redundancy_i \leftarrow redundancy_i + I(X_i; X_k), i \in R$
　　end
end

4.4
Arc Orientation

So far, we have presented several strategies to infer an undirected network from data. The remaining step to be performed consists in the orientation of the inferred arcs. Thus, the question arises, "Can the direction of an arc be inferred in an observational context?" Surprisingly the answer to that question is positive. This can be done in some cases thanks to the previously seen explaining-away effect $X_i \to X_k \leftarrow X_j \Rightarrow I(X_i; X_j | X_k) > I(X_i; X_j)$.

Let us first remark that because of the equality:

$$I(X_i; X_j) - I(X_i; X_j | X_k) = I(X_i; X_k) - I(X_i; X_k | X_j) = I(X_j; X_k) - I(X_j; X_k | X_i) \quad (4.17)$$

the reversal statement of (4.13) is not necessarily true:

$$I(X_i; X_j | X_k) > I(X_i; X_j) \not\Rightarrow X_i \to X_k \leftarrow X_j$$

since $I(X_i; X_j | X_k) > I(X_i; X_j)$ could also imply $X_k \to X_i \leftarrow X_j$ or $X_i \to X_j \leftarrow X_k$. Notwithstanding, given particular configurations of the undirected network, arcs can be oriented thanks to the explaining-away effect. This can be done in two ways:

i) If the variable X_k is linked to X_i and X_j, and an explaining-away effect between them has been detected, then the variable X_k is a common effect of the two other variables. More formally:

$$\left. \begin{array}{c} X_i \leftrightarrow X_k \leftrightarrow X_j \\ I(X_i; X_j | X_k) > I(X_i; X_j) \end{array} \right\} \Rightarrow X_i \rightarrow X_k \leftarrow X_j \qquad (4.18)$$

ii) If the variable X_k is a consequence of X_i and is linked to X_j, and no explaining-away effect occurs between them, then X_j is a consequence of X_k:

$$\left. \begin{array}{c} X_i \rightarrow X_k \leftrightarrow X_j \\ I(X_i; X_j | X_k) \leq I(X_i; X_j) \end{array} \right\} \Rightarrow X_i \rightarrow X_k \rightarrow X_j \qquad (4.19)$$

The two rules (4.18) and (4.19) are an information-theoretic translation of the arc orientation criteria used in [18, 19].

Note that the explaining-away effect $X_i \rightarrow X_k \leftarrow X_j \Rightarrow I(X_i; X_j | X_k) > I(X_i; X_j)$ is true in general, but not always. Consider the following example:

Example 4.3

Disease A can cause the skin to be covered by eczema and disease B can cause the skin have wounds. Let the variable S (skin) have three possible values (*eczema, wounds, normal*). Although disease A and B are two causes of skin injuries S, there is no more information brought by evidence in favor of one of the diseases than the information already given by the state of the skin.

This example illustrates that, in order to observe an explaining-away effect, the two causes should have the same effect on the same target variable [17].

Although directed cycles can represent feed-back and feed-forward effects, there is no suitable joint probability to model these situations. Distributions such as $p(A|B)p(B|C)p(C|A)$ are not well-defined probability distributions, apart from very special cases [11]. Furthermore, the notion of loops is related to dynamics, whereas the probability distribution modeled here comes from samples with no temporal dependencies. Hence, a third rule commonly used to orient arcs in a partially oriented network consists in removing cycles in triplets of variables [35].

In order to guarantee a correct inference of an oriented network, additional assumptions are required. Commonly, the causal Markov and the causal faithfulness assumptions are made [18, 19]. These two properties are defined as follows.

Definition 4.6

The *causal Markov condition* holds if every variable is statistically independent of its noneffects conditional on its direct causes.

Definition 4.7

The *faithfulness condition* holds if the only existing conditional independencies are those specified by the causal Markov condition.

Note, that these two conditions include the assumptions defined above such as causality implies dependency or causal transitivity (for a detailed analysis of these assumptions, see [36].)

Algorithm 4.8 Pseudo-Code of SGS/IC arc Orientation Step for Conditioning Sets of Size One

foreach *triplet $X_i \leftrightarrow X_k \leftrightarrow X_j$ with no link connecting $X_i \leftrightarrow X_j$* **do**
| $I(X_i; X_j | X_k) > I(X_i; X_j) \Rightarrow X_i \rightarrow X_k \leftarrow X_j$
end
while *there remain undirected edges, orient edges subject to* **do**
| avoiding new colliders
| avoiding cycles
end

Algorithm 4.9 Pseudo-Code of the "Remove Cycles" Step of [26] Algorithm

compute the list of cycles in the graph
while *there exists cycles in the graph* **do**
| reverse edge that belongs the highest amount of cycles
| update the list of cycles
end

4.4.1
Assessing Arc Orientation Methods

Similar to the undirected case, the quality of the arc orientation process has to be assessed. For this task, there are three quantities of interest: the number of correctly oriented arcs, the ones that are oriented in the wrong direction, and those arcs that have not been oriented at all. Depending on the problem, different strategies are applied in the literature. If one is interested in the overall performance, the number of correctly oriented arcs is compared to the number of wrongly oriented arcs. The problem with this method is that the more arcs are oriented, the less significant the orientation is (due to the decreasing score of an arc). Another evaluation possibility is a weighted matrix, which is interesting when searching for the directionalities in a local network (e.g., the arcs in a certain "neighborhood" of the target variable). In this case, the orientation of arcs close to the target can be higher rated than that of arcs further away.

When taking into consideration the first possibility, one can use measures analogous to those presented in Section 4.1.3 (F-score or PR curves). In order to compare the true network and the inferred causal network, the notions of precision and recall have to be adopted to the partially oriented networks. In [37], four additional categories to the already introduced quantities *fn*, *tp*, *tn*, and *fp* are defined: *pfn* (partial false-negatives) = *ptp* (partial true-positives) and *ptn* (partial true-negatives) = *pfp* (partial false-positives). The former denote the number of arcs for which $X-Y$ is in the obtained network and either $X \rightarrow Y$ or $Y \rightarrow X$ is the true network. The latter denotes the case when $X-Y$ is in the obtained graph and neither $X \rightarrow Y$ nor $Y \rightarrow X$ are in the original network. Precision and recall are then defined as:

$$rec = \frac{|tp| + \frac{|ptp|}{2}}{|fn| + \frac{|pfn|}{2} + |tp| + \frac{|ptp|}{2}} \quad (4.20)$$

$$pre = \frac{|tp| + \frac{|ptp|}{2}}{|fp| + \frac{|pfp|}{2} + |tp| + \frac{|ptp|}{2}} \quad (4.21)$$

4.5
Conclusions

Transcriptional network inference aims at representing interactions between transcription factors and regulated genes with a graph. Most methods introduced in this chapter are able to distinguish between dependency (e.g., corregulated gene) and causality (e.g., transcription factor) by relying on assumptions such as *causality implies dependency* or *causal sufficiency*. Among inference methods based on information theory, mutual information networks, relying on the matrix of pairwise mutual information, are particularly adapted to large number of variables and low number of samples typically encountered in microarray data. However, these methods only infer an undirected graph. In order to orient the arcs, a second step is required such as methods based on the *explaining-away effect*. Finally, validation measures such as PR curves and *F*-scores have been introduced here in order to assess inferred networks. Many tools and algorithms introduced in this chapter can be tested using the Bioconductor package minet.

References

1 van Someren, E.P., Wessels, L.F.A., Backer, E., and Reinders, M.J.T. (2002) Genetic network modeling. *Pharmacogenomics*, **3**, 507–525.

2 Gardner, T.S. and Faith, J. (2005) Reverse-engineering transcription control networks. *Phys. Life Rev.*, **2**, 65–88.

3 Chow, C. and Liu, C. (1968) Approximating discrete probability

distributions with dependence trees. *IEEE Trans. Inform. Theory*, **14**, 462–467.

4 Cheng, J., Greiner, R., Kelly, J., Bell, D., and Liu, W. (2002) Learning Bayesian networks from data: an information-theory based approach. *Artif. Intell.*, **137**, 43–90.

5 Liang, K. and Wang, X. (2008) Gene regulatory network reconstruction using conditional mutual information. *EURASIP J. Bioinform. Syst. Biol*, 14.

6 Zhao, E., Serpedin, E., and Dougherty, E.R. (2008) Inferring connectivity of genetic regulatory networks using information-theoretic criteria. *IEEE/ACM Trans. Computat. Biol. Bioinform.*, **5**, 262–274.

7 Butte, A.J. and Kohane, I.S. (2000) Mutual information relevance networks: functional genomic clustering using pairwise entropy measurements. *Pac. Symp. Biocomput.*, 415–426.

8 Faith, J.J., Hayete, B., Thaden, J.T., Mogno, I., Wierzbowski, J., Cottarel, G., Kasif, S., Collins, J.J., and Gardner, T.S. (2007) Large-scale mapping and validation of *Escherichia coli* transcriptional regulation from a compendium of expression profiles. *PLoS Biol.*, **5**, 8.

9 Margolin, A.A., Nemenman, I., Basso, K., Wiggins, C., Stolovitzky, G., Dalla Favera, R., and Califano, A. (2006) ARACNE: an algorithm for the reconstruction of gene regulatory networks in a mammalian cellular context. *BMC Bioinformatics*, **7**, S7.

10 Meyer, P.E., Kontos, K., Lafitte, F., and Bontempi, G. (2007) Information-theoretic inference of large transcriptional regulatory networks. *EURASIP J. Bioinform. Syst. Biol. (Special Issue on Information-Theoretic Methods for Bioinformatics)*, 9.

11 Whittaker, J. (1990) *Graphical Models in Applied Multivariate Statistics*, John Wiley & Sons, Inc., New York.

12 Shannon, C.E. (1948) A mathematical theory of communication. *Bell Syst. Tech. J.*, 379–423, 623–656.

13 Butte, A.J., Tamayo, P., Slonim, D., Golub, T.R., and Kohane, I.S. (2000) Discovering functional relationships between RNA expression and chemotherapeutic susceptibility using relevance networks. *Proc. Natl. Acad. Sci. USA*, **97**, 12182–12186.

14 Sokolova, M., Japkowicz, N., and Szpakowicz, S. (2006) Beyond accuracy, F-score and ROC: a family of discriminant measures for performance evaluation. AAAI Workshop on Evaluation Methods for Machine Learning, Boston, MA.

15 Meyer, P.E., Lafitte, F., and Bontempi, G. (2008) Minet: an open source R/Bioconductor package for mutual information based network inference. *BMC Bioinformatics*, 461.

16 Bhadra, S., Bhattacharyya, C., Chandra, N.R., and Mian, S. (2009) A linear programming approach for estimating the structure of a sparse linear genetic network from transcript profiling data. *Algorithms Mol. Biol.*, **4**, 5+.

17 Neapolitan, R.E. (2003) *Learning Bayesian Networks*, Prentice Hall, Englewood Cliffs, NJ.

18 Pearl, J. (2000) *Causality: Models, Reasoning, and Inference*, Cambridge University Press, Cambridge.

19 Spirtes, P., Glymour, C., and Scheines, R. (2001) *Causation, Prediction, and Search*, MIT Press, Cambridge, MA.

20 Meyer, P.E., Schretter, C., and Bontempi, G. (2008) Information-theoretic feature selection using variable complementarity. *IEEE J Spec. Topics Signal Process*, **2**, 91–102.

21 Hwang, K., Lee, J.W., Chung, S., and Zhang, B. (2002) Construction of large-scale Bayesian networks by local to global search. 7th Pacific Rim International Conference on Artificial Intelligence, Tokyo, Japan.

22 Tsamardinos, I., Aliferis, C., and Statnikov, A. (2003) Algorithms for large scale Markov blanket discovery. 16th International FLAIRS Conference, St. Augustine, Florida, USA.

23 Jakulin, A. and Bratko, I. (2003) Quantifying and visualizing attribute interactions. E-print arXiv.cs/0308002v1.

24 Jakulin, A. and Bratko, I. (2004) Testing the significance of attribute interactions.

21st International Conference on Machine Learning, Banff, Canada.

25 Anastassiou, D. (2007) Computational analysis of the synergy among multiple interacting genes. *Mol. Syst. Biol.*, **3**, 83.

26 Bromberg, F. and Margaritis, D. (2009) Improving the reliability of causal discovery from small data sets using argumentation. *J. Mach. Learn. Res.*, **10**, 301–340.

27 Olsen, C., Meyer, P.E., and Bontempi, G. (2009) On the impact of missing values on transcriptional regulatory network inference based on mutual information. *EURASIP J. Bioinform. Syst. Biol.*, 308959.

28 Junker, B.H. and Schreiber, F. (2008) *Analysis of Biological Networks*, Wiley Series in Bioinformatics, Wiley-Interscience, New York.

29 Moret, B.M.E. and Shapiro, H.D. (1991) An empirical analysis of algorithms for constructing a minimum spanning tree. *Lecture Notes Comput. Sci.*, **519**, 400–411.

30 Cover, T.M. and Thomas, J.A. (1990) *Elements of Information Theory*, John Wiley & Sons, Inc., New York.

31 Peng, H., Long, F., and Ding, C. (2005) Feature selection based on mutual information: criteria of max-dependency, max-relevance, and min-redundancy. *IEEE Trans. Pattern. Anal.*, **27**, 1226–1238.

32 Merz, P. and Freisleben, B. (2002) Greedy and local search heuristics for unconstrained binary quadratic programming. *J. Heuristics*, **8**, 1381–1231.

33 Lopes, F.M., Martins, D.C., and Cesar, R.M. (2009) Comparative study of GRNS inference methods based on feature selection by mutual information. IEEE International Workshop on Genomic Signal Processing and Statistics, Minneapolis, MN, USA.

34 Shimamura, T., Imoto, S., Yamaguchi, R., Fujita, A., Nagasaki, M., and Miyano, S. (2009) Recursive regularization for inferring gene networks from time-course gene expression profiles. *BMC Syst. Biol.*, **3**, 41.

35 Meek, C. (1995) Strong completeness and faithfulness in Bayesian networks. 11th Conference on Uncertainty in Artificial Intelligence, Montreal.

36 Zhang, J. and Spirtes, P. (2008) Detection of unfaithfulness and robust causal inference. *Minds Mach.*, **18**, 239–271.

37 Zhang, X., Baral, C., and Kim, S. (2005) An algorithm to learn causal relations between genes from steady state data: simulation and its application to melanoma dataset. *Artif. Intell. Med.*, 524–534.

5
Elucidation of General and Condition-Dependent Gene Pathways Using Mixture Models and Bayesian Networks

Sandra Rodriguez-Zas and Younhee Ko

5.1
Introduction

Multidimensional datasets from microarray experiments have been used to infer gene interaction networks based on the observed gene expression profiles across multiple conditions, treatments, or samples. Results from these studies can aid in the confirmation of previously known pathways and motivate the study of newly uncovered relationships among genes.

A Bayesian network approach is well suited to model the relationships between genes [1–3]. (i) Acyclic direct graphs can be used to describe cause–effect relationships between genes. (ii) This approach has a solid theoretical foundation and offers a probabilistic framework to describe the variation typically observed in microarray experiments. (iii) Bayesian networks can accommodate missing data and incorporate prior knowledge through prior distribution of the parameters [4].

Several applications of Bayesian networks to infer gene networks have been reported [1, 4–7]. However, some implementations of this approach present limitations. (i) Some applications require the transformation of continuous gene expression data into binary or discrete input data [8]. In some cases this transformation could potentially result in biased inference. (ii) Gene network inference typically relies on a large number of microarrays that evaluate gene expression across a wide range of conditions. Nevertheless, most Bayesian network implementations assume simple models to describe the gene coexpression pattern within subnetworks [9]. More flexible models that can accommodate potential changes in the interaction among genes across conditions are necessary.

The integration of mixtures of densities into Bayesian networks offers a solution to model potential changes in gene networks across diverse conditions. This approach was proposed by Davies and Moore [10], and applied to word and social

Applied Statistics for Network Biology: Methods in Systems Biology, First Edition.
Edited by M. Dehmer, F. Emmert-Streib, A. Graber, and A. Salvador.
© 2011 Wiley-VCH Verlag GmbH & Co. KGaA. Published 2011 by Wiley-VCH Verlag GmbH & Co. KGaA.

networks by Newman and Leicht [11]. Ko et al. [12] reported the first application of mixture Bayesian networks to infer a simple histone pathway in yeast. Additional computational, statistical, and biological evaluations of this approach in multiple gene pathways further confirmed the superiority of the technique [13]. Comparisons of the predicted and known pathways and benchmark tests corroborated the outstanding performance of the mixture Bayesian network approach. Subsequent work by Ko et al. demonstrated that the mixture Bayesian network approach also allowed the identification of the best coexpression model supported by each gene in the network, and was well-suited to infer gene networks across a wide spectrum of datasets and pathways [14]. This work demonstrated the flexibility of mixture Bayesian networks to infer condition-dependent networks and this feature is particularly critical in gene expression datasets encompassing multiple conditions [14].

The standard implementations of mixture Bayesian networks identified the pathway best supported by the data [12–14]. This accurate and computationally efficient manner to infer a gene network is also easy to interpret and helpful when a single best network is of interest. However, the single network topology inferred may limit the understanding of gene networks when multiple network topologies have similar probability. To address this scenario, Ko et al. [15] proposed a Markov chain Monte Carlo (MCMC) implementation of the mixture Bayesian network approach. This implementation provided informative posterior distributions of the relationships between genes. Similarly, Grzegorczyk et al. [16] proposed a MCMC approach to infer the posterior distribution of the network structure based on Bayesian networks and a mixture model. The goal of this study is to review the capability of mixture Bayesian networks to advance the prediction of gene pathways and gain more insights into general and condition-dependent gene relationships. The capability of mixture Bayesian networks to address the limitations of existing implementations is demonstrated and various implementations are discussed.

5.2
Methodology

Bayesian gene networks are directed acyclic graphs with nodes representing genes and directed edges representing the relationships between the genes [1, 13]. Given a set of N genes $(g_1, g_2, \ldots, g_j, \ldots, g_N)$ in a Bayesian gene network G, $P(G)$ is the joint probability distribution over all genes in the network. Conditional on the ascendant or parent genes, each gene is independent of all other nondescendant genes in the network following the narrow sense form of the Markov property for a stochastic process. The conditional independence between genes allows the factorization of the joint probability distribution as the product of conditional probabilities. The total network is conceived as a set of N gene subnetworks, each corresponding to a given gene g_j and the corresponding parent gene(s) $a(g_j)$. Ko et al. [13] describe that the likelihood of G is:

$$P(G) = \prod_{j=1}^{N} \frac{P(g_j, a(g_j))}{P(a(g_j))} = \prod_{j=1}^{N} \frac{\prod_{i=1}^{D}\left(\sum_{k=1}^{K_j} \alpha_{jk} f_{jk}(\mathbf{x}_{ji})\right)}{\prod_{i=1}^{D}\left(\sum_{k=1}^{K_j} \alpha_{jk} f_{jk}^*(\mathbf{x}_{ji}^*)\right)}$$

$$= \prod_{j=1}^{N} \frac{\prod_{i=1}^{D}\sum_{k=1}^{K_j} \alpha_{jk} \left(2\pi|\Sigma_{jk}|\right)^{-1/2} \exp\left[-\frac{1}{2}(\mathbf{x}_{ji}-\mu_{jk})^T (\Sigma_{jk})^{-1}(\mathbf{x}_{ji}-\mu_{jk})\right]}{\prod_{i=1}^{D}\sum_{k=1}^{K_j} \alpha_{jk} \left(2\pi|\Sigma_{jk}^*|\right)^{-1/2} \exp\left[-\frac{1}{2}(\mathbf{x}_{ji}^*-\mu_{jk}^*)^T (\Sigma_{jk}^*)^{-1}(\mathbf{x}_{ji}^*-\mu_{jk}^*)\right]}$$

(5.1)

In the gene network scenario, the probability density function for the jth gene conditional on the parent genes ($P(g_j| a(g_j))$) is equal to the ratio between the joint probability density function of the parent and child genes ($P(g_j, a(g_j))$) and the marginal probability density function of the parent genes in the jth subnetwork ($P(a(g_j))$). A Gaussian mixture model represented with a mixture of K_j multivariate Gaussian distributions (f_{jk}), each with a weight α_{jk}, describes the probability density function P of the jth gene subnetwork. In the Gaussian conditional distribution ($P(g_j, a(g_j))/P(a(g_j))$), $\mathbf{x}_{ji}$ is a p_j dimensional vector including the ith gene expression measurement ($i = 1$ to D) of the child and parent genes in the jth subnetwork, and $\mathbf{x}_{ji}^*$ is a p_j dimensional vector including the ith gene expression measurement of the parent genes. The kth component of the mixture is described with a mean vector μ_{jk} of dimension p_j and a variance–covariance matrix Σ_{jk} of dimension $p_j \times p_j$ for $P(g_j, a(g_j))$, and with a mean vector μ_{jk}^* of dimension $p_j - 1$ and a variance–covariance matrix Σ_{jk}^* of dimension $(p_j - 1) \times (p_j - 1)$ for $P(a(g_j))$.

5.2.1
Network Learning Algorithms: Frequentist- and Bayesian MCMC-based algorithms

The learning algorithms encompass the inference of each gene subnetwork structure and the total network, based on the likelihood of the network G. The gene subnetwork inference includes the identification of the subnetwork for each gene (i.e., the parent genes of each child gene), the estimation of the parameters of the mixture models using the expectation-maximization (EM) algorithm, and the identification of the number of mixture components [13]. The total network inference involves the combination of the individual gene subnetworks (exploiting the conditional independence properties) and the removal of cyclic relationships between genes in the overall network [13]. The EM equations follow the derivations of Bilmes [17]. The Bayesian information criterion (BIC) was used to identify the optimal number of mixture components and the total network structure best supported by the data [18].

The identification of the "best" total network structure requires the consideration of alternative structures and is an NP-hard problem because the number of possible candidate structures super-exponentially increases with the number of genes. Ko et al. [13, 14] evaluated two complementary approaches – the sparse candidate algorithm [1] and the simulated annealing method [19] – to infer the total network structure. The sparse candidate algorithm limits the search space of potential parent genes and this strategy can potentially increase the hazard of failing to identify the true network structure. Ko et al. [13] addressed this potential limitation by modifying the sparse candidate algorithm and evaluating additional parent genes for each gene subnetwork with less than the maximum number of parent genes. In addition, a potential drawback of using likelihood-based criteria to propose networks for evaluation is that the algorithm may be confined to a local likelihood maxima and fail to propose a network from the global maximum. Ko et al. [14] used the simulated annealing algorithm to address this situation. At each iteration, this algorithm randomly jumps from the current network structure to a nearby structure formed from the addition, deletion, or reversion of an edge in the current structure. This strategy allows the simulated annealing algorithm to escape from potential local minima by occasionally sampling from less suitable networks and the probability of jumping to another status is empirically adjusted.

In many gene expression experiments, each condition or class studied (e.g., treatments, ages) includes multiple samples. Most approaches to infer networks assume that all the samples in an experiment share the same coexpression patterns between genes. However, departures from this assumption are expected, especially when the conditions under consideration have substantial differences or when substantial sample-to-sample or biological variation is typical in the model (e.g., species) studied. The identification of variation in gene coexpression patterns between conditions or between samples within condition can uncover the plasticity of gene networks within and across conditions, provide more accurate insights into the molecular pathways, offer a more precise classification of samples, and support the effective design of confirmatory studies. Ko et al. [14] demonstrated the capability of mixture Bayesian networks to identify changes in gene pathways across conditions and samples within condition, and to use this information to accurately classify samples within conditions. The integration of mixture Bayesian network and multidimensional reduction algorithms allowed the ascertainment of pathway variation associated with changes in conditions and sample-to-sample variation. The probability that a sample pertained to each mixture coexpression model was used to identify the coexpression model supported by each sample and assess the variation in gene relationships among samples within and across conditions. The estimates of the coexpression model were weighted by the mixture weights and combined into one value per network gene and sample. Using unsupervised learning (*K*-means clustering) and principal component analysis on the mixture parameter estimates, the samples were distributed into groups with similar coexpression patterns. A leave-one-out cross-validation approach was used to assess the reliability of the classification of samples to classes. The comparison of the resulting classes or groups against the conditions studied in the microarray experiment and the distribution of samples within group provided information on the changes in coexpression patterns between conditions.

The integration of mixture Bayesian networks and the EM approach offers the total network best supported by the data. This outcome constitutes a single point estimate of the network. However, in many cases there is no single best network, but a collection of networks that offer similar fit to the data. This is particularly common when the data is fairly uninformative, and consequently the likelihood is flat. In addition, of interest may be the posterior distribution of the network edges instead of a single summary of the network. Ko *et al.* [15] proposed a MCMC method to sample from the posterior distribution of the total network. At each iteration of the Monte Carlo sampler, a candidate network is proposed as a result of adding, deleting, or reverting edges to the previous iteration network from a proposal distribution. The acceptance probability is:

$$A(G_{[t]}, G_{[t-1]}) = \min\left\{1, \left[\frac{P(x|G_{[t]})P(G_{[t]})Q(G_{[t-1]}|G_{[t]})}{P(x|G_{[t-1]})P(G_{[t-1]})Q(G_{[t]}|G_{[t-1]})}\right]\right\}$$
$$= \min\left\{1, \left[\frac{P(x|G_{[t]})}{P(x|G_{[t-1]})}\right]\right\} \quad (5.2)$$

where [t] denotes the iteration number, G denotes the network model, x denotes the gene expression data, $P(x|)$ denotes the likelihood of the data, $P(G)$ denotes the prior probability of the network model, and $Q()$ denotes the proposal distribution. Under the assumption that $P(G)$ has a uniform distribution and $Q()$ is a symmetrical, the prior probability for the network and proposal probability cancel out [20, 21]. A network topology is sampled at each MCMC iteration from Equation (5.2). Ko *et al.* [15] evaluated 10 000 iterations for the chain burn-in and convergence was achieved by 5000 iterations. Estimates of posterior distribution of the network structure were based on up to 300 samples.

The posterior distribution of the network structure offers comprehensive information on the network. However, the summarization of the posterior into a network is challenging. Several methods can be used to summarize the chain of structures into a single network including the consideration of edges that are present in a minimum number of iterations, also known as the importance of an edge [15]. Grzegorczyk *et al.* [16] also proposed an approach based on a mixture model, Bayesian networks, and MCMC to model nonhomogeneous and nonlinear gene regulatory processes. The implementation of the MCMC used an allocation sampler instead of reversible jumping MCMC.

5.3
Applications

5.3.1
Elucidation of Gene Networks

The application of the mixture Bayesian network and sparse candidate approach to three independent datasets – circadian rhythm in honey bees, adherens junction pathways in mouse embryos, and cell cycle in yeast – was reported by Ko *et al.* [13]. All (honey bee circadian rhythm and mouse adherens junction pathways) or the vast majority (yeast cell cycle) of the gene relationships reported were detected. Based on the consistent performance of mixture Bayesian networks across these pathways, only the results for the circadian rhythm pathway will be discussed.

The elucidation of the circadian rhythm pathway in honey bees used gene expression measurements from a microarray experiment including samples of bees from either one of two races (*Apis mellifera mellifera* and *Apis mellifera ligustica*) raised in a colony from either one of two races (*mellifera* and *ligustica*) obtained at six maturation stages. Six genes assigned to the circadian rhythm pathway in the KEGG (Kyoto Encyclopedia of Genes and Genomes) database (http://www.genome.jp/kegg/) were present on the microarray platform and considered for pathway inference. The experimental design allowed the division of the honey bee dataset into four separate bee-colony race subsets. The four subsets were expected to support the similar gene networks allowing the cross-validation of the network predicted on one dataset, on the remaining datasets.

The circadian rhythm gene network predicted by the mixture Bayesian network approach on each of the four bee-host subdatasets is presented in Figure 5.1. All gene relationships predicted by the mixture Bayesian network approach were consistent with known relationships on the KEGG database. Both, direct relationships present in the KEGG database and indirect relationships through intermediate genes not present on the microarray platform were detected by the approach. None of the gene relationships predicted were absent on the KEGG pathway. When each of the four race-colony subsets were used for cross-validation purposes, the mixture Bayesian network consistently detected most gene relationships across the subsets (edge denoted 4/4 or 3/4). The failure to detect all the gene relationships in all four subdatasets was attributed to differences in the genomic, maturity, and physiological profile of each race combination. Data permutation was used to assess the reliability of the gene network predicted using the mixture Bayesian network. Networks were predicted on each of the permuted datasets and compared to the network predicted using the original dataset. None or only one of the edges predicted in the original dataset were present in 39% of the randomized datasets and only 1% of these datasets shared a maximum of five relationships out of the 14 gene relationships identified on the original dataset. These results further confirmed the capability of the mixture Bayesian network approach to accurately infer the topology of the gene network.

5.3.2
Discovery of Condition-Dependent Gene Relationships

The remarkable value of mixture Bayesian networks to uncover variation on the gene coexpression models between conditions or samples within condition was demonstrated by Ko *et al.* [14]. In this study, the mixture Bayesian network approach successfully uncovered variation in the relationship between genes across samples on the starch and sucrose metabolism and circadian rhythm pathways. The implica-

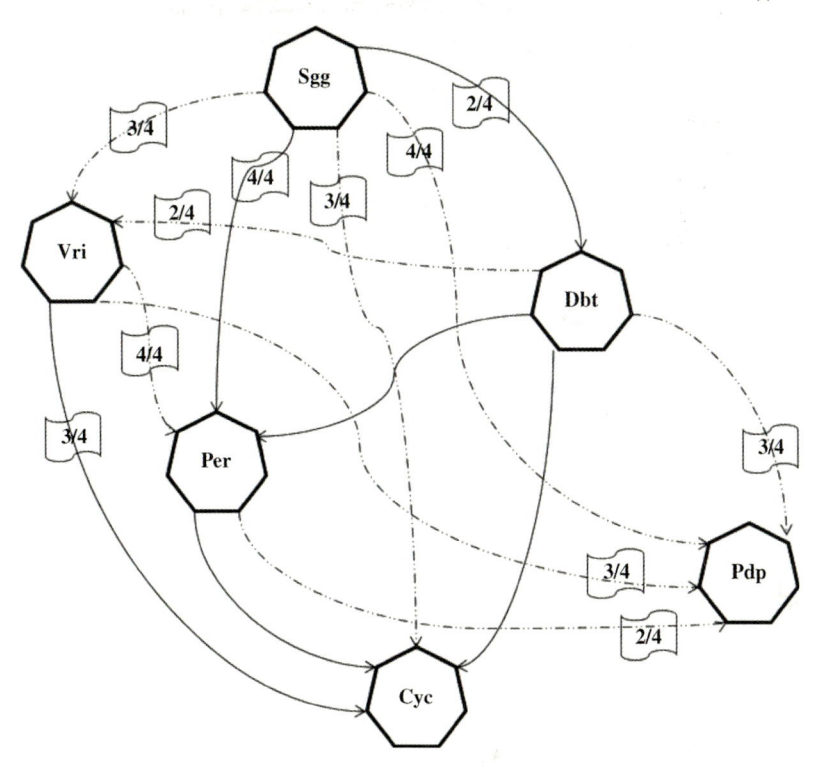

Figure 5.1 Inference of the circadian rhythm gene network. Continuous edges depict direct gene relationships that were predicted and validated in the KEGG pathway. Broken edges depict indirect gene relationships through intermediate genes not present in the data that were predicted and confirmed in the KEGG pathway. Ratios indicate the fraction out of the four cross-validation subsets exhibiting the edge. Annotation: Per = period clock; Cyc = cycle; Vri = vrille; Pdp = pyruvate dehydrogenase phosphatase 1; Sgg = Shaggy; Dbt = double-time. (Adapted from Ko et al. [13].)

tions of this feature of the mixture Bayesian network model are discussed for the starch and sucrose metabolism pathway.

Gene expression data from two microarray experiments that compared six maturation conditions were used to demonstrate the capability of mixture Bayesian networks to reveal changes in gene coexpression patterns across conditions. In typical colonies, young nurse honey bees age and become old forager honey bees. In the first experiment, typical or traditional young nurse (TN) and old forager (TF) honey bees were compared. The second experiment was performed to discriminate changes in gene expression due to age (young versus old) versus behavior (nurse versus forager). Gene expression from honey bees belonging to one of four age-behavior classes: traditional young nurses (YN), nontraditional young ("precocious") foragers (YF), traditional old foragers (OF) and nontraditional old ("overage") nurses (ON) were measured. The mixture Bayesian network approach was well-suited to investigate whether the coexpression patterns vary with age or behavior. Ten genes in the KEGG starch and sucrose metabolism pathway were present in the microarray platform used in the experiments.

The starch and sucrose metabolism gene pathway inferred using the mixture Bayesian approach and simulated annealing is depicted in Figure 5.2 [14]. All except one relationship between genes predicted by this approach were present in the KEGG pathway. The missing relationship involves a hub gene and it is postulated that the other relationships of this gene may have hindered the detection of the missing edge. A mixture of three gene coexpression models was favored for most of the gene subnetworks and thus only the assignment of samples to models for two of the genes, sugarless (*Sgl*) and UDP-glucuronosyltransferase cytological map location 86D (*Ugt86dg*) are presented in Table 5.1. A discussion of the mixture parameter estimates for the subnetwork of *Sgl* highlights the insights into coexpression patterns offered by the mixture Bayesian network. The three coexpression models for *Sgl* (1, 2, and 3) had

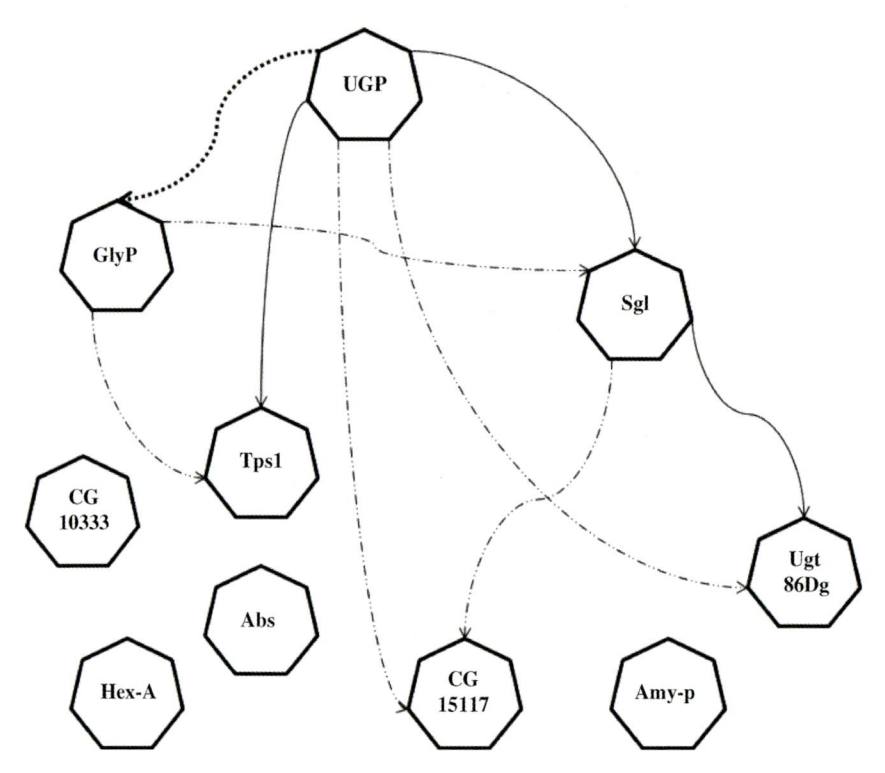

Figure 5.2 Inference of the starch and sucrose metabolism gene network. Continuous edge depict direct gene relationships that were predicted and validated in the KEGG pathway. Broken edges depict indirect gene relationships through intermediate genes not present in the data that were predicted and confirmed in the KEGG pathway. The dotted edge denote a gene relationship predicted with opposite direction to that in the KEGG pathway. Annotation: *Sgl* = sugarless; *abs* = abstract; CG15117 = β-glucuronidase; *Amy-p* = amylase proximal; *Hex-A* = hexokinase A; *Tps1* = trehalose-6-phosphate synthase 1; CG10333 = DmRH19; UGP = UDP glucose-1-phosphate uridylyltransferase; *Ugt86Dg* = UDP-glucuronosyltransferase; *GlyP* = glycogen phosphorylase. (Adapted from Ko et al. [14].)

Table 5.1 Number of samples assigned to either of the three mixture coexpression models for the subnetworks of genes *Sgl* and *Ugt86Dg* in the sucrose and starch metabolism pathway across the six age-behavior conditions (T = traditional, Y = young, O = old, N = nurse, F = forager honey bees) studied.

Condition	Ugt86Dg			Sgl		
	Model 1	Model 2	Model 3	Model 1	Model 2	Model 3
TN	16	0	2	13	1	4
YN	5	0	1	3[a]	0	2
ON	0	5	1	0	5	1
YF	0	3	3	2	4	0
OF	4	0	2	3	2	1
TF	13	0	5	13	1	4

a) Denotes a sample that was assigned with equal probability to two models.

weights of 0.53, 0.19, and 0.28. The correlations between *Sgl* and parent genes UDP glucose-1-phosphate uridylyltransferase (UGP) and glycogen phosphorylase (*GlyP*) were higher in Model 2 (0.90 and 0.81, respectively) and Model 3 (0.72 and 0.81, respectively) compared to Model 1 (0.51 and 0.17, respectively). In addition, the expression level of *Sgl* was lower in Model 1 relative to Models 2 and 3.

All or most of the samples within each of the six conditions considered were assigned to one of the three coexpression mixture models in the *Sgl* and *Ugt86dg* subnetworks in the starch and sucrose metabolism pathway. For instance, the vast majority of the YN, OF, TF, and TN samples were assigned to the same model, meanwhile the majority of the YF and ON were assigned to a different model. Using principal component and *K*-means clustering analysis on the network estimates of all the genes in the pathway uncovered three distinct networks. Nontraditional bees (YF and ON), traditional or old foragers (TF and OF), and traditional or young nurses (TN and YN) were distributed into three different clusters [14]. A leave-one-sample out cross-validation supported the classification of samples into three distinct groups. The mixture Bayesian network approach exposed different coexpression models between the groups and revealed the major changes in gene coexpression patterns associated with age and behavior.

5.3.3
MCMC Mixture Bayesian Network

Ko et al. [15] used MCMC to obtain the posterior distribution of the parameters of the mixture Bayesian network. This approach was used to predict the dorso-ventral axis formation pathway in mouse and the transforming growth factor-β signaling pathway in the honey bee. Results from the first pathway are discussed.

The expression data used to infer the dorso-ventral axis formation pathway was obtained from nine experiments that investigated the effect of multiple toxic agents on mouse embryos. Five genes in the pathway were present in the microarray platform and their expression data was used to infer the pathway. Samples from the posterior distribution of the gene network parameters permitted the computation of gene relationship summaries, the identification of the most likely and most frequent network structure among the structures sampled in the chain, and provided insights into other likely network topologies, not available from the single network estimate provided by the EM approach. The gene relationships depicted in Figure 5.3 were present in more than 50% of the sampled networks. All the gene relationships detected were confirmed in the corresponding KEGG pathway as direct or indirect relationships through intermediate genes in the KEGG database that were not present in the microarray platform. The mixture Bayesian approach using MCMC only failed to detect one edge (between genes *Ph1* and *Rolled*) that was present in the KEGG pathway. In comparison, the simulated annealing implementation of mixture Bayesian networks uncovered the gene relationship missing from the MCMC network and confirmed the other predicted relationships.

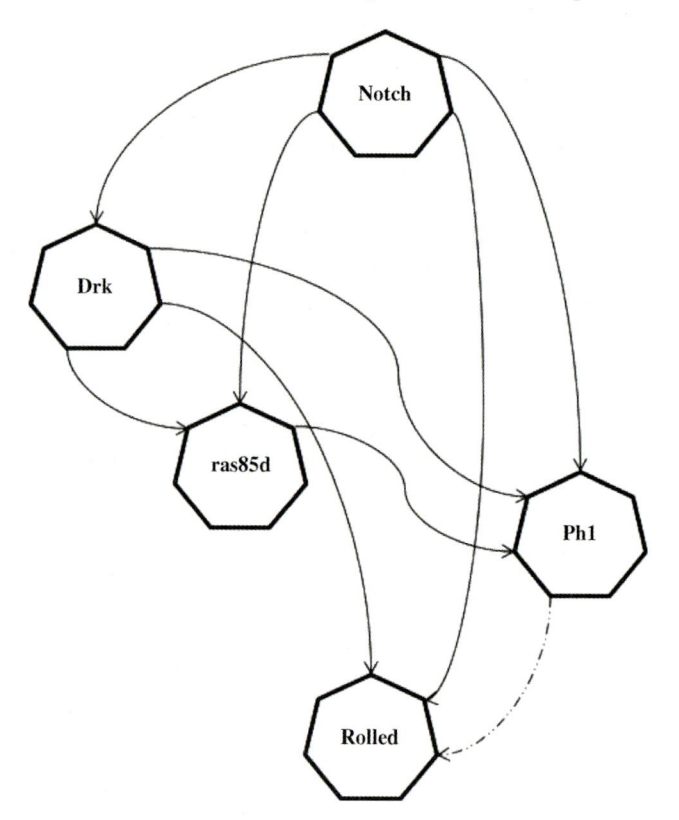

Figure 5.3 Inference of the dorso-ventral axis formation pathway gene network. Continuous edges depict direct gene relationships that were predicted and validated in the KEGG pathway. Broken edges depict indirect gene relationships through intermediate genes not present in the data that were predicted and confirmed in the KEGG pathway. Annotation: *Drk* = downstream of receptor kinase; *ras85d* = Ras oncogene at 85D; *Ph1* = pole hole; *Rolled* = extracellular signal-regulated kinase 1/2. (Adapted from Ko *et al.* [15].)

The similarity of the networks inferred using the MCMC and EM-BIC-simulated annealing approach for both pathways (dorso-ventral axis formation and transforming growth factor-β signaling pathway) disproves the suggestion of Grzegorczyk et al. [16] that the BIC-based mixture Bayesian network approach would be only valid for very large datasets. Likewise, the identification of additional gene relationships by the BIC-simulated annealing approach does not support the generalization put forward by Grzegorczyk et al. [16] that BIC is over-regularized in network inference similarly to other applications. Evaluation of scenarios that encompass a wide range of data sizes, data information content, network sizes, and topology is required to offer conclusive recommendations on the complementary advantages and suitability of the EM-BIC and MCMC implementation of mixture Bayesian networks.

5.3.4
Computational Considerations

Evaluations of the relative advantages of the more flexible and data-driven mixture Bayesian network approach to infer networks in a wide range of scenarios have been reported. The first of these benchmarking exercises confirmed that a mixture of multiple coexpression models offered a better fit to the pathways studied than a single coexpression model [13]. Likewise, Grzegorczyk et al. [16] used synthetic datasets for the Raf–Mek–Erk signaling pathway including 11 protein nodes and demonstrated that mixture descriptor is preferred for datasets simulated using Gaussian mixtures.

An important consideration when developing approaches to infer pathways is the computational time required to predict a network. An evaluation of the computational time required for the prediction of a network across a wide range of number of genes using an EM-BIC mixture Bayesian network approach was undertaken [13]. The computing time required to predict a network depended on the number of genes and the degree of connectivity between the genes studied. The time required by the EM-BIC mixture Bayesian approach supported the effective evaluation of a large number of alternative pathways. Although the MCMC implementation of the mixture Bayesian network offers a more comprehensive depiction of the network topology, the additional time required by this approach (including thousands of burn-in iterations and hundreds or thousands of postburn samples) could hamper the exploration of alternative pathways. A combination of an exploratory phase using the EM-BIC implementation and final inference phase using an MCMC implementation of the mixture Bayesian network approach may be the best alternative.

5.4
Conclusions

A wide range of comparisons between the predicted and known pathways confirmed the outstanding performance of the mixture Bayesian network approach to predict gene relationships. A major advantage of the mixture Bayesian approach was the additional insights gained on general and condition-dependent relationships among genes. Consideration of the mixture model parameter estimates facilitated the interpretation of the inferred networks. The EM-BIC implementation of mixture Bayesian networks provided fast and accurate inference of a single network. The MCMC implementation provided more information through the posterior distribution of all the parameters in the network. Benchmarking tests demonstrated the superiority of the mixture Bayesian network approach that estimates the optimal number of mixture components from the data for each gene subnetwork when compared to nonmixture models and models with a fixed number of components. In addition, the increase in computational requirements with increase network complexity (i.e., number of genes and genes within subnetwork) was moderate, thus facilitating the application of the flexible mixture Bayesian network to a wide range of real-life gene pathway scenarios.

References

1 Friedman, N., Nachman, I., and Pe'er, D. (1999) Learning Bayesian network structure from massive datasets: the "sparse candidate" algorithm, in *Uncertainty in Artificial Intelligence* (eds K.B. Laskey and H. Prade), Morgan Kaufmann, San Francisco, CA, pp. 196–205.

2 Imoto, S., Goto, T., and Miyano, S. (2002) Estimation of genetic networks and functional structures between genes by using Bayesian networks and nonparametric regression. *Pac. Symp. Biocomput.*, 175–186.

3 Rodriguez-Zas, S.L., Ko, Y., Adams, H.A., and Southey, B.R. (2008) Advancing the understanding of the embryo transcriptome coregulation using meta-, functional, and gene network analysis tools. *Reproduction*, **135**, 213–224.

4 Friedman, N., Linial, M., Nachman, I., and Pe'er, D. (2000) Using Bayesian networks to analyze expression data. *J. Comput. Biol.*, **7**, 601–620.

5 Shah, A., Tenzen, T., McMahon, A.P., and Woolf, P.J. (2009) Using mechanistic Bayesian networks to identify downstream targets of the sonic hedgehog pathway. *BMC Bioinformatics*, **10**, 433.

6 Broom, B.M., Rinsurongkawong, W., Pusztai, L., and Do, K.A. (2010) Building networks with microarray data. *Methods Mol. Biol.*, **620**, 315–343.

7 Zhu, J., Chen, Y., Leonardson, A.S., Wang, K., Lamb, J.R., Emilsson, V., and Schadt, E.E. (2010) Characterizing dynamic changes in the human blood transcriptional network. *PLoS Comput. Biol.*, **6**, e1000671.

8 Salzman, P. and Almudevar, A. (2006) Using complexity for the estimation of Bayesian networks. *Stat. Appl. Genet. Mol. Biol.*, **5**, Article 21.

9 Singh, A.K., Elvitigala, T., Cameron, J.C., Ghosh, B.K., Bhattacharyya-Pakrasi, M., and Pakrasi, H.B. (2010) Integrative analysis of large scale expression profiles reveals core transcriptional response and coordination between multiple cellular processes in a cyanobacterium. *BMC Syst. Biol.*, **4**, 105.

10 Davies, S. and Moore, A. (2000) Mix-nets: factored mixtures of Gaussians in Bayesian networks with mixed continuous and discrete variables, in *Uncertainty in Artificial Intelligence* (eds G. Boutiler and M. Goldszmidt), Morgan Kaufmann, San Francisco, CA, pp. 168–175.

11 Newman, M.E. and Leicht, E.A. (2007) Mixture models and exploratory analysis in networks. *Proc. Natl. Acad. Sci. USA*, **104**, 9564–9569.

12 Ko, Y., Zhai, C.-X., and Rodriguez-Zas, S.L. (2007) Inference of gene pathways using Gaussian mixture models. IEEE/BIBM International Conference on Bioinformatics and Biomedicine, Silicon Valley, CA.

13 Ko, Y., Zhai, C.-X., and Rodriguez-Zas, S.L. (2009) Inference of gene pathways using mixture Bayesian networks. *BMC Syst. Biol.*, **3**, 54.
14 Ko, Y., Zhai, C.-X., and Rodriguez-Zas, S.L. (2010) Discovery of gene network variability across samples representing multiple classes. *Int. J. Bioinf. Res. App.*, **6**, 402–417.
15 Ko, Y., Rodriguez-Zas, S.L., and Zhai, C.-X. (2008) An efficient mixture model approach to characterize gene pathways using Bayesian networks. American Statistical Association Conference, Biometrics Section, Alexandria, VA.
16 Grzegorczyk, M., Husmeier, D., Edwards, K.D., Ghazal, P., and Millar, A.J. (2008) Modelling non-stationary gene regulatory processes with a non-homogeneous Bayesian network and the allocation sampler. *Bioinformatics*, **24**, 2071–2078.
17 Bilmes, J.A. (1998) A gentle tutorial of the EM algorithm and its application to parameter estimation for Gaussian mixture and hidden Markov models. Technical Report, International Computer Science Institute and Computer Science Division, Department of Electrical Engineering and Computer Science, UC Berkeley.
18 Schwarz, G. (1978) Estimating the dimension of a model. *Ann. Stat.*, **6**, 461–464.
19 Nicholas, M., Arianna, W.R., Marshall, N.R., Augusta, H.T., and Edward, T. (1953) Equation of state calculations by fast computing machines. *J. Chem. Phys.*, **21**, 1087–1092.
20 Hastings, W.K. (1970) Monte Carlo sampling methods using Markov chains and their applications. *Biometrika*, **57**, 97–109.
21 Metropolis, N., Rosenbluth, A., Rosenbluth, M., Teller, A., and Teller, E. (1953) Equation of state calculations by fast computing machines. *J. Chem. Phys.*, **21** (6), 1087–1092.

6
Multiscale Network Reconstruction from Gene Expression Measurements: Correlations, Perturbations, and "*A Priori* Biological Knowledge"

Daniel Remondini and Gastone Castellani

6.1
Introduction

6.1.1
Complex Networks

In the last decade, physics has been expanding to new research areas. In particular, life-related sciences (ecology, sociology, economics, and last but not least biology) have been showing striking analogies with complex systems arising from various physical areas. Such an approach has happened from both fronts: on the life sciences side, huge amounts of data have become available for detailed analysis, thanks also to the Internet, through which these data are nowadays easily collectable and queryable (e.g., stock market financial series, tables of social relationships from movie copartnerships to e-mail fluxes, high-throughput biological data). On the other side, many physical and mathematical tools that had proven useful in explaining complex phenomena like polymer growth or spin glasses began to spread to other research areas like biological and social sciences in a broad sense.

The common trait of these research fields can be found in the framework of network theory, such that focusing on the relationships among elements allows us to draw general conclusions even though the details of the system are not completely known or easily tractable from a mathematical point of view. Relaxing attention to the details of the specific interaction or element, network theory aims to provide tools for the characterization of a set of relationships, represented as edges or *links*, occurring among similar elements, referred to as vertices or *nodes*.

One of the most powerful approaches to physical systems is statistical mechanics. Many results (for "ideal" gases or solids) have been obtained by considering random interactions between elements of the system, so that a "mean field theory" could be built from the average behavior of the system. The main drawback of this mean field approach (and the actual challenge at the same time) is that complex systems (to which living and life-related systems belong) are often characterized by a nontrivial set of

Applied Statistics for Network Biology: Methods in Systems Biology, First Edition.
Edited by M. Dehmer, F. Emmert-Streib, A. Graber, and A. Salvador.
© 2011 Wiley-VCH Verlag GmbH & Co. KGaA. Published 2011 by Wiley-VCH Verlag GmbH & Co. KGaA.

interactions and a mean field approach can completely miss the point. Moreover, social and biological systems can be considered as constantly far-from-equilibrium systems, since equilibrium for every life-related process equals death, and a continuous influx and efflux of energy and matter is necessary to maintain conditions suitable for life. It is thus quite hard to fit them into equilibrium-based models that we can say constitute the "core" of classical statistical mechanics.

An approach that has received renewed attention, is based on the so-called chemical master equation (CME) that describes the temporal evolution of the probability of having a given number of molecules for each chemical species involved. The discrete probabilistic approach, as with the CME, is attractive because it ensures the correct physical interpretation of fluctuations in the presence of a small number of reacting elements (as compared to continuum approaches such as Langevin and Fokker–Planck formalism [1]) and because it provides an unitary formulation for many biological processes, from chemical reactions to ion channel kinetics. CME theory can be related to predictions on the noise levels in selected biological processes (e.g., during transcription and translation) [2]. In particular, the observation that mRNA is produced in bursts varying in size and time has led to the development of new models capable of better explaining the distributions of synthesized products [3].

The models based on CME can help to characterize the role of noise in the networks reconstruction as well as the role of fluctuation in the enhancement and maintenance of biological functions.

Furthermore, the CME approach allows us to compute all the thermodynamic quantities, including entropy and free energy, with the consequent possibility to characterize the system as a nonequilibrium system if the detailed balance is not satisfied.

One of the greatest contributions that may be given by network theory to the understanding of biological and social systems is that the network architecture may reflect the dynamic processes that led to it. In a pure statistical-physical fashion, different "universality classes" can be sought in order to fit the process we are studying, be it the ask–bid mechanism for a stock, the spreading of a political idea within a society, or the patterns of gene expression following a biological stimulus. Actually, the main limit to this approach is the lack of a complete set of models and formalisms in which every problem can be embedded. This is in fact a challenge inherent to the complexity naturally found in life-related systems. We remark that the features of a network model are peculiar both from a static point of view (e.g., the relation between network topology and the evolutionary model that led to it) and for system dynamics (e.g., the responses to perturbation or the noise features of a stochastic dynamics), and while the first approach is much more developed, dynamics of networks is still a very unexplored field. Recent models of social networks [4] show that the situation can be even more complicated, with node interactions affecting network topology and network topology affecting node interaction dynamics. This is a common paradigm for biological systems at several levels, from nervous and immune system development and functionality to genomics (for a recent review, see [5]).

6.1.2
Gene Interaction Networks from Gene Expression Measurements

Referring to biological systems, microarray technology enables the construction of time series of whole-genome expression, although with a much coarser detail than the aforementioned financial series; other tools for high-throughput measurement of biological observables are also available (e.g., regarding gene copy number variation, single-nucleotide polymorphism frequency, protein–protein interactions, and whole-genome methylation states). Such huge amounts of biological data allow us to revisit biological studies from a more holistic point of view: "systems biology" is a new branch of science (combining biology, physics, statistics, and informatics) that tries to synthesize such information, giving a new perspective to the study of genomic and functional diseases, as well as to the comprehension of cellular activity. Systems biology is massively exploiting network tools to represent and understand these novel datasets, also because biological knowledge collected up to now has been resumed in large networks of biochemical interactions and ontologies of relationships (chemical, functional, or simply structural) whose architecture can be studied by itself, as has been done for simple organisms like yeast or *Caenorhabditis elegans*.

How to extract reliable knowledge from such biological data, in which cellular variability and experimental techniques make them very noisy, and the number of observations is very small compared to the variables to be studied, is a very challenging task by itself. However, the application of "robust" network approaches (e.g., by an appropriate choice of method of network reconstruction) can reveal the emergence of interesting phenomena, even with these limitations in dataset quality and size. Linear Markov models and correlation are common approaches for network reconstruction starting from time-series data. In our research activity, we applied both methods, and we observed that the second was more robust to noise and more sensitive to the information embedded in the data, even with few time points available. We observed a similar behavior in several situations, ranging from the perturbation of a regulatory gene (c-Myc) in a single cell line (rat fibroblasts) to a whole organism (*Drosophila Melanogaster*) in response to a dietary restriction: correlation analysis reveals that cells increase their internal synchronization, expressing (and switching off) a large number of genes at the same time, and such coordinated behavior is even more evident for the top-ranking expressed genes (in terms of statistical significance). It will be interesting to understand how cells deal with such high synchronization, in particular self-organization and external stimulation (long-range mediators like hormones and short-range cell–cell signaling). If we consider higher-scale systems, like people in a crowd or stocks in a financial market, very often correlation is related to catastrophic situations, like riots or market crashes, and thus we could learn a lot about how to control such phenomena by understanding how cells synchronize and desynchronize their activity. The network obtained from such relations (i.e., considering a link between two genes that are coexpressed in time, both for up- and downregulation) shows a structure that conveys information about the time relationships between genes. Figure 6.1 shows the joint distribution of two network parameters (degree connectivity K and betweenness

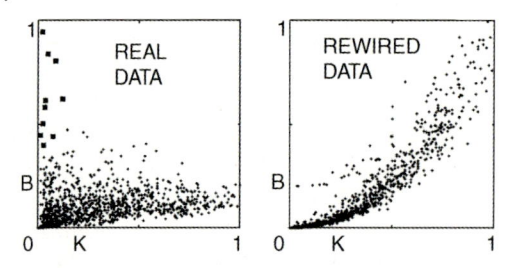

Figure 6.1 Joint distribution of two network parameters (degree connectivity K and BC B) for real data and for randomly rewired data.

centrality (BC) B) for real data and for randomly rewired data, that keeps some of the real data structure but disrupts time relationships, in the case of c-Myc activated rat fibroblast cell lines. Without entering into too "technical" details, it is evident that there are great differences between the two distributions from a global point of view. Network theory can give large-scale information (about global network topology, structure, and modularity), but it can also provide several ways of ranking single nodes or links. In this case, we can observe that a group of nodes (genes) are clearly outstanding, both with respect to the distribution obtained from the rewired dataset and to real data (shown as black squares), their biological meaning is still under investigation, but some of them appear related by an early activation governing the c-Myc transregulation cascade.

6.2
"Perturbation Method"

An emerging approach for the analysis of functional genomics data is based on the so-called "perturbation method," which consists of perturbing the system with external tunable stimuli, and following the changes in the gene interaction network properties as a function of time and perturbation magnitude [6, 7]. Perturbation studies can be conducted within relatively short time scales (from a few hours to a few days) under relatively homogeneous conditions for all the cells/organisms, in order to acquire information on short-time biological processes (e.g., transcriptional cascade reconstruction). However, the same experimental design can be exploited for studies on long time scales related to different classes of biological processes (e.g., characterizing the expression changes due to organismal aging usually requires observations spanning the whole lifespan of the investigated species [8]). Our hypothesis is that the strategy successfully used to analyze classical perturbation studies can be applied to the studies of aging, where the perturbation is the length of the time scale itself, within a time-series framework. This scenario poses a particular challenge from an experimental point of view, since it is hard to observe the most evolved organisms (like humans) for the whole of their lifespan. Moreover, since individual variability (e.g., due to genetic factors) is deeply interlaced with external (environmental) as well as stochastic factors, this could lead to different strategies adopted by different individuals to face aging, thus leading to a "divergence" in gene expression (or methylation or protein–protein interaction) patterns over time.

An observation that is gaining experimental support is based on the detection of global changes in selected groups of genes with a high degree of temporal "synchronization" or "polarization" of gene expression dynamics. Recent experimental and theoretical developments in epigenetics suggest that, for short time-scale phenomena, a possible basis for such synchronization is a global change in chromatin structure, related to concomitant changes of transcription factor activity, resulting in histone modification (acetylation and methylation state). Whether the same hypothesis could be applied also for life-long mechanisms is a very debatable question.

In this chapter we will illustrate c-Myc- and refeeding-induced synchronization in gene expression dynamics as an example of time-series analysis of global gene expression changes. A similar approach can be used in aging research within the assumption that the correlation properties are similar. If these two experimental settings show similar behavior, we can assume that they share basic underlying mechanisms, including global changes of cellular functional circuitry and/or molecular mechanisms such as changes in chromatin structure. In particular, similar global changes of gene expression profiling in classical perturbation experiments and aging will support the hypothesis that the occurrence of "phase transitions" is a general consequence of external perturbations. Our working hypothesis is that a similar process is also taking place in the case of human aging and that it can be captured with the correlation method if applied to time series of gene expression profiling.

Finally, all these methods can be integrated into a systems biology framework, capable of capturing the global and complex character of aging, that can also help to identify individual genes responsible of exceptional phenotypes (centenarians) as well as age-associated changes in complexity, robustness, and frailty.

In this chapter, we review some methods that have been used to reconstruct networks from gene expression data. The methods are strongly data-dependent; in fact, the experimental design often drives the type of analysis, including statistical inference, to be conducted on the data.

Among a plethora of methods that can be used for the reconstruction of a transcriptional network we will describe (i) a general method, based on a correlation matrix, useful for time-series analysis, and (ii) a method based on pathway mapping of probe sets and pathways intersections, using the so-called "*a priori* biological knowledge" giving information on how the probe sets group into functional sets.

6.3
Network Reconstruction by the Correlation Method from Time-Series Gene Expression Data

Time series of gene expression data from high-throughput experiments have been used to infer networks of coexpressed genes. By following the changes in expression at the genomic level, it is possible to identify groups of genes with a similar expression pattern. Most of the techniques currently used in functional

In this chapter, we extend the significance analysis of gene pathways to higher-order structures (i.e., networks of pathways whose intersections contain a significant number of differentially expressed genes). Network structure can reveal the degree of coordination of different biological functions as a consequence of the treatment, as well as the presence of "focal areas" in which groups of genes play central roles. We show examples in which some biological functions (related to specific pathways) are biologically relevant for the studied process, due to their position inside the pathway network. This analysis can be extended to groups of genes at the "interface" between pathways, whose imbalance can affect more than one biological function.

Our approach is aimed at understanding how external perturbations, such as gene activation or tumor induction, can induce in various types of cells, cell lines, or derived tissues, behaviors that can generate, integrate, and respond to dynamic informational cues.

The broad question that we are trying to answer is how a cell converts perturbations of its signaling activity into a "binary," or at least discrete, decision, resulting in the appearance of a given phenotype. Thus, the signaling activity has to be diffused within the cell between and within pathways. A signaling pathway is not a rigid unit, since it can achieve one or more functions with different subsets of its elements. The communication with other pathways, due to the fact that many elements are shared between several pathways, may be captured by looking at those elements belonging to the interface between pathways.

6.5
Examples and Methods of Correlation Network Analysis on Time-Series Data

We considered three datasets of time-course gene expression arrays: (i) conditional, tamoxifen-dependent, activation of the c-Myc proto-oncogene in rat fibroblast cell lines, (ii) genomic response over time to nutrition changes in *Drosophila melanogaster*, and (iii) patterns of gene activity as a consequence of aging occurring over a lifespan time series (25–97 years) sampled from T cells of human donors.

As a first step, we identified a set of genes that significantly respond to the stimuli (keeping in mind that aging for the third database can be seen as a stimulus in a wide sense) by means of statistical analysis, since many genes may not be expressed in specific cell types, or in general may not respond to the perturbation, and thus constitute the background noise to be separated from the signal. Due to the heterogeneous nature and properties of the datasets, we selected the significant genes according to different criteria.

The first dataset (MYC dataset) consists of time-course gene expression arrays based on reconstituted $c\text{-}myc^{-/-}$ rat fibroblast cell lines with the conditionally active, tamoxifen-specific c-Myc–estrogen receptor fusion protein. Binding of tamoxifen to the estrogen receptor domain elicits a conformational change that allows the fusion protein to migrate to the nucleus and to act as a transcription factor. It contains the gene expression data collected after the addition of tamoxifen. Samples were harvested at five time points after the addition of tamoxifen to the culture medium: 0,

2, 4, 8, and 16 h. The entire experiment was repeated on three separate occasions, providing three biological replicates for each gene and time point. Expression profiling was done by using the Affymetrix platform and U34A Gene Chips [48].

For the c-Myc dataset we applied a two-way analysis of variance (ANOVA) (time and treatment as variability factors) to identify genes whose expression pattern was significantly affected by c-Myc activation and that were changing over time. With this method we identified a set of 1191 significant genes out of a total of 8799 [6].

The *D. melanogaster* dataset has been produced according to the following procedures. Newly enclosed virgin females were maintained in yeast-free media until 4 days old and then transferred either to media with yeast (Y-treatment) or to control media without yeast (NY-control). Samples were collected every hour for the following 12 h. Four additional samples were collected prior to refeeding, and arbitrarily designated -4, -3, -2, and -1. Synchronization of the physiological state was obtained by imposing diet restriction in third instars. Affymetrix gene chips were used to measure mRNA abundance at each hour for both Y-treatment and NY-controls. A time-ordered sequence of expression ratios was then computed for each gene in the array [49].

The *D. melanogaster* diet arrays provide a higher resolution dataset and genes were selected via GeneTrace – a statistical method that looks for change points in the time series of the expression ratios between the two cohorts. GeneTrace identified 3519 genes with a significant change point. These results showed that physiological response to nutrient uptake involves a rapid change in transcriptional profile at a global scale and that most of the changes are small (81% of the 3519 ratios smaller than 1.5-fold).

The human aging dataset consists of a set of 25 human male healthy donors, whose lymphocytes were extracted from peripheral blood and hybridized onto comparative arrays containing about 19 000 probes. A Universal Human Reference sample was chosen as one of the channels for each array.

For the human aging dataset we applied one-way ANOVA (with donor age as variability factor) dividing the 25 samples into five age groups, since we selected donors with ages that could be grouped into 10-year intervals separated by at least 5 years (25–35, 40–50, 55–67, 80–80, and above 85). With a highly selective significance threshold ($P < 0.01$ after Benjamini–Hochberg *post hoc* correction) we obtained a set of 768 probe sets selected out of a total of 14 688 probe sets (resulting from the processing of the original arrays).

When c-Myc is activated by tamoxifen, the activity profile of the probe sets clearly changes into a strongly synchronized regime. This is reflected in the histograms of the correlation coefficients for the N-control and T-treatment data sets (Figure 6.2) and in the main parameters of the connectivity distributions obtained from the corresponding adjacency matrices (Table 6.1). The adjacency matrix characterizing the network was obtained by considering only the correlation coefficients whose absolute value exceeded a threshold fixed between 0.95 and 0.99 (we remark that the lower threshold value was higher than the value requested for a $P < 0.05$ statistical significance of the correlation coefficients). The results shown in this chapter were obtained for a threshold equal to 0.98, but similar results held for the [0.95–0.99]

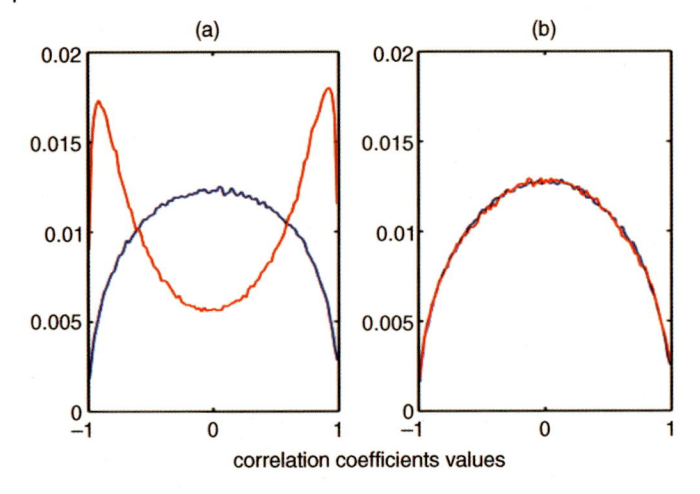

Figure 6.2 Histogram of correlation coefficients of the gene expression time series between genes for the MYC dataset. The red line refers to the perturbed case (tamoxifen-induced c-Myc activity), whereas the blue to the unperturbed one. (a) Perturbation induces a bimodal distribution (i.e., genes tend to be either strongly correlated or anticorrelated, differing significantly from the unperturbed case). (b) correlation coefficient histograms obtained after time reshuffling of the same genes do not show any significant difference.

interval. These coefficients were set equal to 1, producing a symmetric adjacency matrix (diagonal elements were removed). For each gene, connectivity degree k was defined as the total number of genes it was connected to; $k(l) = \sum a_{k,l}$.

For the tamoxifen-induced dataset the number of coefficients close to $+1$ or -1 increases significantly. This finding indicates that many of the 1191 genes, whose expression levels over time were affected by tamoxifen stimulation, became either strongly correlated or anticorrelated.

In Figure 6.3 we show the histogram of the correlation coefficients between all the genes selected with the change point analysis in the *D. melanogaster* dataset. In the NY-controls (top left) the histogram resembles a Gaussian distribution slightly skewed towards positive correlation values. When considering the expression ratio

Table 6.1 Principal network parameters for the control (N) and tamoxifen-induced (T) gene expression datasets.

Network parameters	N	T
k_{min}	0	0
k_{max}	17	99
Mean, k	4.53	23.44
Standard deviation, $\sigma(k)$	2.61	23.97
Skewness $\gamma(k)$	0.89	1.16
Clustering coefficient $c(k)$	0.43	0.45

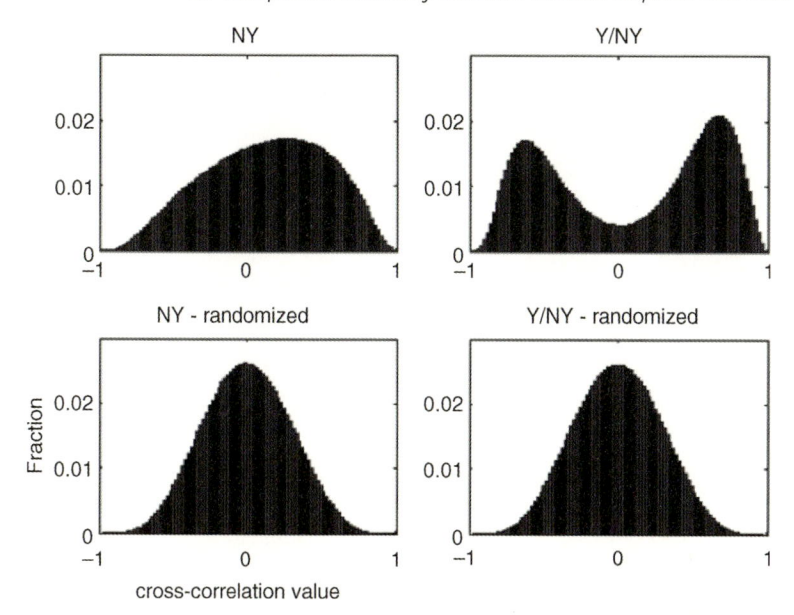

Figure 6.3 Histogram of the correlation coefficients of the gene expression time series between all the genes selected with the change point analysis in the *D. melanogaster* dataset. In the NY-controls (top left) the histogram resembles a Gaussian distribution slightly skewed towards positive correlation values. When considering the expression ratio Y-treatment over NY-control (top right) the distribution becomes bimodal and genes tend to be either strongly correlated or anticorrelated. These results have been validated by reshuffling the time points independently for each gene. In both cohorts this leads to a Gaussian distribution (bottom row).

Y-treatment over NY-control (top right) the distribution becomes bimodal and genes tend to be either strongly correlated or anticorrelated.

We used time reshuffling to test the dependence of the results from the exact time series. By randomly shuffling the time series for each gene separately, time relationships between expression levels are broken, but the mean and standard deviation for each gene are unaltered. Properties of the gene network that truly depend on the expression level dynamics should be significantly affected by a random shuffling in time. The bottom row of Figure 6.3 shows the result for the *D. melanogaster* data. In both cohorts this leads to a Gaussian distribution for the values of the correlation coefficients (Figure 6.3, bottom row). Analogous results have been obtained for the other two datasets (Figures 6.2 and 6.4).

The role of noise was taken into account by analyzing the correlation coefficient distribution obtained from a dataset of randomly generated vectors of the same size as the experimental datasets (e.g., for the MYC dataset: a 5×1191 array of values sampled from the standardized Gaussian distribution [50]). The resulting distribution strictly resembles that obtained with the no-tamoxifen dataset.

It is possible to identify a subnetwork of strongly correlated/anticorrelated genes by selecting the probe sets based on the treatment factor's strength in the ANOVA

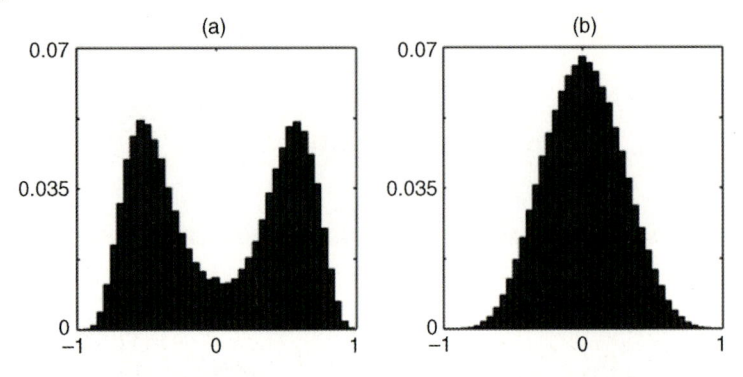

Figure 6.4 Histogram of correlation coefficients of the gene expression time series between genes for the aging dataset. (a) Histogram of the correlation coefficients for the set of 768 probe sets selected with one-way ANOVA, $P < 0.01$. (b) Histogram of the correlation coefficients for a set of 768 probe sets randomly sampled from the whole dataset of 14 688 probe sets. A single-gene time reshuffling applied onto each dataset produces a Gaussian distribution (data not shown).

analysis. This selection process can be characterized as a transition from a unimodal (most genes are uncorrelated) to a bimodal behavior (genes are either correlated or anticorrelated). Figure 6.5 shows how this process occurs by decreasing the cutoff P value (P_{thr}) used to select significant genes in the c-Myc dataset. The different panels show the histogram of the correlation coefficients between the expression values over time from the probe sets in the dataset N (left column) and T (right column) for decreasing P_{thr}. The top row includes all the probe sets used in the analysis ($P_{thr} > 1$); the central row corresponds to an intermediate threshold ($P_{thr} = 0.2$); the bottom row corresponds to the lowest threshold ($P_{thr} = 0.05$). Notice that the transition from unimodal to bimodal is only present in the T dataset, while the N dataset in not affected even when the cutoff P-value is very small. Analogous results have been obtained for the other two datasets.

We have thus observed, in three different perturbation conditions and with different cell types, a relationship between patterns of gene response and patterns of synchronization over time. This may suggest that there are high levels of intracellular communication (at least following macroscopic perturbations) in terms of the transcription machinery (and of gene–gene interactions) that go beyond the genes that may be directly involved (from a functional point of view) in the response to perturbations. As an example, a global regulation mechanism could be provided by three-dimensional chromatin structure (we considered only eukaryotic cells) that may "open" to transcription genes also not directly involved in the mechanisms. For the case of the c-Myc regulation cascade, following our publication it has been shown [51] that c-Myc is directly responsible for the opening of chromatinic regions throughout the genome, thus favoring broad effects in gene transcription regulation.

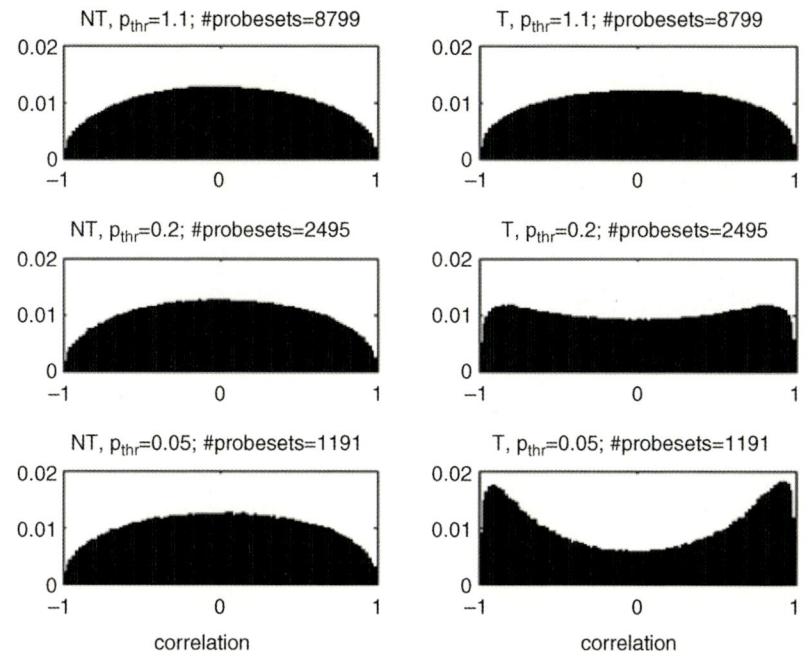

Figure 6.5 Transition from unimodal to bimodal behavior for the MYC dataset. The different panels show the histogram of the correlation coefficients between the expression values over time from the probe sets in the dataset N (left column) and T (right column) for decreasing cutoff P values (P_{thr}). The top row includes all the probe sets used in the analysis ($P_{thr} > 1$); the central row corresponds to an intermediate threshold ($P_{thr} = 0.2$); the bottom row corresponds to the lowest threshold ($P_{thr} = 0.05$).

6.6
Examples and Methods for Pathway Network Analysis

For pathway network analysis, we consider the combined effect of several genes, looking at the ratio of significantly responding genes (with respect to a statistical analysis depending on the specific experimental design) inside a biological pathway. The aim is 2-fold. (i) We achieve a better interpretation of single-gene results, combining them into a higher-level structure with a clearer biological interpretation. (ii) Combining single-gene analysis with a pathway-based statistical analysis, we obtain a filtering of our results that does not rely only on a increase of conservativity (as happens for *post hoc* tests like Bonferroni, Benjamini–Hochberg, and all their variants). Moreover, as explained in the following sections, the single-pathway structure can be embedded in a pathway network in which the interaction between different biological functions (represented by the genes shared between pathways) can also be taken into account.

For a description of this method, we consider two datasets: the c-Myc activation dataset (MYC; as previously described) and a human leukemia dataset.

The second dataset we consider consists of the gene expression measurements as described in [52]. This dataset contains bone marrow samples obtained from acute leukemia patients, that can be classified as acute lymphoblastic leukemia (ALL) and acute myeloid leukemia (AML). The mRNA prepared from bone marrow mononuclear cells was hybridized with Affymetrix Hgu6800 containing probes for 6817 human genes. The experimental design is a comparison between ALL and AML (one factor) on the basis of 6817 probes. The dataset (AML/ALL dataset) contains 72 samples (47 obtained from ALL patients and 25 obtained from AML patients).

6.6.1
Gene Selection and Pathway Grouping

For the MYC dataset, one-way ANOVA was applied to each of the 8799 probe sets to identify those that significantly changed expression level over time. A P-value of 0.05 was chosen as the cutoff significance level. No post hoc correction for multiple testing (i.e., Benjamini–Hochberg, false discovery rate) was applied, since *post hoc* validation is provided by pathway analysis: 765 genes resulted significant, 251 of which are annotated in KEGG and belonged to 142 pathways.

The AML/ALL dataset was analyzed with a linear model with an empirical Bayes method to shrink gene variances (LIMMA, R package) and 1924 genes were found as significantly differentially expressed between the AML and the ALL groups ($P < 0.05$). Among the differentially expressed genes, 801 genes were annotated in the KEGG database.

6.6.2
Pathway Significance and Pathway Network

In order to reconstruct a network, we need to specify both its nodes and links. From a biological point of view, nodes can be defined as groups of genes (such as pathways or ontologies) coding for proteins/peptides with similar functional properties (e.g., ion channels, kinases, phosphatases, and transcription factors), performing similar tasks, or involved in the same biological function. The links between nodes can be drawn in various ways and their definition may also depend on the particular type of experimental design (e.g., temporal correlation or physical interactions of proteins) [53, 54].

We choose to define network nodes as groups of genes belonging to the same pathway as described in the KEGG database. To each node we associate a feature corresponding to the state of the pathway, which can be significantly involved (over-represented), significantly not involved (under-represented), or not significant in the experimental context [55]. The same classification is used for the links between nodes by analyzing the ratio of significant genes at the intersection between the corresponding pathways.

Significance of nodes and links can be assessed within the framework of 2×2 contingency tables (Table 6.2).

Table 6.2 2 × 2 Contingency table.

	Differentially expressed	Not differentially expressed	
G	α	β	N_G
$\bar{G}$	γ	δ	$N_{\bar{G}}$
	S	$\bar{S}$	N

α = number of significant genes G; β = number of not significant genes G; γ = number of significant genes $\bar{G}$; δ = number of not significant $\bar{G}$; $S = \alpha + \gamma$ = number of significant genes in the array; $\bar{S}$ = number of not significant genes in the array; $N_G = \alpha + \beta$ = number of genes G; $N_{\bar{G}} = \gamma + \delta$ = number of genes $\bar{G}$; N = total number of measured genes.

Given a subset G of the N measured genes with N_G genes, α will be differentially expressed while $\beta = N_G - \alpha$ will not. We compare α and β to the number of differentially expressed genes γ and not differentially expressed genes δ not belonging to G.

The statistical significance of the contingency table can be computed in different ways: Fisher exact test, binomial, and χ^2 distribution-based tests [17]. We chose to apply the Fisher exact test because the computation of the hypergeometric distribution is straightforward for tables with both small numbers (arising when testing intersections, see below) and large numbers (arising when testing pathways). The Fisher exact test first computes the probability p^* of the observed 2×2 table by using the hypergeometric distribution with parameters (S, N_G, N):

$$p^* = p(X = \alpha | S, G, N) = \frac{\binom{S}{\alpha}\binom{N-S}{N_G-\alpha}}{\binom{S}{\alpha}}$$

The p-value to reject the null hypothesis (independence of rows and columns in the contingency table) is given by the sum of the probabilities of all the tables with a value lower than S and with the same marginal totals; that is, we consider the cumulative hypergeometric function $P(x \leq S)$:

$$p = \sum_{\substack{i=1, N_G \\ p_i \leq p^*}} \frac{\binom{S}{i}\binom{N-S}{N_G-i}}{\binom{S}{i}}$$

This procedure gives a probability for a two-tailed Fisher test. Distinction between over- or under-representation of the selected group of genes G can then be obtained by comparing the proportion α/NG of differentially expressed genes in G with the proportion of differentially expressed genes S/N on the array. A group G is considered significant if $P \leq 0.05$.

We apply a similar framework to evaluate the significance of all nonempty intersections between two pathways:

$$Q_{ij} = G_i \cap G_j; \quad \forall i, j = 1, \ldots, N; \quad Q_{ij} \neq \{\}$$

The only difference is the definition of the total number of genes N, which is taken to be equal to the total number of genes in the two groups. More precisely, for the group intersection significance analysis, the significant genes in the intersection S_I and the total number of genes N_I in the intersection are compared to the total number of genes Q_{ij} found in the union of the two groups and the number of significant ones, $N_{G_1 \cup G_2}$ and $S_{G_1 \cup G_2}$, respectively.

The reason for this choice is as follows. Suppose we have two groups with 100 genes each and with 50 genes in common. Suppose 60 genes are significant in each group, 30 of which are in the intersection. If 60% call rate is significant for the two sets, it is likely that it will also be for the intersection. However, if we take a random subset of 50 genes from 150 genes in the union of the two groups, we can expect on average a 60% call rate. Hence, a random subset with the same numbers of genes as the original intersection would be likely to be significant. By using the union of the two groups as a background, we increase the requirement for the intersection to be considered as significant and reduce the above problem. Intersections are considered significant if their p-value is lower than 0.05, in which case a link is drawn between the two pathways, either red if it is "significantly involved" or blue if it is "significantly not involved," while if no link is drawn $p > 0.05$.

All the gene groups that we consider are biological pathways defined according to the KEGG annotation, and the mapping between probes and pathways is accomplished by querying the KEGG database via R software (KEGGSOAP package).

Once the significant links and nodes are established, we perform a meta-analysis on the obtained network structure. The aim of this analysis is 2-fold. (i) The network structure (e.g., the presence of subnetworks, clusters, communities) can reveal important biological features (e.g., pathways activating together or non-activating even if they share common genes). (i) Each network element (node or link) can be ranked not only on the basis of its statistical significance (the p-value obtained by the above method), but also considering its centrality in the network. We consider as a centrality measure the BC for each vertex – a parameter that characterizes the degree of "trafficking" through a network element [56]. For a given vertex, BC is proportional to the sum of the shortest paths passing through it.

We remark that BC can be defined analogously for the edges of a network (i.e., our pathway intersections) since it can perform the same count of the number of shortest paths passing through an edge, so we can have information about the relevance of modules (larger than single genes but smaller than whole pathways) involved in the biological mechanism under study, thus obtaining a real multiscale analysis (from single genes to groups of pathways, if they are significantly linked together). The importance of this method will probably increase as soon as a larger number of genes will be included in KEGG pathway network (we remark that the KEGG pathway is updated at least monthly, due to the evolution in the understanding of gene biological function inside metabolic or signaling pathways).

Table 6.3 MYC dataset: statistically significant pathways obtained by the Fisher method.

Pathway	p	Genes	Significant Genes	Under/over representation
Neuroactive ligand–receptor interaction	1.02e-05	223	11	under
RNA polymerase	0.000354	4	4	over
Cytokine–cytokine receptor interaction	0.002103	87	3	under
Pyrimidine metabolism	0.002192	23	9	over
DNA polymerase	0.003769	9	5	over
Aminophosphonate metabolism	0.009321	4	3	over
Cell cycle	0.014101	44	12	over
N-Glycan biosynthesis	0.016643	12	5	over
Jak–STAT signaling pathway	0.019052	69	3	under
Folate biosynthesis	0.025244	9	4	over
Fatty acid metabolism	0.027632	37	10	over
Ether lipid metabolism	0.044526	15	5	over
Glycan structure biosynthesis 1	0.046622	20	6	over

6.6.3 Results

The list of pathways obtained from the Fisher test shows the most significant over- and under-represented pathways in the MYC (Table 6.3) and AML/ALL datasets (Table 6.4). In the MYC dataset three pathways are significantly under-represented: neuroactive ligand receptor interaction, cytokine–cytokine receptor interaction, and Jak–STAT signaling pathway. Among the over-represented pathways are RNA and DNA polymerase, cell cycle, and some metabolic and biosynthetic pathways (pyrimidine, fatty acid and ether lipid metabolism, folate and glycan structures biosynthesis). In the case of the AML/ALL dataset two pathways are found to be under-represented (ribosome and neuroactive ligand receptor interaction); among the over-represented pathways there are cell cycle, many metabolic pathways (glycerophospholipid, galactose, pyrimidine and purine metabolism among other), and a signaling pathway (B cell receptor signaling pathway).

In our case studies (see Figures 6.6 and 6.7) the networks are very small, due to the sparseness of the significant links and nodes, thus very few network elements have nontrivial values of BC. Anyway, the analysis of the MYC network shows the emergence of four main subnetworks (Figure 6.5). These subnetworks are related to different biological functions: the first subnetwork is composed by pathways involved in signaling processes (mitogen-activated protein (MAP) kinase signaling pathway, vascular endothelial growth factor (VEGF) signaling pathway, gonadotropin-releasing hormone (GnRH) signaling pathway), and pathways that are related to the communication between cells and the external environment (regulation of actin cytoskeleton and gap junction). Another interesting subnetwork connects the metabolism with the signaling system, showing links between the peroxisome

Table 6.4 AML/ALL dataset: statistically significant pathways obtained by the Fisher method.

Pathway	p	Genes	Significant genes	Under/over representation
Ribosome	6.70e-05	78	10	under
Cell cycle	0.00152	80	40	over
Glycerophospholipid metabolism	0.003631	28	17	over
Neuroactive ligand–receptor interaction	0.004484	211	51	under
Galactose metabolism	0.008409	21	13	over
B cell receptor signaling pathway	0.008725	49	25	over
Aminoacyl-tRNA biosynthesis	0.015056	20	12	over
Pyrimidine metabolism	0.016126	51	25	over
Purine metabolism	0.02202	90	40	over
Glycerolipid metabolism	0.024062	39	20	over
Leukocyte transendothelial migration	0.0245	75	34	over
Histidine metabolism	0.033768	26	14	over
Nitrobenzene degradation	0.03577	3	3	over
Proteasome	0.040852	27	14	over
Reductive carboxylate cycle (CO_2 fixation)	0.043013	7	5	over
Protein export	0.043013	7	5	over
Aminophosphonate metabolism	0.043363	5	4	over
Nucleotide sugar metabolism	0.043363	5	4	over

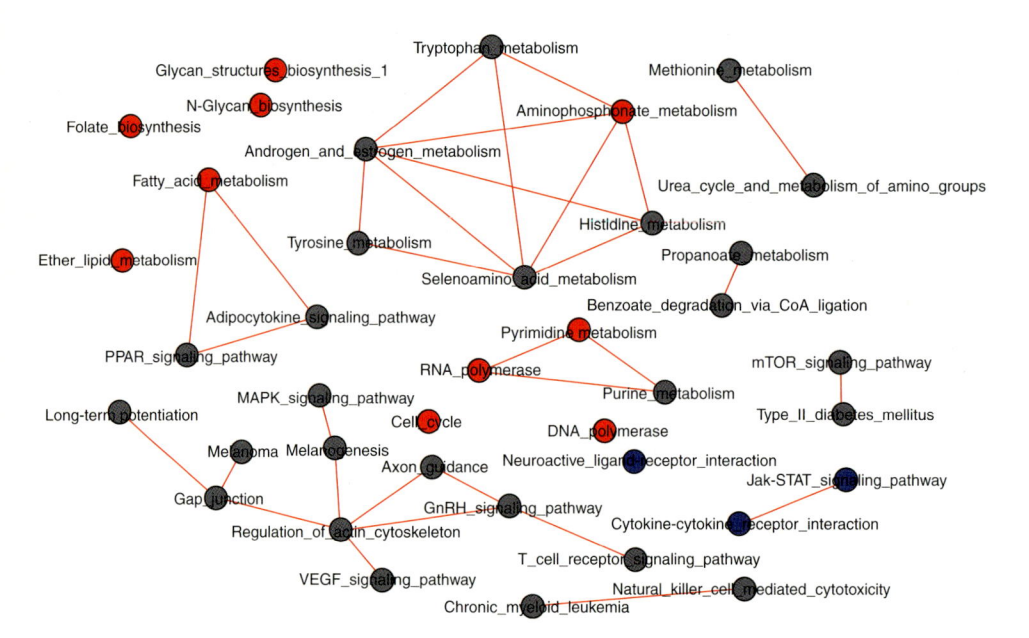

Figure 6.6 Network of pathways for the MYC dataset. Gray circles indicate not significant pathways, red and blue circles indicate significant pathways that are, respectively, over-represented and under-represented. Red links indicate pathways interconnections that are statistically significant and over-represented. The largest pathway subnetworks are clearly related to metabolism, genetic information processing and signaling biological functions.

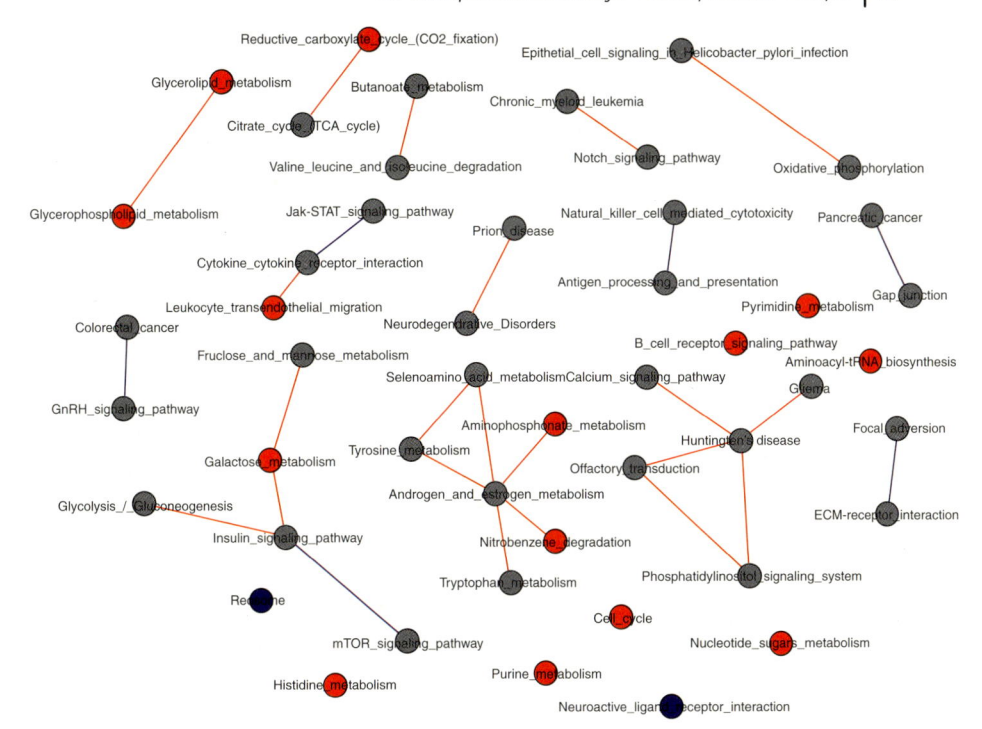

Figure 6.7 Network of pathways for the AML/ALL dataset. Gray circles indicate not significant pathways, red and blue circles indicate significant pathways that are, respectively, over-represented and under-represented. Red lines indicate pathways interconnections (links) that are statistically significant and over-represented, whereas blue lines indicate significant pathways interconnections that are under-represented.

proliferator-activated receptor (PPAR) signaling pathway, adipocytokine signaling pathway, and fatty acid metabolism. A further subnetwork is related to nucleic acids precursor synthesis and nucleic acids polymerization (pyrimidine metabolism, RNA polymerase, and purine metabolism). Another interesting subnetwork contains some basic metabolic pathways (aminophosphonate, tryptophan and tyrosine metabolism, androgen and estrogen metabolism among them).

For the AML/ALL dataset, the pathways network showed in Figure 6.7 evidences a subnetwork connecting signaling and metabolism (insulin signaling pathway, glycolysis/gluconeogenesis, galactose metabolism, fructose and mannose metabolism, and mammalian target of rapamycin (mTOR) signaling pathway that shows an under-represented intersection with insulin signaling pathway). Another interesting subnetwork contains some basic metabolic pathways (aminophosphonate, tryptophan and tyrosine metabolism, androgen and estrogen metabolism, etc.). The presence of a subnetwork involving calcium and phosphatidylinositol signaling, Huntington's disease, glioma, and olfactory transduction can also be noticed.

Single-gene and pathway significance can be represented together by means of a bipartite graph. In a bipartite graph, we have two different types of nodes (in our

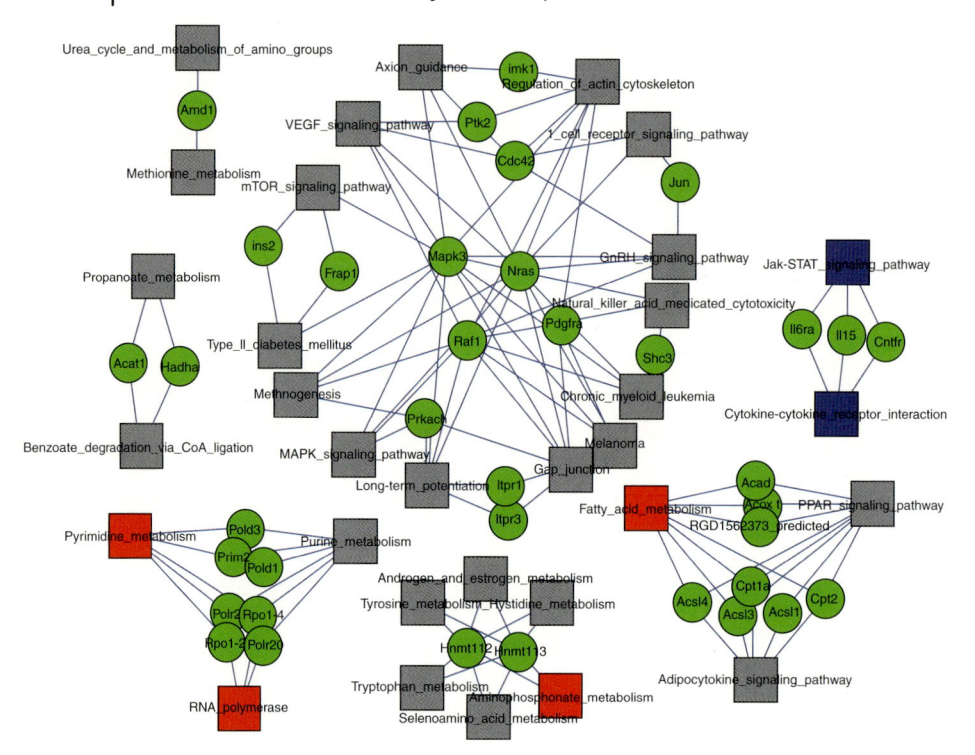

Figure 6.8 Bipartite graph of pathways and genes for the MYC dataset. Pathways are represented by squares and genes by circles. The isolated pathways have been removed so that the graph contains only the connected components of Figure 6.2, evidencing only the significant genes of the significant intersections. Red and blue squares indicate, respectively, significantly over-represented and under-represented pathways whereas the gray indicates not significant pathways. The green tone indicates the degree of pathway membership: from light green (connected to few pathways) to dark green (the hubs).

case, genes and pathways) that can be connected to each other but not between them directly. A one-type graph can be obtained by a "contraction" of the adjacency matrix, producing only a pathway–pathway network or a gene–gene network. The bipartite graph representation highlights the central role of "hub genes" emerging from the pathway network analysis, due to their statistical significance and also to their topological relevance, due to the positioning inside the network. In the MYC dataset (Figure 6.8) the Signaling subnetwork is strongly connected and presumably coordinated by a small number of genes such as MAP kinase III (*Mapk3*), neuroblastoma *ras* oncogene (*Nras*), v-*raf*-1 murine leukemia viral oncogene homolog 1 (*Raf1*), platelet-derived growth factor receptor-α (*Pdgfra*), and cell division cycle 42 homolog (*Cdc42*). The subnetwork containing the basic metabolic pathways (aminophosphonate, tryptophan and tyrosine metabolism, androgen and estrogen metabolism, etc.) shows at its intersections genes belonging to the family of protein

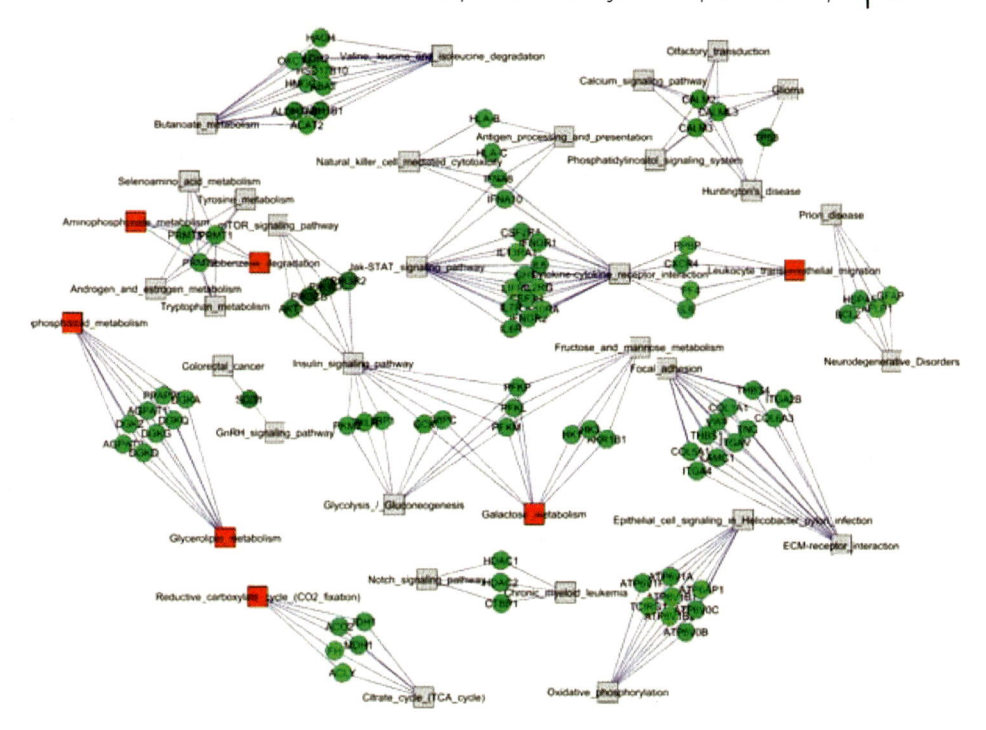

Figure 6.9 Bipartite graph of pathways and genes for the AML/ALL dataset. Pathways are represented by squares and genes by circles. The isolated pathways have been removed so that the graph contains only the connected components of Figure 6.3, evidencing only the significant genes of the significant intersections. Red and blue squares indicate, respectively, significantly over-represented and under-represented pathways, whereas the gray indicates not significant pathways. The green tone indicates the degree of pathway membership: from light green (connected to few pathways) to dark green (the hubs).

arginine methyltransferases (*Hrmt1l2*, *Hrmt1l3*) that are involved in histone modification and chromatin remodeling.

As far as the AML/ALL dataset (Figure 6.9) is concerned, in the subnetwork connecting signaling and metabolism, especially at the intersection between insulin and mTOR signaling pathways, some crucial genes emerge (*PIK3CA*, *PIK3CB*, *PIK3R2*, and *AKT1*).

Similarly to the MYC dataset, the subnetwork that contains some basic metabolic pathways (aminophosphonate, tryptophan and tyrosine metabolism, androgen and estrogen metabolism, etc.) shows hub genes belonging to the family of protein arginine methyltransferases (*PRMT1*, which is a homolog to the rat counterpart *Hrmt1l2* seen in the MYC dataset, *PRMT2*, and *PRMT5*).

It is worth noticing that in the subnetwork involving both calcium and phosphatidylinositol signaling it is possible to highlight some crucial genes, such as calmodulin and the well-known tumor protein p53 (*TP53*).

6.7
Discussion

A global picture of gene expression is greatly enhanced by the use of gene categorization and pathway analysis, but there can be several cases where this approach is not completely satisfactory. It may fail to capture the relationship between the categories and it may discard some important pathways or genes, because it does not take into account their relevance based on their central position. A typical case is that of genes that are at the interface among pathways (as in the case of hubs). With our method we try to overcome these limitations by assigning more relevance to the position occupied by a group of genes in a higher-level structure (i.e., the pathway network) in addition to their statistical significance alone.

The comparison between the pathways listed in Tables 6.3 and 6.4 with the corresponding networks of Figures 6.2 and 6.3 shows how the lists alone cannot grasp the complexity of pathway activation induced by gene expression changes; remarkably, most of the pathways in these networks are not listed in Tables 6.3 and 6.4.

For the MYC dataset, the biggest subnetwork links pathways that are involved in signaling processes (MAP kinases signaling pathway, VEGF signaling pathway, GnRH signaling pathway), and in structural reorganization, communication, and connections (regulation of actin cytoskeleton and gap junction). The bipartite graph shows how this subnetwork is strongly connected and coordinated by a small number of hub genes (*Mapk3*, *Nras*, *Raf1*, *Pdgfra*, *Cdc42*). All these genes are well-known proto-oncogenes involved in proliferation, regulation of growth, cell cycle progression control, and structural reorganization of the cell. They are major c-Myc downstream effectors and they are responsible of the profound effects that c-Myc exerts in cellular physiology: the downregulation of the connections among cells and the connections between cells and the extracellular matrix, cytoskeleton reorganization, and the induction of cell growth and proliferation [57]. The MYC network evidences also a strongly connected component related to basic metabolism, comprising both biosynthetic and catabolic pathways and a small subnetwork related to the synthesis of nucleic acids, which are known to be major targets of c-Myc that upregulates both energy metabolism and biosynthesis needed for growth and proliferation [48, 57].

For the AML/ALL dataset, the pathways network showed in Figure 6.3 evidences a subnetwork connecting signaling and metabolism that underlines how the regulation of energy metabolism may play a key role in the discrimination between the two types of leukemia. It is interesting to note that Insulin and mTOR signaling pathways are known to be involved in AML [58, 59], in particular the crucial genes (*PIK3CA*, *PIK3CB*, *PIK3R2*, and *AKT1*; see Figure 6.7) in the intersection between these two pathways have been recently pointed out as promising novel targets for AML therapy [60].

Among the relevant genes extracted by our method we can notice the well-known tumor protein p53 (*TP53*), involved in a wide variety of cancers, found in the subnetwork involving calcium and phosphatidylinositol signaling systems.

In both datasets the subnetworks containing basic metabolic pathways (aminophosphonate, tryptophan and tyrosine metabolism, androgen and estrogen metabolism, etc.) show at their intersections genes belonging to the family of protein arginine methyltransferases that are involved in histone modification and chromatin remodeling. They have been recently pointed out to have a major role in lymphoid tumors, leukemia, and more generally in cancer [61, 62].

The role of epigenetic modification in cancer induction and differentiation is gaining experimental support and is giving new perspectives on complex cellular processes: our results provide further evidence for the role of c-Myc in promoting oncogenesis through chromatin remodeling [51].

6.8 Conclusions

Our analysis method tries to combine high-throughput experimental procedures and advanced data processing in a general systems biology framework to discover pathway network changes following variation of cellular phenotypes. The use of known pathways, such as those described in the KEGG database, is motivated by the clarity of their biological interpretation, but our method can be applied also to custom-defined pathways or to groups of genes obtained from other methods. This approach can be further generalized by considering different statistical methods for assessing single-gene significance or the significance of single network modules.

This approach leads to an improved biological insight of the single-gene and single-pathway statistical analysis results by adding topological information (BC) related to their intercommunication, represented by a bipartite network. It may improve the comparability of microarray studies, both between different cell types and different perturbations by considering changes in pathway networks instead of single genes.

Moreover, this network-based method highlights the existence of "focal areas" or hub genes that are more likely found at the intersection between pathways. In this way it is possible to reconsider genes on the basis of their central role in the network and not only for their statistical significance. This can be of great importance also considering that the most central genes typically are subjected to very small changes that could be hardly detectable by any single-gene statistical analysis, but can anyway exert great biological effects due to their central role in pathway interconnections and communication. Recently, other authors have been developing methods trying to extract from the data genes that can be biologically relevant even if they are not top ranking in terms of statistical significance [63].

The problem of assessing pathway relevance is considered in a different way by Draghici [64]. In his paper the biological relevance of each pathway is scored both on the basis of a statistical significance test (pathway enrichment analysis) and on other parameters referred to the position of single genes in the pathway. Our method for pathway relevance can be seen as a top-down approach (from a KEGG-based network to single pathways and genes), as much as the method by Draghici is a bottom-up one (from genes to pathways).

It is worthwhile noting how similar problems appear in different branches of science and how similar phenomena are observed at several scales, from human social behavior to single-cell internal signaling, in a self-similar manner reminiscent of the concept of fractality that broke through biology about 20 years ago. A unifying approach as given by network theory allows a "knowledge transfer" between different systems at different levels of detail, from single cells to communities of living organisms.

A deeper insight into network-based mechanisms inside living organisms (from single-cell organization of gene and protein expression, referred to as *interactome*, to multicellular communication strategies like for the immune system) could be helpful in improving the efficiency and robustness of higher-level complex systems, like stock exchange dynamics, the welfare of a nation, immunization strategies, information spreading through society, or (and this could be bad news for common people) marketing and political strategies. Moreover, the possibility of observing biological phenomena from a novel perspective is one of the strongest points in favor of systems biology approaches to biological systems and more generally for network theory methods applied to life-related sciences.

In this chapter we have tried to survey different results obtained by the application of a network paradigm to complex biological systems. Such results are quite varied, both regarding the field of application and the level of detail inside living organisms. The hope is that from this bulk of results a unifying theory will emerge, by exploiting the numerous links between them, leading to a better comprehension of complex systems as a whole and, in particular, when they live far from equilibrium – a very common situation for life sciences, but not yet so common for classical physical sciences.

References

1 van Kampen, N.G. (2007) *Stochastic Processes in Physics and Chemistry*, 3rd edn, Elsevier, Amsterdam.

2 Friedman, N., Cai, L., and Xie, X.S. (2006) Linking stochastic dynamics to population distribution: an analytical framework of gene expression. *Phys. Rev. Lett.*, **97**, 168302.

3 Cai, L., Friedman, N., and Xie, X.S. (2006) Stochastic protein expression in individual cells at the single molecule level. *Nature*, **440**, 358.

4 Holme, P. (2006) Newman nonequilibrium phase transition in the coevolution of networks and opinions. *Phys. Rev. E*, **74**, 056108.

5 Gross, T. and Blasius, B. (2008) Adaptive coevolutionary networks: a review. *J. R. Soc. Interface*, **5**, 259–271.

6 Remondini, D., O'Connell, B., Intrator, N., Sedivy, J.M., Neretti, N., Castellani, G.C., and Cooper, L.N. (2005) Targeting c-Myc-activated genes with a correlation method: detection of global changes in large gene expression network dynamics. *Proc. Natl. Acad. Sci. USA*, **102**, 6902–6906.

7 Milo, R., Shen-Orr, S., Itzkovitz, S., Kashtan, N., Chklovskii, D., and Alon, U. (2002) Network motifs: simple building blocks of complex networks. *Science*, **298**, 824–827.

8 Remondini, D, Salvioli, S., Francesconi, M., Pierini, M., Mazzatti, D.J., Powell, J.R., Zironi, I., Bersani, F., Castellani, G., and Franceschi, C. (2010) Complex patterns of gene expression in human T cells during *in vivo* aging. *Mol. Biosyst.*, **6**, 1983–1992.

9 Soukas, A., Cohen, P., Socci, N.D., and Friedman, J.M. (2000) Leptin-specific patterns of gene expression in white adipose tissue. *Genes Dev.*, **14**, 963–980.

10 Tamayo, P., Slonim, D., Mesirov, J., Zhu, Q., Kitareewan, S., Dmitrovsky, E., Lander, E.S., and Golub, T.R. (1999) Interpreting patterns of gene expression with self-organizing maps: methods and application to hematopoietic differentiation. *Proc. Natl. Acad. Sci. USA*, **96**, 2907–2912.

11 Tavazoie, S., Hughes, J.D., Campbell, M.J., Cho, R.J., and Church, G.M. (1999) Systematic determination of genetic network architecture. *Nat. Genet.*, **22**, 281–285.

12 Toronen, P., Kolehmainen, M., Wong, G., and Castren, E. (1999) Analysis on gene expression data using self-organizing maps. *FEBS Lett.*, **451**, 142–146.

13 Alizadeh, A.A., Eisen, M.B., Davis, R.E., Ma, C., Lossos, I.S., Rosenwald, A., Boldrick, J.C., Sabet, H., Tran, T., Yu, X. et al. (2000) Distinct types of diffuse large B-cell lymphoma identified by gene expression profiling. *Nature*, **403**, 503–511.

14 Bittner, M., Meltzer, P., Chen, Y., Jiang, Y., Seftor, E., Hendrix, M., Radmacher, M., Simon, R., Yakhini, Z., Ben-Dor, A. et al. (2000) Molecular classification of cutaneous malignant melanoma by gene expression profiling. *Nature*, **406**, 536–540.

15 Eisen, M.B., Spellman, P.T., Brown, P.O., and Botstein, D. (1998) Cluster analysis and display of genome-wide expression patterns. *Proc. Natl. Acad. Sci. USA*, **95**, 14863–14868.

16 Ewing, R.M., Ben Kahla, A., Poirot, O., Lopez, F., Audic, S., and Claverie, J.M. (1999) Large-scale statistical analyses of rice ESTs reveal correlated patterns of gene expression. *Genome Res.*, **9**, 950–959.

17 Ross, D.T., Scherf, U., Eisen, M.B., Perou, C.M., Rees, C., Spellman, P., Iyer, V., Jeffrey, S.S., Van de Rijn, M., Waltham, M. et al. (2000) Systematic variation in gene expression patterns in human cancer cell lines. *Nat. Genet.*, **24**, 227–235.

18 Spellman, P.T., Sherlock, G., Zhang, M.Q., Iyer, V.R., Anders, K., Eisen, M.B., Brown, P.O., Botstein, D., and Futcher, B. (1998) Comprehensive identification of cell cycle-regulated genes of the yeast *Saccharomyces cerevisiae* by microarray hybridization. *Mol. Biol. Cell*, **9**, 3273–3297.

19 Chen, T., He, H.L., and Church, G.M. (1999) Modeling gene expression with differential equations. *Pac. Symp. Biocomput.*, 29–40.

20 Friedman, N., Linial, M., Nachman, I., and Pe'er, D. (2000) Using Bayesian networks to analyze expression data. *J. Comput. Biol.*, **7**, 601–620.

21 Matsuno, H., Doi, A., Nagasaki, M., and Miyano, S. (2000) Hybrid Petri net representation of gene regulatory network. *Pac. Symp. Biocomput.*, 341–352.

22 Akutsu, T., Miyano, S., and Kuhara, S. (2000) Inferring qualitative relations in genetic networks and metabolic pathways. *Bioinformatics*, **16**, 727–734.

23 Szallasi, Z. and Liang, S. (1998) Modeling the normal and neoplastic cell cycle with "realistic Boolean genetic networks": their application for understanding carcinogenesis and assessing therapeutic strategies. *Pac. Symp. Biocomput.*, 66–76.

24 Butte, A.J. and Kohane, I.S. (1999) Unsupervised knowledge discovery in medical databases using relevance networks. *Proc. AMIA Symp.*, 711–715.

25 Butte, A.J., Tamayo, P., Slonim, D., Golub, T.R., and Kohane, I.S. (2000) Discovering functional relationships between RNA expression and chemotherapeutic susceptibility using relevance networks. *Proc. Natl. Acad. Sci. USA*, **97**, 12182–12186.

26 Camacho, D., de la Fuente, A., and Mendes, P. (2005) The origin of correlations in metabolomics data. *Metabolomics*, **1**, 53–63.

27 Martins, A.M., Camacho, D., Shuman, J., Sha, W., Mendes, P., and Shulaev, V.

(2004) A systems biology study of two distinct growth phases of *Saccharomyces cerevisiae* cultures. *Curr. Genomics*, **5**, 649–663.

28 Kishino, H. and Waddell, P.J. (2000) Correspondence analysis of genes and tissue types and finding genetic links from microarray data. *Genome Inform. Ser. Workshop Genome Inform.*, **11**, 83–95.

29 Toh, H. and Horimoto, K. (2002) Inference of a genetic network by a combined approach of cluster analysis and graphical Gaussian modeling. *Bioinformatics*, **18**, 287–297.

30 Waddell, P.J. and Kishino, H. (2000) Cluster inference methods and graphical models evaluated on NCI60 microarray gene expression data. *Genome Inform. Ser. Workshop Genome Inform.*, **11**, 129–140.

31 de la Fuente, A., Bing, N., Hoeschele, I., and Mendes, P. (2004) Discovery of meaningful associations in genomic data using partial correlation coefficients. *Bioinformatics*, **20**, 3565–3574.

32 Magwene, P.M. and Kim, J. (2004) Estimating genomic coexpression networks using first-order conditional independence. *Genome Biol.*, **5**, R100.

33 Wille, A., Zimmermann, P., Vranova, E., Furholz, A., Laule, O., Bleuler, S., Hennig, L., Prelic, A., von Rohr, P., Thiele, L. *et al.* (2004) Sparse graphical Gaussian modeling of the isoprenoid gene network in *Arabidopsis thaliana*. *Genome Biol.*, **5**, R92.

34 Dobra, A., Hans, C., Jones, B., Nevins, J.R., and West, M. (2004) Sparse graphical models for exploring gene expression data. *J. Multivariate Analysis*, **90**, 196–212.

35 Schafer, J. and Strimmer, K. (2005) An empirical Bayes approach to inferring large-scale gene association networks. *Bioinformatics*, **21**, 754–764.

36 Hekstra, D., Taussig, A.R., Magnasco, M., and Naef, F. (2003) Absolute mRNA concentrations from sequence-specific calibration of oligonucleotide arrays. *Nucleic Acids Res.*, **31**, 1962–1968.

37 Affymetrix (2005) Guide to Probe Logarithmic Intensity Error (PLIER) Estimation. Technical Note, Affymetrix, Santa Clara, CA.

38 Irizarry, R.A., Hobbs, B., Collin, F., Beazer-Barclay, Y.D., Antonellis, K.J., Scherf, U., and Speed, T.P. (2003) Exploration, normalization, and summaries of high density oligonucleotide array probe level data. *Biostatistics*, **4**, 249–264.

39 Fortunel, N.O., Otu, H.H., Ng, H.H., Chen, J., Mu, X., Chevassut, T., Li, X., Joseph, M., Bailey, C., Hatzfeld, J.A. *et al.* (2003) Comment on "'Stemness': transcriptional profiling of embryonic and adult stem cells" and "a stem cell molecular signature". *Science*, **302**, 393.

40 Subramanian, A., Tamayo, P., Mootha, V.K., Mukherjee, S., Ebert, B.L., Gillette, M.A., Paulovich, A., Pomeroy, S.L., Golub, T.R., Lander, E.S. *et al.* (2005) Gene set enrichment analysis: a knowledge-based approach for interpreting genome-wide expression profiles. *Proc. Natl. Acad. Sci. USA*, **102**, 15545–15550.

41 Manoli, T., Gretz, N., Gröne, H.J., Kenzelmann, M., Eils, R., and Brors, B. (2006) Group testing for pathway analysis improves comparability of different microarray datasets. *Bioinformatics*, **22**, 2500–2506.

42 Ashburner, M., Ball, C.A., Blake, J.A., Botstein, D., Butler, H., Cherry, J.M., Davis, A.P., Dolinski, K., Dwight, S.S., Eppig, J.T. *et al.* (2000) Gene Ontology: tool for the unification of biology. The Gene Ontology Consortium. *Nat. Genet.*, **25**, 25–29.

43 Kanehisa, M. and Goto, S. (2000) KEGG: Kyoto Encyclopedia of Genes and Genomes. *Nucleic Acids Res.*, **28**, 27–30.

44 Beissbarth, T. and Speed, T.P. (2004) GOstat: find statistically over-represented Gene Ontologies within a group of genes. *Bioinformatics*, **20**, 1464–1465.

45 Pandey, R., Guru, R.K., and Mount, D.W. (2004) Pathway Miner: extracting gene association networks from molecular pathways for predicting the biological significance of gene expression microarray data. *Bioinformatics*, **20**, 2156–2158.

46 Goeman, J.J., van de Geer, S.A., de Kort, F., and van Houwelingen, H.C. (2004) A global test for groups of genes: testing

association with a clinical outcome. *Bioinformatics*, **20**, 93–99.
47 Mootha, V.K., Lindgren, C.M., Eriksson, K.F., Subramanian, A., Sihag, S., Lehar, J., Puigserver, P., Carlsson, E., Ridderstråle, M., Laurila, E. et al. (2003) PGC-1alpha-responsive genes involved in oxidative phosphorylation are coordinately downregulated in human diabetes. *Nat. Genet.*, **34**, 267–273.
48 O'Connell, B.C., Cheung, A.F., Simkevich, C.P., Tam, W., Ren, X., Mateyak, M.K., and Sedivy, J.M. (2003) A large scale genetic analysis of c-Myc-regulated gene expression patterns. *J. Biol. Chem.*, **278**, 12563–12573.
49 Gershman, B., Hang, L., Puig, O., Tatar, M., and Garofalo, R.S. (2007) High resolution dynamics of the transcriptional response to nutrition in *Drosophila*: a key role for dFOXO. *Physiol. Genomics*, **29**, 24–34.
50 Remondini, D., Neretti, N., Sedivy, J., Franceschi, C., Milanesi, L., Tieri, P., and Castellani, G.C. (2007) Networks from gene expression time series: characterization of correlation patterns. *Int. J. Bifurcat. Chaos*, **17**, 2477–2483.
51 Knoepfler, P.S., Zhang, X.Y., Cheng, P.F., Gafken, P.R., McMahon, S.B., and Eisenman, R.N. (2006) Myc influences global chromatin structure. *EMBO J.*, **25**, 2723–2734.
52 Golub, T.R., Slonim, D.K., Tamayo, P., Huard, C., Gaasenbeek, M., Mesirov, J.P., Coller, H., Loh, M.L., Downing, J.R., Caligiuri, M.A. et al. (1999) Molecular classification of cancer: class discovery and class prediction by gene expression monitoring. *Science*, **286**, 531–537.
53 Lezon, T.R., Banavar, J.R., Cieplak, M., Maritan, A., and Fedoroff, N.V. (2006) Using the principle of entropy maximization to infer genetic interaction networks from gene expression patterns. *Proc. Natl. Acad. Sci. USA*, **103**, 19033–19038.
54 Hooper, S.D., Boué, S., Krause, R., Jensen, L.J., Mason, C.E., Ghanim, M., White, K.P., Furlong, E.E., and Bork, P. (2007) Identification of tightly regulated groups of genes during *Drosophila melanogaster* embryogenesis. *Mol. Syst. Biol.*, **3**, 72.
55 Draghici, S., Khatri, P., Martins, R.P., Ostermeier, G.C., and Krawetz, S.A. (2003) Global functional profiling of gene expression. *Genomics*, **81**, 98–104.
56 Freeman, L.C. (1977) A set of measures of centrality based on betweenness. *Sociometry*, **40**, 35–41.
57 Dang, C.V., O'Donnell, K.A., Zeller, K.I., Nguyen, T., Osthus, R.C., and Li, F. (2006) The c-Myc target gene network. *Semin. Cancer Biol.*, **16**, 253–264.
58 Doepfner, K.T., Boller, D., and Arcaro, A. (2007) Targeting receptor tyrosine kinase signaling in acute myeloid leukemia. *Crit. Rev. Oncol. Hematol.*, **63**, 215–230.
59 Récher, C., Dos Santos, C., Demur, C., and Payrastre, B. (2005) mTOR, a new therapeutic target in acute myeloid leukemia. *Cell Cycle*, **4**, 1540–1549.
60 Martelli, A.M., Tazzari, P.L., Evangelisti, C., Chiarini, F., Blalock, W.L., Billi, A.M., Manzoli, L., McCubrey, J.A., and Cocco, L. (2007) Targeting the phosphatidylinositol 3-inase/Akt/mammalian target of rapamycin module for acute myelogenous leukemia therapy: from bench to bedside. *Curr. Med. Chem.*, **14**, 2009–2023.
61 Cheung, N., Chan, L.C., Thompson, A., Cleary, M.L., and So, C.W. (2007) Protein arginine-methyltransferase-dependent oncogenesis. *Nat. Cell Biol.*, **9**, 1208–1215.
62 Pal, S., Baiocchi, R.A., Byrd, J.C., Grever, M.R., Jacob, S.T., and Sif, S. (2007) Low levels of miR-92b/96 induce PRMT5 translation and H3R8/H4R3 methylation in mantle cell lymphoma. *The EMBO J.*, **26**, 3558–3569.
63 Chuang, H.Y., Lee, E., Liu, Y.T., Lee, D., and Ideker, T. (2007) Network-based classification of breast cancer metastasis. *Mol. Syst. Biol.*, **3**, 140.
64 Draghici, S., Khatri, P., Tarca, A.L., Amin, K., Done, A., Voichita, C., Georgescu, C., and Romero, R. (2007) A systems biology approach for pathway level analysis. *Genome Res.*, **17**, 1537–1545.

7
Gene Regulatory Networks Inference: Combining a Genetic Programming and H_∞ Filtering Approach

Lijun Qian, Haixin Wang, and Xiangfang Li

7.1
Introduction

One central problem in biology is understanding the regulation of gene expression under different conditions. A gene regulatory network is a collection of DNA segments in a cell that interact with each other and with other substances in the cell, thereby governing the transcription of genes. The correct inference of gene regulatory networks plays a critical role in understanding biological regulation in phenotypic determination and it can affect advanced genome-based therapeutics. In light of the recent development of high-throughput microarray technology, it becomes possible to dissect transcriptional regulatory networks at the genomics level, which are complex and nonlinear in nature. Specifically, the increasing existence of microarray time-series data makes possible the characterization of dynamic nonlinear regulatory interactions among genes. The modeling, analysis, and control of gene regulatory networks are critical for studying cancer evolution, and may serve as the basis for developing new regulatory therapies [1].

As gene regulatory network models are difficult to deduce solely by means of experimental techniques, computational and mathematical methods are indispensable. Owing to the very large number of genes that need to be studied, the relatively small number of datasets available, the noise in the data, and the different natures of the distinct data types, network inference presents great challenges [2].

Much research has been done on gene regulatory network modeling by *linear* differential/difference equations using time-series data (e.g., [3–10], just to name a few). The basic idea is to approximate the combined effects of different genes by means of a weighted sum of their expression levels. In [7], a connectionist model is used to model small gene networks operating in the blastoderm of *Drosophila*. In [3], the concentrations of mRNA and protein are modeled by linear differential equations. A simple form of linear additive functions is suggested by [4], where $dx_i/dt = \sum_{j=1}^{n} w_{ij} x_j$. The degradation rate of gene i's mRNA and environmental effects are assumed to be incorporated in the parameters w_{ij}, and their influence on gene i's expression level x_i is assumed to be linear. A method to obtain a continuous

---

Applied Statistics for Network Biology: Methods in Systems Biology, First Edition.
Edited by M. Dehmer, F. Emmert-Streib, A. Graber, and A. Salvador.
© 2011 Wiley-VCH Verlag GmbH & Co. KGaA. Published 2011 by Wiley-VCH Verlag GmbH & Co. KGaA.

linear differential equation model from sampled time-series data is proposed in [9]. For added biological realism (all concentrations get saturated at some point in time), a sigmoid (squashing) function may be included into the equation. It has been shown that this sort of quasilinear model can be solved by first applying the inverse of the squashing function [5].

In our study, a gene regulatory network is modeled by continuous nonlinear ordinary differential equations (ODEs). Compared to linear models, identification of the nonlinear differential equation model is computationally more intensive and can require more data; however, the range of nonlinear behaviors exhibited by gene regulatory networks can be more thoroughly understood with nonlinear differential equations. In addition, well-established dynamical systems theory is available to characterize the dynamics produced by these models. When more time-series data become available as a result of advances in microarray or other technologies, and assuming continued improvement in computational capability, it can be expected that continuous nonlinear dynamic models will play a critical role in revealing complicated gene behavior.

In general, modeling gene regulatory networks is a nonlinear identification problem. In this chapter, we provide a framework to infer the proposed nonlinear ODE model with noise using time-series data, where genetic programming (GP) and H_∞ filtering are applied. Both synthetic data and experimental data from microarray measurements are used to evaluate the proposed method. Note that although the proposed method is tested only using polynomials as the nonlinear terms, it is expected that it should perform similarly well for other choices of nonlinear terms in the proposed model, dependent of course on sufficient data for more complex nonlinear models.

The remainder of the chapter is organized as follows: Section 7.2 provides a background on noise in biological networks and a review of various modeling methods; the proposed framework and the iterative algorithm are illustrated in Section 7.3; simulation results are given in Section 7.4; and Section 7.5 presents some concluding remarks.

7.2
Background

7.2.1
Noise in Gene Expression

One of the most important properties in gene expression is the stochasticity. In other words, the gene expression process is noisy and fluctuant. Generally, noise sources can be partitioned into two categories [11–14]: *intrinsic noise* and *extrinsic noise*.

- Gene expression is a sequence of biochemical reactions that are inherently stochastic. Those biochemical reactions depend on the molecular events and the difference in the internal states of cells, such as random births, deaths, and

collisions of molecules [15]. The inherent stochasticity in the system is called *intrinsic noise*.
- Variabilities in factors external to the system also contribute the noise. The environment is complicated and the subtle environmental difference may result in fluctuations in gene expression. Those kinds of noise sources are referred to as *extrinsic noise*.

In general, the noise effects of gene expression are the joint effects of intrinsic noise and extrinsic noise.

The concept of intrinsic noise and extrinsic noise has been proved experimentally [13, 15], although it is difficult to distinguish intrinsic noise from extrinsic noise *in vivo* [11]. Experiments have shown that both intrinsic and extrinsic noise contribute substantially to the overall variations. For instance, rapid fluctuations in mRNA could be the source of intrinsic noise [16]. Another experiment showed that noise in the gene expression level caused fluctuations in the protein level in a clonal population of *Escherichia coli* [13]. Extrinsic noise may be the primary source of stochastic fluctuation in gene expression of certain cases; it is, for example, observed in budding yeast [16].

Intrinsic noise is stochastic and inherent in biochemical reactions such as transcription and translation. Its magnitude is proportional to the inverse of system size and its origin is often thermal [17]. Intrinsic noise is the stochastic events during the process of gene expression from the level of promoter binding to mRNA translation to protein degradation. In single gene expression, the intrinsic noise comes from fluctuations generated by stochastic promoter activation, promoter inactivation, and mRNA and protein production and delay [13].

Extrinsic noise is the difference between cells either in the local environment or in the concentration or activity of any factor that affects gene expression. Extrinsic noise is the consequence of the fluctuations in the amounts or states of other cellular components such as molecular species or RNA polymerase. It originates in the random variation of one or more of the externally set control parameters [17, 18]. Those kinds of noise vary from time to time, from cell to cell, and gene sequence such as the number of RNA polymerase or ribosomes, the stage in the cell cycle, the quantity of the protein, mRNA degradation machinery and the cell environment, signals abundance of polymerase, ribosomes growth, and division of the cell [11]. Extrinsic noise can be divided into two subcategories [13]: global noise (fluctuations in the rates of the basic reactions that affect expression of all genes) and gene- or pathway-specific extrinsic noise(e.g., fluctuations in the abundance of a particular transcription factor or stochastic events in a specific signal transduction pathway).

Noise or variation in the process of gene expression may contribute to cell and organism variability [13]. Both the magnitude and the frequency of the noise affect the consequence. Small changes in protein abundance may have dramatic effects on fitness if they persist long enough, whereas large fluctuations in abundance may not have any effect if they occur too frequently to affect a cellular process. The observation that the time scale for intrinsic noise fluctuations is much shorter than that for extrinsic noise suggests that extrinsic noise may affect cellular phenotype more strongly than intrinsic noise, at least in *E. coli* [13]. The presence of stochasticity in gene expression has been confirmed to result in noise in protein abundance, but other sources of noise may result in phenotype variability. Since cellular components interact with one another in complex regulatory networks, the fluctuation in the amount of even a single component may affect the performance of the whole system. The mechanisms through which a natural genetic network can operate reliably despite noisy environments and stochasticity in gene expression are not known, and remain a difficulty challenge in genetic network engineering [16].

7.2.2
Modeling of Gene Regulatory Networks with Noise

Gene expression data raise the possibility for a functional understanding of genome dynamics by means of mathematical modeling. Many models have been proposed in the literature for modeling gene regulatory networks, such as Boolean networks, probabilistic Boolean networks, Bayesian networks, linear additive regulation models, and neural networks [19–23]. In this section, we briefly review some of these models and propose a noisy nonlinear differential equation model for gene regulatory networks.

7.2.2.1 Boolean Networks Model with Noise
The Boolean networks model is a binary model [21, 24–26]. The basic assumption for this model is that a gene has two states: 1 for active and 0 for inactive. A Boolean function is used to describe the influence of other genes on a gene.

Define a set of genes $E = e_1, e_2, \ldots, e_N$, where $e_i \in \{0, 1\}$, $i = 1, 2, \ldots, N$, represents the state of the ith gene and N is the number of genes. The set of Boolean functions is defined as $B = b_1, b_2, \ldots, b_N$. Then the dynamics of a gene regulatory network is determined through a set of discrete equations:

$$e_i(t+1) = b_i[e_1(t), \quad e_2(t), \ldots, e_N(t)], \quad i = 1, 2, \ldots, N \tag{7.1}$$

In [27, 28], different noise levels are considered in the Boolean network model. Random noise is added to the binary data generated by the Boolean networks. The general equation is $e_i(t) = e_i(t) + \rho_\varepsilon$. ρ_ε is added noise. The Boolean network can reduce the error if the gene expression data has a lot of error inside. In [29], a Boolean network model with noise is proposed. The p_ε is defined as the probability of $e_i(t+1) \neq b_i[e_1(t), \quad e_2(t), \ldots, e_N(t)]$, $i = 1, 2, \ldots, N$. Synthetic data are tested in the model because the quality and quantity of the available real data are not sufficient for the proposed model.

7.2.2.2 Bayesian Networks Model with Noise
A Bayesian network is a graph model to estimate a complicated multivariate joint probability distribution through local probabilities [23, 30]. Figure 7.1 shows a $N=5$ nodes Bayesian network. The vertices represent genes or other components. They are random variables. The edges represent the conditional dependence relation and

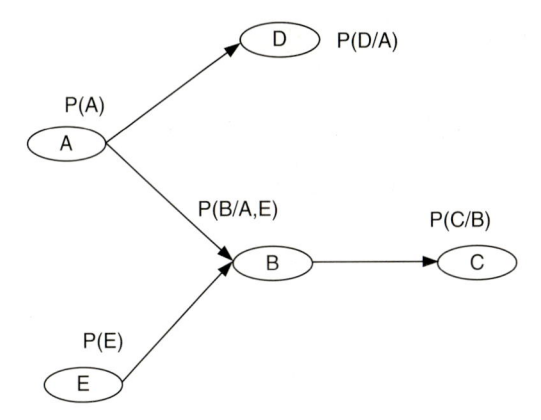

Figure 7.1 Simple Bayesian network structure.

interactions among genes. For the set of parent nodes of a node X_i, a conditional distribution $P(X_i|\text{Parents}(X_i))$ is defined, where $\text{Parents}(X_i)$ denotes the variables corresponding to the direct regulators of the ith node.

In the [28], the dynamic Bayesian network with external noise from time series is introduced. The influence of external noise on the systems dynamics is modeled by flipping the value of a gene X_i at each time step with a probability P_e. The result shows that increasing the value of external noise can reduce the overall performance.

7.2.2.3 Linear Additive Regulation Model with Noise

In this model the expression level of a gene at a certain point can be calculated by the weighted sum of the expression levels of all genes in the network at a previous time point [19]. It may be represented by ODEs:

$$\frac{de_i}{dt} = \sum_{j=1}^{N} w_{ij}e_j + \sum_{k=1}^{K} v_{ik}u_k + \beta_i \qquad (7.2)$$

where e_i is the gene expression level of the ith, N is the number of genes in the gene regulatory network, w_{ij} represents the effect of the jth gene on the ith gene (negative w_{ij} means inhibition while positive w_{ij} represents activation), u_k is the kth external (control) variable, v_{ik} represents the effect of the kth external variable on the ith gene, k is the number of external variables, and β is a bias term. In [31], the noise in the input data is considered in the linear additive regulation model. For each new input data, there is small amount of Gaussian noise with the same standard deviation added.

7.2.2.4 Neural Networks Model with Noise

The neural networks model uses differential equations to describe gene regulatory networks [20]:

$$\tau_i \frac{de_i}{dt} = f\left(\sum_{j=1}^{N} w_{ij}e_j + \sum_{k=1}^{K} v_{ik}u_k + \beta_i\right) - \lambda_i e_i(t) \qquad (7.3)$$

where f is usually a nonlinear function, such as a sigmoidal function, w is the weight matrix, u_k is the kth external (control) variable, v_{ik} represents the effect of the kth external variable on the ith gene, the constant λ_i represents the rate constant of degradation of the gene product i, and β_i represents an external input. Neural networks can be used to assimilate microarray data and construct gene regulatory networks [32]. In [33], a hierarchical Bayesian neural network model was introduced. Two kinds of noise are considered: independent parameters with Gaussian noise and correlated parameters with multivariate normal distribution. In [34], stochastic neural network models are presented for gene regulatory networks. The Poisson random noise is used to represent chance events in the process of synthesis. For expression data with normalized concentrations, exponential or normal random noise are used to generate the synthetic data.

7.2.3
Proposed Nonlinear ODE Model with Noise

In this study, we propose a nonlinear ODE model with noise. Assuming there are N genes of interest and x_i denotes the state (such as the microarray reading) of the ith gene, then the dynamics of the gene regulatory network may be modeled as:

$$\frac{dx_i}{dt} = f_i(x_1, x_2, \ldots, x_N) + v_i \quad i = 1, 2, \ldots, N \tag{7.4}$$

where the nonlinear functions f_i need to be determined from time-series microarray measurements. In this study, we assume the functions (f_i, $\forall i$) are in the form of polynomials:

$$f_i = \sum_{j=1}^{L_i} [(w_{ij} + \mu_{ij}) \, \Omega_{ij}(x_1, x_2, \ldots, x_N)] \quad i = 1, 2, \ldots, N \tag{7.5}$$

where L_i is the number of terms in f_i, w_{ij} are the parameters to be estimated, and $\Omega_{ij}(x_1, x_2, \ldots, x_N)$ is the jth component of the nonlinear function f_i. The polynomials are utilized as universal approximators. In order to mitigate the effect of "the curse of dimensionality," only second-degree polynomials are selected. Note that an advantage of using low-degree polynomial models is that even when there exists some model mismatch, these models may be sufficiently accurate to represent many real systems, and thus are widely utilized in practice [35]. We note that a similar gene regulatory network model has been adopted by [36], but without noise being included in the model. μ_{ij} and v_i are parameter noise and external noise, respectively, and it is assumed that the noise statistics such as the covariance matrices are unknown.

The noisy nature of gene regulatory networks is modeled explicitly in this study. The deterministic model (without noise) corresponds to the nominal case, while the various stochastic effects are included as noise disturbances. For example, there is considerable experimental evidence that indicates the presence of significant stochasticity in transcriptional regulation in both eukaryotes and prokaryotes [37]. The inherent stochasticity of biochemical processes (transcription and translation) is modeled as noise in the parameters (μ_{ij}), which corresponds to the "intrinsic noise" mentioned in the literature [11]. Other effects, such as those from genes not included in the microarray, the amount of RNA polymerase, levels of regulatory proteins, and the effects of mRNA and protein degradation, are modeled by the external noise (v_i) [11]. Previous work has modeled these noise types by Gaussian white noise processes [38]. If the noise statistics are known, then a Kalman filter can be applied to obtain the optimal estimates of the parameters [39]. On the contrary, a robust filter such as a H_∞ filter has to be used to obtain the optimal parameter estimates when the noise statistics such as the covariance matrices are *unknown*.

In this chapter, a two-step procedure is proposed to identify f_i: (i) GP is applied to determine the nonlinear terms and then (ii) the corresponding parameters associated with each term are estimated by H_∞ filtering.

The proposed model includes all the major characteristics of a gene regulatory network: it is nonlinear, dynamic, and noisy. The rationale behind the proposed model are 2-fold: (i) the proposed model is general and sufficiently flexible to include many well-known models and new models yet to be found, and (ii) the noisy nature of gene regulatory networks is modeled explicitly. The deterministic model (without noise) corresponds to the nominal case, while the various stochastic effects are included as noise disturbances. The inclusion of noise also enables the proposed model to provide interpretation of the fact that gene regulatory networks are robust to noise, by which it is meant that the relationships among genes are not greatly affected by small changes caused by noise.

The nonlinear functions (f_i, $\forall i$) need to be identified from time-series microarray measurements such that the identification error is minimized and the simplest model structure is selected. The criteria of selecting f_i are represented by a fitness function and modeling a gene regulatory network becomes a nonlinear optimization problem (minimization of fitness functions).

7.3
Methodology for Identification and Algorithm Description

The task of identifying gene regulatory networks may be considered as an optimization problem. The goal is to minimize the identification error and keep the model as simple as possible, which may be achieved by minimizing the following fitness function:

$$\text{fitness} = \sum_{i=1}^{N} \left[\eta_1 \sum_{k=1}^{M} \left(x_i(k) - x_i^{\text{tar}}(k) \right)^2 + \eta_2 \Gamma_i \right] \tag{7.6}$$

where M is the number of data points, x_i^{tar} is the target time series and x_i is the obtained time series given by the obtained differential equation represented by a GP individual, Γ_i is a penalty term, and $\eta_1 > 0$ and $\eta_2 > 0$ are weights on the estimation error and the model complexity, respectively.

Since it is a global nonlinear optimization problem, a nested optimization structure is adopted, where GP is applied to determine the nonlinear terms (global optimization) while a H_∞ filter is employed to estimate the corresponding parameters for each term (local optimization) in each iteration. Such a decomposition of the problem into a structural part solved by GP and a parameter optimization part solved by H_∞ filtering reduces the complexity significantly and speeds up convergence. The detailed procedures of the proposed iterative algorithm are illustrated in Figure 7.2.

In the study, we have developed a systematic method to infer a gene regulatory network represented by a nonlinear ODE with large dimensionality using rather short length time-series data. We rely on three aspects of our approach to address this issue:

i) The identification problem is decoupled into N subproblems with the ith subproblem focusing on the ith gene. As the time-series data of other genes are fixed (from measurements) when we are focusing on an individual gene, we can solve the identification problem one gene at a time. This approach makes the inference of large gene regulatory networks feasible. Similar decoupling procedures have been used in previous studies, such as the inference of S-system models [41, 42]. In the ith subproblem for the ith gene, the number of parameters needing to be estimated is L_i.

ii) According to a recent result by Sontag [43], $2r+1$ measurements are enough for identification of a set of differential equations with r unknown parameters (if experiments are designed properly, such as the one mentioned in [43]). The $2r+1$ bound is an upper bound. In our case, the minimum number of data points needed to estimate the L_i parameters within the ith subproblem is $2L_i + 1$. For example, the 17 data points in the Yeast Functional Genomics Database (YFGdb; http://yfgdb.princeton.edu/download.html) will allow us to estimate up to eight parameters in the ith equation. Usually, since the gene regulatory network tends to be a sparse network, we do not expect many terms on the right-hand side of the ODE (eight usually being more than enough).

iii) The H_∞ filtering provides optimal estimation with excellent convergence speed. Thus, relatively short-length time-series data is sufficient for the H_∞ filter to converge. In the simulations, we will show that the squared error of the estimation quickly converges close to zero.

7.3.1
GP

GP [44] is a type of evolutionary algorithm. All evolutionary algorithms work with a population of individuals, where each individual may be a solution of the optimization problem. GP operates on a tree structure, which is flexible enough to represent relationships efficiently. The leaves of a tree represent variables or constants, while

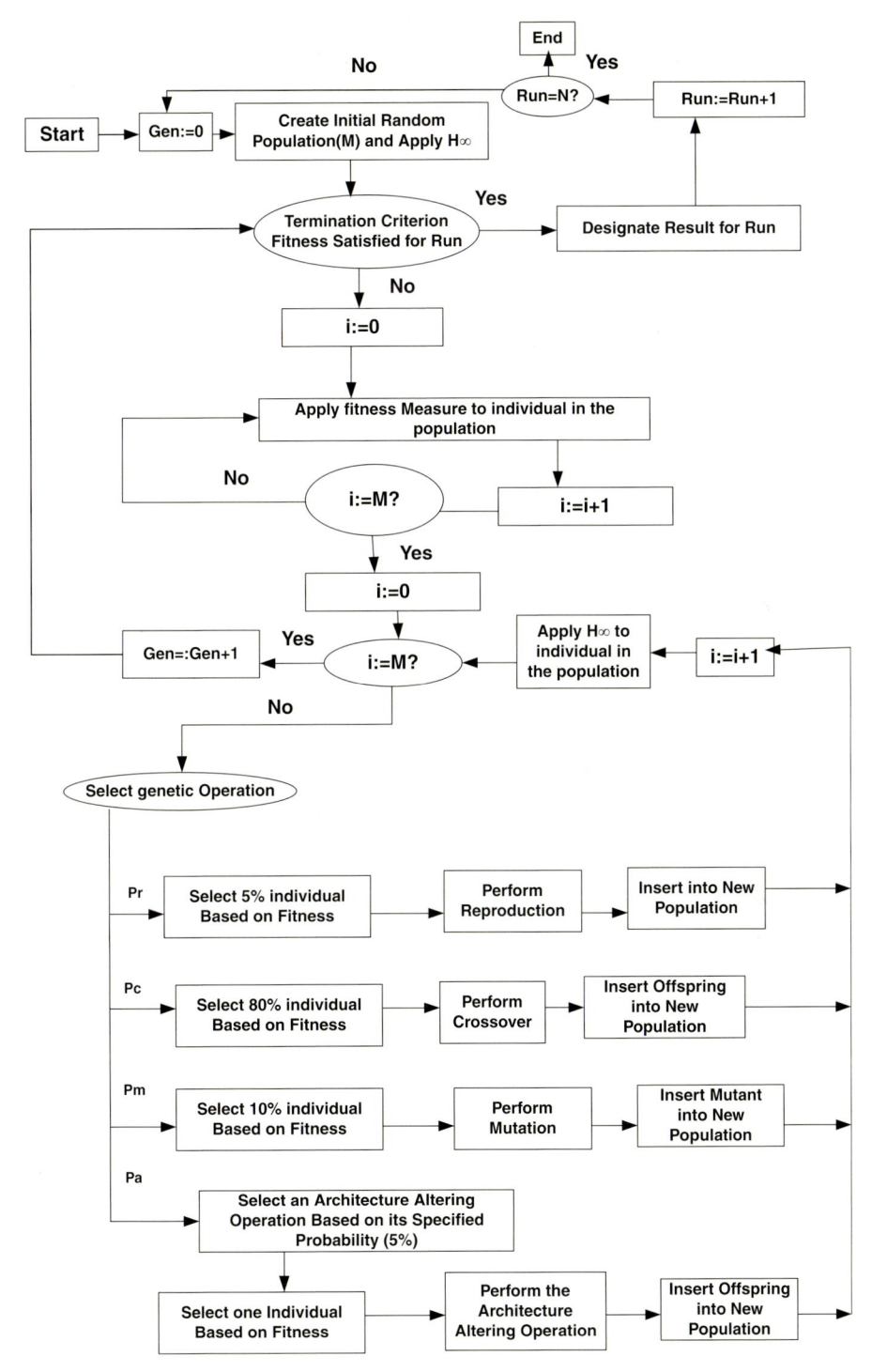

Figure 7.2 GP process with four operations: reproduction, crossover, mutation, and selection. H_∞ filtering is employed to compute the parameters for every generation. (Reproduced with permission from [40], © 2007 IEEE.)

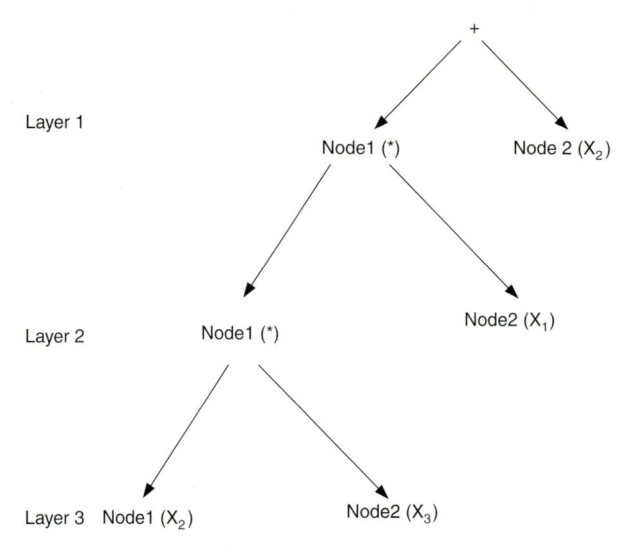

Figure 7.3 Example for the tree structure of the differential equations. (Reproduced with permission from [40], © 2007 IEEE.)

the other nodes implement operators. An example of a tree structure is shown in Figure 7.3, where two operations, multiplication (*) and addition/subtraction (+/−), are used. The corresponding equation is $dx_1/dt = x_2 + x_1 * x_2 * x_3$. Mutation and crossover operations may be performed to generate offspring. Selection of better performing individuals (with smaller fitness value, thus minimizing identification error while favoring the simplest model structure) ensures that the population evolves towards solving the optimization problem.

7.3.2
H_∞ Filter

The development of efficient linear estimation algorithms (e.g., the Kalman filter) has been based mainly on the minimization of the L_2-norm of the corresponding estimation error. This type of estimation assumes that the message-generating process has a known dynamics and that the exogenous inputs have known statistical properties. The well-known Kalman filter offers an optimal filtering algorithm when the system model parameters and power spectral density of the noise are known [45]. However, these assumptions may limit the application of the Kalman filter, because in many situations only approximate signal models are available and/or the statistics of the noise sources are not fully known or are unavailable. Furthermore, the Kalman filter may not be robust against parameter uncertainty of the models [45].

Recent developments in robust estimation have focused on the H_∞ filter [46–48]. The H_∞ filter is designed to guarantee that the operator relating the noise signals to the resulting estimation errors should possess a H_∞-norm less than a prescribed positive value (the noise suppression level) [49]. In the H_∞ filtering, the noise sources can be arbitrary signals with only a requirement of bounded noise. Since H_∞ filtering involves the minimization of the worst possible amplification of the error signal, the goal of the H_∞ filter is to provide an uniformly small estimation error for any processes and measurement noises and any initial states. It is shown that the H_∞ filter is more robust compared with the Kalman filter in terms of model uncertainty and gives better estimates (e.g., in speech processing [50]).

Let the L_i-dimensional vector $w(n)$ denote the state of the system (parameters to be estimated), then the process equation and the measurement equation are

$$w(n) = w(n-1) + \mu(n-1) \tag{7.7}$$

$$d(n) = C(n)w(n) + v(n) \tag{7.8}$$

where $d(n)$ can be calculated as:

$$d(n) = \frac{x(n+1) - x(n)}{\Delta t}$$

C contains all the modules (i.e., $C_i = [\Omega_{i1}\ \Omega_{i2}\ \cdots\ \Omega_{iL_i}]$).

Compared to the Kalman filter that minimizes the variance of the estimation error, the H_∞ filter provides a uniformly small estimation error $e(n) = w(n) - \hat{w}(n)$ with any process and measurement noise. The cost function is given as:

$$J = \frac{\sum_{n=1}^{M} ||w(n) - \hat{w}(n)||_{S(n)}^2}{||w(0) - \hat{w}(0)||_{P(0)^{-1}}^2 + \sum_{n=1}^{M}(||\mu(n)||_{Q_1^{-1}(n)}^2 + ||v(n)||_{Q_2^{-1}(n)}^2)}$$

where $P(0)$, Q_1^{-1}, Q_2^{-1}, and $S(n)$ are positive definite symmetric matrices chosen by the designer based on the performance requirement.

The cost function should be less than a prescribed level, $1/\theta$:

$$\sup J < \frac{1}{\theta}$$

where sup stands for least upper bound.

The implementation of the H_∞ filter is given by:

$$Z(n) = I - \theta S(n) P(n) + C^T(n) Q_2(n)^{-1} C(n) P(n) \tag{7.9}$$

$$K(n) = P(n) Z(n)^{-1} C^T(n) Q_2(n)^{-1} \tag{7.10}$$

$$\hat{w}(n+1) = \hat{w}(n) + K(n)(d(n) - C(n)\hat{w}(n)) \tag{7.11}$$

$$P(n+1) = P(n) Z(n)^{-1} + Q_1(n) \tag{7.12}$$

Note that although the gene regulatory network model itself is nonlinear, the parameter estimation problem is linear given the time-series data.

7.4 Simulation Evaluation

In the simulation study, the proposed scheme (GP plus H_∞ filter) is compared with the approach in [39], where GP and the Kalman filter were combined to deduce the differential equation model and the noise statistics are assumed to be known. Here, the H_∞ filter is used to get better estimates when noise statistics are unknown. We also apply our method to real measurement data from microarray experiments.

7.4.1 Synthetic Data

In this part of the simulation, we use data of a metabolic network, called the E-cell system (a part of the biological phospholipid pathway), that consists of three substances. This network can be approximated as:

$$\dot{x}_1 = -10.32 x_1 x_3$$
$$\dot{x}_2 = 9.72 x_1 x_3 - 17.5 x_2 \quad (7.13)$$
$$\dot{x}_3 = -9.7 x_1 x_3 + 17.5 x_2$$

Here, we apply Runge–Kutta method to calculate the synthetic data, and to add the intrinsic noise and the external noise (both are assumed to be Gaussian white noise).

Since there are three substances in the E-cell system, the tree structure should include a subset of the following terms on the right-hand side of the differential equation: $x_1, x_2, x_3, x_1 x_2, x_1 x_3, x_2 x_3, x_1^2, x_2^2,$ and x_3^2. One thousand individuals are first produced and ranked according to the fitness value; 5% of the individuals with the minimum fitness value are kept for the next generation, 80% individuals performed cross, 10% of individuals perform mutation, and the remaining 5% are for other operations.

We compare our algorithm with the approach in [39] for two different cases. In Case 1, the noise covariance is assumed to be known. The covariance matrices are

$$Q_1 = 10,\ Q_2 = \begin{bmatrix} 0.2 & 0 \\ 0 & 0.2 \end{bmatrix},\ Q_3 = \begin{bmatrix} 0.01 & 0 \\ 0 & 0.01 \end{bmatrix},\ R_1 = 20,\ R_2 = 0.2,\text{ and } R_3 = 0.01.$$

It is assumed that μ_{ij} and v_i are uncorrelated for all i and j. While in Case 2, instead of fixed covariance matrices, it is assumed that the covariance matrices are not known exactly, that is, the covariance matrices of μ_{ij} and v_i are $\tilde{Q}_i = (1+q_i)Q_i$ and $\tilde{R}_i = (1+r_i)R_i$, where q_i and r_i are random variables with $E[q_i] = E[r_i] = 0$, $E[q_i^2] = \sigma_{q_i}^2$, and $E[r_i^2] = \sigma_{r_i}^2$. Variances are given by $\sigma_{q_1}^2 = 10$, $\sigma_{q_2}^2 = 0.2$, $\sigma_{q_3}^2 = 10$, $\sigma_{r_1}^2 = 20$, $\sigma_{r_2}^2 = 0.3$, and $\sigma_{r_3}^2 = 15$. The random variables q_i and r_i are uncorrelated for all i.

The results are summarized in Table 7.1. It is observed that GP plus Kalman filter performs very well when the noise covariance is known, as expected. However, GP plus H_∞ filter outperforms GP plus Kalman filter when the noise covariance is not known exactly. This is also confirmed by the time series shown in Figure 7.4.

Table 7.1 Parameters obtained by GP plus Kalman filter and GP plus H_∞ when noise is present.

	True parameter values	GP + Kalman filter with exact covariance	GP + Kalman filter with uncertain covariance	GP + H_∞ with uncertain covariance
w_{11}	−10.32	−10.34	−7.12	−9.83
w_{21}	9.72	8.87	8.264	9.17
w_{22}	−17.5	−17.42	−16.14	−16.84
w_{31}	−9.72	−9.74	−10.43	−9.03
w_{32}	17.5	17.15	19.78	17.14

The coefficients in the E-Cell model are determined by H_∞ filtering. The convergence of the H_∞ filtering algorithm is an important issue when applying the H_∞ filter to the noisy inputs. The convergence of the H_∞ filter includes the convergence of the estimate $\hat{w}(n)$ and the convergence of the estimation error $\hat{e}(n)$. Figure 7.5 shows the convergence of the H_∞ filter.

7.4.2
Microarray Data

We consider time-series gene expression data corresponding to yeast protein synthesis. Here, the data for three genes (*HAP1*, *CYC7*, and *CYB2*) are picked

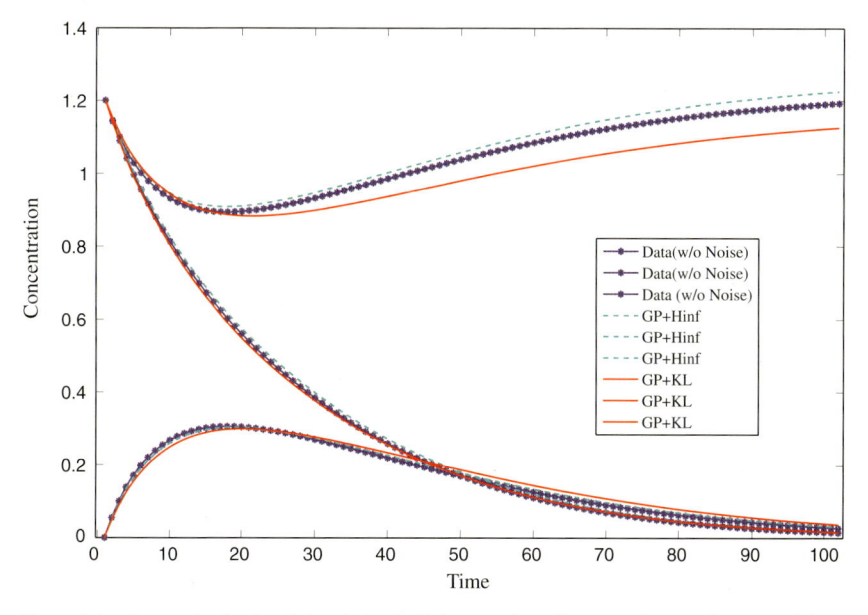

Figure 7.4 Time series for E-cell simulation by Kalman and H_∞ filtering: "data" is the original data without noise; "GP + KL" and "GP + Hinf" are the simulation data from GP plus Kalman filter and GP plus H_∞ filter, respectively. (Reproduced with permission from [40], © 2007 IEEE.)

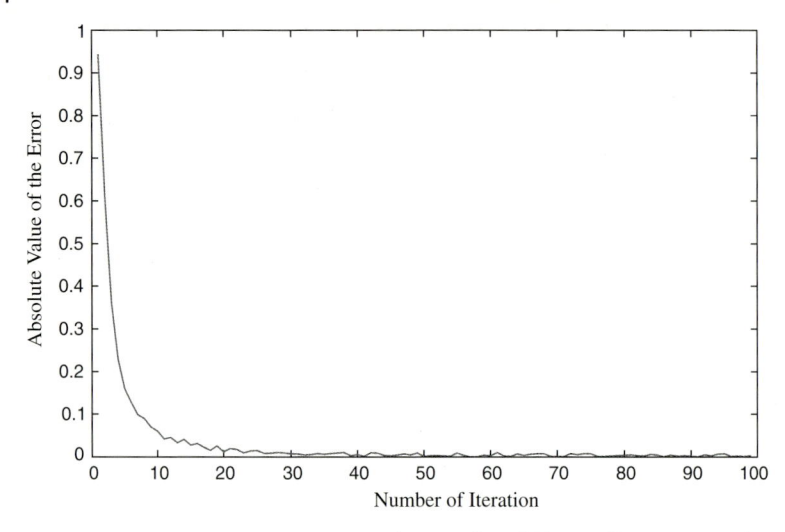

Figure 7.5 Experimental learning curves of the E-cell model by H_∞ filtering. (Reproduced with permission from [40], © 2007 IEEE.)

because the relations among them have been revealed by biological experiments. The trace of the time-series microarray measurement data from the YFGdb is used in this part of the simulation, where 17 sampling data points are provided for each gene by the experiments. The sampling data points are evenly spaced and the observation interval is 10 min. In the simulation, 1000 individuals are produced in each generation. One hundred generations are calculated to reach the minimum fitness values.

The following model is obtained by the proposed algorithm:

$$\begin{aligned}\dot{x}_1 &= (-0.006 + \mu_{11})x_1 + (0.00835 + \mu_{12})x_2 + v_1 \\ \dot{x}_2 &= (-0.3661 + \mu_{21})x_1 + (-0.476 + \mu_{22})x_2 + v_2 \\ \dot{x}_3 &= (-1.6124 + \mu_{31})x_1 + (0.45 + \mu_{32})x_2 + v_3\end{aligned} \quad (7.14)$$

The trajectories for *CYB2*, *HAP1*, and *CYC7* are shown in Figure 7.6.

The obtained relationships among genes are in agreement with biological experimental findings. For example, we observe that *HAP1* represses gene *CYC7* and *CYB2* activates *CYC7*. *HAP1* behaves as a repressor [51]. The interactions among *CYB2*, *HAP1*, and *CYC7* are shown in Figure 7.7.

7.5
Conclusions

Gene expression is an inherently noisy process, because of the stochasticity in molecular processes, such as during transcription and translation (intrinsic noise), and the effect of environmental noise (extrinsic noise). Consequently, the expression levels of genes and proteins in a given cell are continuously fluctuating. As molecular

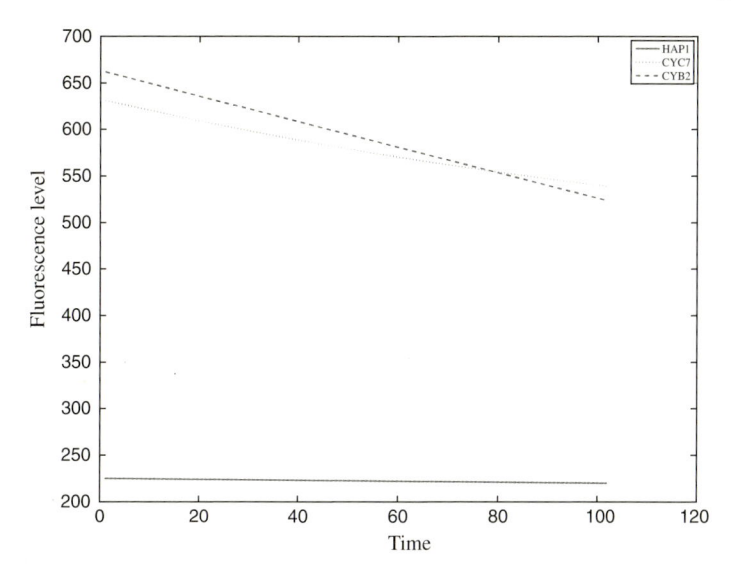

Figure 7.6 The trajectories for *CYB2*, *HAP1*, and *CYC7*. (Reproduced with permission from [40], © 2007 IEEE.)

noise can markedly affect cell behavior, cells have adapted a range of sophisticated mechanisms to control molecular noise [52]. Furthermore, microarray measurements usually contain rather large noise. To make network inference complicated, for all the above-mentioned noises, the statistics of the noise are usually unknown or at least not known exactly.

Apparently, noise cannot be ignored due to its importance in gene regulation. In order to deduce an accurate model of gene regulatory network from noisy datasets, in this study, noise is modeled explicitly in the proposed nonlinear differential equation model of the gene regulatory network. A joint GP and H_∞ filtering approach is proposed to infer the gene regulatory network, where H_∞ filtering provides optimal parameter estimations even when the statistics of the noise are unknown. Simulation results using both synthetic data and microarray datasets demonstrate the effectiveness of the proposed scheme.

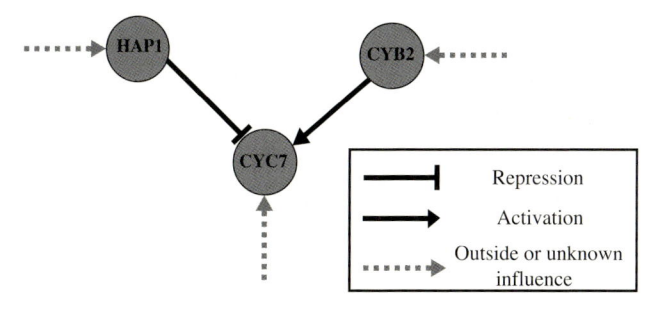

Figure 7.7 The interactions among *CYB2*, *HAP1*, and *CYC7*.

Appendix 7.A: Comparison between the Kalman Filter and H_∞ Filter

The problem of filtering involves estimating the states of a system using past measurements. The development of efficient linear estimation algorithms, such as the Kalman filter, has been based mainly on the minimization of the L_2-norm of the corresponding estimation error. This type of estimation assumes that the message-generating process has a known dynamics and that the exogenous inputs have known statistical properties. The well-known Kalman filter offers an optimal filtering algorithm when the system model parameters and power spectral density of the noise are known. Unfortunately, these assumptions limit the utility of minimum variance estimators in situations where the massage model and/or the noise descriptions are unknown or not known exactly. This fact led to a great interest in minimax estimation and robust filtering.

In the last decade, a measure that is different from the L_2-norm has been introduced for optimal control of linear systems. This measure is the so-called H_∞-norm. The H_∞-norm minimization can be applied not only in control, but also in estimation. The basic idea is to construct a filter that minimize the H_∞-norm of the operator (transfer function) that relates the estimation error with the noise input.

The results of H_∞ control theory of linear systems can be found in [53]. In [54], Bernstein and Haddad consider a steady-state filtering problem where they minimize an upper bound on the H_2-norm of the transfer function from the noise to estimation error while maintaining the H_∞-norm within a certain bound. A more complete solution of both continuous H_∞ filtering and smoothing is given in [55]. In [48], Shen and Deng considered the discrete H_∞ filter using game theory. Shaked and Theodor [46] present a nice tutorial of existing results in H_∞ optimal estimation.

The H_∞-norm of a bounded linear time-invariant operator $p = g(w)$ with the transfer function $G(s) \in RH_\infty$ is defined by:

$$\|G\|_\infty = \sup_{w \in R} \bar{\sigma}[G(jw)] \tag{7.A.1}$$

where $p, w \in L_2$. In single-input single-output case, $\|G\|_\infty$ is just the peak value of $G(jw)$ over all frequencies. The H_∞-norm can be interpreted as an induced norm:

$$\|G\|_\infty^2 = \sup_{\|w\|_2 \neq 0} \frac{\|p\|_2^2}{\|w\|_2^2} \tag{7.A.2}$$

In this brief review, we focus on the continuous-time H_∞ estimation problem. The reader can refer to [48] for all the details of its discrete-time counterpart.

Consider the following linear system:

$$\dot{x}(t) = A(t)x(t) + G(t)w(t) \tag{7.A.3}$$

$$y(t) = C(t)x(t) + v(t) \tag{7.A.4}$$

with the following situations for the initial condition:

i) Initial condition known; without loss of generality, we can assume that $x(0) = 0$.
ii) Initial condition not known exactly; but in the neighborhood of $x(0) = 0$ with 0 as the best *a priori* estimate of $x(0)$.

The states to be estimated are:

$$z(t) = L(t)x(t) \tag{7.A.5}$$

The problem is to obtain an estimate $\hat{z}(t)$ of $z(t)$ using the measurement y. First, let us define the performance index:

$$J_1 := \sup_{0 \neq w \in L_2} \frac{\|z-\hat{z}\|_2^2}{\|w\|_2^2} = \|T_{ed}\|_\infty^2 \; ; \; x(0) = 0 \tag{7.A.6}$$

$$J_2 := \sup_{0 \neq (x_0, w) \in R^n \times L_2} \frac{\|z-\hat{z}\|_2^2}{\|w\|_2^2 + x_0' R x_0} \; ; \quad x(0) = x_0 \tag{7.A.7}$$

where $R = R' > 0$ and T_{ed} denotes the transfer function from the disturbance to the estimation error. We are interested in the following two problems:

- (P1) Find an optimal estimation algorithm that minimize J_1 (J_2), in turn, $\|T_{ed}\|_\infty$, and obtain the resulting $\gamma_0 = \inf_T \|T_{ed}\|_\infty$.
- (P2) Given a scalar $\gamma > 0$, find whether $\gamma \geq \gamma_0$. If this is the case, find an estimator that achieves $\|T_{ed}\|_\infty \leq \gamma$.

The above (P2) is also called the suboptimal H_∞ filtering problem. The solution of (P1) can be approximated as closely as desired by iterating on the values of γ in (P2). The reasons that one usually looks for a solution to the suboptimal H_∞ filtering problem are: (i) finding an optimal solution is often both numerically and theoretically complicated, and (ii) practice it is often not necessary and sometimes even undesirable to design optimal filters, and it is usually much cheaper to obtain suboptimal filters that may also have other nice properties over optimal ones (e.g., lower bandwidth).

Theorem 7.1 *There exists an H_∞ filter such that $J_1 < \gamma^2$ if and only if there exists a symmetric matrix function $P(t)$ for $t \in [0, T]$ that is continuous, differentiable, and the following differential Riccati equation:*

$$\dot{P} = AP + PA^T - PC^TCP + \frac{1}{\gamma^2} PL^TLP + BB^T$$

with $P(0) = 0$. If $P(t)$ exists, one suitable filter for which $J_1 < \gamma^2$ is:

$$\dot{\hat{x}} = A\hat{x} + PC^T(y - C\hat{x}) \tag{7.A.}$$

$$\hat{z} = L\hat{x} \tag{7.A.}$$

with $\hat{x}(0) = 0$.

For the case that initial condition not known, we have the following theorem:

Theorem 7.2 *There exists a H_∞ filter such that $J_2 < \gamma^2$ if and only if there exists a symmetric matrix function $P(t) > 0$ for $t \in [0, T]$ that is continuous, differentiable, and* satisfies the following differential Riccati equation:

$$\dot{P} = AP + PA^T - PC^T CP + \frac{1}{\gamma^2} PL^T LP + BB^T \qquad (7.A.11)$$

with $P(0) = R^{-1}$. Moreover, if $P(t)$ exists, the same filter as in (7.A.9) and (7.A.10) with $P(t)$ now given by (7.A.11) achieves $J_2 < \gamma^2$.

In order to compare between Kalman and H_∞ filter, first let us recall the performance index of the Kalman filter:

$$J := E\{(x-\hat{x})^T M(x-\hat{x})\} = ||z-\hat{z}||_2^2 \qquad (7.A.12)$$

An equivalent statement of the H_∞ filter is: a filter is said to achieve a H_∞ estimation level of γ if the following cost function (performance index):

$$J_1 = ||z-\hat{z}||_2^2 - \gamma^2[||w||_2^2] \qquad (7.A.13)$$

$$J_2 = ||z-\hat{z}||_2^2 - \gamma^2[||w||_2^2 + x'_0 R x_0] \qquad (7.A.14)$$

satisfies $J_1 \leq 0$ and $J_2 \leq 0$ for all $w, v \in L_2[0, T]$ and $x_0 \in R^n$.

From above it is easy to see the differences between the Kalman and H_∞ filter. The Kalman filter only minimizes the estimation error variance, or the L_2-norm of the estimation error. The H_∞ filter also consider the disturbances (noises). The strength of the H_∞ estimation result lies in its superior performance at the peak error range and in its impressive robustness to parameter uncertainty. The minimax structure of the H_∞ filter decides its better performance in the worst case (maximum allowable noise).

The disadvantage of the H_∞ filter is that additional computational complexity is introduced when calculating the H_∞-norm of the transfer function T_{ed}. Note that γ should be calculated through iterations, especially when γ is required to be close to the optimal value γ_0.

Note

A previous version of this chapter was presented as an Invited Paper at the 2007 IEEE Statistical Signal Processing Workshop [40].

References

1 Dougherty, E.R., Datta, A., and Sima, C. (2005) Research issues in genomic signal processing. *IEEE Signal Proc. Mag.*, **22**, 46–68.

2 Sun, N. and Zhao, H. (2009) Reconstructing transcriptional regulatory networks through genomics data. *Stat. Methods Med. Res.*, **18**, 595–617.

3 Chen, T., He, H.L., and Church, G.M. (1999) Modeling gene expression with differential equations. *Pac. Symp. Biocomput.*, 29–40.

4 Yeung, M.K.S., Tegnăr, J., and Collins, J.J. (2002) Reverse engineering gene networks using singular value decomposition and robust regression. *Proc. Natl. Acad. Sci. USA*, **99**, 6163–6168.

5 Weaver, D.C., Workman, C.T., and Stormo, G.D. (1999) Modeling regulatory networks with weight matrices. *Pac. Symp. Biocomput.*, 112–123.

6 D'haeseleer, P., Wen, X., Fuhrman, S., and Somogyi, R. (1999) Linear modeling of mRNA expression levels during CNS development and injury. *Pac. Symp. Biocomput.*, 41–52.

7 Mjolsness, E., Sharp, D.H., and Reinitz, J. (1991) A connectionist model of development. *J. Theor. Biol.*, **152**, 429–453.

8 de Jong, H. (2002) Modeling and simulation of genetic regulatory systems: a literature review. *J. Comput. Biol.*, **9**, 67–103.

9 Tabus, I., Giurcaneanu, C.D., and Astola, J. (2004) Genetic networks inferred from time series of gene expression data. First International Symposium on Control, Communications and Signal Processing, Hammamet.

10 de Hoon, M.J.L., Imoto, S., Kobayashi, K., Ogasawara, N., and Miyano, S. (2003) Inferring gene regulatory networks from time-ordered gene expression data of *Bacillus subtilis* using differential equations. *Pac. Symp. Biocomput.*, 17–28.

11 Swain, P., Elowitz, M., and Siggia, E. (2002) Intrinsic and extrinsic contributions to stochasticity in gene expression. *Proc. Natl. Acad. Sci. USA*, **99**, 12795–12800.

12 Hooshangi, S. and Weiss, R. (2006) The effect of negative feedback on noise propagation in transcriptional gene networks. *Chaos*, **16**, 1–10.

13 Raser, J.M. and O'Shea, E.K. (2005) Noise in gene expression: Origins, consequences, and control. *Science*, **309**, 2010–2013.

14 Tao, Y. (2004) Intrinsic and external noise in an auto-regulatory genetic network. *J. Theor. Biol.*, **229**, 147–156.

15 Lei, J. (2009) Stochasticity in single gene expression with both intrinsic noise and fluctuation in kinetic parameters. *J. Theor. Biol.*, **256**, 485–492.

16 Karn, M., Elston, T.C., Blake, W.J., and Collins, J.J. (2005) Stochasticity in gene expression: from theories to phenotypes. *Nat. Rev. Genet.*, **6**, 451–464.

17 Hasty, J., Pradines, J., Dolnik, M., and Collins, J.J. (2000) Noise-based switches and amplifiers for gene expression. *Proc. Natl. Acad. Sci. USA*, **97**, 2075–2080.

18 Tao, Y. (2004) Intrinsic noise, gene regulation and steady-state statistics in a two-gene network. *J. Theor. Biol.*, **231**, 563–568.

19 D'haeseleer, P., Liang, S., and Somogy, R. (2000) Genetic network inference: from co-expression clustering to reverse engineering. *Bioinformatics*, **16**, 707–726.

20 Keedwell, E., Narayanan, A., and Savic, D. (2002) Modeling gene regulatory data using artificial neural networks. *Proc. IEEE*, **1**, 183–188.

21 Pal, R., Datta, A., and Dougherty, E.R. (2006) Optimal infinite horizon control for probabilistic Boolean networks. *IEEE Trans. Signal Proc.*, **54**, 2375–2387.

22 Ressom, H., Wang, D., Varghese, R.S., and Reynolds, R. (2003) Fuzzy logic-based gene regulatory network. IEEE International Conference on Fuzzy Systems, St Louis, MO.

23 Zou, M. and Conzen, S. (2005) A new dynamic Bayesian network (DBN) approach for identifying gene regulatory networks from time course microarray data. *Bioinformatics*, **21**, 71–79.

24 Lahdesmaki, H., Shmulevich, I., and Yli-Haerja, O. (2004) On learning gene regulatory networks under the Boolean network model. *Mach. Learn.*, **52**, 147–167.

25 Shmulevich, I., Dougherty, E., and Zhang, W. (2002) Gene perturbation and intervention in probabilistic Boolean networks. *Bioinformatics*, **18**, 1319–1331.

26 Zhou, X., Wang, X., and Dougherty, E. (2003) Construction of genomic networks using mutual-information clustering and reversible-jump Markov-chain-Monte-Carlo predictor design. *Signal Process.*, **83**, 745–761.

27 Kim, H., Lee, J.K., and Park, T. (2007) Boolean networks using the chi-square

test for inferring large-scale gene regulatory networks. *BMC Bioinformatics*, **8**, 1–15.
28. Streib, F.E., Dehmer, M., Bakir, G.H., and Mühlhäuser, M. (2005) Influence of noise on the inference of dynamic Bayesian networks from short time series. *Proc. World Acad. Sci.*, **10**, 70–74.
29. Akutsu, T., Miyano, S., and Kuhara, S. (2000) Inferring qualitative relations in genetic networks and metabolic pathways. *Bioinformatics*, **16**, 727–734.
30. Chen, T., Filkov, V., and Skiena, S.S. (2001) Identifying gene regulatory networks from experimental data. *Parallel Comput.*, **27**, 141–162.
31. D'haeseleer, P. and Fuhrman, S. (1999) Gene network inference using a linear, additive regulation model.
32. Xu, R., Hu, X., and Wunsch, D.C. (2004) Inference of genetic regulatory networks with recurrent neural network models. *Int. Conf. of IEEE EMBS*, 2905–2908.
33. Liang, Y. and Kelemen, A.G. (2004) Hierarchical Bayesian neural network for gene expression temporal patterns. *Stat. Appl. Genet. Mol. Biol.*, **3**, 1–24.
34. Tian, T. and Burrage, K. (2003) Stochastic neural network models for gene regulatory networks. *IEEE Evol. Comput.*, **1**, 162–169.
35. Nelles, O. (2001) *Nonlinear System Identification*, Springer, Berlin.
36. Ando, S., Sakamoto, E., and Iba, H. (2002) Evolutionary modeling and inference of gene network. *Inform. Sci.*, **145**, 237–259.
37. Kepler, T. and Elston, T. (2001) Stochasticity in transcriptional regulation: origins, consequences, and mathematical representations. *Biophys. J.*, **81**, 3116–3136.
38. Hasty, J., Pradines, J., Dolnik, M., and Collins, J.J. (2000) Noise-based switches and amplifiers for gene expression. *Proc. Natl. Acad. Sci. USA*, **97**, 2075–2080.
39. Wang, H., Qian, L., and Dougherty, E. (2006) Inference of gene regulatory networks using genetic programming and Kalman filter. IEEE International Workshop on Genomic Signal Processing and Statistics (GENSIPS), College Station, TX.
40. Qian, L. and Wang, H. (2007) Inference of genetic regulatory networks by evolutionary algorithm and H_∞ filtering (Invited Paper). IEEE Statistical Signal Processing Workshop, Madison, WI.
41. Kimura, S., Ide, K., Kashihara, A., Kano, M., Hatakeyama, M., Masui, R., Nakagawa, N., Yokoyama, S., Kuramitsu, S., and Konagaya, A. (2004) Inference of S-system models of genetic networks from noisy time-series data. *Chem-Bio Inform. J.*, **4**, 1–14.
42. Maki, Y. (2002) Inference of genetic network using the expression profile time course data of mouse p19 cells. *Genome Inform.*, **13**, 382–383.
43. Sontag, E.D. (2002) For differential equations with r parameters, $2r+1$ experiments are enough for identification. *J. Nonlinear Sci.*, **12**, 553–583.
44. Koza, J.R. (1992) *Genetic Programming: On the Programming of Computers by Means of Natural Selection*, MIT Press, Cambridge, MA.
45. Grewal, M. and Andrews, A. (1993) *Kalman Filtering: Theory and Practice*, Prentice Hall, Englewood Cliffs, NJ.
46. Shaked, U. and Theodor, Y. (1992) H_∞ optimal estimation: a tutorial. 31st IEEE Conference on Decision and Control, Tucson, AZ.
47. Hassibi, B. and Kailath, T. (1995) H_∞ adaptive filtering. IEEE International Conference on Acoustic, Speech and Signal Processing, Detroit, MI.
48. Shen, X. and Deng, L. (1997) Game theory approach to discrete H_∞ filter design. *IEEE Trans. Signal Process.*, **45**, 1092–1095.
49. Simon, D. (2006) *Optimal State Estimation: Kalman, H Infinity, and Nonlinear Approaches*, John Wiley & Sons, Ltd, Chichester.
50. Shen, X. and Deng, L. (1998) A dynamic system approach to speech enhancement using H-infinity filtering algorithm. *IEEE Trans. Speech Aud. Process.*, **7**, 391–399.
51. Woolf, P. and Wang, Y. (2000) A fuzzy logic approach to analyzing gene expression data. *Physiol. Genomics*, **3**, 9–15.

52 Arias, A.M. and Hayward, P. (2006) Filtering transcriptional noise during development: concepts and mechanisms. *Nat. Rev. Genet.*, **7**, 34–44.

53 Zhou, K., Doyle, J.C., and Glover, K. (1995) *Robust and Optimal Control*, Prentice Hall, Englewood Cliffs, NJ.

54 Bernstein, D.S. and Haddad, W.M. (1989) Steady-state Kalman filtering with an H_∞ error bound. *Syst. Control Lett.*, **12**, 9–16.

55 Nagpal, K.M. and Khargonekar, P.P. (1991) Filtering and smoothing in an H_∞ setting. *IEEE Trans. Automatic Control*, **36**, 152–166.

8
Computational Reconstruction of Protein Interaction Networks

*Konrad Mönks, Irmgard Mühlberger, Andreas Bernthaler, Raul Fechete,
Paul Perco, Rudolf Freund, Arno Lukas, and Bernd Mayer*

8.1
Introduction

Since the identification of the genetic code, hypothesis-driven research has entered molecular biology, and causative reasoning regarding the interplay of genes, transcripts, proteins, and metabolites has become amenable. This research was basically single object (or limited number of objects) driven – studying a particular (or a small number of) gene or gene products under different conditions [1]. To do this, qualitative but also (semi)quantitative experimental procedures have been derived, including the polymerase chain reaction for transcript identification and relative quantification utilizing housekeepers, or mass spectrometry approaches for protein identification based on fingerprints, up to isotope-labeled metabolites for spiking probes allowing quantitative assessment of metabolite concentration. Overall, experimental technologies were derived for addressing all levels of cellular organization from genes up to metabolites, but analyzing a limited number of features in the context of a (more or less precisely defined) biological hypothesis.

The next major breakthrough came about with the sequencing of the human genome, with a first sequence draft published in 2001 [2], allowing (at least theoretically) an assessment of all coding fragments and downstream products. Concomitantly, miniaturization and subsequent parallelization of experimental techniques was developed, culminating in array technologies for assessment of all protein-coding transcripts of a cell [3]. In the same spirit, technologies were developed for parallel assessment of protein [4] and metabolite abundance [5], presently seeing a major step in next-generation sequencing technologies [6].

On the basis of this "omics" revolution the analysis approach also changed from hypothesis-driven research to explorative research [7]: if a major portion of cellular players can be determined for a particular cellular state in parallel no upfront hypothesis is needed, as they may at least in principle be derived by accessing the "complete" omics landscape.

Applied Statistics for Network Biology: Methods in Systems Biology, First Edition.
Edited by M. Dehmer, F. Emmert-Streib, A. Graber, and A. Salvador.
© 2011 Wiley-VCH Verlag GmbH & Co. KGaA. Published 2011 by Wiley-VCH Verlag GmbH & Co. KGaA.

8 Computational Reconstruction of Protein Interaction Networks

gene ID	group A			group B			sign. p-value	fold change
1	104	123	98	67	103	99	no	-
2	27	34	21	534	443	632	yes	up
3	1048	1326	2715	625	613	589	yes	down
4	33	12	19	104	143	99	yes	up
---	---	---	---				---	
n	---	---	---				---	

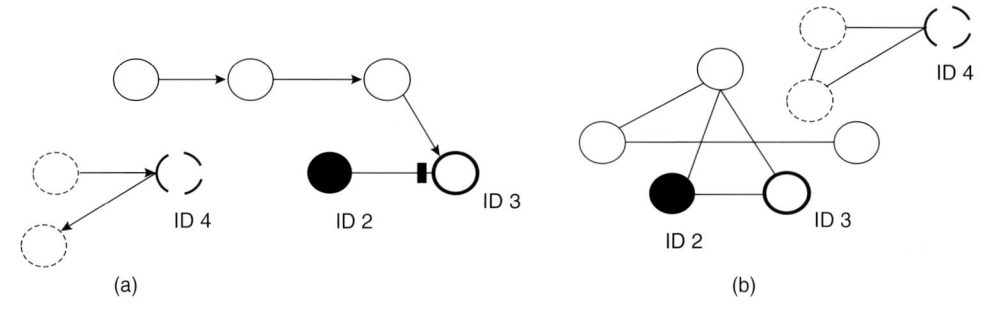

(a)   (b)

Scheme 8.1 Schematic representation of traversing descriptive feature lists into feature dependencies. Result of omics processing is a list of features (given by their gene IDs) and the expression (abundance) values for group A and group B, each holding three samples. Based on the value distributions a test for difference can be conducted for evaluating a significance of difference, further providing the direction of fold change when comparing group A and B. Based on these results relevant features (given in bold) may be interpreted on a directed protein interaction network (a) or an undirected protein interaction network (b).

From this example it becomes clear that mapping omics feature lists onto directed graphs is the preferred method of choice for hypothesis generation resting on relevant features. However, available information on this detailed level is limited: KEGG, for example, provides in total 343 pathways holding 4756 unique genes, which represents only 20% of the total number of about 23 000 human protein-coding genes presently annotated in RefSeq (http://www.ncbi.nlm.nih.gov/RefSeq). For alternative pathway resources the situation is comparable: for PANTHER (www.pantherdb.org) 3138 unique genes are found in more than 165 pathways, for REACTOME (www.reactome.org) the respective numbers are 4076 and 1081. Coverage of genes and proteins is increased for undirected protein interaction networks. OPHID, for example, includes in total more than 20 000 proteins for human.

Next to coverage of genes and proteins, the false-positive and false-negative edges of the given interaction networks have to be respected. Where directed interaction networks usually exhibit higher confidence due to manual inspection and preparation, interactions presented in undirected graphs either hold experimental bias (e.g., when derived from yeast two-hybrid screens [16]) or are predicted based on homology considerations bridging different species.

Independent of considerations regarding the coverage and correctness, the result of interpreting omics features centrally depends on the type of interaction network used: cliques identified on the level of physical interaction networks will be interpreted differently when compared to the analysis on the level of metabolic networks, subsequently resulting in different hypotheses. However, both options for interpreting the omics features may provide valuable insight for hypothesis generation based on the descriptive list of relevant features.

Therefore, in the following we focus on three specifically different types of protein interaction networks: metabolic networks, paralog networks, and physical interaction networks. Our main intention is to derive a procedure for extrapolating from given networks with limited protein coverage to complete networks with reasonable accuracy. Such networks can then be used for specifically analyzing omics features in the context of molecular pathways, in the context of paralogs, and in the context of physical interaction networks.

8.2
Protein Interaction Networks

8.2.1
Network Categories

Recent activities in biological research propose a more differentiated view on biological networks. Lu *et al.* [17], for example, describe the development of an edge-ontology, whereas Alexeyenko and Sonnhammer [18] successfully reconstructed biological networks in a more fine-grained resolution of edge types. STRING (string.embl.de) and other tools, in contrast, annotate edges with similarity scores calculated from omics data [19, 20]. However, reaching an interpretation of omics data on the level of such aggregate biological interactions is difficult. We therefore focus on three well-defined types of biological networks: metabolic, physical, and paralog networks.

8.2.1.1 Metabolic Networks
Metabolic networks are an abstraction of more detailed representations of bundled stoichiometric matrices and reaction kinetics. A recent approach to globally reconstruct the human metabolic network was derived by Duarte *et al.* [21] defining an individual's metabolic network as a function of nutrition, environment, and genetics. The major proteins performing functions in metabolic networks are enzymes and coenzymes, but also features controlling metabolic processes (transcriptional regulation, protein complex formations) are considered as a part of a metabolic network.

Our data source, KEGG, holds process-oriented, manually curated knowledge on more than 343 pathways that execute defined tasks in a cell. These pathways essentially consist of three building blocks: chemical compounds, gene products, and reactions, referred to as relations.

Various procedures can be followed for parsing these pathways into a graph representation. Whereas one possibility is to represent chemical compounds in nodes [22], we prefer to follow the interpretation of Kelley and Ideker and Rapaport *et al.* [23, 24], where the resulting graph is build from enzymes cosharing chemical compounds. Here, the nodes represent enzymes and edges represent substrates and products; if e_1, e_2 are enzymes, an edge exists, if and only if the product of e_1 is a substrate of e_2 or vice versa. As KEGG is designed as a database where gene products refer to clusters of paralogous groups, we selected one representative gene product to avoid introducing paralogous information in the metabolic network. This process resulted in a metabolic network of 2134 nodes and 5450 edges. An example subgraph of a metabolic network is exemplarily depicted in Figure 8.1(a).

8.2.1.2 Paralog Networks

The paralog network represents relationships between paralogous gene products, defined as a set of homologous genes that originate from a common ancestor by gene duplication in one single organism. Paralogs often retain a similar functionality and, in general, also exhibit high sequence similarity. Thus, the relation between functional similarity and local sequence similarity tends to be rather strong. However, considering that original interactions of the duplicated gene are only conserved to a certain extent [25], sequence similarity is a necessary, but not sufficient, condition in order to define a paralog.

As for the metabolic network, the data was automatically retrieved querying the web API provided by the KEGG database. As mentioned above, KEGG integrates information on paralogous groups supported by the subsidiary SSDB database (http://www.genome.jp/kegg/ssdb). The SSDB database utilizes the SSEARCH algorithm [26] followed by a clustering algorithm based on automated clique search [27]. This process resulted in a paralog network of 4208 nodes and 89 854 edges. An subgraph of a paralog network is exemplarily depicted in Figure 8.1(b).

8.2.1.3 Physical Interaction Networks

Physical interaction networks describe direct interaction of molecular entities, mainly driven by noncovalent binding. The Human Proteome Organization (www.hupo.org) designed an exact ontology for the description and documentation of interaction experiments, namely the Proteomics Standard Initiative – Molecular Interaction (PSI-MI) [28]. A subset of this ontology accounts for the description of different types of physical interactions.

The IntAct database [29] at the European Bioinformatics Institute (EBI) stores physical interactions in PSI-MI format. It is manually annotated by expert biologists

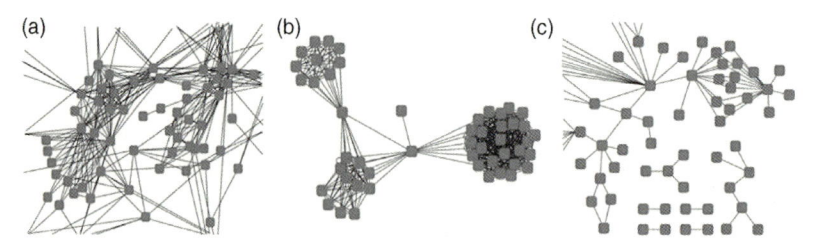

Figure 8.1 Example subgraphs for metabolic (a, edge directionality not included), paralog (b), and physical interaction networks (c).

to a high level of detail, including experimental methods, conditions, and interacting domains, and currently holds more than 200 000 binary interactions on a variety of model organisms and human.

We parsed IntAct for the PSI-MI term MI:0407 (direct interaction), which is defined as: "Interaction that is proven to involve only its interactors." Furthermore, this term is by its parent terms' definition assured to be physically located in the same structural complex. This process resulted in a physical network of 893 nodes and 1104 edges. An example of a physical interaction network is depicted in Figure 8.1(c).

8.2.2
Parameters for Protein Annotation

Various parameters for the characterization of proteins in the context of their participation in interaction networks are of interest, among them the protein abundance in certain tissues, the location where they exert their function, as well as their functional categorization. In the following we present parameters for the characterization of proteins that are then used for deriving network models.

8.2.2.1 Gene Expression Profiles
A major descriptor of a gene is its activity level resulting in expressed mRNA. In our analysis we used the Gene Expression Omnibus (GEO) Body Map dataset GSE7905 [30], holding expression levels for over 15 000 genes in 32 human reference tissues (in the following abbreviated as GEX).

8.2.2.2 Subcellular Location
To be able to conduct their function, proteins have to be localized at their appropriate subcellular compartment. Numerous proteins exhibit their function in the cytoplasm while others need to be located in the cell membrane or are secreted in order to fulfill their function. Immunohistochemistry experiments, among others, are used in order to identify the location of proteins in the cell. Next to these experimental procedures, *in silico* predictions based on the protein sequence can provide information on the subcellular localization of proteins. We used experimentally derived data on subcellular location of proteins as stored in the UniProt database (www.uniprot.org) along with *in silico* predictions based on WoLF PSORT (wolfpsort.org). For each protein a vector of probabilities assigned to "cytoplasm," "cytoskeleton," "endoplasmic reticulum," "extracellular," "Golgi body," "lysosome," "mitochondria," "nuclear," "peroxisome," or "plasma membrane" was computed (in the following denoted as LOC).

8.2.2.3 Gene Annotation
We used GO terms as provided by the Gene Ontology Consortium for functional gene annotation. The Gene Ontology Consortium provides a controlled vocabulary of terms for describing gene products based on their molecular functions, their involvement in certain biological processes, as well as their subcellular localization.

GO terms are structured in the form of a directed acyclic graph and genes or their respective products can be assigned from one to n categories. Assignment of certain genes to GO terms can either be based on experimental evidence or on *in silico* predictions as indicated by the evidence codes of the assignment.

We used the two main GO categories "molecular function" (in the following denoted as GOF) and biological processes (in the following abbreviated as GOP) in our analyses. Associations of genes to GO categories were obtained from the National Center for Biotechnology Information (NCBI; www.ncbi.nlm.nih.gov).

8.2.2.4 Transcription Factors

Transcription factors are proteins binding to specific DNA segments, thus regulating the transcription of genes. Transcription factors rarely act alone and both activating as well as repressing transcription factors are known. Thus, the combination of transcription factors binding to the promoter region and to regulatory regions like enhancers of a specific target gene determine the rate of gene transcription. Next to a number of experimental techniques for the elucidation of the regulatory potential of a transcription factor and a specific target gene, computational methods can be used to predict the regulatory potential of a transcription factor on specific genes. Known binding profiles of transcription factors are used in pattern search algorithms in order to find transcription factor binding sites in the regulatory regions of potential target genes. We used binding matrices of 270 human transcription factors obtained from the JASPAR (jaspar.cgb.ki.se) and TRANSFAC (www.gene-regulation.com) database, and searched for binding sites in the regulatory regions of about 21 000 human genes. Thus, each gene is characterized by a bit string vector of length 270 where each position is representative for a single transcription factor and holds the value 1 if a binding site for the respective transcription factor was found in the regulatory region and 0 otherwise (in the following denoted as TF).

8.2.2.5 MicroRNA

MicroRNAs (miRNAs) are a class of 18- to 24-nucleotide long RNAs. They associate with Argonaute proteins and are mainly involved in post-transcriptional gene regulation. Nearly 800 miRNAs are known in the human genome as stored in the miRBase (www.mirbase.org) and many more are predicted to exist. The main algorithms to predict target genes are TargetScan, miRanda, and PicTar. Based on these predictions a bit string vector can be constructed for each gene where a 1 encodes a predicted regulatory target of a specific miRNA.

8.2.2.6 Pathways

The PANTHER classification system is a resource that classifies genes by their functions based on experimental evidence and evolutionary relationships. The PANTHER resource also holds over 165 manually curated biological pathways where individual components represent single proteins or proteins fulfilling similar biological roles. The PANTHER pathway ontology uses a controlled vocabulary and follows the systems biology markup language (SBML). Four key classes in the ontology include the pathways, the molecule, the reaction, and relationship, as well as the cell type or cellular component class. Based on the assignment of proteins to specific pathways a similarity can be calculated based on joint pathway occurrence (in the following denoted as PPA).

8.2.3
Data Preparation

8.2.3.1 Integration of Data Sources
When confronted with the integration of different data sources, several issues arise, the majority of which originate from the heterogeneity of both biological entity naming and information type. Furthermore, considerations regarding data maintenance are necessary.

The first question concerns the unification of the different namespaces. This is a two-dimensional problem, as the information is not only differentiated on the specific biological entity it refers to, like protein, transcript, or gene, but is also database-specific. For example, the apoptosis regulator BCL2 – as addressed to in the NCBI databases – bears the name ENSG00000171791 at European Molecular Biology Laboratory (EMBL)-EBI. We tackle this issue by using an Abstract Biological Entity (ABE) as a convergence point for the different namespaces. This structure connects the distinct biological views on the entity (protein, gene, etc.), each holding further identifiers originating from different data sources. The granularity of this approach is dictated by the International Protein Index (IPI) database (http://www.ebi.ac.uk/IPI), an abstract identity being uniquely associated to a protein sequence. In the case of splice variants, several such structures may share the same gene names, while holding different protein identifiers. On the gene level we presently integrate the identifiers from NCBI (www.ncbi.nlm.nih.gov) and from ENSEMBL (www.ensembl.org), while on the protein level we integrate the identifiers from NCBI RefSeq (www.ncbi.nlm.nih.gov/RefSeq), and IPI, UniProt, SwissProt and TrEMBL (www.ebi.ac.uk/Databases/protein.html).

Presently, our system holds information on roughly 70 000 different human proteins. However, since much of the available information is gene-specific, or applies only to the most prevalent splice variant of the gene, it makes sense, for most analysis, to use only a subset of all available proteins, that is, approximately 24 000 canonical sequences corresponding to the human protein-coding genes, as defined by the UniProt database in SwissProt.

For ease of maintenance and to ensure data consistency, we have implemented a modular, fully automated workflow. This approach enables us to easily execute full database updates or to incorporate additional data sources and thereby keep pace with the continuously changing informational landscape. For the implementation we use Java as our programming language of choice, while for the persistence layer we employ an object relational mapping (ORM) implemented by Hibernate and use a MySQL database as backend.

8.2.3.2 Obtaining Edge Weights

The ultimate goal of this inquiry is to estimate the likelihood that an arbitrary pair of ABEs (Abstract Biological Entities) is related in a specific context (Scheme 8.2).

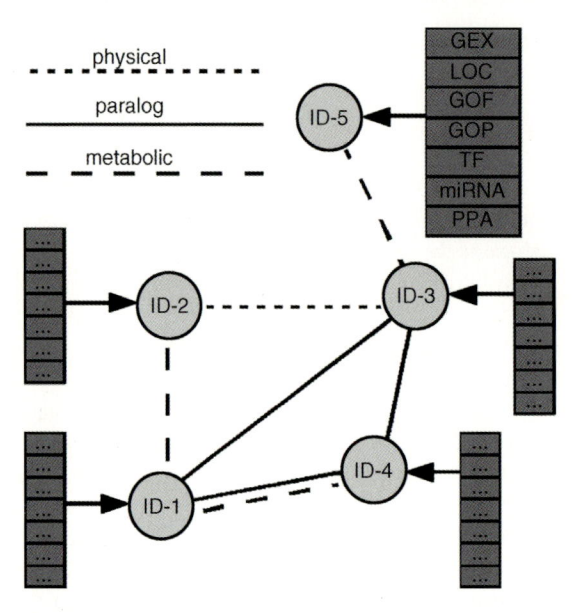

Scheme 8.2 After integrating all parameters employed for protein, gene, or transcript annotation, each ABE can be associated with its corresponding omics data. Furthermore, the relations represented in the incomplete paralog, metabolic, and physical networks are incorporated. The objective is to use both of these sources of information to extrapolate each type of specific network such that a reasonably complete and accurate coverage of the whole human proteome is obtained.

The most straightforward approach to do so is to derive the desired estimate based on a comparison of the annotations of the involved proteins. This can be done individually for each parameter used for annotation, each of which requires a specific procedure for the comparison to be carried out.

Regarding gene expression profiles (GEX), similar abundance levels measured in a variety of tissues indicate a relationship between the ABEs. This relationship could, for example, be a regulatory one in a metabolic context (the products of gene A have a downstream effect on the expression of gene B) or a functional similarity of the corresponding proteins (i.e., they are paralogs), both of which might explain the observed correlation of the expression profiles. As a measure for the similarity on the level of gene expression, we use the Pearson correlation coefficient.

The subcellular location (LOC) associated with a pair of proteins can be seen as providing solely negative evidence for a direct interaction of these proteins: even if both entities are present in the same cellular compartment, it would be rash to conclude that they actually collaborate in a certain process. On the contrary, if two proteins are only found in different regions of the cell, it is reasonable to conclude that a direct interplay of these objects is very unlikely. To measure this likelihood, the inverse of the normalized city-block metric [20] is applied.

At a first glance, the set of gene annotation terms (GOF, GOP) describing each entity suggests a straightforward method to obtain estimates of the probability that a given pair of ABEs is related: the bigger the overlap of both involved sets of terms, the higher the chances that the corresponding genes are actually linked, given that the terms define functional categories or process membership. However, it has to be taken into account that the term's extensions exhibit significant differences. While some are very specific (e.g., "GO:0051714 – positive regulation of cytolysis of cells of another organism"), assigned only to a very limited number of gene products, others are very general (e.g., "GO:0010926 – anatomical structure formation") and apply to a rather elevated number of entities. Thus, we have devised a customized function incorporating the different entropy levels of the terms in order to achieve an adequate likelihood estimate $d_{GO}(X,Y)$ for two entities X,Y (Equation 8.2). Note that in Equation 8.1, T_{sg} designates the number of all genes assigned to the term T or to one of its subcategories, T_{st} the number of terms which stand in an is-a relationship to T, tg the total number of genes in the ontology, and tt the total number of terms in the ontology. Thus, $d_{GO}(X,Y)$ is the sum over $ic(\cdot)$ of all terms $T_{X,Y}$ common to both entities X,Y.

$$ic(T) = \sqrt{\frac{1}{2}\left(\left(\ln\frac{1+T_{sg}}{tg}\right)^2 + \left(\ln\frac{1+T_{st}}{tt}\right)^2\right)} \qquad (8.1)$$

$$d_{GO}(X, Y) = \sum_{T_{X,Y}} ic(T_i) \qquad (8.2)$$

As outlined above, we identified transcription factor binding sites (TF) and thus are able to associate a set of transcription factors with each biological entity. Consequently, the coregulation of genes can be investigated, which in turn serves as an indicator of the genes relatedness. As a measurement of the level of relatedness, the dice coefficient was used.

The same rationale was exploited when deriving a measurement to compare the sets of miRNAs assigned to each ABE (i.e., a similar regulatory context provides evidence for a functional interplay of the regulated genes or their products).

Finally, comparing the pathway data (PPA) linked to a pair of entities obviously allows deducing the chances that these entities are related, given that the co-occurrence of two genes in one or more pathways implies a, at least indirect, relationship. Again, the dice coefficient was used as measure of relationship.

Using these distance functions, pairwise distances can be calculated for all entities. From a graph theoretic point of view, these distances can be modeled as a network in which each edge is associated with a vector of seven weights and whose nodes are constituted by the ABEs. However, in order to be able to calculate these weights, the corresponding parameters have to be known for both entities. The fraction of present entries for each parameter is shown in Table 8.1.

Table 8.1 For each network ("physical," "metabolic," and "paralog") and each data type the percentage of available data per edge is shown. Note that the networks are of different size (i.e., the paralog network comprises 4208 nodes, the metabolic 2134, and the physical 893).

	PPA	GEX	LOC	GOP	GOF	TF	RNA
Physical	6	71	100	94	97	85	82
Metabolic	12	76	100	92	96	87	83
Paralog	9	42	100	92	95	84	77

8.2.3.3 Data Completeness

Given that, as mentioned above, the weight vectors assigned to the network's edges are incomplete, a missing value handling is required in order to further process the accumulated evidence. We opted for a replacement of the missing values with the corresponding mean value, as this method guarantees to minimize the overall deviation from the (unknown) true value.

Clearly, a wide range of other methods to tackle missing values exist (e.g., [31]), like simply removing those edges whose weight vector is incomplete. More sophisticated replacement techniques, such as multiple imputation [32], are typically associated with elevated computational costs and thus scalability becomes an issue. Furthermore, these techniques frequently require a certain minimal level of data completeness in order to achieve reliable results. This minimum level is hardly ever attained when integrating biological data from sources as diverse as those described above.

8.2.3.4 Data Normalization

Before incorporating the obtained edge weights into one single estimate for the probability of the likelihood of a pair of ABEs to be linked (i.e., before a single-weighted network can be reconstructed) the individual estimates have to be pre-processed to allow for a balanced integration. Namely, the mean value was set to zero, the standard deviation to one and the skewness of the weights' distributions was removed. In order to control the shape of the distribution, a suitable adjustment of the mean and the deviation was carried out, followed by a log transformation. The result of this normalization step can be seen for selected parameters in Figure 8.2.

8.2.4
Deriving Models

In order to condense all seven scores calculated for one edge based on the comparison of the individual parameters into a scalar value, the most straightforward approach would be to use a linear model. However, as the key objective is to extrapolate any desired, dedicated network, the known parts of these networks have to be integrated somehow into the reconstruction effort. This very objective can be attained using machine learning algorithms or, more precisely, classification algorithms.

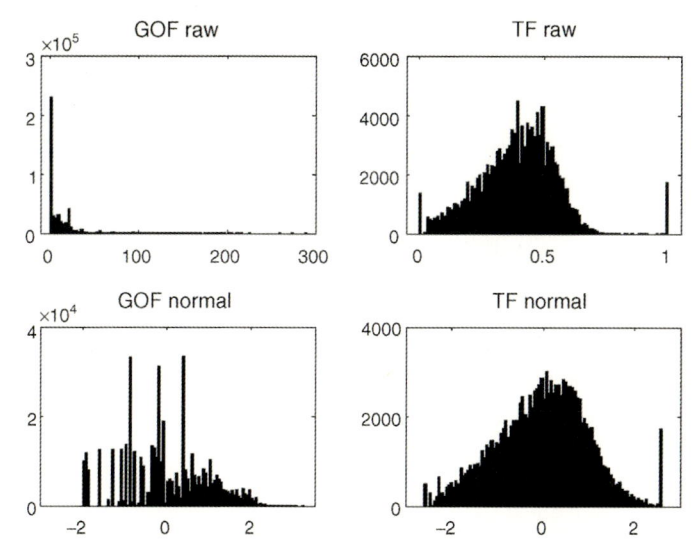

Figure 8.2 Upper row: distribution of the calculated GOF and TF comparison scores before normalization; lower row: respective distributions after normalization. The mean of the normalized distribution is located at 0, the standard deviation equals 1, and the skewness is removed.

8.2.4.1 Basic Considerations

In order to apply classification algorithms, the network has to be transformed into a matrix format (Scheme 8.3).

This can easily be done, given that each edge is characterized by the weight vector obtained from the comparison of the single parameters. Furthermore, to integrate the information given in the form of known portions of a (physical, metabolic, or paralog) network, each edge can be annotated according to its membership in the respective network. Stated in other words, for each edge, it is known if the corresponding entities are related to each other (positives) or not (negatives), taking the reference network (physical, metabolic, or paralog) into account.

The key advantages originating from the use of machine learning techniques are:

- The model derived in the first step of the procedure can be used to predict (any) desired types of relations among biological entities (i.e., a model trained on a metabolic network yields an extrapolation of this very type of network, etc.).
- The model can be applied to any desired set of biological entities, given that these could be annotated sufficiently with the used parameters.
- The obtained predictions are usually more reliable than those provided by less sophisticated methods, such as an equally weighted linear model.

However, several issues have to be handled with care when exploiting the advantages of machine learning algorithms.

First of all, there is a strong bias in the protein data towards interaction and no explicit reporting of noninteracting pairs. The pure absence of an edge in a reference network does not provide any evidence with respect to the nonrelatedness

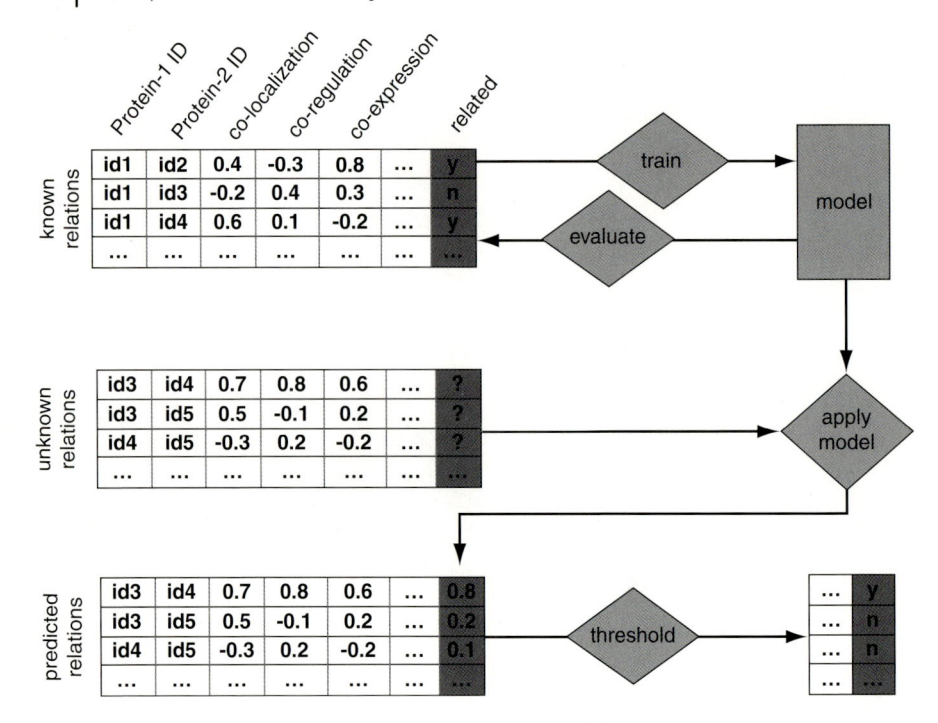

168 | 8 Computational Reconstruction of Protein Interaction Networks

Scheme 8.3 Based on a training set of known relations a model is trained, which can be evaluated on a test set. In order to obtain a faithful estimate of the model's performance, training and test sets have to be disjoint. Furthermore, the model can be applied to a set of unknown relations, which are characterized by the same parameters as the training set. As a result of this application, predictions are obtained (the last column of "predicted relations") that correlate more or less strongly, depending on the model's performance, with the probability that the corresponding biological entities are actually related. A threshold may be defined to obtain binary predictions.

of the respective entities. Consequently, it has to be decided upon which pairs of entities that are not associated in a reference network are believed to be actually not related. Given that biological networks are sparse (i.e., the vast majority of components are not interacting [33, 34]), we decided to randomly select pairs of ABEs. The sample of negatives obtained in this way reflects the actual ratio between related/nonrelated entities (i.e., the edge density) and thus can be considered to be reasonably pure. Another approach to select negatives would be to use the subcellular location (LOC, see above), but this method (i) requires to exclude this parameter from the machine learning procedure, (ii) might introduce a bias with the negative sample, and (iii) could not be proven to be superior to a random selection [18].

Even though biological networks are sparse, in order to maximize the derived model's performance, balanced training samples should be used (i.e., the number of positive instances should equal the number of negative instances). This ratio has been shown to be near optimal in most cases [35], whereas determining the best possible ratio requires extensive experiments, without significantly improving the obtained prediction reliability.

8.2.4.2 Choosing an Algorithm

Given the huge diversity of classification algorithms the question naturally arises which one to pick for a particular network reconstruction task. The key obstacle in this context is formulated in the so-called "no-free-lunch theorem," which roughly states that in terms of predictive performance, no algorithm is in general superior to all the others [36]. However, on a particular dataset, the predictions obtained from different algorithms may vary significantly in their reliability. Consequently, experiments have to be carried out to identify the best-performing algorithm for the given analysis goal.

We carried out a preselection of algorithms for further testing based on the following criteria:

- The induced model's application needs to be reasonably fast, as the number of edges for which predictions are needed is $n \times (n - 1) \times 0.5$, where n is the number of nodes of the network that shall be reconstructed. Given the low dimensionality of our data (i.e., just seven weights for each edge) rather reduced training sets can be used and, thus, the time required to train a models is of no concern in this context.
- The examined algorithms should belong to different algorithm families to increase the chances to obtain predictions that vary in their quality.

For our modeling we utilized C4.5 [37], Random Forests [38], naïve Bayes [39], Bayesian networks [40], support vector machines [41], multilayer perceptron [42], and k-nearest neighbor (k-NN) [43] all implemented in WEKA (http://www.cs.waikato.ac.nz/ml/weka), and linear discriminant analysis as provided in MATLAB (www.mathworks.com).

8.2.5
Validation Procedures

Validation of the obtained results was carried out on two levels: (i) the quality of the derived models was evaluated to allow for identifying the best-performing algorithm and (ii) the topological features of the investigated networks were compared in order to better understand the different network characteristics.

8.2.5.1 Model Performance Evaluation

In order to obtain a reliable estimate of the models performance, 10-fold cross-validation (CV) was used instead of a single split of the known relations into training and test sets (see Scheme 8.3). Now, given that the ratio of positive and negative instances is far from being balanced, maintaining this ratio in the test sets would either enormously increase the overall test set size or lead to an insufficiently small number of positive instances in the test sets. A straightforward way to handle this situation is to work with a balanced test set instead, while using measurements that are either insensitive to the class distribution or that can easily be extrapolated to any desired ratio of positive and negative instances.

An example for the former type of measurement is the receiver operating characteristic (ROC), which is a plot of the true-positive rate (sensitivity) versus the false-positive rate (1 – specificity). To summarize this plot in a scalar value, the area under curve (AUC) is frequently used [44].

8.2.5.2 Network Structure Assessment

To assess the topological properties of the investigated networks, the following measurements were used:

- **Density**. Apart from the average edge density (avg. e-dens; i.e., the number of edges relative to the number of possible edges), we also considered the average edge density within the components (avg. e-dens. in comps.; i.e., the number of edges relative to the number of possible edges within the existing components). While the former is a global measurement, the later is semiglobal.
- **Clustering**. To characterize the level of clustering in the network, we used the average of the cluster coefficients (avg. CC) of all nodes and additionally an alternative cluster coefficient, which is relative to the distribution of the node degree k (avg. CC rel. k) [45]. This second variant reduces the impact of highly connected nodes (e.g., hub proteins) whose neighbors are only sparsely interconnected. Thus, a comparison of both measures serves as an indication for the prevalence of such hubs. Finally, the size of the biggest component relative to the total number of nodes in the network (the index of aggregation (IoA)) was used to asses the overall clustering properties of the network. The first two measures summarize local network properties, whereas the IoA captures a global network property.
- **Connectivity**. To asses the network's connectivity, we used the average shortest-path length (avg. shortest path) between all nodes of the network. Furthermore, for each node the average distance to all its neighbors was calculated and the mean of this distribution was used as an indication of the average centrality of the nodes in the network (mean neighbor distance). As a measurement for the average distance of all nodes from the center of the network, we used the average closeness-centrality. Finally, the number of pairs of nodes which are connected by a path of finite length relative to the number of possible path was taken as an indication of the overall connectivity (% of non-inf. paths). While the first three measurements reflect semiglobal properties, the last one refers to a global characteristic.
- **Degree correlation**. The correlation of edge degrees of adjacent nodes (degree corr.) reflects the dependence of the interconnection scheme on a local node property.

For details on the above-mentioned topological measurements, we refer to [34].

8.3
Characterization of Computed Networks

In this section we outline how the three different networks were extrapolated and present the results obtained on the level of a structural (topological) comparison. Furthermore, the derived model performances are discussed in detail. We will conclude the section with an application of the predicted networks to a list of differentially regulated genes associated with B cell lymphoma in order to exemplify the fundamentally different hypotheses that arise when interpreting such an omics list on a physical, metabolic, or paralog network, respectively.

8.3.1
Evaluation of the Specific Protein–Protein Interactions

The experimental setup we used to extrapolate the given reference networks is straightforward, as depicted in Scheme 8.4.

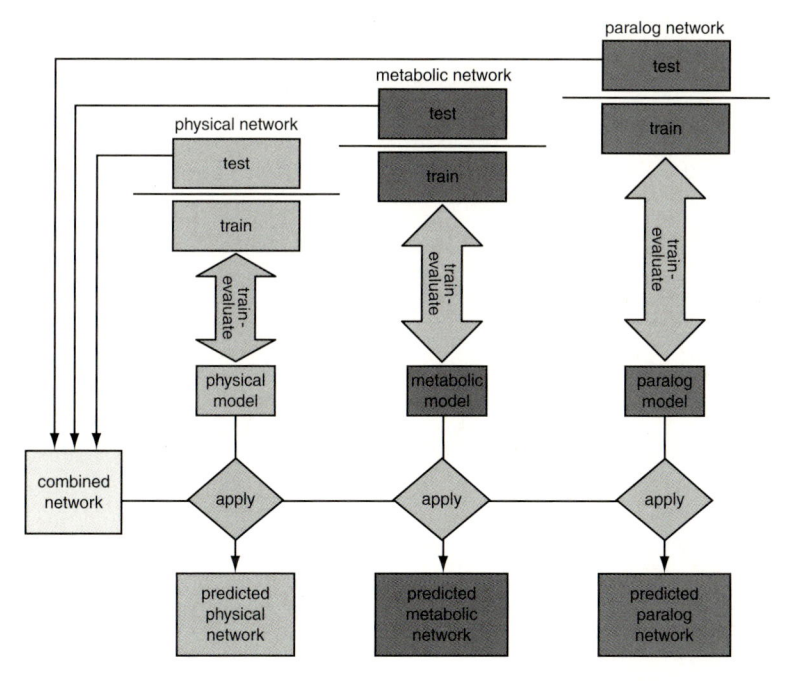

Scheme 8.4 Each reference network (paralog, metabolic, or physical) was split into a training and a test network (training, test). The test networks were combined into one single network, comprising interactions of three different types (i.e., edges representing paralog, metabolic, or physical relationships). Using machine learning algorithms, for each of the training networks models were derived allowing the prediction of the interaction type of interest (paralog model, metabolic model, or physical model). These models were applied to the combined test set, leading to three different predicted networks (pred-para net, pred-meta net, or pred-physical net). Validation was carried out on three different levels. First of all, the topology of the training networks was assessed. Furthermore, the model's performance on the training sets was examined via CV and also measured on the combined test set.

Just notice that, in order to be able to evaluate the correctness of the predicted networks in terms of AUC, we split each reference network and formed a combined network. Thus, all the generated models can be applied to the same set of nodes while allowing validating the obtained predictions against the respective type of relation. This approach allows comparing the model's performances measured via CV on the training set with the performance on the node set of interest, which, ultimately, is the relevant one.

As can be seen in Figure 8.3, all examined networks are very sparse (i.e., they exhibit a very low edge density). However, they differ significantly in all the other calculated measurements. To begin with, the paralog network consists of rather small components (IoA, % of non-inf. paths), whose nodes are heavily interconnected (avg. e-dens. in comps, avg. CC, avg. CC rel. k). The average shortest path length and the mean distance to all neighboring nodes is the lowest of all compared networks, which provides further evidence of the high connectivity within the components. Finally, a degree corr. close to 1 indicates that the topology within one component resembles an "any-to-any" interwiring scheme. In contrast to this, the metabolic network is dominated by one large component (IoA, % of non-inf. paths), whose nodes are only loosely coupled (avg. shortest-path length, mean neighbor dist., avg. e-dens in comps). As the avg. CC shows, some clusters do exist, and the difference between the avg. CC and the avg. CC rel. k indicates the presence of hub proteins, while the degree corr. shows that no dependence between the node degree of related entities can be found. Finally, the physical network reveals an intermediate architecture, composed of small components (IoA, % of non-inf. paths) that are, despite of their size, only sparsely interconnected (e-dens in comps). Both cluster coefficients indicate a very

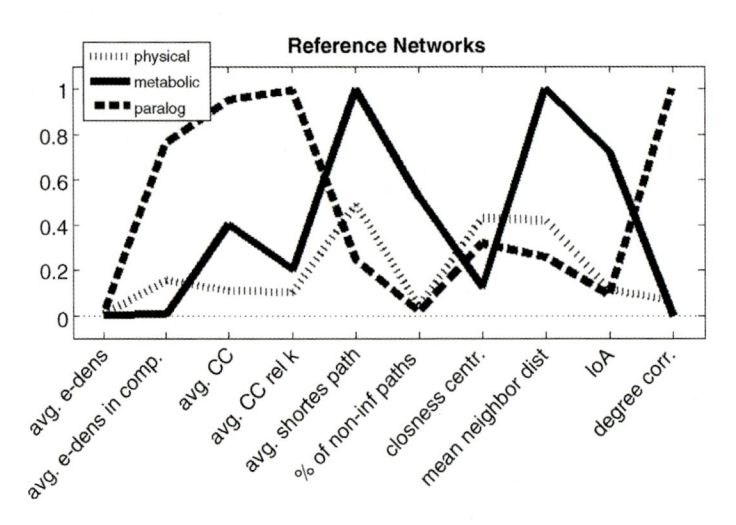

Figure 8.3 Topological profiles of the reference networks. Several measurements were calculated for the reference networks, as indicated on the abscissa. On the ordinate, the corresponding values obtain for the different networks are shown. In order to facilitate the comparison, for each network a connecting line has been drawn for visualizing the networks "topological footprint." The average shortest path length as well as the mean distance between neighboring nodes was rescaled for allowing joint graphical representation.

low tendency to form clusters (avg. CC, avg. CC rel. *k*) and the average shortest path length, together with the mean-neighbor distance, suggest that relatively short pathways do exist, which are, however, still longer than those encountered in the paralog network.

In order to gain a deeper understanding of the impact of data source selection on the accuracy achieved by the reconstructed network, we performed a feature ranking for each of the three investigated network types, as shown in Figure 8.4.

As can be seen, the similarity score based on gene annotation (GOF, GOP) yields rather high AUC scores for the metabolic and the physical network, while providing outstanding results for the prediction of (functional) paralog genes, as expected. On the other hand, the edge weights derived from a comparison of gene expression profiles provide minor performance, questioning the suitability of this parameter for the aim of network reconstruction.

As outlined above, several classification algorithms were compared in order to identify the most simple, best-performing approach (Table 8.2).

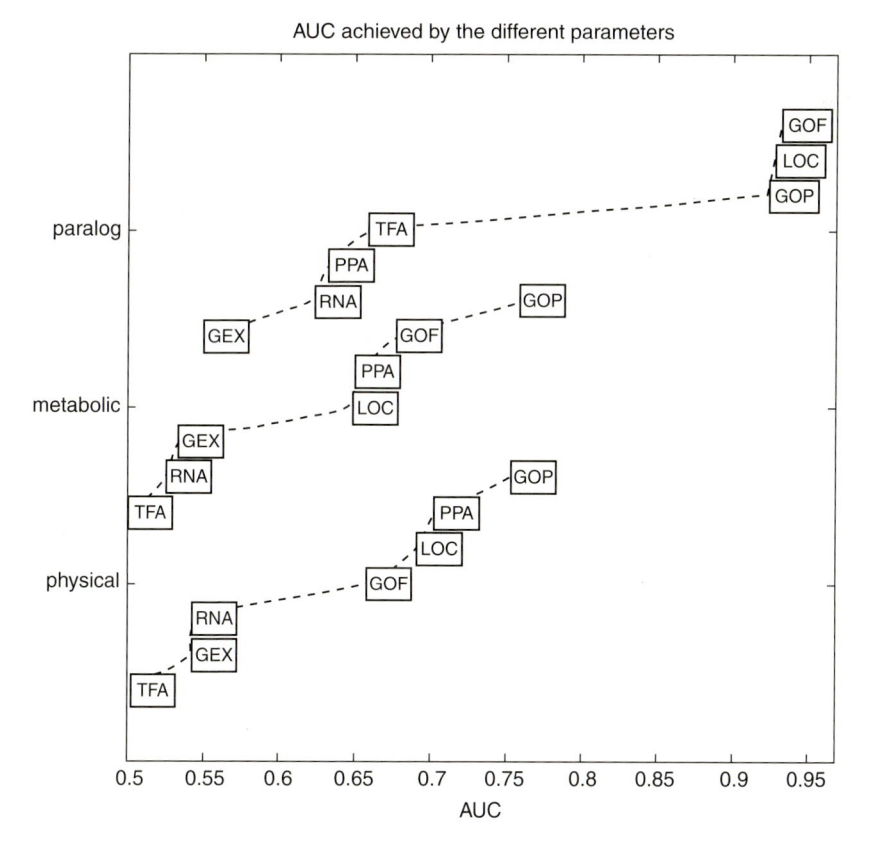

Figure 8.4 The performance of each parameter for each of the reference networks was measured. On the abscissa, the AUC can be seen, whereas each of the lines corresponds to one of the networks, as indicated on the ordinate. Note that the horizontal distances do not have any meaning and were introduced in order to improve legibility.

Table 8.2 For each network type (physical, metabolic, and paralog), the average AUC achieved by the selected algorithms (based on 10-fold CV) is shown, together with the standard deviation in brackets; additionally, the average performance over all classification methods achieved for each network type is shown.

	C4.5	Random forests	Naïve Bayes	Bayesian networks	Support vector machines	NN	20-NN	Linear discriminant analysis	Quadratic discriminant analysis	Average
Physical	0.65 (0.07)	0.71 (0.10)	0.72 (0.08)	0.71 (0.08)	0.67 (0.08)	0.71 (0.1)	0.70 (0.09)	0.75 (0.05)	0.71 (0.06)	0.70 (0.029)
Metabolic	0.79 (0.04)	0.80 (0.03)	0.81 (0.03)	0.83 (0.03)	0.71 (0.02)	0.80 (0.03)	0.78 (0.03)	0.78 (0.01)	0.81 (0.02)	0.79 (0.034)
Paralog	0.97 (0.00)	0.99 (0.00)	0.97 (0.00)	0.98 (0.00)	0.93 (0.00)	0.98 (0.00)	0.98 (0.00)	0.97 (0.00)	0.95 (0.00)	0.96 (0.018)

Evidently, the AUC is largely determined by the network type and the choice of the algorithm has only minor impact (see the standard deviation of the average performance in Table 8.2). Thus, we opted for quadratic discriminant analysis, given that this method minimizes the risk of overfitting while still providing competitive performance results.

Due to the experimental setup depicted in Scheme 8.4, we were able to evaluate the different models' performance on one and the same set of nodes (Figure 8.5).

The results are comparable to those obtained via CV (Table 8.2), indicating the models' stability and thus suggesting that this method can indeed be applied to arbitrary sets of proteins.

8.3.2
Application of the Specific Protein–Protein Interactions

Based on the Oncomine database [46] we selected five transcriptomics studies (ONCO 389, 41, 40, 36, and 863), reported in three publications [47–49], which characterize mantel cell lymphoma, B cell chronic lymphocytic leukemia, and diffuse large B cell lymphoma. After combining the sets of significantly differentially

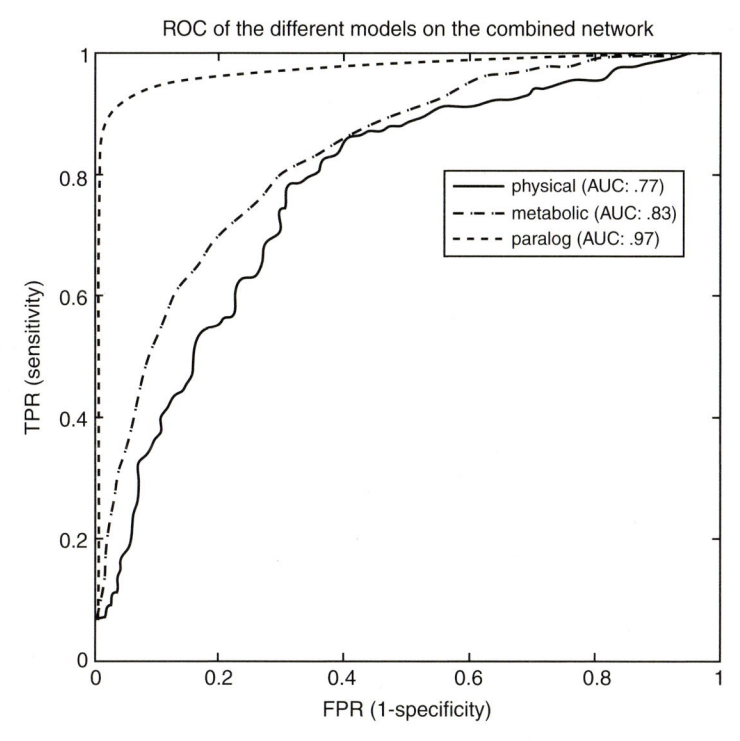

Figure 8.5 Abscissa represents the false-positive rate, whereas the true-positive rate is given on the ordinate. For each specific network model evaluated on the combined network, the ROC is shown.

regulated genes, the resulting list was mapped on the three predicted networks. To illustrate the different interpretations arising on the level of specific networks, we selected a subgraph of each network, comprising approximately the same set of nodes (Figure 8.6).

The biological relevance of the predicted physical interaction partners of the IL-2 receptor γ chain (IL-2RG) is straightforward, since IL-2RG is an important signaling component and is shared by various interleukin (IL) receptors, including IL-4R and IL-7R. Thus, in the physical network, the edge IL-2RG–IL-7R appears to be a true-positive. As the IL-2RG complements the function of the IL-7R, an edge between these two entities in the paralog network is not expected (i.e., the missing edge IL-2RG–IL-7R probably represents a true-negative). Looking at the metabolic network it is exactly the other way around. Metabolic relationships within functional units such as those of IL-2RG and IL-7R are a matter of course. Similar assumptions could be made for the interactions between IL-2RG and IL-4R; however, interestingly, a paralog relationship was predicted and also confirmed by the KEGG database.

Two further nodes in the metabolic as well as in the physical interaction network are members of another class of cytokine receptors: the colony stimulating factor (CSF) receptors CSF3R and CSF2RA. Both of them control the production, differentiation, and function of granulocytes, and are thus involved in similar metabolic processes. A physical interaction between these two factors as predicted in our corresponding network is not reported and may be a false-positive result.

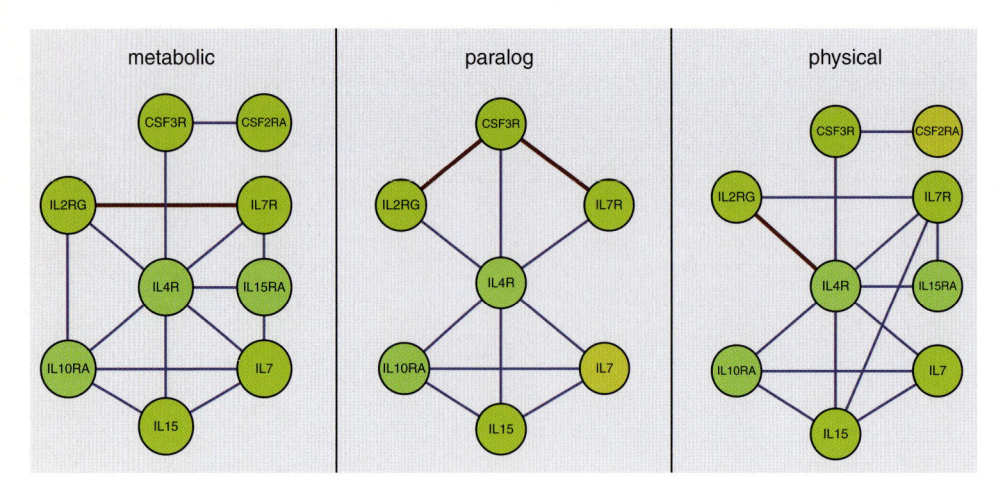

Figure 8.6 Traversing a descriptive omics feature list into context. After mapping a list of differentially regulated B cell lymphoma genes onto the three predicted networks, we selected a subgraph of each network in order to illustrate the benefits arising from the use of dedicated network types. Nodes are identified by corresponding gene symbols; edges that are also represented in the initial datasets KEGG and OPHID are given in bold.

Apart from the cytokine receptors there are two ligands, IL-7 and IL-15, which are part of the selected subgraphs and an edge between them was predicted in all of the three networks. A metabolic as well as a paralog relationship can be explained by their shared function as regulators of lymphoid homeostasis, whereas a physical interaction could not be observed so far (i.e., the predicted edge in the physical network is likely to be a false-positive).

Since IL-2RG is also a subunit of IL-15R and contributes to IL binding, the missing edge between IL-15 and IL-2RG in the predicted physical interaction network has to be taken as a false-negative result.

As can be seen, interpreting and manually validating omics data in dedicated interaction networks provides additional interpretatory options in contrast to non-specific network types. Whereas deciding upon the plausibility of a "somewhat relatedness" between two interactors is hardly possible, the correctness of a predicted interaction of a specific type (e.g., physical, metabolic, paralog) is almost trivial in some cases.

8.4
Conclusions

We outlined an integration process for omics data and demonstrated the suitability of machine learning methods to extrapolate an arbitrary type of interaction network with initially reduced protein coverage to complete interaction networks with reasonable accuracy. As a result we obtained three dedicated networks (metabolic, physical, and paralog), which were exemplarily used to analyze differentially regulated B cell lymphoma genes on a network level. Even though these networks still exhibit a significant number of false-positive/false-negative interactions, the practical benefits arising from the use of dedicated network types for omics data interpretation are evident.

Given that the type of interaction is indicated, manually identifying false-positives/false-negatives becomes feasible and thus a refinement of the raw hypothesis provided by machine learning algorithms is possible. Most importantly, interpreting descriptive omics data in a functional context, such as in a metabolic network, provides insights into the molecular biological processes determining phenotypic observables. Thus, drug targets can be identified based on their position in the network or, alternatively, based on their topological characteristics (e.g., closeness-centrality).

Given the layered, evolutionary complexity of biological systems, it will not be possible to understand them comprehensively on the basis of hypothesis-driven research alone. Likewise, it will not be possible to do so solely through omics studies of genes, proteins, and other molecules in aggregate. The two modes of research are complementary and synergistic. The twin complexities of cancer biology and drug discovery demand such synergy as a basis for the next generation of oncopharmacomic research. Thus, the question on which approach to use might settle in the middle [7].

References

1 Verma, I.M. (2002) Hypothesis-driven science. *Mol. Ther.*, **6**, 1.
2 Lander, E.S. *et al.* (2001) Initial sequencing and analysis of the human genome. *Nature*, **409**, 860–921.
3 Schena, M., Shalon, D., Davis, R.W., and Brown, P.O. (1995) Quantitative monitoring of gene expression patterns with a complementary DNA microarray. *Science*, **270**, 467–470.
4 Anderson, N.L. and Anderson, N.G. (1998) Proteome and proteomics: new technologies, new concepts, and new words. *Electrophoresis*, **19**, 1853–1861.
5 Goodacre, R., Vaidyanathan, S., Dunn, W.B., Harrigan, G.G., and Kell, D.B. (2004) Metabolomics by numbers: acquiring and understanding global metabolite data. *Trends Biotechnol.*, **22**, 245–252.
6 Hert, D.G., Fredlak, C.P., and Barron, A.E. (2008) Advantages and limitations of next-generation sequencing technologies: a comparison of electrophoresis and non-electrophoresis methods. *Electrophoresis*, **29**, 4618–4626.
7 Weinstein, J.N. (2002) "Omics" and hypothesis-driven research in molecular pharmacology of cancer. *Curr. Opin. Pharmacol.*, **2**, 361–365.
8 Perco, P., Rapberger, R., Siehs, C., Lukas, A., Oberbauer, R., and Mayer, B. (2006) Transforming Omics data into context: bioinformatics on genomics and proteomics raw data. *Electrophoresis*, **2**, 2659–2675.
9 ENCODE Consortium (2007) Identification and analysis of functional elements in 1% of the human genome by the ENCODE pilot project. *Nature*, **447**, 799–816.
10 Joyce, A.R. and Palsson, B.O. (2006) The model organism as a system: integrating "omics" data sets. *Nature*, **7**, 198–210.
11 MacKenzie, S. (2001) High-throughput interpretation of pathways and biology. *Drug News Perspect.*, **14**, 54.
12 Westfall, P.H. and Young, S.S. (1993) Resampling-Based Multiple Testing: Examples and Methods for P-Value Adjustment, Wiley Series in Probability and Mathematical Statistics, John Wiley & Sons, Inc., New York.
13 Tusher, V.G., Tibshirani, R., and Chu, G. (2001) Significance analysis of microarrays applied to the ionizing radiation response. *Proc. Natl. Acad. Sci. USA*, **24**, 5116–5121.
14 Bryne, J.C., Valen, E., Tang, M.H., Marstrand, T., Winther, O., da Piedade, I., Krogh, A., Lenhard, B., and Sandelin, A. (2008) JASPAR, the open access database of transcription factor-binding profiles: new content and tools in the 2008 update. *Nucleic Acid Res.*, **36**, D102–D106.
15 Brown, K.R. and Jurisica, I. (2005) Online predicted human interaction database. *Bioinformatics*, **21**, 2076–2082.
16 Gietz, R.D., Triggs-Raine, B., Robbins, A., Graham, K., and Woods, R. (1997) Identification of proteins that interact with a protein of interest: applications of the yeast two-hybrid system. *Mol. Cell. Biochem.*, **172**, 67–79.
17 Lu, L.J., Sboner, A., Huang, Y.J., Lu, H.X., Gianoulis, T.A., Yip, K.Y., Kim, P.M., Montelione, G.T., and Gerstein, M.B. (2007) Comparing classical pathways and modern networks: towards the development of an edge ontology. *Trends Biochem. Sci.*, **32**, 320–331.
18 Alexeyenko, A. and Sonnhammer, E.L. (2009) Global networks of functional coupling in eukaryotes from comprehensive data integration. *Genome Res.*, **19**, 1107–1116.
19 Jensen, L.J., Kuhn, M., Stark, M., Chaffron, S., Creevey, C., Muller, J., Doerks, T., Julien, P., Roth, A., Simonovic, M., Bork, P., and von Mering, C. (2009) STRING 8 – a global view on proteins and their functional interactions in 630 organisms. *Nucleic Acids Res.*, **37**, D412–D416.
20 Bernthaler, A., Mühlberger, I., Fechete, R., Perco, P., Lukas, A., and Mayer, B. (2009) A dependency graph approach for analysis of differential gene expression profiles. *Mol. Biosyst.*, **5**, 1720–1731.

21 Duarte, N.C., Becker, S.A., Jamshidi, N., Thiele, I., Mo, M.L., Vo, T.D., Srivas, R., and Palsson, B. (2007) Global reconstruction of the human metabolic network based on genomic and bibliomic data. *Proc. Natl. Acad. Sci. USA*, **104**, 1777–1782.

22 Jeong, H., Tombor, B., Albert, R., Oltvai, Z.N., and Barabási, A.-L. (2000) The large-scale organization of metabolic networks. *Nature*, **407**, 651–654.

23 Kelley, R. and Ideker, T. (2005) Systematic interpretation of genetic interactions using protein networks. *Nat. Biotechnol.*, **23**, 561–566.

24 Rapaport, F., Zinovyev, A., Dutreix, M., Barillot, E., and Vert, J.-P. (2007) Classification of microarray data using gene networks. *BMC Bioinformatics*, **8**, 35.

25 Hughes, A.L. and Friedman, R. (2005) Gene duplication and the properties of biological networks. *J. Mol. Evol.*, **61**, 758–764.

26 Pearson, W.R. (1996) Effective protein sequence comparison. *Methods Enzymol.*, **266**, 227–258.

27 Minowa, Y., Katayama, T., Nakaya, A., Goto, S., and Kaneshisa, M. (2003) Classification of protein sequences into paralog and ortholog clusters using sequence similarity profiles of KEGG/SSDB. *Genome Inform.*, **14**, 528–530.

28 Orchard, S. *et al.* (2007) The minimum information required for reporting a molecular interaction experiment (MIMIx). *Nat. Biotechnol.*, **25**, 894–898.

29 Kerrien, S. *et al.* (2007) IntAct – open source resource for molecular interaction data. *Nucleic Acid Res.*, **35**, D561–D565.

30 Barrett, T. *et al.* (2007) NCBI GEO: mining tens of millions of expression profiles – database and tools update. *Nucleic Acids Res.*, **35**, D760–D765.

31 Donders, A.R., van der Heijden, G.J., Stijnen, T., and Moons, K.G. (2006) Review: a gentle introduction to imputation of missing values. *J. Clin. Epidemiol.*, **59**, 1087–1091.

32 Barnard, J. and Meng, X.L. (1999) Applications of multiple imputation in medical studies: from AIDS to NHANES. *Stat. Methods Med. Res.*, **8**, 17–36.

33 Barabsi, A.-L. and Oltvai, N.Z. (2004) Network biology: understanding the cell's functional organization. *Nat. Rev. Genet.*, **5**, 101–113.

34 Platzer, A., Perco, P., Lukas, A., and Mayer, B. (2007) Characterization of protein-interaction networks in tumors. *BMC Bioinformatics*, **8**, 224.

35 Provost, F. and Weiss, G.M. (2001) The effect of class distribution on classifier learning: an empirical study. Technical Report, Department of Computer Science, Rutgers University.

36 Wolpert, D.H. (1996) The lack of *a priori* distinctions between learning algorithms. *Neural Comput.*, **8**, 1341–1390.

37 Quinlan, R. (1993) *C4.5: Programs for Machine Learning*, Morgan Kaufmann, San Mateo, CA.

38 Breiman, L. (2001) Random forests. *Mach. Learn.*, **1**, 5–32.

39 John, G. and Langley, P. (1995) Estimating continuous distributions in Bayesian classifiers. Eleventh Conference on Uncertainty in Artificial Intelligence, Montreal.

40 Enrique, C., Gutiérrez, J.M., and Hadi, A.S. (1997) Learning Bayesian networks, in *Expert Systems and Probabilistic Network Models, Monographs in Computer Science*, Springer, Berlin, pp. 481–528.

41 Shawe-Taylor, J. and Cristianini, N. (2000) *Support Vector Machines and Other Kernel-Based Learning Methods*, Cambridge University Press, Cambridge.

42 Widrow, B. and Lehr, M.A. (1990) 30 years of adaptive neural networks: perceptron, Madaline and backpropagation. *Proc. IEEE*, **78**, 1415–1442.

43 Aha, D.W., Kibler, D., and Albert, M.K. (2005) Instance-based learning algorithms. *Mach. Learn.*, **6**, 37–66.

44 Fawcett, T. (2005) An introduction to ROC analysis. *Pattern Recogn. Lett.*, **27**, 861–874.

45 Soffer, S.N. and Vazquez, A. (2005) Network clustering coefficient without degree-correlation biases. *Phys. Rev. E*, **71**, 057101.

46 Rhodes, D.R., Yu, J., Shanker, K., Deshpande, R., Ghosh, D., Barrette, T., Pandey, A., and Chinnaiyan, A.M. (2004) Oncomine: a cancer microarray database

and integrated data-mining platform. *Neoplasia*, **6**, 1–6.

47 Rosenwald, A. *et al.* (2001) Relation of gene expression phenotype to immunoglobulin mutation genotype in B-cell chronic lymphocytic leukemia. *J. Exp. Med.*, **194**, 1639–1647.

48 Rosenwald, A. *et al.* (2002) The use of molecular profiling to predict survival after chemotherapy for diffuse large-B-cell lymphoma. *N. Engl. J. Med.*, **346**, 1937–1947.

49 Zhan, F. *et al.* (2002) Global gene expression profiling of multiple myeloma, monoclonal gammopathy of undetermined significance, and normal bone marrow plasma cells. *Blood*, **99**, 1745–1757.

Part Three
Analysis of Gene Networks

9
What if the Fit is Unfit? Criteria for Biological Systems Estimation Beyond Residual Errors

Eberhard O. Voit

9.1
Introduction

The analysis of biological networks has made enormous strides in recent years. In the context of static networks, which do not change over short periods of time, new biological techniques have begun to permit the characterization of very large interaction maps (e.g., among proteins), and computational graph theory has been the tool of choice for analyzing and interpreting what these maps entail. Complementing these activities has been the exploration of dynamically changing, regulated biological systems. On the experimental side, these efforts have enormously benefitted from the astounding advances in high-throughput biology at the genomic, proteomic, metabolic, and physiological levels. On the analytical side, the procedures and results of this new field of experimental systems biology have first been supported by a rapidly expanding repertoire of bioinformatics tools that permit the storage, retrieval, and analysis of very large datasets. More recently, the bioinformatics tools have become tightly interwoven with analytical and simulation techniques that are at the heart of the emerging field of computational systems biology.

At this point in time, it is no longer a real challenge to simulate large linear or even nonlinear systems in the form of algebraic or differential equations. It has also become feasible to simulate hybrid systems that contain continuous and discrete events, stochastic effects, and delays (e.g., [1, 2]). The simplicity with which we can perform large-scale simulations is in stark contrast to the overwhelming challenges we face much earlier in any biological systems analysis, namely when the biological phenomenon of interest is to be translated into a mathematical or computational model. This translation task may be subdivided into three aspects. The first concerns the acquisition of data. While biology is producing high-quality data in large quantities, these data are not always of the type and completeness that elucidate all aspects of the biological phenomenon from sufficiently many angles to construct a mathematical model. For instance, models of dynamic processes in the brain are hampered by the extremely difficult access to specific, restricted neuronal areas in living organisms [3]. This aspect of data availability will without doubt continue to

Applied Statistics for Network Biology: Methods in Systems Biology, First Edition.
Edited by M. Dehmer, F. Emmert-Streib, A. Graber, and A. Salvador.
© 2011 Wiley-VCH Verlag GmbH & Co. KGaA. Published 2011 by Wiley-VCH Verlag GmbH & Co. KGaA.

improve throughout the foreseeable future and we will not discuss issues of data generation in this chapter. The other two components of the translation from biology into a computational construct are distinct, but closely related to each other. The first is the determination of suitable mathematical descriptions for all relevant details of the biological phenomenon, while the second is the identification of numerical values for the parameters in these descriptions [4]. These two fundamental tasks of computational systems biology are the focus of this chapter.

Before we discuss details, challenges, methods, and pitfalls associated with the construction of biological systems models, we should ask why such an effort appears to be worth our while. One might begin by pre-empting a widespread critique of modeling, namely that models merely recreate, often in a much abstracted and simplified fashion, what "real" biologists had known all along and in greater detail. So what, if a model produces results similar to those observed? Modelers are sometimes stunned by this critique, because it is certainly not a trivial matter to write computer code that fits a large collection of biological data well. The truth behind this (mis-)conception is that a well-fitting model is a necessary but not sufficient condition for greater things to come. Indeed, without further analysis, exploration, explanation, or prediction, an accurate fit by itself does not earn the modeler much more than bragging rights.

In generic terms, the construction of a model is worthwhile if the model is able to answer specific questions or helps decide between acceptance and rejection of a hypothesis. Such a hypothesis may take many different forms. It may be qualitative in a sense that one is primarily interested in whether some key variable in the system responds to a specific input with an increase or a decrease. It may be semiquantitative if one is interested in the rough extent of the response and it is quantitative or numerical if the model is supposed to show the correct value of the affected key variable. What level of accuracy is needed in a model result depends on the questions asked, on the effort one is willing to invest in the modeling effort, and on many issues associated with the biological phenomenon itself and with the model.

9.2
Model Design

The challenge of converting a biological system into a mathematical structure requires the specification of functions that describe all pertinent processes, as well as the identification of suitable parameter values. The selection of process descriptions is by no means trivial. Granted, there are situations where a function can be inferred from the type and mechanism of the process. For example, there is good reason to choose an exponential function for the description of the growth of a small bacterial population, because the process is biologically driven by repeated cell doubling. However, such cases of mechanism-based model selection are actually rare and even traditional choices like a Michaelis–Menten rate function for an enzyme-catalyzed reactions are not without troubling questions, because deep underlying assumptions like homogeneity of the medium and free movement of enzymes and substrates are usually not satisfied *in vivo* [5]. Physics affords us with a rich repertoire of proven formulations for fundamental characteristics like forces and energy, but in biology these fundamental aspects are convoluted and often mixed together in a complicated manner. As an example, consider the process of gene expression, which involves the opening of the DNA strands, the right spatial and temporal availability and action of transcription factors, and the complex process of transcription into RNA. It is simply impossible to reduce this collective event into mechanistic pieces that permit elementary, physics-based representations.

Faced with similar challenges, engineers typically resort to linear approximations. These are very convenient, because there are stringent rules for their design, as well as for their analysis. Indeed, the repertoire of analytical and computational methods for linear systems is huge. The problem with linear approaches in biology is that most phenomena are genuinely nonlinear. They saturate or oscillate in a stable fashion, show switches, and sometimes appear to be chaotic. Reducing their dynamics to linear functions would not permit a proper analysis of these features. At the same time, the number of nonlinear functions is infinite and there are no guidelines as to which of these might be optimal or even appropriate descriptions of biological processes. A useful alternative is a nonlinear approximation. The first idea presumably coming to mind might be a second-order (quadratic) approximation, but this choice actually turns out to be rather inconvenient for later analyses [6]. Instead, it has proven beneficial to approximate biological processes with linear functions in logarithmic coordinates. This procedure is mathematically sound, as it directly adheres to the tenets of Taylor's theory and leads to nonlinear descriptions that can capture all types of responses, including different types of oscillations and chaos [7, 8]. Besides, these representations have desirable properties for mathematical and computational analysis. The concept of linearization in logarithmic coordinates is the core of biochemical systems theory (BST) [9, 10], which has been documented in several hundred articles and book chapters; book-length descriptions include [11–14].

BST comes in two main variants. In the generalized mass action (GMA) formulation, every process is represented with one product of power-law functions. For instance, in the simple branched pathway with two feedback signals that is shown in Figure 9.1, the equation for X_2 can be formulated directly as:

$$\dot{X}_2 = \gamma_{21} X_1^{f_{21}} - \gamma_{22} X_2^{f_{221}} - \gamma_{23} X_2^{f_{222}} X_4^{f_{24}} \tag{9.1}$$

where the γ parameters denote rate constants, which can take any non-negative values, while f_{21}, f_{221}, f_{222}, and f_{24} are kinetic orders that may take any real values. As X_1 is the substrate of the production reaction, f_{21} is positive. By contrast, f_{24} is negative, because it represents the inhibitory signal exerted by X_4. In general, a GMA system always has the format:

$$\dot{X}_i = \gamma_{i1} \prod_{j=1}^{n+m} X_j^{f_{ij1}} \pm \gamma_{i2} \prod_{j=1}^{n+m} X_j^{f_{ij2}} \pm \cdots \pm \gamma_{ik} \prod_{j=1}^{n+m} X_j^{f_{ijk}} \pm \cdots \quad i=1,\ldots,n \tag{9.2}$$

In addition to the dynamically changing variables, the system may also contain independent variables, $X_{n+1}, \ldots, X_{n+m}$, which affect the system, but are not affected

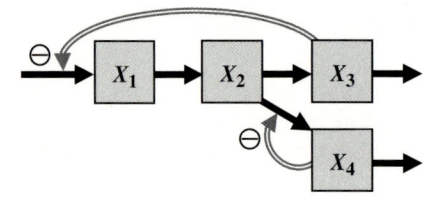

Figure 9.1 Generic pathway with one branch and two feedback signals.

by the system. In many cases, these variables are constant during a given mathematical experiment.

In the alternative S-system formulation, all processes entering a variable or pool are collectively represented with a single product of power-law functions that contains all variables affecting the collection of fluxes. Similarly, all processes leaving a variable or pool are collectively represented with a single product of power-law functions that contains all variables affecting the collection of fluxes. Revisiting the pathway in Figure 9.1, the only difference to the GMA formulation occurs for the degradation of X_2, which is now represented with only one term that contains both X_2 and X_4. Using the conventional parameter names for S-systems, the equation for X_2 is therefore:

$$\dot{X}_2 = \alpha_2 X_1^{g_{21}} - \beta_2 X_2^{h_{22}} X_4^{h_{24}} \tag{9.3}$$

All other equations are the same as before, with the minor deviation of traditionally different names for the parameters. Accounting again for independent variables, the generic S-system format is:

$$\dot{X}_i = \alpha_i \prod_{j=1}^{n+m} X_j^{g_{ij}} - \beta_i \prod_{j=1}^{n+m} X_j^{h_{ij}} \quad i = 1, \ldots, n \tag{9.4}$$

BST models have a number of advantages over *ad hoc* formulations. First, in order to formulate the system equations one does not have to know the true mechanisms governing the phenomenon of interest. As long as it is clear how the system is "connected" (i.e., which component affects which other components), either as a source of material or as the source of a regulatory signal, it is a straightforward procedure to set up a symbolic model. Here, "symbolic" means that we formulate the equations of the model, but that we do not know what the values of their parameter are; examples are Equations 9.1–9.4. The second advantage of BST models is that each parameter, even if it does not yet have a specific value, has a clearly defined meaning. An example is f_{24}, which exclusively represents the inhibition by X_4 of the conversion between X_2 and X_4. Similarly, it is immediately clear into which parameter a particular feature of a biological system has to be translated. This one-to-one relationship between biological aspects and parameters is very helpful both during the model design and the interpretation of results from the model analysis. The third advantage of the BST formulation is that the particular format has convenient mathematical properties. The main disadvantage of BST is that there is no guarantee that the model will capture all relevant features of the biological system of interest with sufficient accuracy. While the Taylor approximation is mathematically guaranteed to match any data close to a chosen operating point, this guarantee extends only over a very small range. Outside this range, which is of unknown size, the model may or may not fit well. BST shares this feature with all other models in biology.

BST models are just one option for nonlinear representations, but they are especially powerful as initial default models when not much is known about the details of the biological system. Specifically, since the construction of the symbolic equations is essentially automatic, the model design challenge is reduced to determining optimal parameter values with which the model matches the observed data well. This determination, while still difficult, is to be seen in comparison with the task of setting up a model with unspecified functions that are supposed to capture the dynamics of complex and ill-characterized processes. Many methods are available for parameter estimation purposes (for a recent review, see [4]), but none of them works perfectly or even satisfactorily in cases of moderately large biological systems.

9.3
Concepts and Challenges of Parameter Estimation

Methods of parameter estimation for biochemical systems fall into two broad categories, which are directly tied to the available types of data (e.g., see chapter 5 in [14]). In the past, the data almost always consisted of kinetic features associated with a particular step in the biochemical pathway system. Such features included the K_M of the enzyme, sometimes a flux rate or V_{max}, a dissociation constant, or some other kinetic characteristic. Also, once in a while steady-state values for the variables or fluxes of the system were available. Given such data, the "bottom-up" strategy of parameter estimation consisted of formulating each step symbolically and optimizing the parameter values such that they matched the alleged shape of the reaction step. Subsequently, all "local" descriptions (of individual reaction steps) were merged into a system of differential equations that described the entire pathway, the equations were integrated, and the numerical solutions of the system were compared against additional observation data. As the comparison typically led to numerical inconsistencies, one had to go back to the individual process representations many times and adjust functions or parameter values. This iterated reformulation and data matching could easily take months if not years. The vast majority of all existing biochemical models have been estimated with methods of this type; a very detailed example is [15].

Recent advances in molecular biology have begun to offer an alternative "top-down" approach to parameter estimation. The data here consist of measurements of metabolite concentrations at many subsequent time points, usually following some stimulus. The estimation now occurs in the "opposite" direction. Namely, one attempts to determine parameter values such that the solutions of the differential equations match the observed time-series data. The local parameters, which earlier were the starting point, are now the result of the top-down estimation. This type of an estimation task is known in mathematics and computing as an inverse problem, for which many algorithms are available. However, these algorithms tend to fail for moderately complicated biological systems [4].

As the identification of parameter values is a severe bottleneck of systems biology, enormous efforts have been dedicated to the development of general and specific techniques. One crucial issue is the fact that essentially all estimation algorithms are iterative. They use some method to determine a candidate set of parameter values, solve the differential equations with these parameter values, compare the solution with the observed data, and evoke some method for improving the parameter set for the next iteration. As an immediate consequence, the set of differential equations must be solved thousands of times and while each solution may be relatively fast, their collection can become prohibitively long. To circumvent this particular issue, two independent groups [16, 17] proposed almost 30 years ago the smoothing of the raw data and the interpretation of the slopes of the time course for each metabolite as estimates for the differentials on the left-hand sides of the differential equations (Figure 9.2). Thus, according to this method, each differential equation is replaced with a set of K algebraic equations, where K is the number of time points where the metabolite concentrations and slopes are measured or estimated. Each of the algebraic equations contains on the left-hand side the estimated slope value $S(t_k)$ at a given time point t_k and on the right-hand side the expression given by the differential equation and evaluated at t_k. Pursuing this strategy, estimating the parameters $p_1, \ldots, p_M$ in the differential equation:

$$\dot{X}_i = f_i(X_1, X_2, \ldots, X_n; p_{i1}, \ldots, p_{iM_i}) \tag{9.5}$$

becomes a task involving a system of K algebraic equations of the form:

$$S_i(t_k) \approx f_i(X_1(t_k), X_2(t_k), \ldots, X_n(t_k); p_{i1}, \ldots, p_{iM_i}), \quad i=1,\ldots,n, \quad k=1,\ldots,K \tag{9.6}$$

As the estimation is based on observed time courses, all quantities $X_i(t_k)$ and $S_i(t_k)$ are known or estimated, and the only unknowns are the parameter values of $p_1, \ldots, p_M$.

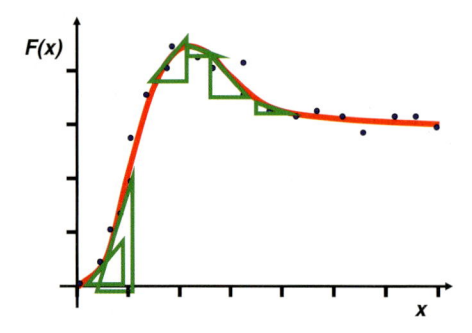

Figure 9.2 Generic univariate dataset (blue dots), a smoothing function (red), and examples of estimated slopes (indicated by green triangles) along the time course. The slopes thus estimated are used as surrogates of the differentials in the differential equations of the systems model.

This slope estimation strategy has two significant advantages. First, the need for solving differential equations is eliminated and the estimation is done instead with a purely algebraic system of equations. This conversion is very consequential, because in excess of 95% of the estimation time for a system of differential equations is spent on numerically integrating the equations, and this percentage can approach 100% [18]. Second, each time course is "decoupled" from the others, because it can be addressed independently of all other time courses. Intriguingly, if this method of slope estimation and decoupling is applied to Lotka–Volterra models, the initially nonlinear parameter estimation task becomes a simple linear regression [19]. Similarly, applied to S-systems within BST, the nonlinear estimation can be converted into an iteration of linear regression tasks [20] (see also [21]).

Clearly, the computational speed-up of the slope estimation and decoupling strategy is very appealing, but one might wonder about it statistical rigor. For a long time, the method was seen as such a convenient shortcut that it would outweigh possible concerns of statistical bias. The argument was that the solution obtained with this method could at least be used as a starting point for more conventional and possibly less biased estimations. However, Brunel recently showed that the procedure is asymptotically normal, consistent, and indeed statistically sound [22].

As the optimization of parameters in algebraic equations is much easier than in differential equations, one might be tempted to assume that the estimation task is essentially solved. However, this is not always the case. The slope estimation and decoupling strategy consists of two key steps – the smoothing of the data, which is necessary for the estimation of slopes, and the parameter estimation of the systems of nonlinear algebraic equations. The smoothing step is typically achieved with splines [23], although more sophisticated methods have also been proposed for this purpose [24, 25]. The choice of a spline or another smoother necessarily requires a decision regarding the degree of desired smoothness, and this decision cannot be made with total objectivity, because the smoothness of the data – or the lack thereof – depends on the processes governing the system from which the data were obtained, including experimental noise. At the same time it is quite evident that the degree of smoothing will affect the second step of slope estimation. No matter which method is chosen, smoothing incurs an approximation error, which is generally larger for smoother splines that are of lower order and consist of fewer pieces. If more spline pieces or higher-order splines are used, the approximation error is generally lower, but the appearance of the smoothing function is "bumpier" (Figure 9.3). This bumpiness may be due to time courses with many true ups and downs or to experimental noise in the data. If the latter is the case, a bumpy smoother simply tries to mimic the noise, the slopes along the smoothed time course increase and decrease to an unreasonable extent, and the subsequent parameter estimation results in a larger residual error. Thus, if the data contain even moderate noise, the smoothing error and the parameter estimation error are inversely related to each other, and the two must be weighed against each other. Addressing this issue with statistical rigor, Ramsay et al. developed algorithms that optimize weights associated with the two types of error [26]. A remaining question in this context is to what degree

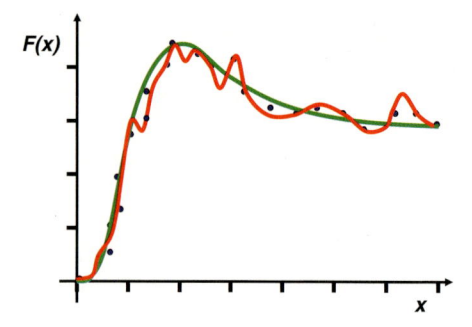

Figure 9.3 Different degrees of smoothing incur different residual errors. A smooth interpolation (green) generally results in a larger residual error with respect to the data (blue dots), whereas the more detailed smoother (red) is associated with a smaller error but a "bumpier" appearance, which may cause problems with the slope estimation and decoupling technique.

different extents of smoothing may lead to a loss of biological information, especially if the true "nature" of the data is unknown.

Overall, the slope estimation and decoupling strategy makes the estimation task simpler and much faster, but it does not solve all problems. We discuss some remaining issues in the following.

9.3.1
Typical Parameter Estimation Problems

The problems typically encountered in the estimation of parameter values fall into two classes. In the first class, the algorithm simply does not find a suitable solution and, as a consequence, the residual error is unacceptably high. In the second class, an algorithm does find a solution, but something is not quite right with this solution. We discuss the different cases one by one.

9.3.1.1 Data Fit is Unacceptable
If no satisfactory fit can be found, the reasons may be manifold. In a relatively clear-cut case, the algorithm does not converge at all, reaches the maximally permitted number of iterations, or produces a fit that is obviously very different from the observed data. In such cases, the foremost (although not necessarily only) reasons are likely of a technical nature. It might be that the computer- or user-suggested initial guesses for all parameters are simply so bad that the algorithm does not reach a basin of attraction surrounding the optimal solution. It is also possible that the algorithm is attracted to an unacceptable local minimum.

A distinctly different cause for not yielding a good solution may be that the alleged functions in the model are so far from the truth that the algorithm cannot determine a satisfactory solution. In contrast to purely technical issues, the result in this case is often "some" fit that, however, is clearly not optimal. For instance, if one attempts to fit a Michaelis–Menten function to a sigmoidal time course, it is clear that the initial

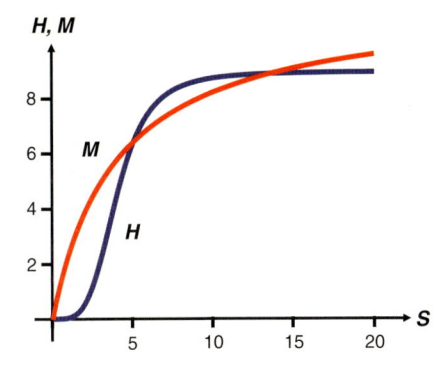

Figure 9.4 An algorithm tries to fit a sigmoidal Hill function (*H*; blue) with a Michaelis–Menten function (*M*; red) that simply does not have sufficient shape flexibility. As a consequence, no adequate fit can be reached.

phase cannot be matched appropriately, because the structure of the Michaelis–Menten model is not equipped to capture S-shaped datasets (Figure 9.4). In simple cases like the one described, the problem is easily detected and diagnosed. However, this analysis is not so readily accomplished in a high-dimensional parameter space. Of course, all combinations of the above causes may be encountered. Many studies have focused on these and other technical issues.

9.3.1.2 Differently Structured Candidate Models are Difficult to Compare

Within the realm of linear regression, methods have been developed for assessing the relative worth of an additional parameter. Specifically, objective criteria exist, based on residual errors and numbers of parameters and data points, for deciding the superiority of one of two candidate fits where one involves M and the other one $M + 1$ parameters [27]. For nonlinear estimation tasks, such comparisons are much more difficult, especially if different model structures are involved. Surprisingly, there is even ambiguity in the number of parameters. As an example, consider a Hill function of the type:

$$V(S) = \frac{V_{max} S^2}{K_M^2 + S^2} \tag{9.7}$$

which has a sigmoidal shape as shown in Figure 9.4. One will easily recognize K_M and V_{max} as parameters, but should the Hill coefficient (i.e., the power associated with S) be counted, even if it is *a priori* set equal to 2? After all, it clearly affects the shape of the function and if 2 did not fit, we could easily change it to a different value. It is difficult to find an objective criterion accounting for this issue. Thus, it is in general not a trivial matter to compare two fits, such as one with a Michaelis–Menten model (which is a special case with the Hill coefficient equaling 1) and one with a Hill model with a fixed or tunable Hill coefficient. Similarly, it is difficult to compare fits with a Hill model and a logistic model, which can both capture sigmoidal processes equally well but have different mathematical structures (Figure 9.5).

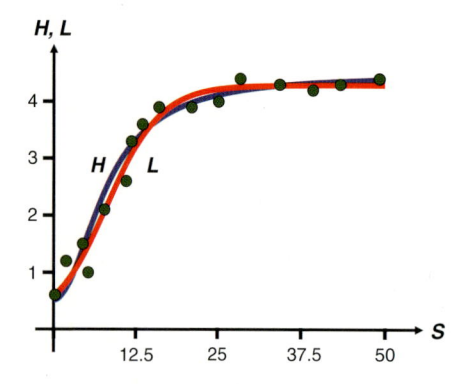

Figure 9.5 Data (green symbols) following a sigmoidal trend are equally well represented by a Hill function (H; blue; $H(S) = 4S^2/(8^2 + S^2) + 0.5$) or a logistic function (L; red; $L(S) = 4.3/[1 + \exp(-0.24 \cdot (S-8))]$).

9.3.1.3 The Fit is Acceptable, But...

In spite of all technical challenges, one often obtains a fit that is good, as judged by visual inspection and an acceptable sum of squared residual errors (SSE), and it might appear that the problem is solved. In some cases this is the case, but a good SSE should not be taken as the sole criterion.

The best-known situation is convergence to a local minimum. Even if an optimization algorithm returns an acceptable fit, there is no guarantee that the fitted model is truly the best option. It is easily possible that the solution corresponds to a local minimum and that other solutions, maybe far away, are even better. However, the algorithm may not necessarily find these superior solutions because they are separated from the current solution by domains of parameter sets that correspond to drastically inferior model fits. Thus, any time the algorithm attempts to move toward the global optimum, it hits the separating areas of high errors, deems the direction futile, and searches in other parts of the parameter space. A partial, but not always effective solution to the problem is repeated optimization with different initial guesses for some or all parameters.

A slightly different situation occurs if the identified optimum is surrounded by a large area of solutions with very similar errors. In the simplest cases, these "almost optimal" solution sets form slightly distorted ellipses in a higher-dimensional space, but this is not necessarily so. Recent years have seen quite a bit of attention dedicated to what is now called "sloppy" solutions (e.g., [28, 29]). One could argue that the optimized solution is still a tiny bit better than its neighbors in the sloppy set and that it should therefore be preferred over all other candidates. However, in some cases the residual errors within the sloppy set are so similar that a slight change in just one data point, which could easily correspond to experimental noise, would identify a different solution as optimal. Thus, one should not discard the range of parameters surrounding the optimized solution. In a way, sloppy solution sets are not necessarily a cause for concern, because the corresponding models are quite similar in their fits to the data. In fact, it might be possible to simplify the optimized solution by setting certain parameters equal to zero, if the sloppy set includes this setting as a legitimate possibility. The *SSE* would very slightly increase, but arguments of simplicity might favor this increase in exchange for a simpler model structure. Intriguingly, if the sloppy set permits positive, zero, as well as negative values for a certain parameter, the interpretation of the corresponding models differs. For instance, within the context of BST, a negative kinetic order is to be interpreted as an inhibitory signal, while a positive kinetic order suggests an activating or augmenting influence and a value of zero corresponds to an unimportant role of the parameter. This result poses the interesting question of whether sloppy solutions of this type are computational artifacts or whether inter-individual variation could go so far as to allow activation in one organism and inhibition in another [30]. At present, this question cannot be answered with confidence. Finally, while Ockham's razor might suggest that the simpler solution is to be preferred, biology has presented us in many cases with solutions that initially seem more complex than necessary. Later we may find that the more complex solutions are preferable because of improved robustness or other higher-order features.

Somewhat related to sloppy solutions is the situation where the data are not plentiful of informative enough to allow a precise determination of all parameter values. A very simple example arises if two parameters p_1 and p_2 always appear in the model in the same constellation, such as the ratio p_1/p_2. Clearly, by multiplying the two parameters with a nonzero factor, the ratio is unchanged. Turning the argument around, a search algorithm cannot find a unique solution in this situation, but only one where the ratio fits well. In these cases, the solution is "structurally non-identifiable," which entails that infinitely many combinations of parameter values can yield solutions with the same *SSE*. It is recommended to remove these identifiability issues, for instance, with methods of model reduction [31].

Quite a different concern with an apparently good fit is the occasional identification of parameter values that are impossible or unreasonable from a biological point of view. The most obvious case is an optimized value that turns out to be negative, although the parameter must be positive. Examples include rate constants and K_M values. If this situation occurs, the optimization should be redone with corresponding constraints on the parameters. A more subtle variation of this issue is a resulting model that is unstable or extremely sensitive. It may happen in this case that a very small percent change in a parameter value would lead to dramatic changes in an important system feature such as the steady state. It is rare that such a high sensitivity is realistic in biology. Whether the situation is caused by a faulty parameter or by the misidentification or omission of some process or regulatory signal cannot be said without further analysis.

Another common and generic issue with an apparently good fit is often overlooked at first and becomes an inconvenient surprise later. This issue is the frequent inability of the parameterized model to predict responses to new stimuli in an adequate fashion [32]. Thus, the original data are matched quite nicely and the model traverses the cloud of data points seemingly fine. It may also be able to predict responses of the model throughout a modest time period beyond the measured time points. However, if new observations are to be modeled, the model may woefully fail. How is that

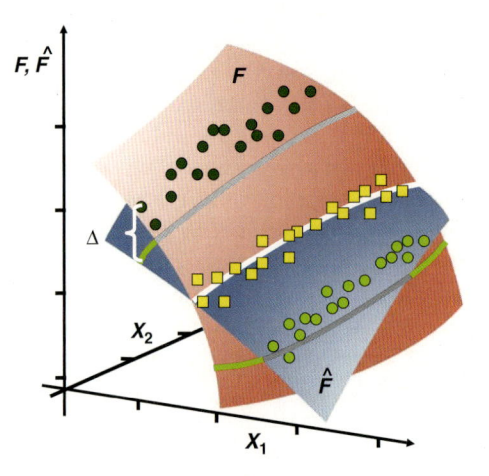

Figure 9.6 Inability of a model to represent new data. In this generic example, the true function of a flux $F(X_1, X_2)$ (red surface) is modeled with the function $\hat{F}(X_1, X_2)$ (blue surface), based on one dataset (yellow squares). The fit (white line) is quite good. However, since F and $\hat{F}$ only intersect in the (white) model fit and otherwise diverge, extrapolated fits to other data (light and dark green circles) are no longer accurate and lead to significant residual errors (exemplified with Δ).

possible? One explanation is that the fitted curve constitutes the intersection of two (usually curved) surfaces – one describing the model function with optimized parameter values and the other representing the true dynamics of the biological system (see Figure 9.6). If the data points are close to this intersection, the fit not only appears to be good, but actually is so. However, analyzing new data means moving along the true surface and away from the intersection, and it becomes significant that the two surfaces diverge. As a result, if the extrapolation differs substantially from the original data, the model is no longer able to capture it. Again, several root causes for this situation are possible. One is the compensation of error among different terms within a system of equations. As an example, consider an equation with two terms, of which one is modeled quite badly. It is not difficult to imagine that, at least in certain situations, the other term could compensate for the error. In fact by also being modeled badly, the overall results can be surprisingly good. At first glance, it appears that two wrongs could indeed make a right. However, the error compensation usually no longer holds for new situations where the system variables in the two terms of the equation change to different degrees. As consequence, the fit to the extrapolated dataset can be unacceptably bad. To some degree, the method of dynamic flux estimation can remedy the extrapolation problem [32]. However, this method requires ideal conditions that are seldom satisfied and therefore requires additional information, which is not always easy to obtain [33, 34].

Finally, a genuine challenge with the otherwise appealing slope estimation and decoupling technique is that dependencies among equations are *a priori* ignored. For instance, if the same parameter appears in two equations, the decoupling causes independent estimations and unless countermeasures are put into place, the same parameter might be identified with two different values [21].

9.3.1.4 Needed: A Better Fit! Or Not?

The previous sections have demonstrated that even an apparently good fit may not be the ultimate solution, because, for instance, it may not hold up in extrapolations. If so, what should be done? This is a complex question, and as in so many cases the answer is: it depends. The need to search for other solutions depends on the purpose of the model and the availability of appropriate data.

In some cases, even symbolic models, in which parameters are not at all numerically defined, can yield valuable insights. In fact, if insights can be gained for arbitrary parameter values, they are usually more general than results obtained for certain numerical parameter combinations. A beautiful example is the exploration of design principles [35]. In this line of research one asks what the role of a particular feature of a biological system is. For instance, one might observe a feedback inhibition signal in a pathway system and ask why it is there. According to the method of controlled mathematical comparisons (MCMC), one sets up two essentially identical systems models in parallel; however, one represents the observed signal and the other one misses the signal of interest [36]. The responses of these two models are compared with respect to performance criteria, such as robustness and response time. Studying a whole roster of such features and results, one system design is ultimately declared superior, either in general or within a certain environment. Many such comparisons have been performed without the specification of numerical parameter values, while other comparisons required the definition of relevant ranges of parameters. Using this MCMC strategy, Savageau proposed general rules for the regulation of bacterial gene circuits that have held up in all cases tested so far [37, 38]. These rules were independent of specific parameter values and identified the superior circuit designs primarily based on demands exerted by the environmental of the bacteria. Other examples included more complex gene circuit designs and a variety of other system structures [39–44]. In the context of signaling cascades, the structural design and performance demands even helped determine the ranges of effective parameter values [45].

Trusting in the observation that accurate parameter values are often not as critical as the correct model structure, Alves *et al.* exhaustively evaluated likely parameter ranges on discrete grid points, thereby yielding sufficiently good, although maybe not optimal solutions [46]. Parameter ranges can also be restricted by biological and clinical constraints, and coarse solutions can be refined by means of simulations. In spite of the rather uncertain nature of this process, the model results can be surprisingly strong. For example, in an effort to construct a model of dopamine metabolism in the human brain, Qi *et al.* collected semiqualitative input from clinicians, biochemists, and toxicologists regarding the relative concentrations of relevant metabolites and the flux split ratios at diverging branch points in the pathway system [3]. This information turned out to be sufficient for setting up complex pathway models with coarse parameter values. Even though theses models did not capture the precise numerical features of the biological system, they helped explain the inner workings of the system. For instance, Qi *et al.* used these models to perform exhaustive analyses of root causes leading to Parkinson's disease and schizophrenia [3, 47, 48]. Models of this type can also indicate whether the responses to targeted alterations occur in the right direction. Thus, even if the parameter values are coarse, some types of valuable insights may be gained.

Alas, there are situations where the parameterization should be as good and reliable as possible. For instance, it is becoming possible to develop health and disease models, which could be used for designing specific, and maybe even personalized, treatment strategies [49]. Clearly such strategies need to be quantitative and accurate enough to allow at least modest extrapolations from the normal state. These models should also permit reliable predictions of what might happen to an individual if s/he is or is not treated for some abnormality. One might be tempted to discard such predictions into the future, because our current models are simply too inexact. However, even short-term predictions can be extremely beneficial. For instance, a 5-min warning that a critical care patient is diverging from the normal trajectory might be sufficient to initiate efficient countermeasures [50]. Such short time horizon seems to be within reach of computational models, even if they are based on relatively coarse approximations.

If the fit is good with respect to the SSE of one data set, but fails in extrapolations, the situation is dire, because there is no general diagnostic tool for identifying where the problem lies or whether the reason of failure is a combination of several problems. To some degree, the modeling process needs to be restarted in such a case. Of course, the process does not really start at the very beginning, because a reasonable model for at least one situation is now available and this model may serve as a starting point or as a constraint for further model development. One might also expect both, greater reliability of the model and insight into the true nature of the biological processes if more and ideally diverse datasets are available that cover a greater portion of the space of variation in the system variables. Nevertheless, the difficult question at this point is whether the chosen functions in the model are appropriate or not. If they are, renewed parameter estimation with additional data might lead to a better numerical model implementation, but if they are not, new functions need to be determined. This task can be very difficult, because the structure of these functions is *a priori* unclear and because one usually does not even have data regarding the individual functions, but only regarding the entire dynamics of the system, which is governed by numerous processes simultaneously [34].

One strategy is to use biological insights to find functions that might be appropriate representations of individual processes. Accordingly, one designs a model with more complex, biologically relevant functions, and fits it to the original data and to the new data at the same time. The original, simpler model may be evoked as a constraint in a sense that the new function should coincide with the original model function (and possibly its slopes) with respect to the original data. In principle, this strategy seems to make sense, but in reality it often requires considerable effort [34]. Ultimately, this strategy leads to the question of when enough is enough. Is it really desirable to develop a very complicated process description that now captures two datasets? It is known that even slight overparameterization tends to lead to extrapolation problems and the more parameters are involved in a functional description, the more severe the potential problem is.

An alternate strategy is an extension of the concept of generic approximations. Experience has shown that extensions toward higher-order Taylor terms become mathematically cumbersome and while they naturally fit better, the facility of their analysis is compromised [6]. Instead, one may employ a piecewise approximation [51]. This strategy is straightforward in principle, but realistically requires the determination of breakpoints, at which one approximation is substituted with another. For univariate functions, the breakpoints can almost be determined by visual inspection or with a simple algorithm, but the determination is much harder for multivariate functions. For linear and power-law functions, algorithms have been developed that automate this process [52, 53]. The result is an approximation, consisting of a certain number of pieces, throughout which the overall SSE is within desired bounds. The actual estimation of parameter values for these piecewise approximations requires considerably more data and of course the number of parameter values grows with each added piece. Nevertheless, this strategy is relatively unbiased and therefore may offer a first default in situations where biology does not suggest candidate functions or where such candidates are so complicated [54] that there is hardly a chance that parameters could be estimated without serious overfitting.

9.4
Conclusions

The goal of parameter estimation is obvious – find numerical values that render a model optimal for the representation of a biological system. While clear in principle, the task is often convoluted in practice. In the past, most data were coarse and scarce, and simple model fits had to be considered adequate. With recent advances in molecular biology, the situation has changed and some data are so good now that it is difficult to excuse bad fits.

As parameter estimation is so important, many groups have devoted substantial effort to it. In most cases, these investigations focused on the substantial technical challenges associated with the task and on algorithmic improvements. In addition to distinctly different optimization methods, which include regression, simulated annealing, and numerous variants of genetic, colony, and swarm algorithms, it has turned out that the estimation of slopes and the subsequent conversion of differential into algebraic equations is very beneficial. While these methodological and algorithmic improvements have made parameter estimation manageable in principle, they have not solved all problems. There are still issues with slow convergence or the trapping of algorithms in local minima.

Beyond the well-known technical issues, even apparently good fits should be subjected to additional muster. Criteria for such additional tests should be the reasonableness of the numerical values of all parameters, model stability, sensitivity, and robustness, and the ability to provide good fits in extrapolations. Possible causes for models to fail these criteria are plentiful. Particularly hideous among them is the compensation of fitting errors within and among the system equations, because such compensation is not always easy to diagnose and remediate. The recent method of dynamic flux estimation [32] is able to help, but only under ideal conditions, and it will be necessary to expand such methods toward more realistic conditions.

If a fit with a good SSE is obtained, but extrapolations cause problems, the entire model structure may have to be revisited, including the choice of functions, the numbers and types of parameters, and the availability of data. Clearly, several replicates of time-series data and data obtained under different conditions will allow better estimations, along with their statistical analyses and interpretations. Ultimately, one must consider the purpose of the model and judge the real need for accurate parameter values. Indeed, a simple model with fewer parameters is often more robust and less sensitive to overparameterization, yet may yield as much, if not more insight.

Acknowledgments

The author expresses his gratitude to Yun Lee for help with one of the figures. This work was supported in part by a Molecular and Cellular Biosciences Grant (MCB-0958172; E.O. Voit, PI) from the National Science Foundation, a grant from the University System of Georgia (E.O. Voit and Y. Xu, PIs), a grant from the National Institutes of Health (2 RO1 GM063265-09; Y. Hannun, PI) and a Biological Energy Science Center (BESC) grant from the Department of Energy (P. Gilna, PI). Any opinions, findings, and conclusions or recommendations expressed in this material are those of the authors and do not necessarily reflect the views of the sponsoring institutions.

References

1 Chen, L., Qi-Wei, G., Nakata, M., Matsuno, H., and Miyano, S. (2007) Modelling and simulation of signal transduction in an apoptosis pathway by using times Petri nets. *J. Biosci.*, **32**, 113–127.

2 Wu, J. and Voit, E. (2009) Hybrid modeling in biochemical systems theory by means of functional Petri nets. *J. Bioinform. Comput. Biol.*, **7**, 107–134.

3 Qi, Z., Miller, G.W., and Voit, E.O. (2008) Computational systems analysis of dopamine metabolism. *PLoS ONE*, **3**, e2444.

4 Chou, I.-C. and Voit, E.O. (2009) Recent developments in parameter estimation and structure identification of biochemical and genomic systems. *Math. Biosci.*, **219**, 57–83.

5 Savageau, M.A. (1995) Enzyme kinetics *in vitro* and *in vivo*: Michaelis–Menten revisited, in *Principles of Medical Biology* (ed. E.E. Bittar), JAI Press, Greenwich, CT, pp. 93–146.

6 Cascante, M., Sorribas, A., Franco, R., and Canela, E.I. (1991) Biochemical systems theory: increasing predictive power by using second-order derivatives measurements. *J. Theor. Biol.*, **1991**, 521–535.

7 Savageau, M.A. and Voit, E.O. (1987) Recasting nonlinear differential equations as S-systems: a canonical nonlinear form. *Math. Biosci.*, **87**, 83–115.

8 Voit, E.O. (1993) S-system modeling of complex systems with chaotic input. *Environmetrics*, **4**, 153–186.

9 Savageau, M.A. (1969) Biochemical systems analysis. I. Some mathematical properties of the rate law for the component enzymatic reactions. *J. Theor. Biol.*, **25**, 365–369.

10 Savageau, M.A. (1969) Biochemical systems analysis. II. The steady-state solutions for an *n*-pool system using a power-law approximation. *J. Theor. Biol.*, **25**, 370–379.

11 Savageau, M.A. (1976) *Biochemical Systems Analysis: A Study of Function and Design in Molecular Biology*, Addison-Wesley, Reading, MA, p. 379.

12 Voit, E.O. (ed.) (1991) *Canonical Nonlinear Modeling. S-System Approach to Understanding Complexity*, Van Nostrand Reinhold, New York, p. 365.

13 Torres, N.V. and Voit, E.O. (2002) *Pathway Analysis and Optimization in Metabolic Engineering*, Cambridge University Press, Cambridge.

14 Voit, E.O. (2000) *Computational Analysis of Biochemical Systems: A Practical Guide for Biochemists and Molecular Biologists*, Cambridge University Press, Cambridge, p. 530.

15 Alvarez-Vasquez, F., Sims, K.J., Hannun, Y.A., and Voit, E.O. (2004) Integration of kinetic information on yeast sphingolipid metabolism in dynamical pathway models. *J. Theor. Biol.*, **226**, 265–291.

16 Varah, J.M. (1982) A spline least squares method for numerical parameter estimation in differential equations. *SIAM J. Sci. Stat. Comput.*, **3**, 28–46.

17 Voit, E.O. and Savageau, M.A. (1982) Power-law approach to modeling biological systems; III. Methods of analysis. *J. Ferment. Technol.*, **60**, 233–241.

18 Voit, E.O. and Almeida, J. (2004) Decoupling dynamical systems for pathway identification from metabolic profiles. *Bioinformatics*, **20**, 1670–1681.

19 Voit, E.O. and Chou, I.C. (2010) Parameter estimation in canonical biological systems models. *Int. J. Syst. Synth. Biol.*, **1** (1), 1–19.

20 Chou, I.C., Martens, H., Voit, E.O. (2006) Parameter estimation in biochemical systems models with alternating regression. *Theor. Biol. Med. Model*, **3**, 25.

21 Vilela, M., Chou, I.C., Vinga, S. et al. (2008) Parameter optimization in S-system models. *BMC Syst. Biol.*, **2**, 35.

22 Brunel, N.J.-B. (2008) Parameter estimation of ODE's via nonparameteric estimators. *Electr. J. Stat.*, **2**, 1242–1267.

23 Seatzu, C. (2000) A fitting based method for parameter estimation in S-systems. *Dyn. Syst. Appl.*, **9**, 77–98.

24 Eilers, P.H. (2003) A perfect smoother. *Anal. Chem.*, **75**, 3631–3636.

25 Vilela, M., Borges, C.C., Vinga, S. et al. (2007) Automated smoother for the numerical decoupling of dynamics models. *BMC Bioinformatics*, **8**, 305.

26 Ramsay, J.O., Hooker, G., Cao, J., and Campbell, D. (2007) Parameter estimation for differential equations: a generalized smoothing approach. *J. Roy. Stat. Soc. B*, **69**, 741–796.

27 Neter, J. and Wasserman, W. (1974) *Applied Linear Statistical Models*, Irwin, Homewood, IL.

28 Gutenkunst, R.N., Casey, F.P., Waterfall, J.J., Myers, C.R., and Sethna, J.P. (2007) Extracting falsifiable predictions from sloppy models. *Ann. NY Acad. Sci.*, **1115**, 203–211.

29 Gutenkunst, R.N., Waterfall, J.J., Casey, F.P. et al. (2007) Universally sloppy parameter sensitivities in systems biology models. *PLoS Comput. Biol.*, **3**, 1871–1878.

30 Vilela, M., Vinga, S., Mattoso Maia, M.A.G., Voit, E.O., and Almeida, J. (2009) Identification of neutral sets of biochemical systems models from time series data. *BMC Syst. Biol.*, **3**, 47.

31 Raue, A., Kreutz, C., Maiwald, T. et al. (2009) Structural and practical identifiability analysis of partially observed dynamical models by exploiting the profile likelihood. *Bioinformatics*, **25**, 1923–1929.

32 Goel, G., Chou, I.-C., and Voit, E.O. (2008) System estimation from metabolic time series data. *Bioinformatics*, **24**, 2505–2511.

33 Voit, E.O., Goel, G., Chou, I.-C., and da Fonseca, L. (2009) Estimation of metabolic pathway systems from different data sources. *IET Syst. Biol.*, **3**, 513–522.

34 Goel, G. (2009) Dynamic Flux Estimation: a novel framework for metabolic pathway analysis. Dissertation, Department of Biomedical Engineering. Georgia Institute of Technology, Atlanta, GA.

35 Savageau, M.A. (2001) Design principles for elementary gene circuits: elements, methods, and examples. *Chaos*, **11**, 142–159.

36 Savageau, M.A. (1985) A theory of alternative designs for biochemical control systems. *Biomed. Biochim. Acta*, **44**, 875–880.

37 Savageau, M.A. (1998) Demand theory for gene regulation: quantitative development of the theory. *Genetics*, **149**, 1665–1676.

38 Savageau, M.A. (1998) Demand theory for gene regulation: quantitative application to the lactose and maltose operons of *Escherichia coli*. *Genetics*, **149**, 1677–1691.

39 Coelho, P.M., Salvador, A., and Savageau, M.A. (2009) Quantifying global tolerance of biochemical systems: design implications for moiety-transfer cycles. *PLoS Comput. Biol.*, **5**, e1000319.

40 Igoshin, O.A., Price, C.W., and Savageau, M.A. (2006) Signalling network with a bistable hysteretic switch controls developmental activation of the sigma transcription factor in *Bacillus subtilis*. *Mol. Microbiol.*, **61**, 165–184.

41 Alves, R. and Savageau, M.A. (2000) Effect of overall feedback inhibition in unbranched biosynthetic pathways. *Biophys. J.*, **79**, 2290–2304.

42 Hlavacek, W.S. and Savageau, M.A. (1996) Rules for coupled expression of regulator and effector genes in inducible circuits. *J. Mol. Biol.*, **255**, 121–139.

43 Irvine, D.H. and Savageau, M.A. (1985) Network regulation of the immune response: alternative control points for suppressor modulation of effector lymphocytes. *J. Immunol.*, **134**, 2100–2116.

44 Schwacke, J.H. and Voit, E.O. (2004) Improved methods for the mathematically controlled comparison of biochemical systems. *Theor. Biol. Med. Model.*, **1**, 1.

45 Schwacke, J.H. and Voit, E.O. (2007) Concentration-dependent effects on the rapid and efficient activation of MAPK. *Proteomics*, **7**, 890–899.

46 Alves, R., Herrero, E., and Sorribas, A. (2004) Predictive reconstruction of the mitochondrial iron–sulfur cluster assembly metabolism. II. Role of glutaredoxin Grx5. *Proteins Struct. Funct. Bioinformat.*, **57**, 481–492.

47 Qi, Z., Miller, G.W., and Voit, E.O. (2008) A mathematical model of presynaptic dopamine homeostasis: implications for schizophrenia. *Pharmacopsychiatry*, **41**, S89–S98.

48 Qi, Z., Miller, G.W., and Voit, E.O. (2009) Computational analysis of determinants of dopamine dysfunction. *Synapse*, **63**, 1133–1142.

49 Voit, E.O. and Brigham, K.L. (2008) The role of systems biology in predictive health and personalized medicine. *Open Path J.*, **2**, 68–70.

50 Lu, Y., Burykin, A., Deem, M.W., and Buchman, T.G. (2009) Predicting clinical physiology: a Markov chain model of heart rate recovery after spontaneous breathing trials in mechanically ventilated patients. *J. Crit. Care*, **24**, 347–361.

51 Savageau, M.A. (2001) Design principles for elementary gene circuits: elements, methods, and examples. *Chaos*, **11**, 142.

52 Ferrari-Trecate, G. and Muselli, M. (2002) A new learning method for piecewise linear regression. *Lecture Notes Comput. Sci.*, **2415**, 444–449.

53 Machina, A., Ponosov, A., and Voit, E.O. (2010) Automated piecewise power-law modeling of biological systems. *J. Biotechnol.*, **149**, 154–165.

54 Peskov, K., Goryanin, I., and Demin, O. (2008) Kinetic model of phosphofructokinase-1 from *Escherichia coli*. *J. Bioinform. Comput. Biol.*, **6**, 843–867.

10
Machine Learning Methods for Identifying Essential Genes and Proteins in Networks
Kitiporn Plaimas and Rainer König

10.1
Introduction

Defining essential genes or their corresponding proteins enables identifying potential drug targets and may also provide an understanding of the minimal requirements for a synthetic cell. However, experimentally assaying the essentiality of coding genes is resource intensive and not feasible for all organisms, specifically if they are pathogenic, such as *Salmonella* or *Staphylococcus*. Thus, computational methods supporting the prediction of essential genes are needed for directed drug development. Finding essential genes and the proteins they encode can be facilitated by investigating their function with respect to other genes and proteins. Cellular networks can be put up to analyze these interactions systematically. A cellular network can conceptually be divided into three parts: the metabolic network, the signaling network, and the transcriptional regulatory network. Metabolism is the best observed and described part of the network. Metabolic reactions typically involve enzymatic conversion and mass flow of small molecules (e.g., sugars) that have been studied for several decades using enzyme kinetics and tracer experiments. In contrast, knowledge about signaling interactions is much less established on a general level, and models are often obtained from the functional context and potential wiring/rewiring aspects. Finally, the transcriptional regulatory network contains the less conserved topology. It adapts broadly and often very dynamically to the physiological situation.

To identify drug targets, typically, the cellular network is analyzed by removing a node in the network model mimicking specific drug treatment that inhibits the corresponding protein. A single node in the network needs then to be characterized by estimating the robustness of the network when this node has been discarded. Several network topology-based features have been described to identify essential genes *in silico* for various organisms [1–6]. We will explain them in depth in the following sections. A single feature describing the topology may often not yield a good essentiality estimate and intelligently combining these features can yield a far more comprehensive model. We will briefly explain the method of machine learning that can do the job of combining these features. Machine learning techniques, such as decision trees, neural networks, or Support Vector Machines, are used to predict which genes are essential and which are nonessential. The machines learn from experimentally determined information about the essentiality of a group of genes and can apply their "knowledge" to predict the essentiality for other, new genes. They are trained to analyze the relationship between various gene characteristics using attributes (also called "features" or "descriptors" in the following) derived from the cellular networks, and also enable the easy integration of sequence information and information from other high-throughput data, like gene expression. In Section 10.2, we describe how to construct cellular networks, Section 10.3 explains useful and commonly used network descriptors which can serve as features for the machines, Section 10.4 outlines the basic concept of machine learning and how to apply it to our needs, and, finally, in Section 10.5, we review some successful examples reported in the literature.

10.2
Definitions and Constructions of the Network

The terms "network" and "graph" will be used synonymously in the following. A graph G (V, E) consists of vertices u, $v \in V$ and edges $(u, v) \in E$ connecting these vertices. Edges (u, v) can be undirected or directed. Directed edges are represented by ordered pairs of nodes (u, v) and lead from source u to sink v. They are graphically depicted by arrows. Undirected edges are represented by unordered pairs of nodes (u, v), and are depicted by a line between vertices u and v. They are used if information about the direction is lacking or not needed. Bidirectionality between vertices u and v can be represented by two edges – one leading from u to v and one in the opposite direction.

Metabolic networks are represented as bipartite graphs consisting of two disjoint sets of vertices $m \in M$ and $r \in R$ representing metabolites and reactions [7]. Directed edges lead from the substrates of a reaction to the reaction and from the reaction to its products. Doing this for every reaction yields a network that consists of alternating nodes of metabolites and reactions. For some applications a reaction-based representation is needed in which the vertices of the network are the reactions and edges are set if a product of one reaction is the substrate of the other. Similarly, in a metabolite-based representation, the vertices are the metabolites that are connected by reactions (Figure 10.1). Often, ubiquitous metabolites like water, oxygen, ATP, and cofactors are discarded to model only the most relevant metabolic fluxes. Until now, no optimal and standardized method exists for the construction of a cellular metabolic network [8, 9].

Regulatory networks can be constructed by linking transcription factors and their regulating genes [10]. The interaction information for this can be experimentally

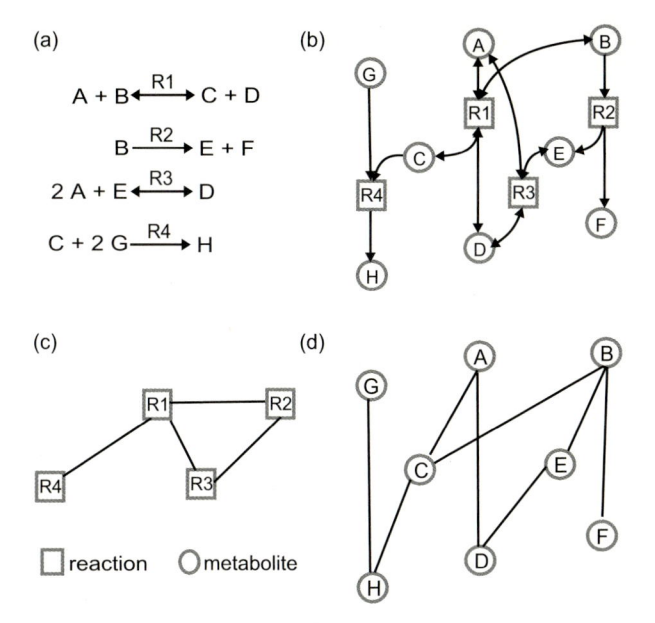

Figure 10.1 Representations of metabolic networks. Four representations of metabolic reactions are shown: (a) stoichiometric equations, (b) bipartite graph with rectangles as reactions (or enzymes) and circles as metabolic compounds (c) reaction-based representation of the metabolic network, and (d) metabolite-based representation.

inferred from chromatin immunoprecipitation (ChIP) or ChIP in high-throughput using microarrays (ChIP-Chip) [11] and next-generation sequencing techniques (ChIP-Seq). Constructing the signaling network is much more demanding and several different approaches have been reported. In the simplest and most commonly used case, data of known protein–protein interactions (PPIs) is used as edges forming an undirected graph [12]. One of the most elaborated approaches was suggested by Kohn, in which a detailed signaling flow was constructed similar to maps for electronic circuits [13].

10.3
Network Descriptors

In the following, the most relevant network descriptors (features) are explained for estimating the essentiality of nodes in a network. Figure 10.2 shows example networks to illustrate the explained features. The section is subdivided into two parts. Section 10.3.1 describes node features for undirected graphs and can be used for any of the networks (metabolism, transcription regulatory, and signaling). Section 10.3.2 explains features that have been specifically designed for metabolism (they may need some modification if employed for other networks).

(a)   (b)   (c)

(d)   (e)

□ reaction   ○ metabolite

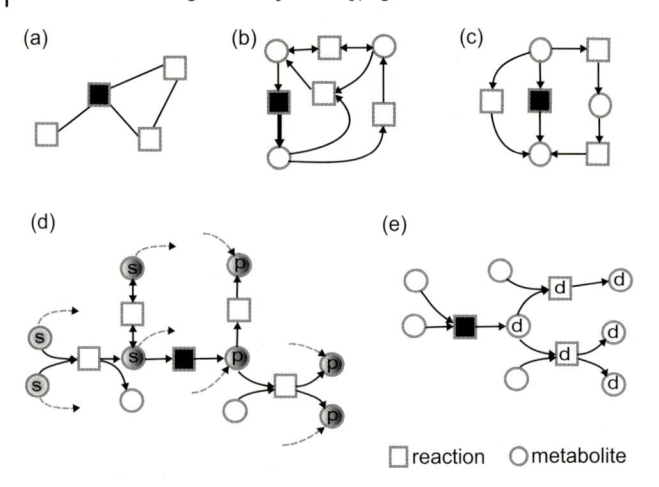

Figure 10.2 Network examples to illustrate the topological features. Circles represent metabolites, rectangles represent reactions, dark arrows represent directions of the metabolic flux, and dark lines represent links between two neighboring reactions. Dark rectangles represent the observed reactions. (a) Example of an undirected graph (reaction-based representation of the metabolic network) for computing *degree*, *clustering coefficient*, and *centrality* features. The observed node has a degree of 3 and a clustering coefficient of 1/3. The observed node is placed central and more pathways pass through the node (two out of six) compared to the other depicted nodes. Therefore, its betweenness centrality is higher in comparison to the other nodes (observed node: 2; other nodes: 0). (b–e) Representation of a bipartite graph for metabolism. (b) The observed reaction is a chokepoint. Enzymes are chokepoints if they exclusively consume or produce a certain metabolite as depicted here for the filled reaction being the only reaction producing the lower left metabolite. (c) The example shows alternative pathways for the observed reaction. There are two alternative paths to produce the product of the observed reaction and therefore the number of deviations is 2. Their lengths are one and two reactions, respectively. Thus, the length of the shortest path is 1 and the average of alternative path lengths is 1.5. (d) Reactions nearby the observed reaction and its upstream substrates (S) and downstream products (P). Dash-line arrows represent possible alternative pathways to consume substrates S for producing products P. The producibility gives the percentage of products of the considered reaction that can be produced from the substrates via alternative pathways. (e) Damaged reactions and compounds (d in rectangles and circles, respectively) when removing the observed reaction.

10.3.1
Network Topological Features for Undirected Graphs

As mentioned above, a network may be represented as an undirected graph $G(V, E)$ that consists of a set of nodes V and a set of edges E. Each node $i \in V$ represents a unique cellular entity such as enzymes, genes, and proteins, while each edge $(i,j) \in E$ represents an observed interaction between two nodes i and j. Regulatory networks may be constructed by linking transcription factors and the genes they regulate. Signaling networks may be constructed using information about PPIs. To construct an undirected graph for metabolism, the network representation of a reaction-pair network can be used instead of a bipartite graph.

10.3.1.1 Connectivity

The connectivity is given by the number of neighboring nodes of a node. It is also called the *degree* of a node (see Figure 10.2a). In other terms, the degree k of a node $v \in V$ is defined as the number of edges between v and its adjacent nodes.

10.3.1.2 Clustering Coefficient

The clustering coefficient is used to estimate the local density of the network. The clustering coefficient (CC) of a node v is defined as the ratio of the number of connecting edges m_v among all neighbors of v and the total number of all edges among them that could be possible:

$$CC(v) = \frac{2m_v}{k_v(k_v-1)} \tag{10.1}$$

where k_v is the degree of node v.

10.3.1.3 Centrality Measures

In the context of cellular networks, descriptors for *node centrality* are quite powerful for describing the essentiality of the node. They describe not only the impact of the node on its direct vicinity, but also the contribution of a node to the global structure of the network. The simplest of all centrality measures is the degree k, as mentioned above, describing the local vicinity of the node. The following centrality measures consider the entire network:

- **Betweenness centrality** (BW) is the frequency of a node to be part of the shortest paths connecting all pairs of nodes [14]. Betweenness centrality $BW(v)$ for node v is given by:

$$BW(v) = \sum_{i \neq j \neq v \in V} \frac{d_{ij}(v)}{d_{ij}} \tag{10.2}$$

in which d_{ij} is the number of shortest paths from node i to node j and $d_{ij}(v)$ is the number of shortest paths from i to j that pass through node v. The sum comprises all pairs (i, j) of nodes of the network.

- **Closeness centrality** (CN) is defined by the inverse of the average length of the shortest paths from node v to all the other nodes in the network:

$$CN(v) = \frac{n-1}{\sum_{i \neq v, i \in V} d_{vi}} \tag{10.3}$$

in which n is the number of nodes in the network [14].

- **Eccentricity** is the longest distance from the given node v to any other node [15]. The eccentricity-centrality (EC) is the reciprocal of the eccentricity:

$$EC(v) = \frac{1}{\max_{i \neq v, i \in V}(d_{vi})} \tag{10.4}$$

- **Eigenvector centrality** (*EV*) is based on the assumption that the utility of a node is determined by the utility of the neighboring nodes [16]. It scores a node higher if it is connected to high-scoring nodes. It is the principal eigenvector of the adjacency matrix of the network. Let x_i denote the score (eigenvector centrality $EV(i)$) of a node i. Let A_{ij} be the adjacency matrix of the network, such that $A_{ij} = 1$ if there is an edge between nodes i and j, and $A_{ij} = 0$ otherwise. For node i, the eigenvector centrality score is proportional to the average of the eigenvector centrality scores of the neighbors of i:

$$x_i = \frac{1}{\lambda} \sum_{j \in Neighbor(i)} x_j = \frac{1}{\lambda} \sum_{j=1}^{n} A_{ij} x_j \qquad (10.5)$$

in which $Neighbor(i)$ is the set of neighboring nodes of node i, n is the total number of nodes, and λ is a constant. This leads directly to the well-known eigenvector equation, $Ax = \lambda x$. Normally, there are different eigenvalues λ for which an eigenvector solution exists. According to the Perron–Frobenius theorem only the eigenvector of the largest eigenvalue is feasible for being used for the eigenvector centrality [17].

10.3.2
Network Topological Features for a Bipartite Graph of Metabolic Networks

As metabolic networks are constructed by connecting reactions and metabolites as a bipartite graph, some special and specific properties of a node in the graph can be considered. Metabolites are nutrients or, in general, compounds that need to be synthesized or catabolized by enzymes of a cell. For identifying drug target enzymes, the network topology features are computed for reaction nodes. Let $G(V, E)$ represent a directed bipartite graph where V consists of two disjoint sets of vertices M and R representing metabolites and reactions, respectively [7]. Each edge connects vertices from M to R and vice versa, representing directed edges leading from the substrates of a reaction to the reaction and from the reaction to its products (see Figure 10.1b).

The *in-degree* and *out-degree* of a node $v \in V$ in directed graphs is defined by the number of incoming edges and outgoing edges, respectively:

$$\begin{aligned} d_{in}(v) &= |\{u \in V | \exists (u, v) \in E\}| \\ d_{out}(v) &= |\{w \in V | \exists (v, w) \in E\}| \end{aligned} \qquad (10.6)$$

10.3.2.1 Stoichiometric Properties

Reactions can be reversible or irreversible and this characteristic of *reversibility* is also used to describe nodes in metabolism. The *number of substrates* and the *number of products* correspond to the number of different metabolites that are needed for the given reaction (substrates S) and that are produced (products P), respectively. Note that for node $v \in R$ the number of substrates $NS(v) = d_{in}(v)$ and the number of products $NP(v) = d_{out}(v)$.

10.3.2.2 Chokepoints

A reaction that is the only reaction that consumes or produces a certain metabolite in the metabolic network is considered to be a chokepoint [5, 6] (see Figure 10.2b). This feature may make it irreplaceable. The chokepoint $CP(v)$ for a node $v \in R$ is given by:

$$CP(v) = true \begin{cases} \text{if } s \in substrate(v) \text{ and } d_{out}(s) = 1, \text{ or} \\ \text{if } p \in product(v) \text{ and } d_{in}(p) = 1 \end{cases} \quad (10.7)$$

where $substrate(v)$ and $product(v)$ are the sets of substrates and products of reaction v, respectively.

10.3.2.3 Load Scores

Load scores are defined to detect hotspots in the network, and are based on the ratio of the number of shortest paths passing through a reaction and the number of nearest neighbor links attached to it [5]. This ratio is compared to the average load value in the network. The load score $LS(v)$ of a node $v \in R$ is given by:

$$LS(v) = \ln \left[\frac{d_{ij}(v)/k_v}{\sum_{i,j \in M, t \in R} d_{ij}(t) / \sum_{t \in R} k_t} \right] \quad (10.8)$$

where k_v is the degree of reaction v and $d_{ij}(v)$ is the number of shortest paths from metabolite i to metabolite j that pass through v.

10.3.2.4 Deviations

Several descriptors have been established for estimating the feasibility of possible flux deviations if the node under observation is discarded. These features describe possible alternative pathways from substrates of the knocked-out reaction to its products. To calculate these descriptors, a breadth-first search algorithm is used to simulate a qualitative metabolic flux from the substrates to the products of the observed reaction when the reaction is discarded. The network without the observed reaction will also be called a "mutated network" in the following. The feasibility of the alternative paths is analyzed. A reaction is more likely to be essential for survival when basically the mutated network cannot yield the products of the reaction from its substrates or if the mutated network has difficulties reaching the products. As an example, the following algorithm has been proven to be useful to investigate this:

i) All metabolites acting as input nodes (substrates) and output nodes (products) of the knocked-out reaction are selected. The set of substrates S defines the input nodes and the set of products P defines the output nodes. To get a broader list of available substrates, the substrates of the reactions upstream of S and the products of the reactions downstream of P are added to the sets S and P, respectively. Substrates of reactions that have at least one of the substrates S as a substrate are also added to S. Further, substrates of reactions that have a metabolite out of P as a substrate are also added to S.
ii) Reactions are selected that use compounds of S as substrates.

iii) These selected reactions and their products are incorporated into the list of discovered reactions and products. The products are set as newly available metabolites in the network.
iv) Steps (ii) and (iii) are repeated until no further reactions can be identified.
v) Metabolites of P that cannot be produced are counted (unreachable products P).

After finishing the process, the number of products of the observed reaction that could be produced within the mutated network yields two features: a quality feature defining if all products could be produced and the percentage of products that could not be produced. Furthermore, from this we can compute the number of (pairwise different) deviations, their average lengths, and the length of the shortest deviation (see Figure 10.2).

10.3.2.5 Damage in Global Networks

The "damage" estimates the number of potentially affected metabolites (number of damaged compounds, see Figure 10.2e) and reactions (number of damaged reactions) downstream of the knocked-out reaction [4]. For irreversible reactions it can be calculated using the following procedure for the observed reaction v (if the reaction is reversible, the following procedure is done for both directions, and the resulting damaged compounds and reactions are put together):

i) All the metabolites that are produced by v are deleted.
ii) All the reactions for which at least one substrate is missing are deleted.
iii) All the metabolites that are produced by the missing reactions are deleted.
iv) Steps (ii) and (iii) are repeated until no further nodes are deleted.

All deleted metabolites and reactions are collected and counted, yielding the feature values for damaged metabolites and damaged reactions, respectively. Combined with the knowledge of chokepoints and alternative pathways, the definition of damage can be applied to define further features such as the number of damaged chokepoints and damaged nodes that cannot be produced by alternative paths [18, 19].

10.4
Machine Learning

All these features describe network topology and can support finding an essential node in the network. However, these features yield much stronger predictions when combined. There exists a variety of methods to combine such descriptors. We will show a supervised machine learning approach. For this, we describe very briefly how (supervised) machine learning works in principle and we refer the reader to the literature for a more detailed description [20, 21].

Machine learning algorithms employ complex data patterns to predict a given class that the data belongs to. In our case the class is the property of a node to be essential or not. Machine learning algorithms have been successfully applied to various fields such as natural language processing, search engines, computer vision, and bioinformatics [22]. Two main approaches exist in machine learning: supervised learning (*classification*) and unsupervised learning (*clustering*). Supervised learning establishes a function that maps objects to classes using a set of attributes or features, whereas unsupervised learning categorizes a set of objects without predefined classes to discover similarities between objects. We use the first approach. The machine learning algorithm or classifier needs prior knowledge about a set of objects. These objects comprise values for their descriptors and a class label. In our case the object is a node in the network and its descriptors are, for example, if the object is a chokepoint, its connectivity, and so on. The class label of the object is its property to be essential or not. The classifier then "learns" from a given dataset for which the class labels are known, which in our case is the information if a node is essential or not, and may have been observed experimentally by a knock-out study for the coding gene of the protein. After learning, the classifier is applied to superimpose the class labels from the given descriptors (features) of new objects for which the class labels are not known.

Mathematically, the classifier finds a function f that maps the feature vector x_i to the correct class C_i for every (known) object i:

$$f : x_i \in X^N \to C_i \in \{-1, +1\}, \quad i = \{1, \ldots, m\} \tag{10.9}$$

where m is the number of objects the classifier is fed for learning, X^N is the feature space (space of all possible combinations for all feature values), and -1 and $+1$ are the two class labels (e.g., -1 denotes nonessential and $+1$ denotes essential).

Ample methods exist for finding a suitable function, which we will not explain here (see, e.g., [20, 21]). Finding a mapping function is called "learning" or "training." After training, the classifier needs to be validated (i.e., tested for how good its function is). For this, an estimate needs to be done that has general applicability. Estimating the performance on the data the classifier has learned would lead to an overestimation of its performance. Therefore, another dataset (called the validation set) with known class labels is used and the learned function of the classifier is applied to predict these labels. The predictions are then compared to the true values and a prediction rate (accuracy) can be calculated saying how good the prediction was (percentage of correct predictions). This gives an estimate how the classifier performs when applied to data for which the class labels are not known. After this performance estimation, the classifier is applied to new data. For these data objects, only the feature values are known and the classifier then predicts to which class each object belongs using this information. In our case, these may be nodes for which no experimental data about their essentiality exists. Often the amount of objects with known class labels is limited. Splitting these objects into a set of training data and validation data would even further limit the amount of data to learn a proper mapping function. To circumvent this problem, a cross-validation can be applied. The simplest cross-validation is a leave-one-out cross-validation. For this, except for one, all objects with known class labels are taken for learning and the classifier is validated with the object that was taken out for learning. As this does not give a very precise performance estimate, the whole procedure is repeated for every object (with a known class label) to be left out and the performance is estimated by taking the mean prediction performance of all these validations. This can of course also be done by leaving out more than one object.

10.5
Some Examples of Applications

10.5.1
Validating an Experimental Knock-Out Screen

Screening microorganisms for drug targets usually starts with a genome-wide knock-out screen of all open reading frames. Positive hits are knock-out mutants showing considerably reduced viability and proliferation or are lethal when passed into a medium of physiological conditions. The detected genes and the corresponding enzymes can then be considered as candidates for further, more detailed investigations. However, these high-throughput screens are error-prone and can produce many unwanted false-positives (i.e., knock-out was found to be essential, but it is not essential) and false-negatives (knock-out was found to be nonessential, but it is essential). To detect these false results, a machine learning approach has been set up and successfully applied to *Escherichia coli* in a case study [18]. Apart from the topological features described above, some other features were also used, such as information about gene homology to support the classification with the information that a knocked-out gene may have homologs in the genome that can replace the function of the knocked-out gene. The aim was to validate a genome-wide knock-out screen from a public repository. The machine learning system was trained and validated by the Keio Collection, which is a comprehensive experimental dataset comprising phenotypic outcomes from single knock-out mutants of about every open reading frame of *E. coli* [23]. For estimating the performance, a leave-one-out cross-validation was applied, and achieved an accuracy of 93% and a precision of 90% (percentage of true-positives out of all positive/essential predictions). Note that these performance estimations are based on the experimental data and only give a good estimate if most of the experimentally received class labels are true (which is presumed in this study and the other studies described below). The predictions of the classifier were then used to detect errors in the experimental knock-out screen. A difference between the experimental data and *in silico* predictions may either be due to an erroneous prediction of the algorithm *or* an error of the knock-out experiment. Thus, genes that were predicted as false-positives and false-negatives were experimentally reinvestigated. Five out of the six selected genes in the false-positive list were found to be not correctly knocked-out and nine out of 33 genes of the false-negatives where found to be not essential. Concluding, this analysis could well reduce the number of false experimentally obtained class labels, specifically for the positive predictions.

10.5.2
Training with Data from One Organism to Predict Essential Genes for Another Organism

Another very challenging application is to train the machine with one organism (e.g., *E. coli* using the Keio Collection) and predict essential nodes for another organism for which the experimental high-throughput screen may be to expensive (e.g., for a pathogen that needs elaborate control and isolation conditions in the laboratory). Plaimas *et al.* [19] used all together five experimental datasets of high-throughput knock-out screens including two sets of different laboratories for *E. coli* and two sets of different studies for *Pseudomonas aeruginosa*. Similar to *E. coli* in the gut, *P. aeruginosa* is an abundant bacterium in the soil. Additionally, they employed a smaller, noncomprehensive dataset of *Salmonella typhimurium* in which 53 knockouts were described to be essential. First, they trained machines with the knock-out dataset and the network of *E. coli*, and validated them with the data from *Pseudomonas*. Then they did it vice versa (i.e., training the machines with the data from *Pseudomonas* and validating them with the *E. coli* datasets). Note that the network descriptors were compiled for each organism and its enzymes separately. The class labels from the organisms were assigned to the descriptors of the genes from the according organism. With this they achieved an average accuracy of 84% and an average precision of 73%. These values were less than in the previous example, but it should be noted that this task is much harder. Finally, they used the trained machines from all four datasets (two from *E. coli* and two from *Pseudomonas*) to predict essential enzymes of *Salmonella*. The results were compared to the experimental dataset of *Salmonella* and 27 genes were found to be consistent. An analysis of all predictions and the experimental dataset revealed some interesting potential drug target candidates such as enzymes of the non-melavonate pathway, which is an alternative pathway for cholesterol biosynthesis. The non-melavonate pathway is not found in the human host, which makes it very attractive for designing a specific treatment. Some of the enzymes of this pathway are known targets for other pathogenic microorganisms (see [19] for more details).

In another study, Hwang *et al.* [24] analyzed PPI networks and gene sequences of *E. coli* and *S. cerevisiae*. They analyzed various network topologies, including degree, clustering coefficient, betweenness, and closeness centrality. They also performed an organism cross-wise prediction using *S. cerevisiae* and *E. coli*. After training the classifier with *S. cerevisiae*, they yielded a prediction performance for *E. coli* of 72% precision and 72% sensitivity. Vice versa, using the classifier trained with *E. coli*, the prediction performance for *S. cerevisiae* was 75% precision and 64% sensitivity.

10.5.3
Further Reported Investigations

Topology features of PPI networks of *E. coli* and *S. cerevisiae* have been used for other machine learning investigations [24–26, 28]. In the study by Acencio *et al.* [28], an integrated network of gene interactions comprising (physical) protein, metabolic and transcriptional regulation interactions of *S. cerevisiae* was constructed and investigated for defining topological features. Not only the topological features but also features describing the cellular localization and biological process from Gene Ontology (www.geneontology.org) were used to train the machines. To define the class labels for training and validation they used a comprehensive dataset from a public repository [29]. They compared class labels for 96% of the genome (i.e., for all but 514 open reading frames). The performance was measured by calculating the area under the curve (AUC) for the receiver operating characteristic (ROC) curves (which in brief is an average of specificities and sensitivities for a large range of stringency thresholds). The best result was an AUC of 81% when the machine was trained with all features and 77% when trained only with the topological features, emphasizing that the topological features contributed considerably to the performance. Using the best classifier (all features), 44 out of 514 genes were predicted to be essential. When scanning the literature, they found that nine of these 44 genes have been previously described to be essential. In addition to this, they discovered that the number of protein interactions (or degree), nuclear localization, and number of regulating transcription factors are important factors determining gene essentiality. Manimaran *et al.* [25] used degree, closeness centrality, and betweenness centrality features, and analyzed PPI networks of *E. coli*. To support classification, they additionally used 61 gene expression datasets from microarrays in which a large variety of different treatments were applied. They put up 61 "conditional" PPI networks by restricting the networks to nodes that were expressed at the respective condition, and yielded a sensitivity of 84% and a specificity of 96% [27].

Seringhaus *et al.* [30] predicted gene essentiality in organisms with sequenced genomes but without other genome-scale experimental data using a weighted combination of features derived from sequence data alone. A classifier was developed using data from *S. cerevisiae* and tested on the closely related yeast species *Saccharomyces mikatae*. The performance of the classifier in *S. cerevisiae* was assessed by a cross-validation and demonstrated a precision of 69%. The trained machine was used to predict essential genes in *S. mikatae*. The predicted essential genes were verified by comparing homology to essential genes in *S. cerevisiae* and some of those genes were tested experimentally.

10.6
Conclusions

Defining drug targets is a challenging task. Many experiments are based on an essentiality screen of genes to identify the associated enzymes as possible drug targets. Employing a machine learning algorithm that combines descriptors of network topology for predicting essential genes is a novel and in the meanwhile widely used approach. It can support experimental genome-wide knock-out screens by validating them as shown in Section 10.5.1. The method aims at finding out points that substantially weaken the network when discarded. The machine learning system can be used to integrate features of proteins describing the topology of their network-embedding in respect to their estimated need and usefulness in the network. Additionally, these machines enable adding other information, such as functional genomics properties. A methodologically very challenging task is to employ the classifier across different organisms (e.g., using the essentiality screen of one organism to infer the information to another). Section 10.5.2 illustrated classifiers that were trained with essentiality information for genes of one organism (e.g., *E. coli, S. cerevisiae*) and employed to predict essential genes of another organism (*S. mikatae, Pseudomonas, Salmonella*). These predictions do not depend on essentiality information of the query organism for which the predictions are made, but solely on the network features (and genomic information) from the query organism.

An intelligent combination of these features may be seen as an alternative approach to the established methods of flux balance analyses (FBAs) and elementary flux modes (EFM). These methods were not addressed here as ample of literature can found elsewhere [33–37]. However, FBAs need detailed growth and nutrient information [31], and the EFM method needs an in-depth refinement of the metabolic network to assign internal nodes and external nodes to reduce the computational complexity [32]. For pathogens it is often hard to define these environmental parameters, which are complex and changeable (e.g., as for intestinal infections). The machine learning approach described in this chapter can, in principle, handle various environment conditions without detailed specification, but of course may benefit from experimental essentiality screens that are done in similar environmental conditions as the application (e.g., in oxygen-deprived conditions when mimicking the environmental conditions of the gut). Until now, most studies investigated predicting essential nodes for the same organism and inferring essentiality information from one organism to another organism has been done only for closely related microorganisms. It will be a challenge to open this up to a wider variety of organisms for training and application, and also to use the descriptors for other applications, specifically for multicellular organisms and human cells (e.g., to predict potential driver mutations in cancer or to predict host factors for viruses).

References

1 Fatumo, S., Plaimas, K., Mallm, J.P., Schramm, G., Adebiyi, E., Oswald, M., Eils, R., and König, R. (2009) *Infect. Genet. Evol.*, **9**, 351.

2 Feist, A.M., Henry, C.S., Reed, J.L., Krummenacker, M., Joyce, A.R., Karp, P.D., Broadbelt, L.J., Hatzimanikatis, V., and Palsson, B.O. (2007) *Mol. Syst. Biol.*, **3**, 121.

3 Jeong, H., Mason, S.P., Barabasi, A.L., and Oltvai, Z.N. (2001) *Nature*, **411**, 41.

4 Lemke, N., Heredia, F., Barcellos, C.K., Dos Reis, A.N., and Mombach, J.C. (2004) *Bioinformatics*, **20**, 115.

5 Rahman, S.A. and Schomburg, D. (2006) *Bioinformatics*, **22**, 1767.

6 Yeh, I., Hanekamp, T., Tsoka, S., Karp, P.D., and Altman, R.B. (2004) *Genome Res.*, **14**, 917.

7 König, R. and Eils, R. (2004) *Bioinformatics*, **20**, 1500.

8 Feist, A.M., Herrgard, M.J., Thiele, I., Reed, J.L., and Palsson, B.O. (2009) *Nat. Rev. Microbiol.*, **7**, 129.
9 Nikoloski, Z., Grimbs, S., May, P., and Selbig, J. (2008) *J. Theor. Biol.*, **254**, 807.
10 Luscombe, N.M., Babu, M.M., Yu, H., Snyder, M., Teichmann, S.A., and Gerstein, M. (2004) *Nature*, **431**, 308.
11 Cawley, S., Bekiranov, S., Ng, H.H., Kapranov, P., Sekinger, E.A., Kampa, D., Piccolboni, A., Sementchenko, V., Cheng, J., Williams, A.J. et al. (2004) *Cell*, **116**, 499.
12 Chuang, H.Y., Lee, E., Liu, Y.T., Lee, D., and Ideker, T. (2007) *Mol. Syst. Biol.*, **3**, 140.
13 Kohn, K.W. (1999) *Mol. Biol. Cell.*, **10**, 2703.
14 Estrada, E.V. (2006) *Proteomics*, **6**, 35.
15 Koschützki, D. and Schreiber, F. (2004) German Conference on Bioinformatics, Bielefeld.
16 Bonacich, P. (1972) *J. Math. Soc.*, **2**, 113.
17 Bonacich, P. (1987) *Am. J. Soc.*, **92**, 1170.
18 Plaimas, K., Mallm, J.P., Oswald, M., Svara, F., Sourjik, V., Eils, R., and König, R. (2008) *BMC Syst. Biol.*, **2**, 67.
19 Plaimas, K., Eils, R., and König, R. (2010) *BMC Syst. Biol.*, **4**, 56.
20 Mitchell, T. (1997) *Machine Learning*, McGraw-Hill, New York.
21 Witten, H. and Frank, E. (2005) *Data Mining: Practical Machine Learning Tools and Techniques*, 2nd edn, Morgan Kaufmann, San Francisco, CA.
22 Tarca, A.L., Carey, V.J., Chen, X.W., and Romero, R., and Draghici, S. (2007) *PLoS Comput. Biol.*, **3**, 6.
23 Baba, T., Ara, T., Hasegawa, M., Takai, Y., Okumura, Y., Baba, M., Datsenko, K.A., Tomita, M., Wanner, B.L., and Mori, H. (2006) *Mol. Syst. Biol.*, **2**, 1.
24 Hwang, Y.C., Lin, C.C., Chang, J.Y., Mori, H., Juan, H.F., and Huang, H.C. (2009) *Mol. Biosyst.*, **5**, 1672.
25 Manimaran, P., Hegde, S.R., and Mande, S.C. (2009) *Mol. Biosyst.*, **5**, 1936.
26 Gustafson, A.M., Snitkin, E.S., Parker, S.C., DeLisi, C., and Kasif, S. (2006) *BMC Genomics*, **7**, 265.
27 Hegde, S.R., Manimaran, P., and Mande, S.C. (2008) *PLoS Comput. Biol.*, **4**, 1.
28 Acencio, M.L. and Lemke, N. (2009) *BMC Bioinformatics*, **10**, 290.
29 Giaever, G., Chu, A.M., Ni, L., Connelly, C., Riles, L., Veronneau, S., Dow, S., Lucau-Danila, A., Anderson, K., Andre, B. et al. (2002) *Nature*, **418**, 387.
30 Seringhaus, M., Paccanaro, A., Borneman, A., Snyder, M., and Gerstein, M. (2006) *Genome Res.*, **16**, 1126.
31 Schuetz, R., Kuepfer, L., and Sauer, U. (2007) *Mol. Syst. Biol.*, **3**, 119.
32 Dandekar, T., Moldenhauer, F., Bulik, S., Bertram, H., and Schuster, S. (2003) *Biosystems*, **70**, 255.
33 Duarte, N.C., Herrgard, M.J., and Palsson, B.O. (2004) *Genome Res.*, **14**, 1298.
34 Edwards, J.S., Ibarra, R.U., and Palsson, B.O. (2001) *Nat. Biotechnol.*, **19**, 125.
35 Edwards, J.S. and Palsson, B.O. (2000) *BMC Bioinformatics*, **1**, 1.
36 Schuster, S., Dandekar, T., and Fell, D.A. (1999) *Trends Biotechnol.*, **17**, 53.
37 Schuster, S., Fell, D.A., and Dandekar, T. (2000) *Nat. Biotechnol.*, **18**, 326.

11
Gene Coexpression Networks for the Analysis of DNA Microarray Data
Matthew T. Weirauch

11.1
Introduction

Biology has classically been dominated by a reductionist philosophy, and years of research are a testament to the success of this approach. The emergence and rapid improvement of sequencing technologies has resulted in increasing evidence that cellular functions are rarely carried out in isolation and instead often involve the concerted interaction of numerous cellular components. A key challenge now facing biologists is achieving an understanding of this complicated and dynamic system.

Recent years have seen an explosion in the development of high-throughput techniques for globally monitoring various aspects of gene activity. Using these new technologies, it is now possible to identify novel associations between genes at a higher resolution than ever before; for example, it will soon be feasible to map the entire set of protein interactions for any organism. The availability of these genome-wide datasets offers unprecedented opportunities for discovering novel cellular attributes from a systems perspective and enhances our ability to accurately predict gene function on a global scale.

Among the more popular recent high-throughput technologies is the DNA microarray [1, 2]. Microarrays allow the expression levels of thousands of genes to be monitored simultaneously and so provide a global picture of the cell's transcriptional activities across multiple conditions. Prior to their development, the determination of when and where a gene is transcribed was a laborious and time-consuming endeavor. Microarrays have opened the door to a new realm of biological exploration, including the elucidation of genes involved in specific processes such as the cell cycle [3] and development [4], the assessment of the impact of genetic and chemical perturbations [5, 6], and the identification of disease-related genes [7].

The sheer volume of data generated by microarray studies requires the development of advanced statistical computational analysis tools. In this chapter, I discuss a subclass of these methods that are based on graph theory. Specifically, the studies and techniques discussed herein compare all pairs of gene expression patterns based on a similarity score and pairs exceeding a threshold are linked to create a network, which

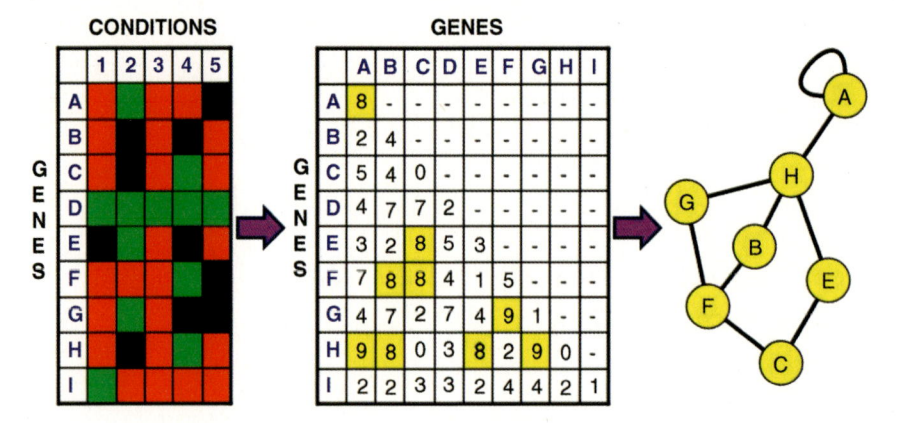

Figure 11.1 Overview of the network construction process. The heatmap on the left indicates data values for a set of nine genes across five conditions. Here, red corresponds to a high value, black a medium value, and green a low value. In the first step, all gene pairs are compared based on a scoring function that measures similarity in gene activity. In the resulting matrix, each row and column correspond to one gene, with each matrix entry representing the score for the corresponding gene pair. In this example, the scoring system is symmetric, so scores are only shown for the lower diagonal of the matrix. A threshold is then chosen using some criteria (such as significance with respect to permutation tests) and all gene pairs are identified that exceed this threshold (highlighted in yellow). Finally, a network is constructed with genes for nodes and edges drawn between gene pairs whose score exceeds the threshold.

provides the basis for subsequent analyses (Figure 11.1). Networks created in this manner from microarray data are often referred to as gene coexpression networks (GCNs) and aid in the identification of genes that share similar expression patterns across a variety of experimental conditions. Genes displaying such patterns are often controlled by the same transcriptional regulatory program, functionally related, or members of the same pathway or protein complex.

11.2
Background

11.2.1
Gene Transcription

The cell is a highly complicated system that must respond to a variety of changing internal and external conditions in order to perform its myriad functions. Incredibly, almost all of the information necessary to maintain this system is encoded in genomic DNA. The fundamental unit of information in DNA is the gene – a sequence that contains the blueprints for proteins and noncoding RNAs, which together perform the majority of cellular functions. Owing to the constant need for gene products, the cell is constantly transcribing genes. Although some genes are more or less continuously transcribed, most are transcribed according to need. Transcriptional regulation of a gene involves a fine-tuned response to a wide range of inner- and extracellular conditions, and is largely achieved by proteins known as transcription factors. Transcription factors regulate gene transcription by binding specifically to genomic DNA sequences, often at locations proximal to the genes that they control. Upon binding, transcription factors can increase or decrease the rate of transcription of the genes that they regulate, through mechanisms such as recruiting or blocking members of the basal transcriptional machinery, or by affecting local chromatin structure. Deciphering the "regulatory code" of the genome (when and how genes are turned on and off) is a major goal of biology, and microarrays aid in this goal by measuring the expression levels of genes under different cellular conditions.

11.2.2
DNA Microarrays

A single microarray experiment examines the expression levels of thousands of genes under a single experimental condition. Most studies involve multiple microarray experiments covering a range of conditions. For example, microarray studies often compare the expression levels of genes across different growth conditions, across different developmental stages, or in tissue samples with and without a particular disease.

Several experimental platforms have been developed for performing microarray experiments [8]. Although the various platforms differ in their design details, at their most basic level all microarrays consist of a grid of DNA sequences, or probes. Often, probes correspond to known genomic features, such as genes or intergenic regions. In a microarray experiment, samples of complementary DNA or complementary RNA are hybridized to probes and the relative abundance of each sequence in the sample is quantified by automated detection of fluorescence. A gene's expression profile is often represented as a vector of numerical values, with each value corresponding to a probe's expression level in one microarray experiment, and is usually visualized in a "heatmap" (Figure 11.1, left), such that highly expressed probes are red and lowly expressed probes are green.

Most studies include both biological replicates (which control for biological variability such as differences between patients or cell lines) and experimental replicates (which control for technical variability within an experiment such as variation between array locations). Replicates are often averaged into a single score for a probe, and expression levels are normalized to account for the inherent noise and variability of microarrays. There is a rich literature on microarray data normalization and the reader is referred to reviews specifically covering this topic for more information [9, 10]. Usually, data is normalized such that values near zero indicate no change in the probe's expression level relative to controls, with positive and negative values indicating increased and decreased levels of expression, respectively. After normalization, further filtering is performed, such as removing probes whose expression levels do not significantly vary across conditions.

A typical microarray study will report a list of genes that are *differentially expressed* across conditions [11], using methods such as analysis of variance (ANOVA) [12], gene set enrichment analysis (GSEA) [13], or LIMMA [14]. A differentially expressed gene has significantly different expression levels between two (or more) conditions. For example, differential expression is frequently used to identify genes that are highly expressed in a particular disease state (as opposed to normal, nondisease states) [7]. In the following sections, I focus on network-based methods, which instead identify relationships based on gene *coexpression*. Relationships based on coexpression capture the overall tendency for a pair or group of genes to have similar expression levels across conditions (as opposed to being high in one set of conditions and low in another).

11.2.3
Networks

Microarrays generate abundant volumes of data, creating the need for sophisticated methods for data mining, interpretation, and representation. One popular data structure that has emerged for this task is the network. Major advantages to network-based methods include their robustness to large amounts of data, the availability of graph-theoretic methods for data mining, and ease of visual interpretation. Networks provide an intuitive way to convey both global and local aspects of large amounts of data. Consequentially, networks have seen extensive use in many different fields, and have been used, for example, in the modeling of the World Wide Web, social interactions, and flight patterns [15].

Networks offer a natural way to model interactions between genes, with nodes representing genes and edges representing various interactions types inferred from different data sources [16]. Network-based methods have been applied to a wide variety of problems in biology, including the display of protein interactions [17], transcription factor binding site discovery [18], and the modeling of genetic interactions [19]. Likewise, several tools have been developed for the visualization and analysis of biological networks, including Cytoscape [20], VisAnt [21], and tYNA [22]. For a review of biological network applications, the reader is referred to [23]. In the following sections, I focus on the application of networks to the study of gene coexpression relationships inferred from DNA microarray experiments.

11.3
Construction of GCNs

A GCN connects pairs of genes (nodes) that are significantly coexpressed across conditions (microarray experiments) in a study. As such, the first step in creating a GCN is to score all pairs of gene vectors. The second step is to choose a score threshold and connect all gene pairs whose scores exceed this value. The resulting GCN can be either unweighted (interactions are binary: present or absent) or weighted (e.g., probabilities representing the likelihood of a given link, as in [24]).

In this section, I discuss various scoring schemes and review methods for choosing a score threshold. In general, networks can be either directed or undirected. Here, I focus on undirected networks, which indicate mutual relationships of coexpression, but do not necessarily indicate causality. Several methods have been developed for the inference of directed regulatory networks from expression data generated from gene disruption or overexpression [25–27]; for a review of directed network methods, the reader is referred to a recent review by Markowetz and Spang [28].

11.3.1
Data Format and Representation

A single expression study contains multiple microarray experiments, each *assaying* (examining) a specific cellular condition. The resulting data is often represented in matrix format, with rows corresponding to probes (genes) and columns representing experiments (conditions):

$$M = \begin{bmatrix} M_{1,1} & M_{1,2} & \cdots & M_{1,N} \\ M_{2,1} & M_{2,2} & \cdots & \vdots \\ \vdots & \vdots & \ddots & \vdots \\ M_{G,1} & M_{G,2} & \cdots & M_{G,N} \end{bmatrix}$$

Here, matrix M contains a total of G rows and N columns, representing the number of genes and the number of conditions, respectively. Each entry in the matrix corresponds to the expression level of one gene in a single condition. Each row represents the expression level of a single gene across all assayed conditions, and is often referred to as a *gene vector*. Typically, a global picture of a microarray dataset is provided in the form of a heatmap (Figure 11.1, left), which is often clustered in order to identify groups of genes and conditions that display similar patterns (see Section 11.5.3.1). The expression matrix serves as the input data for all subsequent network-based analyses, of which the first step is the calculation of pairwise gene coexpression scores.

11.3.2
Calculating Pairwise Gene Scores

11.3.2.1 Overview
Several methods exist for comparing the expression profiles of gene pairs. In this section, I focus on four popular scoring schemes: Euclidean distance, mutual information, Pearson's correlation coefficient, and Spearman's rank correlation. Although other methods have been proposed [29], the majority of studies to date use one of the four methods described here. Each scoring method captures unique aspects of the data. Euclidean distance measures the geometric distance between two vectors, and so takes into account both the direction and the magnitude of the gene vectors. Mutual information measures how much knowing the expression levels of one gene reduces the uncertainty about the expression levels of another. Pearson's

correlation coefficient measures the tendency of two vectors to increase or decrease together, thus quantifying their overall correspondence. Spearman's rank correlation is identical to Pearson's correlation, but operates on expression ranks instead of expression values, and so is more robust to outliers. Each method offers unique advantages and disadvantages for the identification of coexpression relationships, and I discuss these in detail below.

11.3.2.2 Euclidean Distance

The Euclidean distance between two gene vectors measures their geometric distance, taking into account both the direction and the magnitude of the vectors. This scoring scheme is therefore not appropriate when the absolute levels of functionally related genes are expected to be highly different. For example, the Euclidean distance might not be effective at identifying a transcription factor along with its targets if the overall levels of the transcription factor are comparatively lower across assayed conditions. Furthermore, if two genes have consistently low expression levels but are otherwise randomly correlated, they might still appear close in Euclidean space. However, Euclidean distance can be useful for detecting genes in a common pathway that respond differently from the rest of the genome. For example, if a cell is subjected to an environmental stress such as heat shock, a dramatic transcriptional increase might be expected for genes involved in heat shock response.

The Euclidean distance, D, between two gene vectors X and Y is calculated as:

$$D(X, Y) = \sqrt{\sum_{j=1}^{N} (X_j - Y_j)^2}$$

where N is the number of conditions in the study. Note that because this is a distance measure, similar gene vectors have values close to zero.

11.3.2.3 Mutual Information

Mutual information (I) is an information-theoretic quantity that expresses the mutual dependence between two random variables. In other words, it measures the degree to which knowledge about one random variable reduces the entropy of another random variable. One advantage to mutual information is that it can detect non-linear statistical relationships. Multiple studies have demonstrated the utility of mutual information for the construction of gene coexpression networks [30, 31]. The mutual information, I, of two gene vectors X and Y is calculated as:

$$I(X, Y) = S(X) + S(Y) - S(X, Y)$$

where $S(V)$ is the entropy of any discrete variable V and is calculated as:

$$S(V) = \sum_{j=1}^{N} P(V_j) \log P(V_j)$$

where $P(V_j)$ is the probability of assuming the given value. In practice, $P(V_j)$ can be calculated from microarray data by converting the continuous expression values contained in matrix M to discrete values by binning, and calculating the overall frequencies (and conditional frequencies) of each bin.

11.3.2.4 Pearson's Correlation Coefficient

Pearson's correlation coefficient (often denoted as r) measures the tendency of two gene vectors to take on values above and below each of their average levels in a coordinated fashion. As the measure is relative to the gene's own average level, it is useful for detecting similarity between genes that may have different absolute levels of expression. For example, Pearson's correlation coefficient can identify cases where a transcriptional activator and its target genes' expression levels rise and fall in synchrony even if the targets have a more extreme level of expression. Since its introduction to the field of microarray analysis [32], Pearson's correlation coefficient has arguably been the most popular method for the comparison of gene vectors.

The correlation, r, between two gene vectors X and Y is calculated as:

$$r(X, Y) = \frac{\sum_{j=1}^{N} (X_j - \overline{X})(Y_j - \overline{Y})}{\sqrt{\sum_{j=1}^{N} (X_j - \overline{X})^2 \sum_{j=1}^{N} (Y_j - \overline{Y})^2}}$$

where $\overline{X}$ and $\overline{Y}$ are the mean expression levels of X and Y, respectively. The value of r ranges from -1 (perfect anticorrelation) to 1 (perfect correlation), with values near 0 indicating no correlation.

In practice, the assumption made by the Pearson's correlation of an underlying normal distribution across experimental conditions is usually approximately valid. In cases where the data does not follow a normal distribution, a nonparametric method such as Spearman's rank correlation should be used (see below). Popular variations on r include an uncentered version, which does not subtract off the means, and the absolute value of r (or, alternatively, r^2). The uncentered Pearson's correlation coefficient is appropriate when actual expression values are more important than the overall trend of the gene vectors to have a similar shape.

The absolute r (or, alternatively, r^2) is capable of also capturing *negative correlation*, which can occur when a transcriptional repressor's expression levels are inversely proportional to the levels of its targets [33]. However, a substantial complication of interpreting networks constructed from anticorrelation is that they do not exhibit the same transitivity properties as networks constructed from standard correlation measures. For example, if gene A is anticorrelated with gene B and gene B is anticorrelated with gene C, A and C are not necessarily anticorrelated. This characteristic can complicate downstream analyses, such as the identification of gene modules as dense subgroups of interconnected genes (see Section 11.5.3).

11.3.2.5 Spearman's Rank Correlation Coefficient

Spearman's rank correlation coefficient (often denoted ρ) is calculated in an identical manner to r, with the exception that each gene vector is first transformed to a set of ranks by assigning the highest expression level a rank of 1, and the lowest a rank of N.

Due to this rank transformation, ρ is less sensitive to extreme values in the data, as outliers are scaled down upon being mapped to rank space. In cases where there are no ranking ties, ρ can be calculated as:

$$\rho(X, Y) = 1 - \frac{6\sum_{j=1}^{N} D_j^2}{N(N^2 - 1)}$$

where D_j is the difference between the ranks of the corresponding values of X and Y.

11.3.3
Choosing a Threshold

Once a scoring measure has been chosen and all pairs of genes have been scored, the next step is to choose a score threshold and create a GCN by linking all gene pairs with scores exceeding this threshold. In this section, I discuss various strategies for choosing a threshold.

11.3.3.1 Simple Thresholding Methods

The simplest thresholding method is to choose a score cutoff. This method can be useful when a clear interpretation of the scoring scheme is available. For example, several studies use a Pearson correlation coefficient cutoff of 0.50 as a cutoff, corresponding to a "moderate" correlation, or 0.80, corresponding to a "strong correlation" [34–37]. Alternatively, the top k scoring gene pairs can be chosen (where k is any number), a strategy that can be useful for practical reasons (e.g., only having sufficient time or resources to test the top k predictions), or in cases where a "level playing field" is required (e.g., when determining the overlap of a gene coexpression network derived from one tissue to one derived from another tissue).

11.3.3.2 Fisher's Z-Transformation

An alternative to choosing a raw threshold is to choose a cutoff corresponding to a desired significance level. Fisher's Z-transformation assigns a significance estimate to Pearson's correlation coefficient based on the number of conditions across which it is calculated. Such a strategy can be useful because the number of conditions tested in a microarray study can vary considerably. For example, a study involving the knockdown of a specific gene may include only a handful of conditions, while a clinical study may include hundreds of patient samples. Intuitively, a Pearson correlation of 0.9 computed over four conditions should score worse than a correlation of 0.9 computed over 100 conditions, as high correlation levels can be achieved across a small number of conditions by chance. The Fisher Z-transform assigns higher significance to genes with strong correlations across a greater number of samples, and is defined as:

$$Z(X, Y) = \frac{(\sqrt{N-3})}{2} \ln\left(\frac{1 + r(X, Y)}{1 - r(X, Y)}\right)$$

where N and r are as defined above. An advantage of this method is that the resulting distribution approximately follows a standard normal distribution, allowing the desired significance threshold to be chosen using the inverse normal distribution. Several studies have used Fisher's Z for the construction of coexpression networks, including [38].

11.3.3.3 Permutation-Based Thresholds

Often, the optimal threshold for network construction is unknown. In such cases, it can be useful to also obtain an empirically based significance estimate. Such estimates are achieved by comparing to a randomly permuted expression data matrix. One standard permutation procedure is to randomly swap columns within each gene vector (e.g., within each row). This process maintains the distribution of expression levels for each gene, but breaks any relationships that might be present across conditions. Likewise, values can be swapped within each column, which maintains the expression level distribution within each condition, but breaks any relationships across genes. Frequently, permuted data is created by permuting both the rows and columns of the matrix.

Once the permuted data matrix is created, all pairs of gene vectors are then compared using the chosen scoring measure (as above) and a distribution is built from these scores. This distribution can then be used to gage the significance of a real score using a variety of methods. For example, an estimate of the statistical significance can be obtained for any "real" score using a Z-score transformation:

$$Z(X, Y) = \frac{r(X, Y) - \mu}{\sigma}$$

where r is the Pearson correlation as defined above, and μ and σ are the mean and standard deviation across all pairs of genes in the permuted matrix, respectively. Permutation-based methods have been used for choosing coexpression network cutoffs in several studies [31, 39].

Alternatively, the random score distribution can be used to convert scores to a false discovery rate (FDR):

$$FDR(X, Y) = \frac{\sum_{a=1}^{G} \sum_{b=a+1}^{G} I(r(R_a, R_b) > r(X, Y))}{G(G-1)/2}$$

where I is the indicator function, R is the randomly permuted expression data matrix, and G is the number of genes. FDRs are useful in cases where normality assumptions do not hold or when it is necessary to control for the expected rate of false predictions.

11.3.3.4 Other Methods

A handful of other methods have been proposed for choosing a cutoff for coexpression network construction. One popular method is to choose a threshold based on prior biological knowledge [37, 40–42]. For example, if a particular biological process is under investigation, a threshold can be chosen such that the majority of its constitutive genes are connected by coexpression links. Alternatively, a threshold can be chosen based strictly on the topology of the underlying network. For example, the number of connections can be gradually decreased until the network's clustering coefficient is optimized with respect to its randomized counterpart [43]. Others have suggested methods based on random matrix theory [44] and choosing a correlation threshold resulting in the minimum network density [45].

11.4
Integration of GCNs with Other Data

The wide availability of DNA microarray data has yielded unparalleled insights into global gene expression patterns. Despite the clear utility of microarray-based studies, each experiment is subject to the usual setbacks associated with high-throughput technologies, including experimental noise and study-specific biases. One way to address issues arising from individual datasets is to gain confidence in predictions by combining results across studies, a field of statistics often referred to as *meta-analysis*. Direct comparison of expression levels across different microarray platforms and laboratories often reveals low overall reproducibility [46, 47], although recent studies suggest that such comparisons are now becoming more feasible as microarray technologies mature [48, 49]. Meta analysis-based methods attempt to find commonalities across datasets, which often provide stronger predictions than those made based on individual datasets.

In this section, I discuss recent studies and techniques that use meta-analysis techniques to combine multiple biological datasets in a network framework. Such methods can be broadly classified into three basic approaches (Figure 11.2). In the first approach, the combined network is formed from the union of links taken from individual datasets (Figure 11.2(1)). Methods based on this approach achieve high sensitivity at the cost of specificity. In general, union-based approaches should not be used when combining multiple noisy datasets (such as networks built from individual microarray studies), as the resulting network will likely be too densely connected to be of any practical use. Such approaches are instead better applied when multiple higher-confidence datasets are being combined (e.g., see Section 11.4.2.1). A second approach is to restrict to links supported by multiple interaction types (Figure 11.2(2)). Although higher confidence can be gained when restricting to interactions supported by multiple sources, such strategies can still suffer when at least two data sources are highly noisy or when the data sources are complementary to each other. Accordingly, recent studies have focused on probabilistic-based methods for network integration. These methods weight the various data sources based on their ability to predict various attributes (e.g., shared function) and choose a threshold based on the weighted combination (Figure 11.2(3)). Such methods take advantage of the inherent differences in qualities between various data sources and have recently seen a surge in popularity [50].

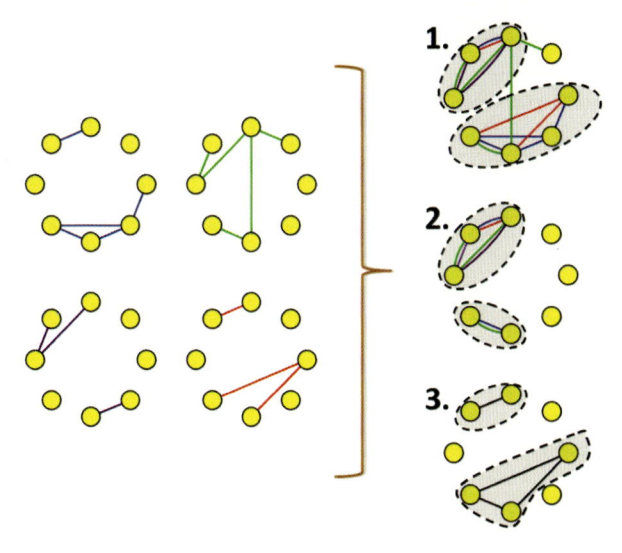

Figure 11.2 Overview of network-based data integration approaches. Three categories of data integration approaches are depicted for combining four different networks (left) into one of three integrated networks (right). For each network, nodes represent genes, with nodes in the same position representing the same gene in different networks. Different edge colors can represent different microarray studies, or different types of interactions (such as protein and genetic interactions). Three categories of integrated networks are depicted on the right: (1) union-based, (2) intersection-based, and (3) probabilistic-based. Dashed ovals indicate groups of genes (*gene modules*) that can be inferred from each integrated network (see text for discussion).

11.4.1
Integration of Multiple Expression Datasets

The results of numerous studies demonstrate that genes that are coexpressed across multiple conditions are likely to be members of the same pathway or protein complex [6, 51, 52]. However, the identification of gene coregulation relationships from the expression patterns of a single experiment is complicated by the presence of overlapping pathways and noise in the data. Therefore, although microarrays offer valuable global insight into gene expression patterns, specific predictions often come with a high false-positive rate. Recent studies have taken advantage of the increasing amount of microarray data by identifying genes that are coregulated across multiple conditions in multiple studies. Such strategies have been aided by a recent increased movement towards systematic cataloging of expression datasets in databases such as the Gene Expression Omnibus (GEO) [53], Array Express [54], and the Stanford Microarray Database [55]. Most microarray databases require standardized formats for the description of experimental conditions, which enables the use of automated methods for identifying studies assaying specific conditions. In this section, I review methods that combine the results of multiple expression datasets using a network-based framework.

11.4.1.1 Integrating Data within a Species

In the early years of microarray technology, there were only a handful of available datasets, rendering meta-analysis based techniques largely unsuitable. Early microarray studies focused mainly on cellular responses to specific scenarios such as the presence and absence of nutrients [56], the activation of a specific transcription factor [57], or comparisons between normal and diseases states [58]. The utility of analyzing many cellular conditions was initially demonstrated in a study in *Saccharomyces cerevisiae* assaying 300 cellular conditions corresponding to individual gene knockdowns and various chemical treatments [6]. Around the same time, another important study assayed a range of cellular challenges such as heat shock or exposure to drugs [5]. Examining gene expression profiles across a diverse range of conditions allowed both groups to use clustering strategies to assign putative functional roles to hundreds of genes. The identification of groups of genes with similar expression patterns enabled the prediction of gene function on a global scale using the "guilt by association" principle, which posits that a gene that is linked to a set of genes involved in a particular process is likely to also be involved in that process [59]. A key insight of these studies was that a greater variety of assayed conditions results in stronger predictions of gene function and in this way they paved the way for subsequent meta-analysis techniques for combining results across studies.

The first work to combine expression data from many studies provided the first global picture of the expression program of a multicellular organism, the nematode *Caenorhabditis elegans* [60]. In this study, the authors calculated Pearson correlation coefficients between all gene pairs and applied a spring-embedded layout algorithm to create a map in which coexpressed genes are located proximal to each other on the *xy* plane, with large groups of genes corresponding to broad functional classes. By demonstrating the utility of combining data from several experiments for gene functional prediction, the authors opened the door for future meta-analysis based studies of gene expression data. To date, gene coexpression "meta-networks" have been constructed for several other organisms, including *Escherichia coli* [61], *S. cerevisiae* [38], mouse [62–64], and human [42, 65–67].

11.4.1.2 Integrating Data across Species

The incredible diversity of life stemmed from a single common ancestor through the process of evolution. A major driving force in evolution is thought to be the interplay between transcription factors and the genes that they regulate [68]. The result of this interplay is a fine-tuning of the cellular response to changing conditions via the alteration of the expression patterns of the thousands of genes contained in a genome. The effect of these alterations is that new genes can be recruited to or dropped from a particular regulatory program, or new regulatory programs can be created. By comparing coexpression relationships across species, new insights can be gained into how specific pathways have evolved and what makes an organism unique [69–71]. Likewise, conservation of coexpression of genes across multiple species implies that their coregulation might provide a selective advantage and are therefore likely to be involved in a similar function. For these reasons, the identification of both conserved and species-specific gene coexpression relationships have been active areas of research.

Early studies compared gene expression levels in *C. elegans* and *S. cerevisiae*, revealing a surprisingly small number of conserved gene coexpression patterns that tended to involve genes whose products form stable protein complexes [72]. Subsequent studies showed that orthologs with conserved expression patterns are likely to be members of the same pathway and can be used to predict gene function [52, 73]. The former study reported the interesting finding that in certain cases, coexpression is conserved, but specific expression patterns are not.

The first study to examine global coexpression patterns across species identified pairs of genes that are coexpressed over thousands of microarray experiments in human, fly, worm, and yeast [51]. In this study, gene pairs displaying significant coexpression across multiple species were linked and used to identify conserved genetic modules corresponding to a diverse range of processes. In addition to providing one of the first conserved functional maps of the cell, this work demonstrated the utility of comparing multiple expression datasets across species for the elucidation of gene function and regulation, a theme frequently revisited in subsequent works [34, 73–75]. Recent studies have used conserved coexpression to predict other cellular attributes, including protein interactions [76] and the involvement of a gene in particular diseases [77].

11.4.2
Integration of Heterogeneous Data Sources

Recent years have seen a rapid increase in the development of high-throughput experimental techniques for the measurement of various cellular aspects, including protein interactions, genetic interactions, and gene perturbation phenotypes [23]. A major focus of recent computational analyses has been the development of methods for combining these data sources in order to gain a more holistic view of the cell [78]. Although different in implementation details, most of these methods take advantage of the fact that different data sources provide complementary perspectives of cellular activities, while at the same time filtering out the experimental noise inherent to individual high-throughput studies [79–81]. In general, such approaches can be broadly classified into those that aggregate networks constructed from individual data sources, and those that combine separate networks into a single network of pairwise gene scores (usually under some probabilistic framework). In the following sections, I provide a brief overview of these two broad classes of data integration and conclude with an overview of methods for the integration of microarray data with other particular data sources.

11.4.2.1 Union and Intersection-Based Methods
One class of network aggregation methods involves simply taking the union of links across networks (Figure 11.2(1)). The combination of different interaction types into a single network in this manner allows for the determination of relationships between the constitutive data types. Although conceptually simple, network

aggregation has proven to be a powerful method. Initial network aggregation-based studies identified pairs of genes with relationships supported by multiple data sources, including protein–protein, gene coexpression, and gene cophenotype interactions [82–84]. An alternative strategy is to find subnetworks supported by multiple interaction types that do not necessarily overlap [85]. Such approaches take advantage of datasets that provide orthogonal, but complementary information, such as synthetic lethal genetic interactions and protein interactions [86]. Other studies have used network aggregation to determine complex relationships between different interaction types [87, 88] and for the characterization of human disease states [89, 90].

11.4.2.2 Probabilistic Methods

Despite the clear utility of the simple aggregation-based methods described above, recent studies have focused on more advanced methods for the integration of data sources into a single probabilistic network. Probabilistic networks can be thought of as representing an average functional similarity across different tissue types, developmental stages, and conditions, such that the final network represents all possible associations across a wide range of states [50]. Several advantages are afforded by taking a probabilistic view of gene associations, including the implicit incorporation of uncertainty and the ability to account for gene pleiotropy.

Probabilistic methods for data integration weight data sources before combining them into a single network, in the process accounting for differences in accuracy. To illustrate, consider the four networks depicted in Figure 11.2, with the red network representing high confidence interactions, the blue network representing medium confidence interactions, and the green and purple networks representing low confidence interactions. A probabilistic method can incorporate these interaction strengths, such that a single red interaction is sufficient for presence in the final network (Figure 11.2(3)). Conversely, any other link type is insufficient on its own for inclusion in the final network, and is only included when a medium strength (blue) link is supported by at least one other interaction type (Figure 11.2(3)). In this manner, probabilistic methods represent a logical compromise between union-based approaches (Figure 11.2(1)) and intersection-based approaches (Figure 11.2(2)).

When combining different sources of information in a probabilistic framework, an important issue is how to weight the various data types. Most methods weight each dataset based on its ability to capture known functional relationships (e.g., genes known to participate in the same biological process). One popular framework is the Bayesian network – a graph-based model of joint multivariate probability distributions that captures conditional independence between variables [91]. Due to their ability to handle noise and describe complex stochastic processes, Bayesian networks have been widely applied for the construction of coexpression networks since their original introduction to the field [92]. Subsequent studies have applied Bayesian networks for the integration of multiple data sources into a single functional network in yeast [93–95] and other eukaryotic species [96].

A second popular Bayesian-based method for data integration is the log likelihood method of Marcotte et al. [97]. In this approach, each experiment is evaluated based on its ability to capture known functional relationships using a log-likelihood score (*LLS*):

$$LLS = \ln\left(\frac{P(L|E)/\sim P(L|E)}{P(L)/\sim P(L)}\right)$$

where $P(L/E)$ and $\sim P(L/E)$ are the frequencies of linkages (L) observed in the given experiment (E) between annotated genes operating in the same pathway and in different pathways, respectively. Likewise, $P(L)$ and $\sim P(L)$ represent the prior expectations (the total frequency of linkages between all annotated cofunctional genes and noncofunctional genes, respectively). A final score for each gene pair is obtained by summing scores across all studies, often while factoring in the degree of independence between experiments. Since their introduction, log-likelihood-based methods have been applied for the construction of integrated networks for a variety of model organisms, including yeast [97, 98], worm [99, 100], and fly [101].

Although most methods weight data sources based on their ability to capture known biological relationships, alternative approaches have been suggested. In particular, approaches that do not require a training set are particularly attractive because only a limited number of (potentially biased) functional associations are known. A recent study presented the POINTILLIST method, which combines data sources into a single network without the need for a training set, instead weighting datasets by their overall statistical power [102]. Although the method's utility is demonstrated by its application to a yeast galactose utilization network, its scalability to genome-wide datasets remains an issue.

11.4.2.3 Integration of Expression Data with Specific Other Data Types

In addition to general data integration strategies, several methods have been specifically tailored to the integration of specific data sources. Such methods take advantage of known relationships between data sources to address specific biological questions. In this section, I briefly review three types of data that are frequently combined with microarray data: transcription factor binding site predictions, genetic variation, and protein interactions.

One of the earliest applications of microarray data integration was for the identification of putative transcription factor binding sites (and their corresponding motifs) in genomic DNA. Such methods operate on the principle that the coexpression relationships learned from microarrays are often indicative of coregulation relationships [103]. Early methods identified groups of coregulated genes in a microarray study using clustering techniques (see Section 11.5.3.1), and subsequently searched for enriched sequence motifs in their genomic upstream (promoter) regions [104–108]. Other methods operate in the opposite direction, identifying groups of genes with similar upstream transcription factor binding sites and restricting to those with coherent expression signatures [40, 109]. Recent methods focus on the simultaneous clustering of microarrays and transcription factor binding site enrichment data [110–112].

A second class of data integration methods combines microarray data with data monitoring genetic variation. The expression levels of most genes are thought to be heritable, thus making them amenable to genetic analyses seeking links between genotypes and phenotypes. By simultaneously measuring the expression levels of thousands of genes at a time, microarrays provide a quantitative genome-wide phenotype associated with a specific experimental condition. Recently, the field of "genetical genomics" has seen a surge in popularity [113–116]. In this framework, mRNA levels and DNA marker data are collected in tissue samples, and the chromosomal regions that affect expression levels are determined by quantitative trait loci (QTL) analysis. Using this data, a direct link can be established between genotype and phenotype, enabling the study of complex diseases and traits [117, 118]. Gene networks provide a natural choice for studies combining microarray and genetic variation data, and several groups have used network-based approaches to this end [119–121].

Recent advances in experimental techniques such as yeast two-hybrid assays [122] and tandem affinity purification [123] have resulted in a rapid increase of protein interaction data paralleling that of microarrays. Genes sharing a common function have an increased tendency to have similar expression patterns, and to encode proteins that interact. Accordingly, several groups have demonstrated a tendency for genes with interacting protein products to also be coexpressed [124–126], resulting in the development of methods for the integration of microarray data with protein interaction data. In one of the earliest studies, Ideker *et al.* identified "active subnetworks" – connected regions of the protein interaction network encoded by genes displaying significant changes in expression over similar experimental conditions [127]. Subsequent studies have used similar strategies to classify protein complexes as permanent or transient [128], to identify tissue-specific interactions [129], and to create a "time-dependent interaction network" by overlaying cell cycle expression data [130]. Additional applications of protein interaction and expression data integration include gene module identification [131, 132] and gene functional prediction [133].

11.5
Analysis of GCNs

Once a GCN has been constructed, techniques are available to mine information from the thousands of interactions it contains, many of which are based on ideas borrowed from the field of graph theory. Taken together, these approaches offer complementary information on network structure and function, each providing unique insights into complex cellular systems. In this section, I discuss three popular categories of methods for extracting biological predictions from gene coexpression networks. In the first section, I discuss the identification of genes with many interaction partners, often referred to as *network hubs* (Figure 11.3(1)). Next, I discuss methods for the discovery of *network motifs*, overrepresented subgraph structures such as interaction triangles [134] or feedback/feedforward loops [88].

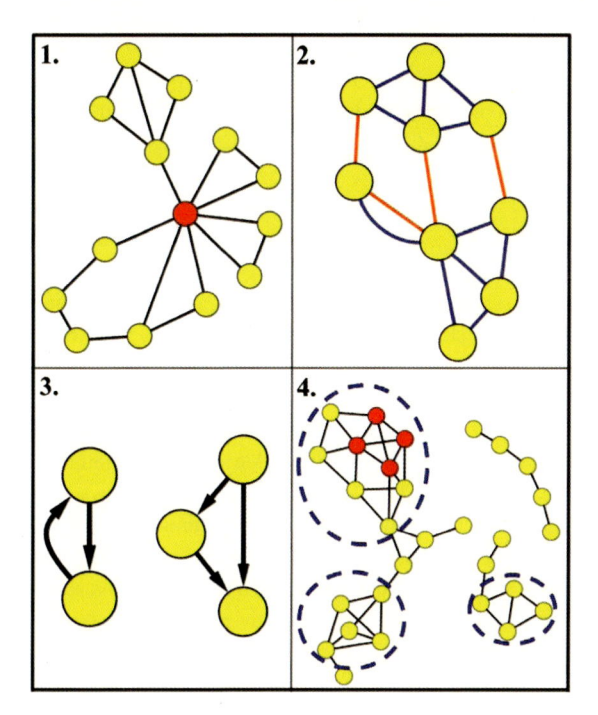

Figure 11.3 Overview of network mining techniques. (1) Discovery of network hubs. Here, a single network hub gene, which has an excess of links with respect to other genes in the network, is depicted in red. (2 and 3) Identification of over-represented network motifs. Over-represented network motifs can be identified by comparison to their frequency of occurrence in randomly permuted networks. It can be observed in (2) that there is an excess of triangle motifs induced by the blue edges. In (3), two additional network motifs are depicted: the feedback loop (left) and the feedforward loop (right). (4) Identification of densely connected subnetworks. Using graph theory-based applications, it is possible to identify densely connected subregions of a network, or gene modules (dashed blue ovals). Nodes indicated in red are members of a clique of size four (see text for discussion).

The final section outlines approaches for the detection of densely connected subregions of a network (often referred to as *gene modules*), as depicted in Figure 11.3(4).

11.5.1
Network Hubs

The majority of biological networks have a *scale-free* distribution, meaning that most genes have a small number of links, with only a few hub genes having many links [39, 135–137]. Such a distribution rarely arises in random networks and is thought to provide beneficial attributes to the overall system, including robustness to accidental node (gene) failure and a shorter average distance between any two nodes [138]. Due to their importance to the overall network structure, the hub genes of many biological networks have been shown to have an increased propensity to be essential [139, 140], to have orthologs in other organisms [141], to be highly conserved throughout evolution [16], and to be involved in complex diseases [90]. In particular, these characteristics are specifically present in GCNs [34, 39].

11.5.2
Network Motifs

One means of globally characterizing a network is by examining its local subgraph interaction patterns. Local subgraph structures that are enriched in a network compared to its randomized counterparts are referred to as *network motifs* [134]. Network motifs can include any number of nodes, with the simplest pattern consisting of three nodes linked in the form of a triangle. For example, the blue triangles depicted in Figure 11.3(2) might represent a network motif due to their high frequency of occurrence, while orange triangles would not.

Network motifs represent simple building blocks of complex systems and so might offer insight into the structural design principles of networks. An advantage to network motifs is that they can be used to assign putative roles to genes based solely on local network topology. Some network motifs are directly interpretable from a biological perspective, such as the feedback and feedforward loops depicted in Figure 11.3(3). Both motifs play important roles in gene transcriptional regulation [142, 143], and their over-representation in the transcriptional regulatory networks of multiple diverse species is further indicative of their biological relevance [144]. Recent studies have explored such three-way interaction motifs in GCNs [145, 146]. Discovery of network motifs in integrated networks has offered insights into relationships between different interaction types, while demonstrating that higher-order "network themes" encompassing multiple occurrences of network motifs are also enriched in biological networks [87, 88].

11.5.3
Gene Modules

Often, the goal of a microarray experiment is to identify groups of genes with similar expression levels across conditions. Gene modules identified in this manner might represent distinct regulatory programs, and often correspond well with pathways and protein complexes [63, 147–149]. For example, the strongest signal in a microarray study is often the clustering (grouping together based on similar expression profiles) of genes encoding ribosomal subunits [5]. Mounting evidence suggests a modular nature to the organization of cellular systems and several studies have accordingly begun to explore cellular processes at the level of gene modules as opposed to a single gene at a time [150–152]. A wide range of methods have been developed for the detection of gene modules using microarray data. These methods can be broadly classified into those that operate directly on the microarray data matrix (gene vector clustering methods) and those that take a GCN as input (network-based methods).

11.5.3.1 Gene Vector Clustering Methods for the Identification of Gene Modules

Microarray studies contain thousands of data points, and therefore require computational methods for visualization and identification of patterns in the data. Often, groups of putatively coregulated genes are identified by the clustering of their expression patterns. Gene vector clustering methods operate directly on the microarray data and so are often more easily interpretable due to their direct correspondence to the expression data matrix. Due to their popularity, I here give a brief summary of a handful of frequently used methods. The reader is referred to other reviews [153–155] for an in depth discussion of these and other non-network-based clustering methods.

One of the earliest methods for microarray data analysis was based on hierarchical clustering [32]. In hierarchical clustering, relationships between genes are represented by a tree whose branch lengths reflect the degree of similarity between their gene vectors, as assessed by a scoring method such as one previously discussed (e.g., Pearson's correlation coefficient). Clusters are formed from the tree by iteratively grouping the closest pair of nodes and replacing them by a single node representing their set average. Other popular clustering methods include *K*-means clustering [108], a partitional method that minimizes the overall within-cluster sum of squared distances from the cluster mean, and self-organizing maps [156], which represent clusters as nodes in a graph that are mapped into a higher dimensional space (initially at random) and iteratively adjusted to fit the data. Methods that identify genes that cluster across a subset of conditions are referred to as *two-way clustering* or *biclustering* methods [41, 157–159]. Such methods are effective at capturing genes that are coregulated only under a specific subset of conditions, and are capable of identifying genes that are involved in multiple pathways [160].

The above gene vector clustering methods all operate solely on the microarray data itself, and so can be classified as unsupervised methods. Supervised methods are a separate class that incorporates prior biological knowledge, such as other interaction types [161] or known biological pathways [162]. Additionally, several methods combine microarray clustering with experimental or computational transcription factor binding site data [40, 104, 111, 112]. Supervised clustering methods are a useful way to combine microarray data with other information, but they can also limit *de novo* discovery by focusing on already established biological relationships. Furthermore, supervised methods are highly dependent on the quality of the data on which they are trained [163]. Nevertheless, supervised clustering methods have proven useful for the discovery of novel associations between genes and known biological processes.

11.5.3.2 Network-Based Methods for the Identification of Gene Modules

Network-based clustering methods identify sets of genes that are more densely linked to each other than expected by chance, given the overall link density of the network. Densely connected subnetworks represent groups of genes that are mutually and tightly coexpressed with each other, and are therefore likely to represent gene coregulation modules. Frequently used statistics for module detection include subnetwork *connectivity*, the number of links connecting a set of nodes to one another divided by the total possible number of such links, and the *clustering coefficient*, the degree of connectivity exhibited by the neighbors of a node.

A major advantage to network-based clustering approaches is their ability to tolerate missing data. Namely, it is not necessary for all of the members of a gene module to be tested in all of the conditions in which they are coexpressed, provided that subsets of the genes are connected densely enough to be considered significantly connected. Furthermore, despite the fact that many genes take part in multiple processes [164], most standard non network-based clustering algorithms only assign genes to a single cluster. Network-based methods allow overlap between gene coregulation groups.

11.5.3.2.1 Connected Components The simplest way to partition a network into subnetworks is through the identification of its *connected components*. A connected component is a maximal set of nodes such that any pair of nodes can be reached by at least one path. To illustrate, there are a total of three connected components in the network depicted in Figure 11.3(4) (left, upper right, and lower right). An advantage to connected components is that they can be easily identified via a breadth-first or depth-first search. However, the majority of biological networks are not sufficiently modular for connected components to be useful in practice, with many consisting of only one large connected component. As a result, connected components can often be further broken down into subnetworks (e.g., the leftmost connected component in Figure 11.3(4) can be broken into the two strongly connected components indicated with dashed circles). A further disadvantage to connected components is that a path between a set of nodes does not guarantee a high degree of link density, as evidenced by the connected component depicted in the upper right of Figure 11.3(4).

11.5.3.2.2 Cliques and Approximate Cliques An alternative approach that ensures a high link density is the identification of network *cliques*. A clique is a set of nodes that are fully connected to each other. For example, the red nodes depicted in Figure 11.3 (4) represent a clique of size four. By definition, a clique containing k nodes will contain $k*(k-1)/2$ links (all possible node pair combinations). It can be seen that a clique of size k (referred to as a *k-clique*) also includes multiple cliques of size $k-1$, $k-2, \ldots, 1$, so algorithms for the detection of *maximal cliques* are often employed in practice. The problem of identifying all maximal cliques in a network is computationally expensive, so advanced methods such as the exact branch-and-bound algorithm employed by Cliquer [165] or the vertex cover extraction-based method of Voy et al. [166] must be employed for their detection.

Given the formulation of the problem of finding densely connected subnetworks, the identification of maximal cliques clearly represents one possible solution. However, in practice, high-throughput data sources like microarrays have a high false-negative rate, so the requirement of full connectivity is usually too stringent. Many methods therefore relax the maximal connectivity requirement and instead search for *approximate cliques*, which allow a certain number of missing links [167, 168].

11.5.3.2.3 Advanced Module Identification Methods Somewhere between connected components and cliques lies a middle ground of connectivity that is optimal for identifying high confidence gene modules while allowing for noise in the data. The number of possible subnetworks grows exponentially with the number of nodes in a network, so heuristic techniques are required for approximate solutions to the problem of identifying densely connected subnetworks. At the time of this writing, over 25 methods have been published, so a thorough review here is infeasible given space constraints. Although there have been attempts to evaluate some methods [169, 170], no one method clearly outperforms the others. Evaluations of module identification methods are hindered by the fact that no clear objective mathematical criteria exist for evaluation purposes. Furthermore, individual methods often have a wide range of possible parameter settings, which frequently are dataset specific. In the future, a formal evaluation of available methods along the lines of those performed for protein structure prediction [171] and transcription factor motif finding [172] would likely be beneficial. In lieu of such an evaluation, I here discuss a handful of popular, publicly available methods, and refer the reader to a review by Sharan et al. [173] for further information.

Most methods for the identification of subnetworks operate in a "bottom-up" fashion. In bottom-up methods, densely connected "seed" regions are initially identified, and are subsequently expanded and merged based on some statistical measure of overall link connectivity [174]. One of the original methods for subnetwork detection in biological data is the MCODE (Molecular COmplex DEtection) algorithm [147]. MCODE weights all nodes by local neighborhood density and identifies densely connected seed regions, which are subsequently modified by adding or removing nodes based on a connectivity criterion. MCODE has been widely applied for the detection of complexes in protein interaction networks, and is available as a default plugin for the Cytoscape network visualization and analysis tool [20].

An alternative method, MODES (Mining Overlapping DEnse Subgraphs), is available as a command-line utility [148]. MODES first identifies dense subnetworks using a modified version of the HCS (Highly Connected Subgraphs) algorithm [175] that allows for the detection of overlapping modules. Each subnetwork is then collapsed into a single node and HCS is run again on the condensed network. Finally, each subnetwork is pruned based on a connectivity criterion and the algorithm iterates from the beginning until no new subnetworks are identified. MODES is particularly useful for meta-analysis of a large number of datasets, due to its command-line implementation and relatively fast operating time.

CFinder [176] uses the Clique Percolation Method [177] to locate the k-clique percolation clusters of a network (all nodes that can be reached via adjacent k-cliques from each other). The authors define two k-cliques as adjacent if they share exactly k-nodes. Due to its focus on clique detection, CFinder is best suited for the quick detection of very dense subnetworks, particularly when they reside in relatively sparse networks.

Finally, the NetworkBlast algorithm [178] calculates likelihood ratio scores for a set of genes by comparing their fit to a protein complex model versus a random model. A greedy search algorithm is used to detect high-scoring modules based on this criterion. NetworkBlast can operate on either a single network or on a pair of networks in different species and is available online.

11.5.3.2.4 Methods Incorporating Multiple Networks

A handful of methods have been developed for the detection of modules containing genes that are densely connected in multiple networks. Such methods take advantage of the complementary nature of data sources generated by different high-throughput sources and can also be useful for the identification of genetic modules conserved across species. Statistical-Algorithmic Method for Bicluster Analysis (SAMBA) [159] is a biclustering framework that models data using a bipartite graph linking genes to various properties, identifying modules as heavy subgraphs. An alternative method, mining COherent DENSE subgraphs (CODENSE) [148] identifies modules using two networks: a summary network consisting of edges that frequently occur in the input networks and a second-order network containing edges that frequently occur in the same input networks. Due to the fact that the slowest step (subnetwork identification) is always performed only on two networks (the summary and second-order networks), CODENSE achieves a fast operating time, regardless of the number of input networks. The preceding methods are unsupervised, in the sense that they operate on the data without any prior biological knowledge. Alternatively, several supervised methods have been developed for the directed search for genes related to a query gene (or gene group) of interest using microarray data [28, 162, 179–181]. Such methods are useful for the identification of novel gene associations to a particular regulatory program or biological process.

11.5.3.2.5 Annotation of Gene Modules

Due to the modular nature of cellular networks, the use of gene modules for the assignment of gene function is now in widespread practice. Subnetworks identified in networks are often used to assign genes to a putative biological function [151]. Unannotated genes that share a module with genes with a known function can be assigned that function based on the "guilt by association" principle. Often, controlled ontologies such as Gene Ontology (GO) [182], KEGG (Kyoto Encyclopedia of Genes and Genomes) [183], or GenMAPP [184] are used for this purpose by determining the significance of the overlap of the genes in a module with the genes in each ontology category using the hypergeometric distribution:

$$P(X \geq n_b) = \sum_{i=n_b}^{n_s} \frac{\binom{n_c}{n_b}\binom{n_s-n_c}{N-n_b}}{\binom{n_s}{N}}$$

where n_b is the number of genes present in both the module and the category, n_s is the size of the module, n_c is the size of the category, and N is the number of genes assigned to at least one category. Each module is then assigned a putative annotation based on its strongest p-value.

11.6
GCNs for the Study of Cancer

Cancer is largely a disease of mis-regulation and many studies have used microarray analysis in an effort to help unravel the complex mechanisms underlying the transition from normal to cancerous states (reviewed in [185–187]). Most studies focus on the identification of differentially expressed genes, such as genes that are highly expressed in tumor states compared to normal tissues [188–191]. An alternative approach is to identify genes that are differentially coexpressed (e.g., a group of genes that are coexpressed in normal tissues, but not in tumor tissues), which might represent regulatory programs disrupted in cancer [35, 39, 192, 193]. Recent work has focused on identifying pathways displaying abhorrent cancer expression patterns [194–198] and on the development of meta-analysis methods for combining microarray results across studies [199–201]. Such approaches have been greatly aided by databases such as Oncomine [202], which contains data from thousands of microarrays examining a broad range of cancer subtypes, progression stages, and patient backgrounds. A handful of studies have taken network-based approaches to help identify genes and groups of genes involved in the transformation of normal tissues to a malignant stage [30, 203–205]. Network frameworks are particularly attractive for this task because they enable the discovery of alterations in transcriptional network connectivity in disease states.

The ultimate goal of these studies is the determination of biomarkers and specific therapeutic targets for every subtype of cancer. Despite the success of the studies mentioned above, cancer is a highly complicated disease and at present there are no validated clinical tests based on microarray applications [206, 207]. In the future, large scale collaborative efforts such as The Cancer Genome Atlas [208] offer promise for increasing our understanding of the molecular basis of cancer.

11.7
Conclusions

In this chapter, I have presented a summary of network-based methods for the analysis of gene expression data generated by DNA microarrays. GCNs provide an intuitive means for the visualization and investigation of complex interactions between thousands of genes at a time. The network framework is well-suited for integration with other sources of data, and enables the application of graph theoretic-based algorithms. Analysis of GCNs offers insights into global systems properties, while at the same time providing the basis for specific gene regulatory and functional hypotheses.

GCNs are a convenient and useful method for the analysis of microarray data. However, it is important to emphasize the importance of examining the original expression data upon the identification of interesting gene associations, as genes can display strong correlations for a variety of superfluous reasons. For example, if many similar experimental conditions are examined, it might be expected that a substantial portion of the genome might display similar expression patterns. Furthermore, genes with expression levels that peak together in a single condition and are otherwise expressed at low levels will have strong correlations. By examining the original expression data (e.g., in the form of a heatmap), such cases can easily be separated from biologically relevant predictions.

As high-throughput technologies mature, the need for computational methods will likewise continue to increase. Although new sequencing technologies such as 454, Solexa, and SOLiD offer unprecedented opportunities, they also produce staggering amounts of data. Likewise, microarray studies are beginning to include more samples (e.g., larger patient cohorts) and cover a wider range of species. In the future, the application of meta-analysis methods for large-scale data integration such as network alignment [178, 209–211] will likely help shed light on the similarities and differences between a wide range of species, tissues, and disease states.

Acknowledgments

Thank you to Brooke E. Crowley and Alex G. Williams for helpful comments on an earlier draft of this manuscript.

References

1 Brown, P.O. and Botstein, D. (1999) Exploring the new world of the genome with DNA microarrays. Nat. Genet., 21, 33–37.

2 Schena, M., Heller, R.A., Theriault, T.P., Konrad, K., Lachenmeier, E., and Davis, R.W. (1998) Microarrays: biotechnology's discovery platform for functional genomics. Trends Biotechnol., 16, 301–306.

3 Spellman, P.T., Sherlock, G., Zhang, M.Q., Iyer, V.R., Anders, K., Eisen, M.B., Brown, P.O., Botstein, D., and Futcher, B. (1998) Comprehensive identification of cell cycle-regulated genes of the yeast Saccharomyces cerevisiae by microarray hybridization. Mol. Biol. Cell, 9, 3273–3297.

4 Jiang, M., Ryu, J., Kiraly, M., Duke, K., Reinke, V., and Kim, S.K. (2001) Genome-wide analysis of developmental and sex-regulated gene expression profiles in Caenorhabditis elegans. Proc. Natl. Acad. Sci. USA, 98, 218–223.

5 Gasch, A.P., Spellman, P.T., Kao, C.M., Carmel-Harel, O., Eisen, M.B., Storz, G., Botstein, D., and Brown, P.O. (2000) Genomic expression programs in the response of yeast cells to environmental changes. Mol. Biol. Cell, 11, 4241–4257.

6 Hughes, T.R., Marton, M.J., Jones, A.R., Roberts, C.J., Stoughton, R., Armour, C.D., Bennett, H.A., Coffey, E., Dai, H., He, Y.D., Kidd, M.J., King, A.M., Meyer, M.R., Slade, D., Lum, P.Y., Stepaniants, S.B., Shoemaker, D.D., Gachotte, D., Chakraburtty, K., Simon, J., Bard, M., and Friend, S.H. (2000) Functional discovery via a compendium of expression profiles. Cell, 102, 109–126.

7 Chung, C.H., Bernard, P.S., and Perou, C.M. (2002) Molecular portraits and the family tree of cancer. Nat. Genet., 32 (Suppl.), 533–540.

8 Sasik, R., Woelk, C.H., and Corbeil, J. (2004) Microarray truths and consequences. J. Mol. Endocrinol., 33, 1–9.

9 Quackenbush, J. (2002) Microarray data normalization and transformation. Nat. Genet., 32 (Suppl.), 496–501.

10 Smyth, G.K. and Speed, T. (2003) Normalization of cDNA microarray data. Methods, 31, 265–273.

11 Gusnanto, A., Calza, S., and Pawitan, Y. (2007) Identification of differentially expressed genes and false discovery rate in microarray studies. *Curr. Opin. Lipidol.*, **18**, 187–193.

12 Kerr, M.K., Martin, M., and Churchill, G.A. (2000) Analysis of variance for gene expression microarray data. *J. Comput. Biol.*, **7**, 819–837.

13 Subramanian, A., Tamayo, P., Mootha, V.K., Mukherjee, S., Ebert, B.L., Gillette, M.A., Paulovich, A., Pomeroy, S.L., Golub, T.R., Lander, E.S., and Mesirov, J.P. (2005) Gene set enrichment analysis: a knowledge-based approach for interpreting genome-wide expression profiles. *Proc. Natl. Acad. Sci. USA*, **102**, 15545–15550.

14 Smyth, G.K. (2004) Linear models and empirical Bayes methods for assessing differential expression in microarray experiments. *Stat. Appl. Genet. Mol. Biol.*, **3**, Article 3.

15 Goldenberg, A., Zheng, A.X., Fienberg, S.E., and Airoldi, E.M. (2009) A survey of statistical network models. *Found. Trends Mach. Learn.*, **2**, 1.

16 Barabasi, A.L. and Oltvai, Z.N. (2004) Network biology: understanding the cell's functional organization. *Nat. Rev. Genet.*, **5**, 101–113.

17 Li, S., Armstrong, C.M., Bertin, N., Ge, H., Milstein, S., Boxem, M., Vidalain, P.O., Han, J.D., Chesneau, A., Hao, T., Goldberg, D.S., Li, N., Martinez, M., Rual, J.F., Lamesch, P., Xu, L., Tewari, M., Wong, S.L., Zhang, L.V., Berriz, G.F., Jacotot, L., Vaglio, P., Reboul, J., Hirozane-Kishikawa, T., Li, Q., Gabel, H.W., Elewa, A., Baumgartner, B., Rose, D.J., Yu, H., Bosak, S., Sequerra, R., Fraser, A., Mango, S.E., Saxton, W.M., Strome, S., Van Den Heuvel, S., Piano, F., Vandenhaute, J., Sardet, C., Gerstein, M., Doucette-Stamm, L., Gunsalus, K.C., Harper, J.W., Cusick, M.E., Roth, F.P., Hill, D.E., and Vidal, M. (2004) A map of the interactome network of the metazoan *C. elegans*. *Science*, **303**, 540–543.

18 Pritsker, M., Liu, Y.C., Beer, M.A., and Tavazoie, S. (2004) Whole-genome discovery of transcription factor binding sites by network-level conservation. *Genome Res.*, **14**, 99–108.

19 Tong, A.H., Lesage, G., Bader, G.D., Ding, H., Xu, H., Xin, X., Young, J., Berriz, G.F., Brost, R.L., Chang, M., Chen, Y., Cheng, X., Chua, G., Friesen, H., Goldberg, D.S., Haynes, J., Humphries, C., He, G., Hussein, S., Ke, L., Krogan, N., Li, Z., Levinson, J.N., Lu, H., Menard, P., Munyana, C., Parsons, A.B., Ryan, O., Tonikian, R., Roberts, T., Sdicu, A.M., Shapiro, J., Sheikh, B., Suter, B., Wong, S.L., Zhang, L.V., Zhu, H., Burd, C.G., Munro, S., Sander, C., Rine, J., Greenblatt, J., Peter, M., Bretscher, A., Bell, G., Roth, F.P., Brown, G.W., Andrews, B., Bussey, H., and Boone, C. (2004) Global mapping of the yeast genetic interaction network. *Science*, **303**, 808–813.

20 Killcoyne, S., Carter, G.W., Smith, J., and Boyle, J. (2009) Cytoscape: a community-based framework for network modeling. *Methods Mol. Biol.*, **563**, 219–239.

21 Hu, Z., Snitkin, E.S., and DeLisi, C. (2008) VisANT: an integrative framework for networks in systems biology. *Brief. Bioinform.*, **9**, 317–325.

22 Yip, K.Y., Yu, H., Kim, P.M., Schultz, M., and Gerstein, M. (2006) The tYNA platform for comparative interactomics: a web tool for managing, comparing and mining multiple networks. *Bioinformatics*, **22**, 2968–2970.

23 Zhu, X., Gerstein, M., and Snyder, M. (2007) Getting connected: analysis and principles of biological networks. *Genes Dev.*, **21**, 1010–1024.

24 Zhang, B. and Horvath, S. (2005) A general framework for weighted gene co-expression network analysis. *Stat. Appl. Genet. Mol. Biol.*, **4**, Article17.

25 Kanabar, P.N., Vaske, C.J., Yeang, C.H., Yildiz, F.H., and Stuart, J.M. (2009) Inferring disease-related pathways using a probabilistic epistasis model. *Pac. Symp. Biocomput.*, 480–491.

26 Markowetz, F., Kostka, D., Troyanskaya, O.G., and Spang, R. (2007) Nested effects models for high-dimensional phenotyping screens. *Bioinformatics*, **23**, i305–312.

27 Mehra, S., Hu, W.S., and Karypis, G. (2004) A Boolean algorithm for reconstructing the structure of regulatory networks. *Metab. Eng.*, **6**, 326–339.

28 Markowetz, F. and Spang, R. (2007) Inferring cellular networks–a review. *BMC Bioinformatics*, **8** (Suppl. 6), S5.

29 Kim, R.S., Ji, H., and Wong, W.H. (2006) An improved distance measure between the expression profiles linking co-expression and co-regulation in mouse. *BMC Bioinformatics*, **7**, 44.

30 Basso, K., Margolin, A.A., Stolovitzky, G., Klein, U., Dalla-Favera, R., and Califano, A. (2005) Reverse engineering of regulatory networks in human B cells. *Nat. Genet.*, **37**, 382–390.

31 Butte, A.J. and Kohane, I.S. (2000) Mutual information relevance networks: functional genomic clustering using pairwise entropy measurements. *Pac. Symp. Biocomput.*, 418–429.

32 Eisen, M.B., Spellman, P.T., Brown, P.O., and Botstein, D. (1998) Cluster analysis and display of genome-wide expression patterns. *Proc. Natl. Acad. Sci. USA*, **95**, 14863–14868.

33 Zeng, T. and Li, J. (2010) Maximization of negative correlations in time-course gene expression data for enhancing understanding of molecular pathways. *Nucleic Acids Res.*, **38**, e1.

34 Bergmann, S., Ihmels, J., and Barkai, N. (2004) Similarities and differences in genome-wide expression data of six organisms. *PLoS Biol.*, **2**, E9.

35 Reverter, A., Ingham, A., Lehnert, S.A., Tan, S.H., Wang, Y., Ratnakumar, A., and Dalrymple, B.P. (2006) Simultaneous identification of differential gene expression and connectivity in inflammation, adipogenesis and cancer. *Bioinformatics*, **22**, 2396–2404.

36 Sanoudou, D., Haslett, J.N., Kho, A.T., Guo, S., Gazda, H.T., Greenberg, S.A., Lidov, H.G., Kohane, I.S., Kunkel, L.M., and Beggs, A.H. (2003) Expression profiling reveals altered satellite cell numbers and glycolytic enzyme transcription in nemaline myopathy muscle. *Proc. Natl. Acad. Sci. USA*, **100**, 4666–4671.

37 Zhou, X., Kao, M.C., and Wong, W.H. (2002) Transitive functional annotation by shortest-path analysis of gene expression data. *Proc. Natl. Acad. Sci. USA*, **99**, 12783–12788.

38 Huttenhower, C., Hibbs, M., Myers, C., and Troyanskaya, O.G. (2006) A scalable method for integration and functional analysis of multiple microarray datasets. *Bioinformatics*, **22**, 2890–2897.

39 Carter, S.L., Brechbuhler, C.M., Griffin, M., and Bond, A.T. (2004) Gene co-expression network topology provides a framework for molecular characterization of cellular state. *Bioinformatics*, **20**, 2242–2250.

40 Bar-Joseph, Z., Gerber, G.K., Lee, T.I., Rinaldi, N.J., Yoo, J.Y., Robert, F., Gordon, D.B., Fraenkel, E., Jaakkola, T.S., Young, R.A., and Gifford, D.K. (2003) Computational discovery of gene modules and regulatory networks. *Nat. Biotechnol.*, **21**, 1337–1342.

41 Ihmels, J., Friedlander, G., Bergmann, S., Sarig, O., Ziv, Y., and Barkai, N. (2002) Revealing modular organization in the yeast transcriptional network. *Nat. Genet.*, **31**, 370–377.

42 Ucar, D., Neuhaus, I., Ross-MacDonald, P., Tilford, C., Parthasarathy, S., Siemers, N., and Ji, R.R. (2007) Construction of a reference gene association network from multiple profiling data: application to data analysis. *Bioinformatics*, **23**, 2716–2724.

43 Elo, L.L., Jarvenpaa, H., Oresic, M., Lahesmaa, R., and Aittokallio, T. (2007) Systematic construction of gene coexpression networks with applications to human T helper cell differentiation process. *Bioinformatics*, **23**, 2096–2103.

44 Luo, F., Yang, Y., Zhong, J., Gao, H., Khan, L., Thompson, D.K., and Zhou, J. (2007) Constructing gene co-expression networks and predicting functions of unknown genes by random matrix theory. *BMC Bioinformatics*, **8**, 299.

45 Aoki, K., Ogata, Y., and Shibata, D. (2007) Approaches for extracting practical information from gene co-expression networks in plant biology. *Plant Cell Physiol.*, **48**, 381–390.

46 Bammler, T., Beyer, R.P., Bhattacharya, S., Boorman, G.A., Boyles, A., Bradford, B.U., Bumgarner, R.E., Bushel, P.R., Chaturvedi, K., Choi, D., Cunningham, M.L., Deng, S., Dressman, H.K., Fannin, R.D., Farin, F.M., Freedman, J.H., Fry, R.C., Harper, A., Humble, M.C., Hurban, P., Kavanagh, T.J., Kaufmann, W.K., Kerr, K.F., Jing, L., Lapidus, J.A., Lasarev, M.R., Li, J., Li, Y.J., Lobenhofer, E.K., Lu, X., Malek, R.L., Milton, S., Nagalla, S.R., O'Malley, P., Palmer, J.V.S., Pattee, P., Paules, R.S., Perou, C.M., Phillips, K., Qin, L.X., Qiu, Y., Quigley, S.D., Rodland, M., Rusyn, I., Samson, L.D., Schwartz, D.A., Shi, Y., Shin, J.L., Sieber, S.O., Slifer, S., Speer, M.C., Spencer, P.S., Sproles, D.I., Swenberg, J.A., Suk, W.A., Sullivan, R.C., Tian, R., Tennant, R.W., Todd, S.A., Tucker, C.J., Van Houten, B., Weis, B.K., Xuan, S., and Zarbl, H. (2005) Standardizing global gene expression analysis between laboratories and across platforms. *Nat. Methods*, **2**, 351–356.

47 Tan, P.K., Downey, T.J., Spitznagel, E.L. Jr, Xu, P., Fu, D., Dimitrov, D.S., Lempicki, R.A., Raaka, B.M., and Cam, M.C. (2003) Evaluation of gene expression measurements from commercial microarray platforms. *Nucleic Acids Res.*, **31**, 5676–5684.

48 Griffith, O.L., Pleasance, E.D., Fulton, D.L., Oveisi, M., Ester, M., Siddiqui, A.S., and Jones, S.J. (2005) Assessment and integration of publicly available SAGE, cDNA microarray, and oligonucleotide microarray expression data for global coexpression analyses. *Genomics*, **86**, 476–488.

49 Yauk, C.L. and Berndt, M.L. (2007) Review of the literature examining the correlation among DNA microarray technologies. *Environ. Mol. Mutagen.*, **48**, 380–394.

50 Fraser, A.G. and Marcotte, E.M. (2004) A probabilistic view of gene function. *Nat. Genet.*, **36**, 559–564.

51 Stuart, J.M., Segal, E., Koller, D., and Kim, S.K. (2003) A gene-coexpression network for global discovery of conserved genetic modules. *Science*, **302**, 249–255.

52 van Noort, V., Snel, B., and Huynen, M.A. (2003) Predicting gene function by conserved co-expression. *Trends Genet.*, **19**, 238–242.

53 Barrett, T., Troup, D.B., Wilhite, S.E., Ledoux, P., Rudnev, D., Evangelista, C., Kim, I.F., Soboleva, A., Tomashevsky, M., and Edgar, R. (2007) NCBI GEO: mining tens of millions of expression profiles – database and tools update. *Nucleic Acids Res.*, **35**, D760–D765.

54 Parkinson, H., Sarkans, U., Shojatalab, M., Abeygunawardena, N., Contrino, S., Coulson, R., Farne, A., Lara, G.G., Holloway, E., Kapushesky, M., Lilja, P., Mukherjee, G., Oezcimen, A., Rayner, T., Rocca-Serra, P., Sharma, A., Sansone, S., and Brazma, A. (2005) ArrayExpress – a public repository for microarray gene expression data at the EBI. *Nucleic Acids Res.*, **33**, D553–D555.

55 Demeter, J., Beauheim, C., Gollub, J., Hernandez-Boussard, T., Jin, H., Maier, D., Matese, J.C., Nitzberg, M., Wymore, F., Zachariah, Z.K., Brown, P.O., Sherlock, G., and Ball, C.A. (2007) The Stanford Microarray Database: implementation of new analysis tools and open source release of software. *Nucleic Acids Res.*, **35**, D766–D770.

56 Wodicka, L., Dong, H., Mittmann, M., Ho, M.H., and Lockhart, D.J. (1997) Genome-wide expression monitoring in *Saccharomyces cerevisiae*. *Nat. Biotechnol.*, **15**, 1359–1367.

57 Harkin, D.P., Bean, J.M., Miklos, D., Song, Y.H., Truong, V.B., Englert, C., Christians, F.C., Ellisen, L.W., Maheswaran, S., Oliner, J.D., and Haber, D.A. (1999) Induction of GADD45 and JNK/SAPK-dependent apoptosis following inducible expression of BRCA1. *Cell*, **97**, 575–586.

58 Heller, R.A., Schena, M., Chai, A., Shalon, D., Bedilion, T., Gilmore, J., Woolley, D.E., and Davis, R.W. (1997) Discovery and analysis of inflammatory disease-related genes using cDNA microarrays. *Proc. Natl. Acad. Sci. USA*, **94**, 2150–2155.

59 Wolfe, C.J., Kohane, I.S., and Butte, A.J. (2005) Systematic survey reveals general

applicability of "guilt-by-association" within gene coexpression networks. *BMC Bioinformatics*, **6**, 227.

60 Kim, S.K., Lund, J., Kiraly, M., Duke, K., Jiang, M., Stuart, J.M., Eizinger, A., Wylie, B.N., and Davidson, G.S. (2001) A gene expression map for *Caenorhabditis elegans*. *Science*, **293**, 2087–2092.

61 Faith, J.J., Hayete, B., Thaden, J.T., Mogno, I., Wierzbowski, J., Cottarel, G., Kasif, S., Collins, J.J., and Gardner, T.S. (2007) Large-scale mapping and validation of *Escherichia coli* transcriptional regulation from a compendium of expression profiles. *PLoS Biol.*, **5**, e8.

62 Dobrin, R., Zhu, J., Molony, C., Argman, C., Parrish, M.L., Carlson, S., Allan, M.F., Pomp, D., and Schadt, E.E. (2009) Multi-tissue coexpression networks reveal unexpected subnetworks associated with disease. *Genome Biol.*, **10**, R55.

63 Freeman, T.C., Goldovsky, L., Brosch, M., van Dongen, S., Maziere, P., Grocock, R.J., Freilich, S., Thornton, J., and Enright, A.J. (2007) Construction, visualisation, and clustering of transcription networks from microarray expression data. *PLoS Comput. Biol.*, **3**, 2032–2042.

64 Guan, Y., Myers, C.L., Lu, R., Lemischka, I.R., Bult, C.J., and Troyanskaya, O.G. (2008) A genomewide functional network for the laboratory mouse. *PLoS Comput. Biol.*, **4**, e1000165.

65 Huang, Y., Li, H., Hu, H., Yan, X., Waterman, M.S., Huang, H., and Zhou, X.J. (2007) Systematic discovery of functional modules and context-specific functional annotation of human genome. *Bioinformatics*, **23**, i222–229.

66 Lee, H.K., Hsu, A.K., Sajdak, J., Qin, J., and Pavlidis, P. (2004) Coexpression analysis of human genes across many microarray data sets. *Genome Res.*, **14**, 1085–1094.

67 Prieto, C., Risueno, A., Fontanillo, C., and De las Rivas, J. (2008) Human gene coexpression landscape: confident network derived from tissue transcriptomic profiles. *PLoS ONE*, **3**, e3911.

68 Carroll, S.B. (2005) Evolution at two levels: on genes and form. *PLoS Biol.*, **3**, e245.

69 Khaitovich, P., Hellmann, I., Enard, W., Nowick, K., Leinweber, M., Franz, H., Weiss, G., Lachmann, M., and Paabo, S. (2005) Parallel patterns of evolution in the genomes and transcriptomes of humans and chimpanzees. *Science*, **309**, 1850–1854.

70 Lelandais, G., Vincens, P., Badel-Chagnon, A., Vialette, S., Jacq, C., and Hazout, S. (2006) Comparing gene expression networks in a multi-dimensional space to extract similarities and differences between organisms. *Bioinformatics*, **22**, 1359–1366.

71 Rifkin, S.A., Kim, J., and White, K.P. (2003) Evolution of gene expression in the *Drosophila melanogaster* subgroup. *Nat. Genet.*, **33**, 138–144.

72 Teichmann, S.A. and Babu, M.M. (2002) Conservation of gene co-regulation in prokaryotes and eukaryotes. *Trends Biotechnol.*, **20**, 407–410; discussion 410.

73 Lefebvre, C., Aude, J.C., Glemet, E., and Neri, C. (2005) Balancing protein similarity and gene co-expression reveals new links between genetic conservation and developmental diversity in invertebrates. *Bioinformatics*, **21**, 1550–1558.

74 Chen, C., Weirauch, M.T., Powell, C.C., Zambon, A.C., and Stuart, J.M. (2007) A search engine to identify pathway genes from expression data on multiple organisms. *BMC Syst. Biol.*, **1**, 20.

75 Tamada, Y., Bannai, H., Imoto, S., Katayama, T., Kanehisa, M., and Miyano, S. (2005) Utilizing evolutionary information and gene expression data for estimating gene networks with Bayesian network models. *J. Bioinform. Comput. Biol.*, **3**, 1295–1313.

76 Ramani, A.K., Li, Z., Hart, G.T., Carlson, M.W., Boutz, D.R., and Marcotte, E.M. (2008) A map of human protein interactions derived from co-expression of human mRNAs and their orthologs. *Mol. Syst. Biol.*, **4**, 180.

77 Ala, U., Piro, R.M., Grassi, E., Damasco, C., Silengo, L., Oti, M., Provero, P., and Di Cunto, F. (2008) Prediction of human disease genes by human-mouse conserved coexpression analysis. *PLoS Comput. Biol.*, **4**, e1000043.

78 Troyanskaya, O.G. (2005) Putting microarrays in a context: integrated analysis of diverse biological data. *Brief Bioinform.*, **6**, 34–43.

79 Bader, G.D., Heilbut, A., Andrews, B., Tyers, M., Hughes, T., and Boone, C. (2003) Functional genomics and proteomics: charting a multidimensional map of the yeast cell. *Trends Cell Biol.*, **13**, 344–356.

80 Deng, M., Sun, F., and Chen, T. (2003) Assessment of the reliability of protein–protein interactions and protein function prediction. *Pac. Symp. Biocomput.*, 140–151.

81 Sprinzak, E., Sattath, S., and Margalit, H. (2003) How reliable are experimental protein–protein interaction data? *J. Mol. Biol.*, **327**, 919–923.

82 Gunsalus, K.C., Ge, H., Schetter, A.J., Goldberg, D.S., Han, J.D., Hao, T., Berriz, G.F., Bertin, N., Huang, J., Chuang, L.S., Li, N., Mani, R., Hyman, A.A., Sonnichsen, B., Echeverri, C.J., Roth, F.P., Vidal, M., and Piano, F. (2005) Predictive models of molecular machines involved in *Caenorhabditis elegans* early embryogenesis. *Nature*, **436**, 861–865.

83 Marcotte, E.M., Pellegrini, M., Thompson, M.J., Yeates, T.O., and Eisenberg, D. (1999) A combined algorithm for genome-wide prediction of protein function. *Nature*, **402**, 83–86.

84 Walhout, A.J., Reboul, J., Shtanko, O., Bertin, N., Vaglio, P., Ge, H., Lee, H., Doucette-Stamm, L., Gunsalus, K.C., Schetter, A.J., Morton, D.G., Kemphues, K.J., Reinke, V., Kim, S.K., Piano, F., and Vidal, M. (2002) Integrating interactome, phenome, and transcriptome mapping data for the *C. elegans* germline. *Curr. Biol.*, **12**, 1952–1958.

85 Byrne, A.B., Weirauch, M.T., Wong, V., Koeva, M., Dixon, S.J., Stuart, J.M., and Roy, P.J. (2007) A global analysis of genetic interactions in *Caenorhabditis elegans*. *J. Biol.*, **6**, 8.

86 Kelley, R. and Ideker, T. (2005) Systematic interpretation of genetic interactions using protein networks. *Nat. Biotechnol.*, **23**, 561–566.

87 Yeger-Lotem, E., Sattath, S., Kashtan, N., Itzkovitz, S., Milo, R., Pinter, R.Y., Alon, U., and Margalit, H. (2004) Network motifs in integrated cellular networks of transcription-regulation and protein–protein interaction. *Proc. Natl. Acad. Sci. USA*, **101**, 5934–5939.

88 Zhang, L.V., King, O.D., Wong, S.L., Goldberg, D.S., Tong, A.H., Lesage, G., Andrews, B., Bussey, H., Boone, C., and Roth, F.P. (2005) Motifs, themes and thematic maps of an integrated *Saccharomyces cerevisiae* interaction network. *J. Biol.*, **4**, 6.

89 Chuang, H.Y., Lee, E., Liu, Y.T., Lee, D., and Ideker, T. (2007) Network-based classification of breast cancer metastasis. *Mol. Syst. Biol.*, **3**, 140.

90 Tuck, D.P., Kluger, H.M., and Kluger, Y. (2006) Characterizing disease states from topological properties of transcriptional regulatory networks. *BMC Bioinformatics*, **7**, 236.

91 Pearl, J. (1985) Bayesian networks: a model of self-activated memory for evidential reasoning. Seventh Conference of the Cognitive Science Society, University of California, Irvine, CA.

92 Friedman, N., Linial, M., Nachman, I., and Pe'er, D. (2000) Using Bayesian networks to analyze expression data. *J. Comput. Biol.*, **7**, 601–620.

93 Jansen, R., Yu, H., Greenbaum, D., Kluger, Y., Krogan, N.J., Chung, S., Emili, A., Snyder, M., Greenblatt, J.F., and Gerstein, M. (2003) A Bayesian networks approach for predicting protein–protein interactions from genomic data. *Science*, **302**, 449–453.

94 Myers, C.L. and Troyanskaya, O.G. (2007) Context-sensitive data integration and prediction of biological networks. *Bioinformatics*, **23**, 2322–2330.

95 Troyanskaya, O.G., Dolinski, K., Owen, A.B., Altman, R.B., and Botstein, D. (2003) A Bayesian framework for

combining heterogeneous data sources for gene function prediction (in *Saccharomyces cerevisiae*). *Proc. Natl. Acad. Sci. USA*, **100**, 8348–8353.

96 Alexeyenko, A. and Sonnhammer, E.L. (2009) Global networks of functional coupling in eukaryotes from comprehensive data integration. *Genome Res.*, **19**, 1107–1116.

97 Lee, I., Date, S.V., Adai, A.T., and Marcotte, E.M. (2004) A probabilistic functional network of yeast genes. *Science*, **306**, 1555–1558.

98 Lee, I., Li, Z., and Marcotte, E.M. (2007) An improved, bias-reduced probabilistic functional gene network of baker's yeast, *Saccharomyces cerevisiae*. *PLoS ONE*, **2**, e988.

99 Lee, I., Lehner, B., Crombie, C., Wong, W., Fraser, A.G., and Marcotte, E.M. (2008) A single gene network accurately predicts phenotypic effects of gene perturbation in *Caenorhabditis elegans*. *Nat. Genet.*, **40**, 181–188.

100 Zhong, W. and Sternberg, P.W. (2006) Genome-wide prediction of *C. elegans* genetic interactions. *Science*, **311**, 1481–1484.

101 Costello, J.C., Dalkilic, M.M., Beason, S.M., Gehlhausen, J.R., Patwardhan, R., Middha, S., Eads, B.D., and Andrews, J.R. (2009) Gene networks in *Drosophila melanogaster*: integrating experimental data to predict gene function. *Genome Biol.*, **10**, R97.

102 Hwang, D., Rust, A.G., Ramsey, S., Smith, J.J., Leslie, D.M., Weston, A.D., de Atauri, P., Aitchison, J.D., Hood, L., Siegel, A.F., and Bolouri, H. (2005) A data integration methodology for systems biology. *Proc. Natl. Acad. Sci. USA*, **102**, 17296–17301.

103 Allocco, D.J., Kohane, I.S., and Butte, A.J. (2004) Quantifying the relationship between co-expression, co-regulation and gene function. *BMC Bioinformatics*, **5**, 18.

104 Conlon, E.M., Liu, X.S., Lieb, J.D., and Liu, J.S. (2003) Integrating regulatory motif discovery and genome-wide expression analysis. *Proc. Natl. Acad. Sci. USA*, **100**, 3339–3344.

105 Haverty, P.M., Hansen, U., and Weng, Z. (2004) Computational inference of transcriptional regulatory networks from expression profiling and transcription factor binding site identification. *Nucleic Acids Res.*, **32**, 179–188.

106 Liu, X., Brutlag, D.L., and Liu, J.S. (2001) BioProspector: discovering conserved DNA motifs in upstream regulatory regions of co-expressed genes. *Pac. Symp. Biocomput.*, 127–138.

107 Roth, F.P., Hughes, J.D., Estep, P.W., and Church, G.M. (1998) Finding DNA regulatory motifs within unaligned noncoding sequences clustered by whole-genome mRNA quantitation. *Nat. Biotechnol.*, **16**, 939–945.

108 Tavazoie, S., Hughes, J.D., Campbell, M.J., Cho, R.J., and Church, G.M. (1999) Systematic determination of genetic network architecture. *Nat. Genet.*, **22**, 281–285.

109 Bussemaker, H.J., Li, H., and Siggia, E.D. (2001) Regulatory element detection using correlation with expression. *Nat. Genet.*, **27**, 167–171.

110 Gao, F., Foat, B.C., and Bussemaker, H.J. (2004) Defining transcriptional networks through integrative modeling of mRNA expression and transcription factor binding data. *BMC Bioinformatics*, **5**, 31.

111 Segal, E., Yelensky, R., and Koller, D. (2003) Genome-wide discovery of transcriptional modules from DNA sequence and gene expression. *Bioinformatics*, **19** (Suppl. 1), i273–i282.

112 Xu, X., Wang, L., and Ding, D. (2004) Learning module networks from genome-wide location and expression data. *FEBS Lett.*, **578**, 297–304.

113 Cheung, V.G. and Spielman, R.S. (2002) The genetics of variation in gene expression. *Nat. Genet.*, **32** (Suppl.), 522–525.

114 Darvasi, A. (2003) Genomics: gene expression meets genetics. *Nature*, **422**, 269–270.

115 Jansen, R.C. and Nap, J.P. (2001) Genetical genomics: the added value from segregation. *Trends Genet.*, **17**, 388–391.

116 Li, J. and Burmeister, M. (2005) Genetical genomics: combining genetics with gene expression analysis. *Hum. Mol. Genet.*, **14**, R163–R169.

117 Hubner, N., Wallace, C.A., Zimdahl, H., Petretto, E., Schulz, H., Maciver, F., Mueller, M., Hummel, O., Monti, J., Zidek, V., Musilova, A., Kren, V., Causton, H., Game, L., Born, G., Schmidt, S., Muller, A., Cook, S.A., Kurtz, T.W., Whittaker, J., Pravenec, M., and Aitman, T.J. (2005) Integrated transcriptional profiling and linkage analysis for identification of genes underlying disease. *Nat. Genet.*, **37**, 243–253.

118 Sweet-Cordero, A., Tseng, G.C., You, H., Douglass, M., Huey, B., Albertson, D., and Jacks, T. (2006) Comparison of gene expression and DNA copy number changes in a murine model of lung cancer. *Genes Chromosomes Cancer*, **45**, 338–348.

119 Chesler, E.J., Lu, L., Shou, S., Qu, Y., Gu, J., Wang, J., Hsu, H.C., Mountz, J.D., Baldwin, N.E., Langston, M.A., Threadgill, D.W., Manly, K.F., and Williams, R.W. (2005) Complex trait analysis of gene expression uncovers polygenic and pleiotropic networks that modulate nervous system function. *Nat. Genet.*, **37**, 233–242.

120 Schadt, E.E., Lamb, J., Yang, X., Zhu, J., Edwards, S., Guhathakurta, D., Sieberts, S.K., Monks, S., Reitman, M., Zhang, C., Lum, P.Y., Leonardson, A., Thieringer, R., Metzger, J.M., Yang, L., Castle, J., Zhu, H., Kash, S.F., Drake, T.A., Sachs, A., and Lusis, A.J. (2005) An integrative genomics approach to infer causal associations between gene expression and disease. *Nat. Genet.*, **37**, 710–717.

121 Schadt, E.E., Molony, C., Chudin, E., Hao, K., Yang, X., Lum, P.Y., Kasarskis, A., Zhang, B., Wang, S., Suver, C., Zhu, J., Millstein, J., Sieberts, S., Lamb, J., GuhaThakurta, D., Derry, J., Storey, J.D., Avila-Campillo, I., Kruger, M.J., Johnson, J.M., Rohl, C.A., van Nas, A., Mehrabian, M., Drake, T.A., Lusis, A.J., Smith, R.C., Guengerich, F.P., Strom, S.C., Schuetz, E., Rushmore, T.H., and Ulrich, R. (2008) Mapping the genetic architecture of gene expression in human liver. *PLoS Biol.*, **6**, e107.

122 Yu, H., Braun, P., Yildirim, M.A., Lemmens, I., Venkatesan, K., Sahalie, J., Hirozane-Kishikawa, T., Gebreab, F., Li, N., Simonis, N., Hao, T., Rual, J.F., Dricot, A., Vazquez, A., Murray, R.R., Simon, C., Tardivo, L., Tam, S., Svrzikapa, N., Fan, C., de Smet, A.S., Motyl, A., Hudson, M.E., Park, J., Xin, X., Cusick, M.E., Moore, T., Boone, C., Snyder, M., Roth, F.P., Barabasi, A.L., Tavernier, J., Hill, D.E., and Vidal, M. (2008) High-quality binary protein interaction map of the yeast interactome network. *Science*, **322**, 104–110.

123 Krogan, N.J., Cagney, G., Yu, H., Zhong, G., Guo, X., Ignatchenko, A., Li, J., Pu, S., Datta, N., Tikuisis, A.P., Punna, T., Peregrin-Alvarez, J.M., Shales, M., Zhang, X., Davey, M., Robinson, M.D., Paccanaro, A., Bray, J.E., Sheung, A., Beattie, B., Richards, D.P., Canadien, V., Lalev, A., Mena, F., Wong, P., Starostine, A., Canete, M.M., Vlasblom, J., Wu, S., Orsi, C., Collins, S.R., Chandran, S., Haw, R., Rilstone, J.J., Gandi, K., Thompson, N.J., Musso, G., St Onge, P., Ghanny, S., Lam, M.H., Butland, G., Altaf-Ul, A.M., Kanaya, S., Shilatifard, A., O'Shea, E., Weissman, J.S., Ingles, C.J., Hughes, T.R., Parkinson, J., Gerstein, M., Wodak, S.J., Emili, A., and Greenblatt, J.F. (2006) Global landscape of protein complexes in the yeast *Saccharomyces cerevisiae*. *Nature*, **440**, 637–643.

124 Ge, H., Liu, Z., Church, G.M., and Vidal, M. (2001) Correlation between transcriptome and interactome mapping data from *Saccharomyces cerevisiae*. *Nat. Genet.*, **29**, 482–486.

125 Hahn, A., Rahnenfuhrer, J., Talwar, P., and Lengauer, T. (2005) Confirmation of human protein interaction data by human expression data. *BMC Bioinformatics*, **6**, 112.

126 Ideker, T., Thorsson, V., Ranish, J.A., Christmas, R., Buhler, J., Eng, J.K., Bumgarner, R., Goodlett, D.R., Aebersold, R., and Hood, L. (2001) Integrated genomic and proteomic analyses of a systematically perturbed metabolic network. *Science*, **292**, 929–934.

127 Ideker, T., Ozier, O., Schwikowski, B., and Siegel, A.F. (2002) Discovering regulatory and signalling circuits in molecular interaction networks. *Bioinformatics*, **18** (Suppl. 1), S233–S240.

128 Jansen, R., Greenbaum, D., and Gerstein, M. (2002) Relating whole-genome expression data with protein–protein interactions. *Genome Res.*, **12**, 37–46.

129 Bossi, A. and Lehner, B. (2009) Tissue specificity and the human protein interaction network. *Mol. Syst. Biol.*, **5**, 260.

130 de Lichtenberg, U., Jensen, L.J., Brunak, S., and Bork, P. (2005) Dynamic complex formation during the yeast cell cycle. *Science*, **307**, 724–727.

131 Komurov, K. and White, M. (2007) Revealing static and dynamic modular architecture of the eukaryotic protein interaction network. *Mol. Syst. Biol.*, **3**, 110.

132 Tornow, S. and Mewes, H.W. (2003) Functional modules by relating protein interaction networks and gene expression. *Nucleic Acids Res.*, **31**, 6283–6289.

133 Karaoz, U., Murali, T.M., Letovsky, S., Zheng, Y., Ding, C., Cantor, C.R., and Kasif, S. (2004) Whole-genome annotation by using evidence integration in functional-linkage networks. *Proc. Natl. Acad. Sci. USA*, **101**, 2888–2893.

134 Milo, R., Shen-Orr, S., Itzkovitz, S., Kashtan, N., Chklovskii, D., and Alon, U. (2002) Network motifs: simple building blocks of complex networks. *Science*, **298**, 824–827.

135 Jeong, H., Tombor, B., Albert, R., Oltvai, Z.N., and Barabasi, A.L. (2000) The large-scale organization of metabolic networks. *Nature*, **407**, 651–654.

136 Uetz, P., Giot, L., Cagney, G., Mansfield, T.A., Judson, R.S., Knight, J.R., Lockshon, D., Narayan, V., Srinivasan, M., Pochart, P., Qureshi-Emili, A., Li, Y., Godwin, B., Conover, D., Kalbfleisch, T., Vijayadamodar, G., Yang, M., Johnston, M., Fields, S., and Rothberg, J.M. (2000) A comprehensive analysis of protein–protein interactions in *Saccharomyces cerevisiae*. *Nature*, **403**, 623–627.

137 Wuchty, S. (2001) Scale-free behavior in protein domain networks. *Mol. Biol. Evol.*, **18**, 1694–1702.

138 Albert, R., Jeong, H., and Barabasi, A.L. (2000) Error and attack tolerance of complex networks. *Nature*, **406**, 378–382.

139 Han, J.D., Bertin, N., Hao, T., Goldberg, D.S., Berriz, G.F., Zhang, L.V., Dupuy, D., Walhout, A.J., Cusick, M.E., Roth, F.P., and Vidal, M. (2004) Evidence for dynamically organized modularity in the yeast protein–protein interaction network. *Nature*, **430**, 88–93.

140 Jeong, H., Mason, S.P., Barabasi, A.L., and Oltvai, Z.N. (2001) Lethality and centrality in protein networks. *Nature*, **411**, 41–42.

141 Krylov, D.M., Wolf, Y.I., Rogozin, I.B., and Koonin, E.V. (2003) Gene loss, protein sequence divergence, gene dispensability, expression level, and interactivity are correlated in eukaryotic evolution. *Genome Res.*, **13**, 2229–2235.

142 Mangan, S., Zaslaver, A., and Alon, U. (2003) The coherent feedforward loop serves as a sign-sensitive delay element in transcription networks. *J. Mol. Biol.*, **334**, 197–204.

143 Pan, G., Li, J., Zhou, Y., Zheng, H., and Pei, D. (2006) A negative feedback loop of transcription factors that controls stem cell pluripotency and self-renewal. *FASEB J.*, **20**, 1730–1732.

144 Hinman, V.F., Nguyen, A.T., Cameron, R.A., and Davidson, E.H. (2003) Developmental gene regulatory network architecture across 500 million years of echinoderm evolution. *Proc. Natl. Acad. Sci. USA*, **100**, 13356–13361.

145 Watkinson, J., Liang, K.C., Wang, X., Zheng, T., and Anastassiou, D. (2009) Inference of regulatory gene interactions from expression data using three-way mutual information. *Ann. NY Acad. Sci.*, **1158**, 302–313.

146 Zhang, J., Ji, Y., and Zhang, L. (2007) Extracting three-way gene interactions from microarray data. *Bioinformatics*, **23**, 2903–2909.

147 Bader, G.D. and Hogue, C.W. (2003) An automated method for finding molecular complexes in large protein interaction networks. *BMC Bioinformatics*, **4**, 2.

148 Hu, H., Yan, X., Huang, Y., Han, J., and Zhou, X.J. (2005) Mining coherent dense subgraphs across massive biological networks for functional discovery. *Bioinformatics*, **21** (Suppl. 1), i213–i221.

149 Kharchenko, P., Church, G.M., and Vitkup, D. (2005) Expression dynamics of a cellular metabolic network. *Mol. Syst. Biol.*, **1**, **2005**, 0016.

150 Gavin, A.C., Aloy, P., Grandi, P., Krause, R., Boesche, M., Marzioch, M., Rau, C., Jensen, L.J., Bastuck, S., Dumpelfeld, B., Edelmann, A., Heurtier, M.A., Hoffman, V., Hoefert, C., Klein, K., Hudak, M., Michon, A.M., Schelder, M., Schirle, M., Remor, M., Rudi, T., Hooper, S., Bauer, A., Bouwmeester, T., Casari, G., Drewes, G., Neubauer, G., Rick, J.M., Kuster, B., Bork, P., Russell, R.B., and Superti-Furga, G. (2006) Proteome survey reveals modularity of the yeast cell machinery. *Nature*, **440**, 631–636.

151 Hartwell, L.H., Hopfield, J.J., Leibler, S., and Murray, A.W. (1999) From molecular to modular cell biology. *Nature*, **402**, C47–C52.

152 Rives, A.W. and Galitski, T. (2003) Modular organization of cellular networks. *Proc. Natl. Acad. Sci. USA*, **100**, 1128–1133.

153 D'Haeseleer, P. (2005) How does gene expression clustering work? *Nat. Biotechnol.*, **23**, 1499–1501.

154 Kaufman, L. and Russeeuw, P. (1990) *Finding Groups in Data: An Introduction to Cluster Analysis*, John Wiley & Sons, Inc., New York.

155 Thalamuthu, A., Mukhopadhyay, I., Zheng, X., and Tseng, G.C. (2006) Evaluation and comparison of gene clustering methods in microarray analysis. *Bioinformatics*, **22**, 2405–2412.

156 Tamayo, P., Slonim, D., Mesirov, J., Zhu, Q., Kitareewan, S., Dmitrovsky, E., Lander, E.S., and Golub, T.R. (1999) Interpreting patterns of gene expression with self-organizing maps: methods and application to hematopoietic differentiation. *Proc. Natl. Acad. Sci. USA*, **96**, 2907–2912.

157 Cheng, Y. and Church, G.M. (2000) Biclustering of expression data. *Proc. Int. Conf. Intell. Syst. Mol. Biol.*, **8**, 93–103.

158 Getz, G., Levine, E., and Domany, E. (2000) Coupled two-way clustering analysis of gene microarray data. *Proc. Natl. Acad. Sci. USA*, **97**, 12079–12084.

159 Tanay, A., Sharan, R., Kupiec, M., and Shamir, R. (2004) Revealing modularity and organization in the yeast molecular network by integrated analysis of highly heterogeneous genomewide data. *Proc. Natl. Acad. Sci. USA*, **101**, 2981–2986.

160 Wu, L.F., Hughes, T.R., Davierwala, A.P., Robinson, M.D., Stoughton, R., and Altschuler, S.J. (2002) Large-scale prediction of *Saccharomyces cerevisiae* gene function using overlapping transcriptional clusters. *Nat. Genet.*, **31**, 255–265.

161 Hanisch, D., Zien, A., Zimmer, R., and Lengauer, T. (2002) Co-clustering of biological networks and gene expression data. *Bioinformatics*, **18** (Suppl. 1), S145–S154.

162 Owen, A.B., Stuart, J., Mach, K., Villeneuve, A.M., and Kim, S. (2003) A gene recommender algorithm to identify coexpressed genes in *C. elegans*. *Genome Res.*, **13**, 1828–1837.

163 Jansen, R. and Gerstein, M. (2004) Analyzing protein function on a genomic scale: the importance of gold-standard positives and negatives for network prediction. *Curr. Opin. Microbiol.*, **7**, 535–545.

164 Gasch, A.P. and Eisen, M.B. (2002) Exploring the conditional coregulation of yeast gene expression through fuzzy k-means clustering. *Genome Biol.*, **3**, RESEARCH0059.

165 Niskanen, S. and Östergård, P.R.J. (2003) Cliquer user's guide, version 1.0. Technical Report T48, Communications Laboratory, Helsinki University of Technology, Espoo.

166 Voy, B.H., Scharff, J.A., Perkins, A.D., Saxton, A.M., Borate, B., Chesler, E.J., Branstetter, L.K., and Langston, M.A. (2006) Extracting gene networks for low-dose radiation using graph theoretical algorithms. *PLoS Comput. Biol.*, **2**, e89.

167 Spirin, V. and Mirny, L.A. (2003) Protein complexes and functional modules in molecular networks. *Proc. Natl. Acad. Sci. USA*, **100**, 12123–12128.

168 Uno, T. (2007) An efficient algorithm for enumerating pseudo cliques, in *Algorithms and Computation* (ed. T. Tokuyama), Springer, Berlin, pp. 402–414.

169 Brohee, S. and van Helden, J. (2006) Evaluation of clustering algorithms for protein–protein interaction networks. *BMC Bioinformatics*, **7**, 488.

170 Newman, M.E. and Girvan, M. (2004) Finding and evaluating community structure in networks. *Phys. Rev. E*, **69**, 026113.

171 Moult, J. (2005) A decade of CASP: progress, bottlenecks and prognosis in protein structure prediction. *Curr. Opin. Struct. Biol.*, **15**, 285–289.

172 Tompa, M., Li, N., Bailey, T.L., Church, G.M., De Moor, B., Eskin, E., Favorov, A.V., Frith, M.C., Fu, Y., Kent, W.J., Makeev, V.J., Mironov, A.A., Noble, W.S., Pavesi, G., Pesole, G., Regnier, M., Simonis, N., Sinha, S., Thijs, G., van Helden, J., Vandenbogaert, M., Weng, Z., Workman, C., Ye, C., and Zhu, Z. (2005) Assessing computational tools for the discovery of transcription factor binding sites. *Nat. Biotechnol.*, **23**, 137–144.

173 Sharan, R., Ulitsky, I., and Shamir, R. (2007) Network-based prediction of protein function. *Mol. Syst. Biol.*, **3**, 88.

174 Girvan, M. and Newman, M.E. (2002) Community structure in social and biological networks. *Proc. Natl. Acad. Sci. USA*, **99**, 7821–7826.

175 Hartuv, E. and Shamir, R. (2000) A clustering algorithm based on graph connectivity. *Inform. Process Lett.*, **76**, 175–181.

176 Adamcsek, B., Palla, G., Farkas, I.J., Derenyi, I., and Vicsek, T. (2006) CFinder: locating cliques and overlapping modules in biological networks. *Bioinformatics*, **22**, 1021–1023.

177 Derenyi, I., Palla, G., and Vicsek, T. (2005) Clique percolation in random networks. *Phys. Rev. Lett.*, **94**, 160202.

178 Kalaev, M., Smoot, M., Ideker, T., and Sharan, R. (2008) NetworkBLAST: comparative analysis of protein networks. *Bioinformatics*, **24**, 594–596.

179 Dhollander, T., Sheng, Q., Lemmens, K., De Moor, B., Marchal, K., and Moreau, Y. (2007) Query-driven module discovery in microarray data. *Bioinformatics*, **23**, 2573–2580.

180 Hibbs, M.A., Hess, D.C., Myers, C.L., Huttenhower, C., Li, K., and Troyanskaya, O.G. (2007) Exploring the functional landscape of gene expression: directed search of large microarray compendia. *Bioinformatics*, **23**, 2692–2699.

181 Mostafavi, S., Ray, D., Warde-Farley, D., Grouios, C., and Morris, Q. (2008) GeneMANIA: a real-time multiple association network integration algorithm for predicting gene function. *Genome Biol.*, **9** (Suppl. 1), S4.

182 Ashburner, M., Ball, C.A., Blake, J.A., Botstein, D., Butler, H., Cherry, J.M., Davis, A.P., Dolinski, K., Dwight, S.S., Eppig, J.T., Harris, M.A., Hill, D.P., Issel-Tarver, L., Kasarskis, A., Lewis, S., Matese, J.C., Richardson, J.E., Ringwald, M., Rubin, G.M., and Sherlock, G. (2000) Gene Ontology: tool for the unification of biology. The Gene Ontology Consortium. *Nat. Genet.*, **25**, 25–29.

183 Kanehisa, M., Araki, M., Goto, S., Hattori, M., Hirakawa, M., Itoh, M., Katayama, T., Kawashima, S., Okuda, S., Tokimatsu, T., and Yamanishi, Y. (2008) KEGG for linking genomes to life and the environment. *Nucleic Acids Res.*, **36**, D480–484.

184 Salomonis, N., Hanspers, K., Zambon, A.C., Vranizan, K., Lawlor, S.C., Dahlquist, K.D., Doniger, S.W., Stuart, J., Conklin, B.R., and Pico, A.R. (2007) GenMAPP 2: new features and resources for pathway analysis. *BMC Bioinformatics*, **8**, 217.

185 Macgregor, P.F. (2003) Gene expression in cancer: the application of microarrays. *Expert Rev. Mol. Diagn.*, **3**, 185–200.

186 Nielsen, T.O. (2006) Microarray analysis of sarcomas. *Adv. Anat. Pathol.*, **13**, 166–173.

187 Segal, E., Friedman, N., Kaminski, N., Regev, A., and Koller, D. (2005) From signatures to models: understanding cancer using microarrays. *Nat. Genet.*, **37** (Suppl.), S38–S45.

188 Chen, X., Cheung, S.T., So, S., Fan, S.T., Barry, C., Higgins, J., Lai, K.M., Ji, J.,

Dudoit, S., Ng, I.O., Van De Rijn, M., Botstein, D., and Brown, P.O. (2002) Gene expression patterns in human liver cancers. *Mol. Biol. Cell*, **13**, 1929–1939.

189 Golub, T.R., Slonim, D.K., Tamayo, P., Huard, C., Gaasenbeek, M., Mesirov, J.P., Coller, H., Loh, M.L., Downing, J.R., Caligiuri, M.A., Bloomfield, C.D., and Lander, E.S. (1999) Molecular classification of cancer: class discovery and class prediction by gene expression monitoring. *Science*, **286**, 531–537.

190 Ramaswamy, S. and Golub, T.R. (2002) DNA microarrays in clinical oncology. *J. Clin. Oncol.*, **20**, 1932–1941.

191 Rhodes, D.R., Kalyana-Sundaram, S., Mahavisno, V., Barrette, T.R., Ghosh, D., and Chinnaiyan, A.M. (2005) Mining for regulatory programs in the cancer transcriptome. *Nat. Genet.*, **37**, 579–583.

192 Choi, J.K., Yu, U., Yoo, O.J., and Kim, S. (2005) Differential coexpression analysis using microarray data and its application to human cancer. *Bioinformatics*, **21**, 4348–4355.

193 Kostka, D. and Spang, R. (2004) Finding disease specific alterations in the co-expression of genes. *Bioinformatics*, **20** (Suppl. 1), i194–i199.

194 Bild, A.H., Yao, G., Chang, J.T., Wang, Q., Potti, A., Chasse, D., Joshi, M.B., Harpole, D., Lancaster, J.M., Berchuck, A., Olson, J.A. Jr, Marks, J.R., Dressman, H.K., West, M., and Nevins, J.R. (2006) Oncogenic pathway signatures in human cancers as a guide to targeted therapies. *Nature*, **439**, 353–357.

195 Edelman, E.J., Guinney, J., Chi, J.T., Febbo, P.G., and Mukherjee, S. (2008) Modeling cancer progression via pathway dependencies. *PLoS Comput. Biol.*, **4**, e28.

196 Efroni, S., Schaefer, C.F., and Buetow, K.H. (2007) Identification of key processes underlying cancer phenotypes using biologic pathway analysis. *PLoS ONE*, **2**, e425.

197 Segal, E., Friedman, N., Koller, D., and Regev, A. (2004) A module map showing conditional activity of expression modules in cancer. *Nat. Genet.*, **36**, 1090–1098.

198 Setlur, S.R., Royce, T.E., Sboner, A., Mosquera, J.M., Demichelis, F., Hofer, M.D., Mertz, K.D., Gerstein, M., and Rubin, M.A. (2007) Integrative microarray analysis of pathways dysregulated in metastatic prostate cancer. *Cancer Res.*, **67**, 10296–10303.

199 Grutzmann, R., Boriss, H., Ammerpohl, O., Luttges, J., Kalthoff, H., Schackert, H.K., Kloppel, G., Saeger, H.D., and Pilarsky, C. (2005) Meta-analysis of microarray data on pancreatic cancer defines a set of commonly dysregulated genes. *Oncogene*, **24**, 5079–5088.

200 Rhodes, D.R., Barrette, T.R., Rubin, M.A., Ghosh, D., and Chinnaiyan, A.M. (2002) Meta-analysis of microarrays: interstudy validation of gene expression profiles reveals pathway dysregulation in prostate cancer. *Cancer Res.*, **62**, 4427–4433.

201 Warnat, P., Eils, R., and Brors, B. (2005) Cross-platform analysis of cancer microarray data improves gene expression based classification of phenotypes. *BMC Bioinformatics*, **6**, 265.

202 Rhodes, D.R., Kalyana-Sundaram, S., Mahavisno, V., Varambally, R., Yu, J., Briggs, B.B., Barrette, T.R., Anstet, M.J., Kincead-Beal, C., Kulkarni, P., Varambally, S., Ghosh, D., and Chinnaiyan, A.M. (2007) Oncomine 3.0: genes, pathways, and networks in a collection of 18,000 cancer gene expression profiles. *Neoplasia*, **9**, 166–180.

203 Bredel, M., Bredel, C., Juric, D., Harsh, G.R., Vogel, H., Recht, L.D., and Sikic, B.I. (2005) Functional network analysis reveals extended gliomagenesis pathway maps and three novel MYC-interacting genes in human gliomas. *Cancer Res.*, **65**, 8679–8689.

204 Butte, A.J., Tamayo, P., Slonim, D., Golub, T.R., and Kohane, I.S. (2000) Discovering functional relationships between RNA expression and chemotherapeutic susceptibility using relevance networks. *Proc. Natl. Acad. Sci. USA*, **97**, 12182–12186.

205 Liu, C.C., Chen, W.S., Lin, C.C., Liu, H.C., Chen, H.Y., Yang, P.C., Chang, P.C., and Chen, J.J. (2006) Topology-based cancer classification and related

206 pathway mining using microarray data. *Nucleic Acids Res.*, **34**, 4069–4080.
206 Garber, K. (2004) Genomic medicine. Gene expression tests foretell breast cancer's future. *Science*, **303**, 1754–1755.
207 Gruvberger-Saal, S.K., Cunliffe, H.E., Carr, K.M., and Hedenfalk, I.A. (2006) Microarrays in breast cancer research and clinical practice–the future lies ahead. *Endocr. Relat. Cancer*, **13**, 1017–1031.
208 Cancer Genome Atlas Research Network (2008) Comprehensive genomic characterization defines human glioblastoma genes and core pathways. *Nature*, **455**, 1061–1068.
209 Berg, J., Lassig, M., and Wagner, A. (2004) Structure and evolution of protein interaction networks: a statistical model for link dynamics and gene duplications. *BMC Evol. Biol.*, **4**, 51.
210 Kelley, B.P., Sharan, R., Karp, R.M., Sittler, T., Root, D.E., Stockwell, B.R., and Ideker, T. (2003) Conserved pathways within bacteria and yeast as revealed by global protein network alignment. *Proc. Natl. Acad. Sci. USA*, **100**, 11394–11399.
211 Sharan, R. and Ideker, T. (2006) Modeling cellular machinery through biological network comparison. *Nat. Biotechnol.*, **24**, 427–433.

12
Correlation Network Analysis and Knowledge Integration
Thomas N. Plasterer, Robert Stanley, and Erich Gombocz

12.1
Introduction

Nature understands how to optimize network structures. From molecules involved in information storage (genes), processing (mRNA, proteins), and execution (proteins, metabolites) up to regulatory motifs, pathways, functional modules, and hierarchical networks [1] that further assemble to build cells, tissues, individuals, and communities, network structures employed at multiple levels are all around us. These associations exists across multiple levels, spanning a single-photon conformational retinol switch in the rhodopsin photoreceptor [2] up to regulation of the oxygen/carbon dioxide balance in the biosphere by photosynthetic phytoplankton [3]. Relationships among components can be highly complex and span multiple levels of organization but, amazingly, this configuration of parts and associations functions – for the most part – due to our 4-billion-year evolutionary experiment in trial and error.

While Nature is comfortable with this complexity, human researchers tend to reduce these networks to component parts and associations to attempt to understand these systems [4]. Cognizant of this cautionary note, networks of many distinct, heterologous node and edge types can be exploited in the study of biological systems. Bioinformaticists assemble networks based on biological properties they wish to elucidate and communicate to biologists. As such, these networks are generally assembled from direct experimental observations or computationally predicted, synthesized from one or more experiment. Examples of experimentally derived networks include protein–protein interaction networks [5–10] where network edges are derived generally from yeast two-hybrid experiments or coimmunoprecipitation or "pull-down" experiments and gene regulatory networks derived from chromatin immunoprecipitation experiments [11]. Examples of computationally predicted networks include homology networks [12] and promoter motif predicted gene regulatory networks [13, 14]. In many cases, networks that are a combination of experimental and computational relationships are used in practice. This is especially likely for biological pathways – a type of network assembled from multiple lines of

Applied Statistics for Network Biology: Methods in Systems Biology, First Edition.
Edited by M. Dehmer, F. Emmert-Streib, A. Graber, and A. Salvador.
© 2011 Wiley-VCH Verlag GmbH & Co. KGaA. Published 2011 by Wiley-VCH Verlag GmbH & Co. KGaA.

experimental evidence where relationships are assumed to be valid in closely related species even if never observed directly. This inference is made on the basis of sequence similarity between constituent parts of the pathway across multiple, closely related organisms. In many cases this is now the general practice for experimental annotation in systems biology, where experimental results across multiple "omics" platforms (transcriptomics, proteomics, metabolomics) are mapped against biological pathways and literature-derived biological networks [15–17]. Correlation networks are a hybrid approach, where mathematical associations between measured analytes drive initial network edge creation and additional edges are added that describe known relationships from biological knowledge bases and literature [18–20].

12.2
Systems Biology Data Quandaries

Heterogeneity of data types and relationships and multiple level of organization of biological networks is one challenge in elucidating biological systems. An additional complication exists in handling the vast number of objects profiled and the vast number of relationships between those objects, both from a computational and human knowledge perspective. A recent systems biology project in idiosyncratic hepatotoxicity measured 33 573 analytes (genes, proteins, metabolites) in both liver tissues and plasma, and generated 202 354 769 correlations among all measured analytes [21]. Determining where to focus attention in this large, multilayered network represents a key challenge in understanding systems biology experiments.

Network heterogeneity and size in systems biology are daunting alone, but a further complication exists in where underlying data is localized and where knowledge bases useful to explain network associations can be brought to bear. Even within the same organization it is common for one group to conduct transcriptomics experiments, another proteomics and another pharmacodynamics, and yet another for clinical examinations. These groups frequently do not speak the same language nor are they facile with the understanding of domain-specific knowledge and techniques held within each silo. This problem is compounded when there is a need to share experimental results and interpretations across multiple organizations.

12.3
Semantic Web Approaches

Data heterogeneity across multiple tiers, data volume, and data silos are three great challenges confronting systems biology, and, by extension, gains in personalized medicine promised by systems biology. One possible solution to surmount these obstacles has been pioneered with the development of the semantic web [22]. The original concept was what Tim Berners-Lee intended for the internet, basically a highly connected network that could be simply accessed and utilized by any computing device, and not simply a way of describing images, tables, and so on, for human consumption. This required the development of three technologies: a common data language useable for multiple software agents, sets of statements that translate domain-specific terms into common terms, and rules for reasoning across these terms. The common language is called the resource description framework (RDF). Within this framework all data is described with a unique universal resource identifier (URI) as is any link between data. The basic building block within RDF is two pieces of data and the relationship between them. This is termed a triple and is a trivial node–edge–node relationship. The sets of definitions are known as ontologies, a common one being the Web Ontology Language (OWL [23, 24]). Finally, inferencing engines have been constructed to use both the ontologies and triples stored within RDF to derive novel associations.

These approaches fit well with the challenges in systems biology. Data heterogeneity across multiple tiers can be addressed with ontologies specific to the problem space. Data volume, once a challenge in this space, seems to be less of a concern as RDF "triplestore" databases, such as Garlik's 4Store (www.4store.org), Franz's Allegrograph (http://www.franz.com/agraph/allegrograph/), Open Link's Virtuoso (virtuoso.openlinksw.com), and Oracle's Spatial 11G (http://www.oracle.com/technology/tech/semantic_technologies/pdf/semantic%20technologies%20in%20oracle%20spatial%2011g%20datasheet.pdf) have become available. As RDF and OWL are open standards, where all nodes and edges are described with URIs, data silos are also less of a problem, depending upon the availability of the resources to the public.

Combining correlation networks with semantic data integration offers the possibility of maximizing experimental measurements with current biological knowledge. This will be the focus of the rest of this chapter.

12.4
Correlation Network Analysis

Correlation networks are another type of experimentally derived network. In these networks, nodes represent gene transcripts, proteins, metabolites, or other experimental observations measured in a given experiment. Edges represent the mathematical association between nodes, across all subjects in an experiment, typically created only if a significance threshold is reached for that association. An example is shown in Figure 12.1.

Typically linear correlation measures, such as the Pearson correlation [25], are used to determine correlation coefficients. The quantitative measure of correlation for a set of data is denoted by r. It is assumed that this measure is an estimate of the unobserved true correlation, ϱ, in the entire population for a given study. When r is close to $+1$, the two analytes correlate well, in the sense that when the level of the first increases, so does the level of the second. When r is close to -1, the two analytes anticorrelate well, in the sense that when the level of the first increases, the level of the

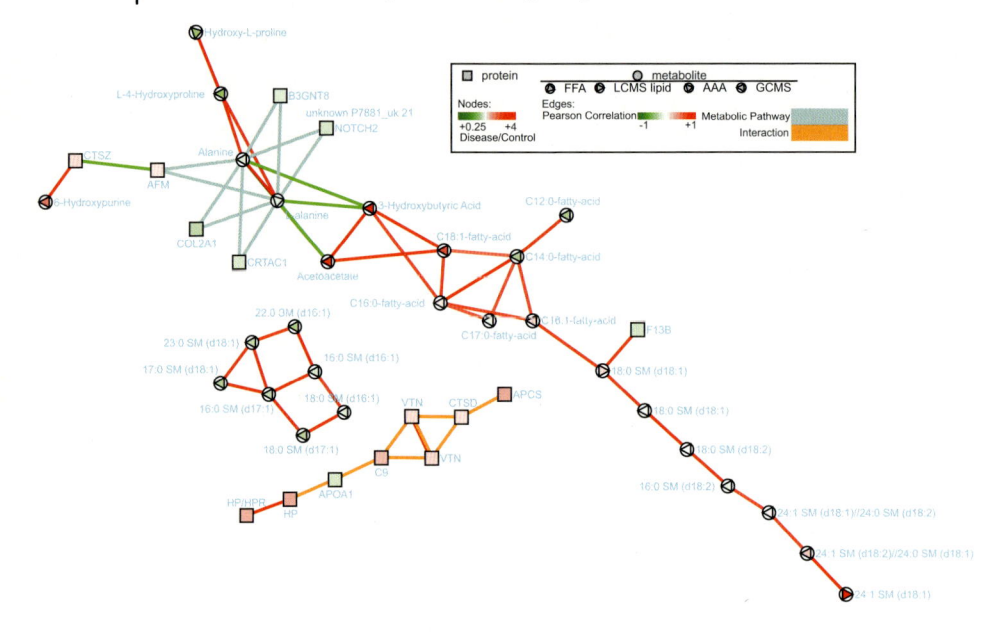

Figure 12.1 Correlation network example. In this representation node shapes denote the experimental platform used and node colors denote the expression change between disease over control. Edge colors denote correlations between analytes (red, green), interacting analytes (orange), or presence in the same metabolic pathway (blue).

second decreases. When r is close to 0, the two analytes are said to be uncorrelated and their scatter plot will show no trend. An illustration of a typical analyte association is shown in Figure 12.2 and the analyte scatter plot in Figure 12.3.

Rank-transformed correlation measures, such as the Spearman correlation [28], are also employed, but results are generally similar. The key distinction of a correlation approach is that the associations are driven by abundance levels of individual subjects, not an aggregate metric (mean, median) of a cohort. This allows inferencing of the type: as abundance levels of analyte X rise, abundance levels of analyte Y also rise (or fall). Figure 12.2 is a great illustration of this; the unordered patient index number comparing serum high-density lipoprotein (HDL) levels (red) with plasma Apolipoprotein A1 (ApoA1) levels (blue) shows a strong association – as HDL rises (or falls) so does ApoA1. Figure 12.3 rearranges this observation into a scatter plot with a corresponding Pearson correlation coefficient of 0.85 ($p \leq 0.001$).

Relevance networks [29] are a close neighbor of correlation networks, employing mutual information as the metric of similarity instead of correlation. Mutual information has the potential to find nonlinear relationships between analytes (Figure 12.4D), but does not have the ability to distinguish between positive and negative associations (Figure 12.4A and B). Mutual information can also produce different results in cases where we expect the same degree of association between variables (Figure 12.4A and C).

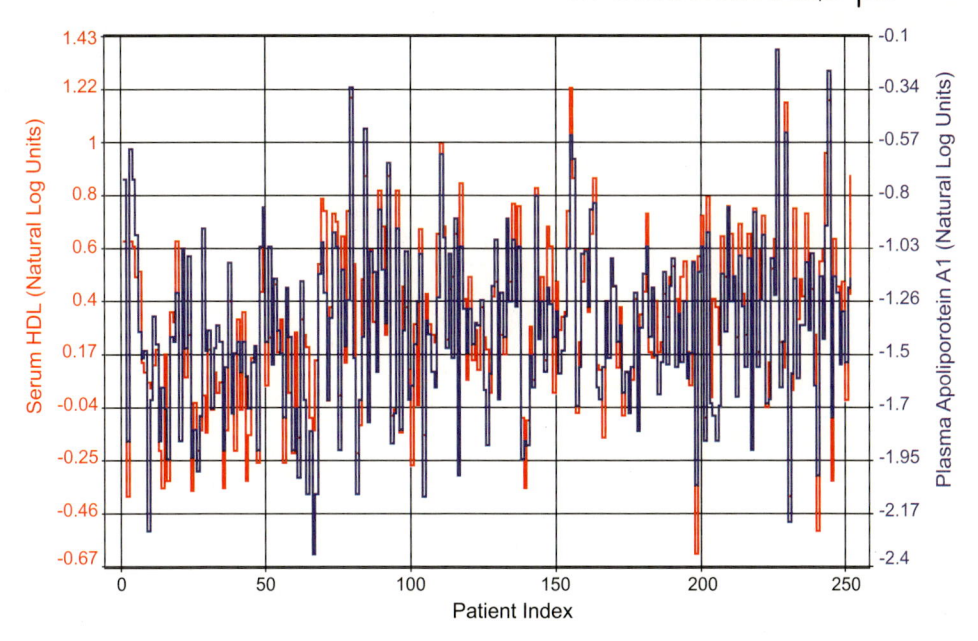

Figure 12.2 Relationship between serum HDL and plasma ApoA1. Across 250 patients measured there was a high association between these two analytes. Patient samples were part of the Copenhagen City Hospital Heart Study or Copenhagen General Population Study [26, 27].

While the relationship among many biological analytes may not be linear across the entire range of expression, linear approximations have proven to be useful in many cases [30, 31].

12.4.1
Selecting Nodes and Edges for Networks

Perhaps the most important consideration in assembling correlation networks is where the threshold for node and edge inclusion needs to be established. If this threshold is too high, important relationships will be omitted from consideration. If too low, visual inspection becomes nearly impossible. While much network structure can be evaluated without the need to explicitly draw and visualize, many of the key hypotheses behind network generation in the first place benefit greatly from domain expert interpretation and oversight. Context and network objectives should really dominate the setting of thresholds rather than dogma around p-values and false discovery. One example of a tightly targeted network occurs among patients in a heart failure study. Here, using a targeted cross-omics plasma panel of 12 analytes measured 184 patients, using Spearman's method, the authors found significant correlation between matrix metalloproteinase-2 and atrial natriuretic peptide of 0.39 ($p \leq 0.001$ [32]). At the other extreme is the Liver Toxicity Biomarker Study (LTBS), a project in idiosyncratic hepatotoxicity. In this project 33 573 analytes (genes, proteins, metabolites) were measured in both liver tissues and plasma and

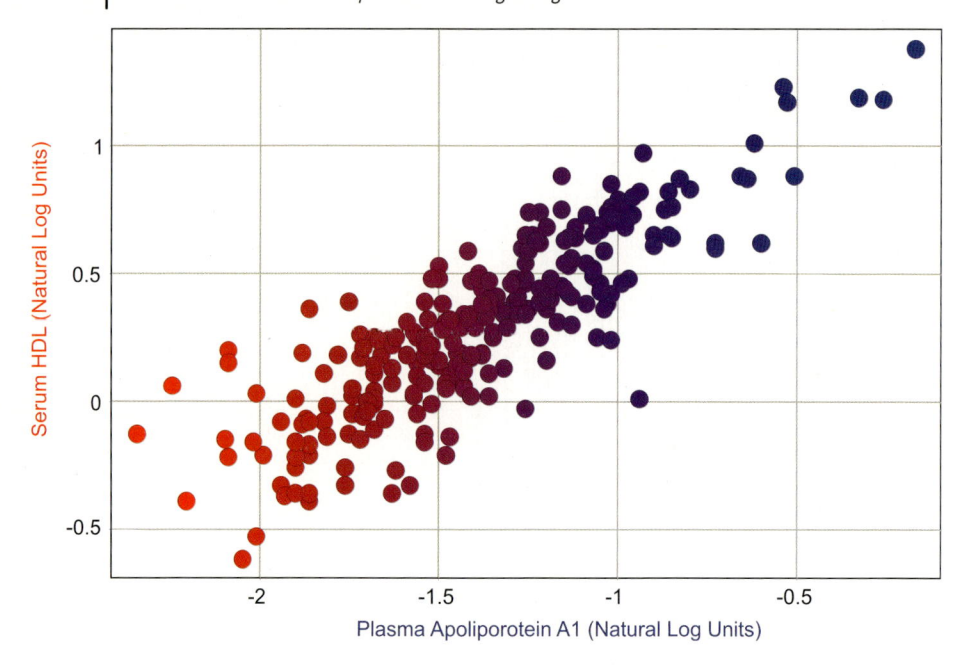

Figure 12.3 Scatter plot relationship between serum HDL and plasma ApoA1. These analytes had a correlation coefficient of 0.85 in a patient population from the Copenhagen City Hospital Heart Study or Copenhagen General Population Study [26, 27]. The red to blue scale is associated with analyte abundance levels.

generated 202 342 406 correlations among all measured analytes [21]. Thresholds chosen to begin data visualization were $|r| \leq 0.8$ and $p \leq 0.05$, and these still generated 1 343 514 correlations – a reduction of 99.3% – to evaluate further. Practical operating procedures need to be employed to make biological observations and generate new hypotheses in this environment.

First, one needs to determine if experimentally observed nodes need to be included in a network. If the inclusion of a node only adds noise to a network it should be excluded. Generally, if the experimental platform variability exceeds the analyte biological variability, these measurements cannot be considered to be robust and these analytes should not be included in either traditional pathway analysis or correlation analysis. In the case of the LTBS study great differences were seen in the coefficients of variation (CVs) for analytical platforms in liver (Table 12.1).

The discovery protein, lipid and amino acid platforms had CVs of less than 20% for more than 89% of measured analytes, while the polar analyte and mRNA analysis platform (completed on an Affymetrix Rat Genome 230 version 2.0 gene chip) had CVs less than 20% for about 75% of their measured analytes. The lower CVs seen among the first three platforms would make analytes measured among them generally better candidates for correlation analysis than the polar and mRNA platforms, although any node inclusion threshold should be determined by the specific analyte CV rather than the platform CV.

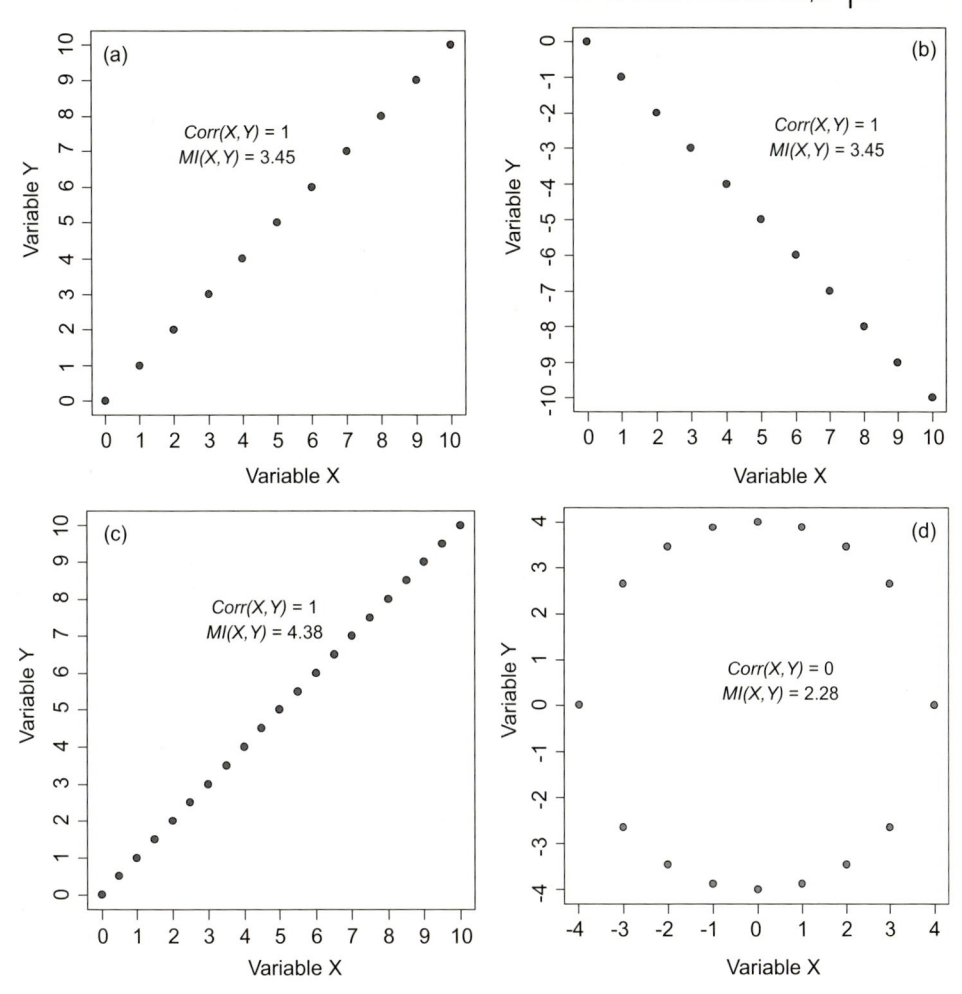

Figure 12.4 Illustrations of the properties of correlation coefficients (Corr) and mutual information (MI) scores, Coefficients for each method appear in each panel.

Table 12.1 CVs for analytical platforms applied to the liver samples [21].

CV range (%)	mRNA analysis	Discovered proteins	Lipids	Amino acids	Polar analytes
0–10	11 957	630	108	34	103
>10–20	10 935	142	15	3	138
>20–30	4219	1	9	0	51
>30–50	2933	0	4	0	21
>50	1054	0	1	0	6
NA	1	529	—	1	—
Total	31 099	1302	137	38	319

12.4.2
Distributions of Correlation Statistics

After generating the covariance matrix between all observed analytes the distribution of correlation statistics can help to determine proper thresholds for inclusion of correlation edges in a correlation network. A recent study in a juvenile neurological disease had 828 total analytes measured in plasma (561 proteins and 267 metabolites) which gave correlation distributions shown in Figures 12.5 and 12.6.

Clearly, the number of analytes affects the number of correlations within an omics modality. When scaled, 0.68% of the metabolite–metabolite correlations (red bars in Figure 12.5) exceed thresholds of $|r| \geq 0.8$ and $p \leq 0.05$, while this is 1.27% for protein–protein correlations. This figure is nearly double for the proteins, perhaps reflecting greater biological association among the protein data. This may also reflect how proteins were counted in this experiment – each wet isoforms is counted as a distinct entity – and these instances of the same protein class are generally highly correlated.

Missingness can also confound selecting correlation edges for networks. A number of correlations had between four and 22 analyte pairs for calculating a correlation (Figure 12.6). There is a clear relationship between the number of subject pairs and the potential for a robust correlation (i.e., you pay a large penalty for not measuring analytes in correlation networks). For example, 5.18% of all four-subject

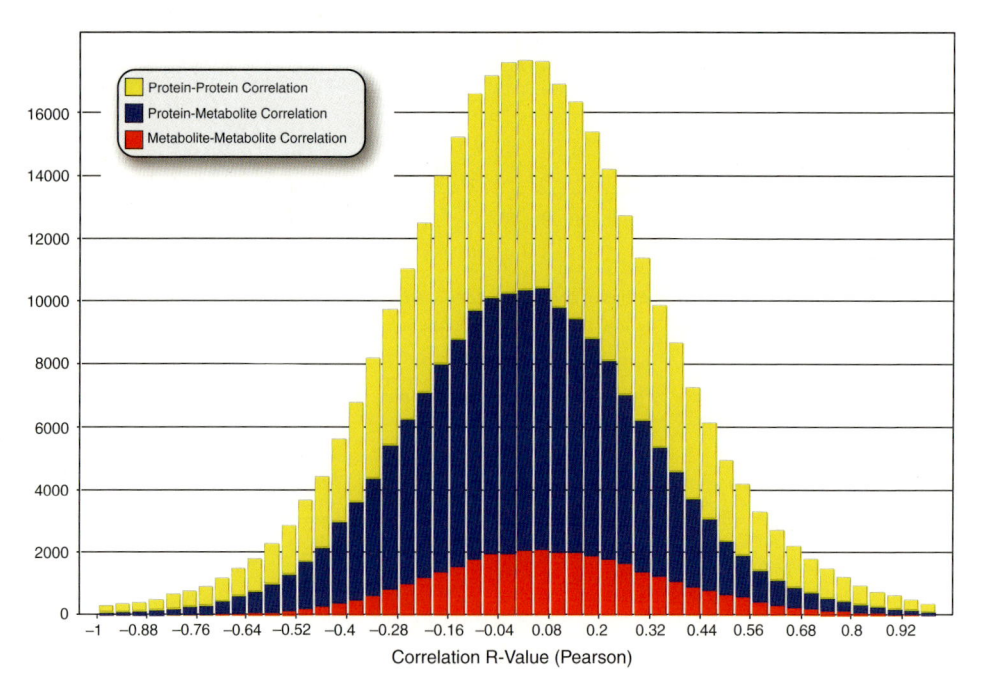

Figure 12.5 Distribution of correlation coefficients (r-value) in plasma; 561 proteins and 267 metabolites were used to generate the covariance matrix.

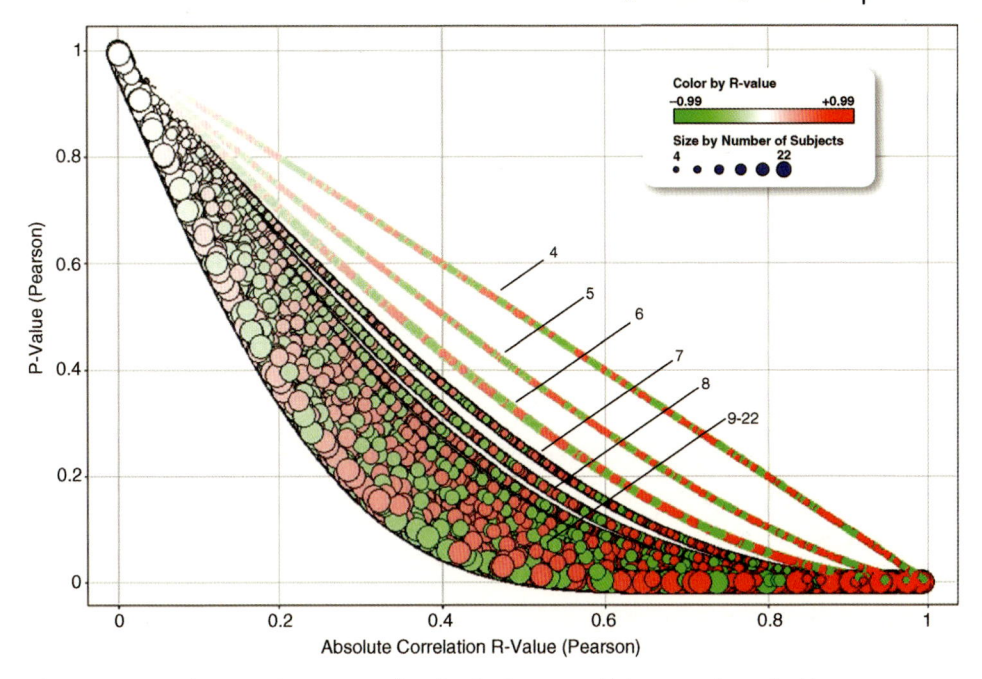

Figure 12.6 Correlation *p*-value versus *r*-value plot. Analyte pairs with lower numbers of subjects have higher *p*-values and lower absolute *r*-values.

pair correlations were significant (with thresholds of $|r| \geq 0.8$ and $p \leq 0.05$) while only 0.29% of all 22-subject pair correlations were significant. However if the only threshold considered is a $p \leq 0.05$ the number of significant correlations does not change for four-subject correlations but increases to 10.24% for 22-subject correlations.

The clear implication of this is that higher numbers of subject pairs allows correlations with lower correlation coefficients to be considered significant and evaluated in correlation networks. As this setting is testing multiple hypotheses (i.e., testing for significance of more than 1000 analytes simultaneously) a false discovery rate (FDR) correction is appropriate. If these correlation *p*-values are corrected, allowing for a 10% FDR, then only 0.018% (two) of the four-subject correlations are declared significant and 0.76% (465) of the 22-subject correlations are declared significant. Correlation network methodologies and the robustness of correlation approaches depends greatly on the number of subjects available, which typically precludes its utility in small studies, as cost is also correlated with the number of subjects.

12.5
Knowledge Annotation for Networks

After decisions have been made on what relationships are to be pursued for biological contextualization (i.e., placing the relationships into their known biological contexts), the next step is to annotate these networks.

There are many different knowledge bases and techniques for annotating networks, some of which are available commercially (e.g., Sentient Suite; IO Informatics, Berkeley, CA, www.io-informatics.com) or as plugins within Cytoscape [33–35], an open source bioinformatics network visualization platform (e.g., MiMI [36]). The most common implementation seen in industry and academia is to (i) download public and private knowledge bases; (ii) convert them into a relational database framework, if not already in this format; (iii) construct a common vocabulary for analyte synonyms; (iv) develop a rules system to match measured analytes to specifically defined relationships; (v) annotate networks and display and interact with networks; and (vi) update knowledge bases, as needed. This process requires a lot of internal informatics resources, and is redundantly addressed throughout many commercial and academic locations. Annotating networks from a shared, publically available resource solves the database curation and redundancy problem and is a critical advance of semantic web approaches.

This process is highlighted from a recent use case in the High-Risk Plaque (HRP) initiative (http://www.hrpinitiative.com/hrpinit/).

12.5.1
HRP and the Paired-Plaque Study

The HRP initiative began as a precompetitive consortium in January 2007 to research and advance the understanding, recognition, and management of HRP – a leading cause of myocardial infarction. While there is a relationship between blood cholesterol levels, arterial plaque burden, vessel stenosis, and ensuing myocardial infarction, many patients suffering heart attacks did not have elevated cholesterol nor was narrowing seen in coronary arteries. For these patients it is likely that a once-stable plaque became vulnerable, due to local stresses or inflammation, and ruptured causing a blood clot to block a coronary artery and subsequently lead to a heart attack [37].

The HRP consortium, made up of members from the pharmaceutical, medical imaging, biotechnology, and insurance industry, commissioned a series of prospective and retrospective studies to examine mechanism of plaque progression and rupture, and prediction of heart attack. These studies employed a wide range of measurement modalities, from molecular measurements in plaque tissues to carotid ultrasound measurements of arterial plaque burden. Semantic approaches were used in both studies, for data analysis to annotation to visualization and interpretation.

One study within the HRP framework, the Plaque Biology Study, was an effort to elucidate changes in the process of plaque biology from a stable, but vulnerable, plaque to a ruptured plaque. Atherosclerotic plaque is quite a heterogeneous tissue. In a 1-cm plaque sample from a typical patient, one segment may contain a stable, hardened section while a neighboring segment may be softened and inflamed, and quite vulnerable to sudden rupture, frequently leading to an ensuing atherothrombotic event.

The Plaque Biology Study was based on a collection of 26 carotid atherosclerotic plaque samples where each sample, derived from a single patient, could be characterized with both a sample region (American Heart Association grade III–V) and a ruptured region (American Heart Association grade VI). These two segments from the same patient were compared in ensuing statistical analysis for differences in the two plaque types using proteomics, transcriptomics and metabolomics methodologies (http://www.hrpinitiative.com/hrpinit/research/plaque-biology.jsp).

Within this study, 23 862 analytes were measured (1330 proteins, 348 metabolites, 22 184 transcripts) across both stable and ruptured plaque types. This generated a correlation matrix of 19 703 968 associations (correlations were only calculated on the set of 2896 transcripts whose expression differed significantly between stable and ruptured plaque phenotypes), 634 744 significant associations in stable plaque (significance was assigned to correlations with a $p \leq 0.05$ and an absolute $r \geq 0.8$), and 119 920 in ruptured plaque samples. This large dataset provided data volume and data tier problems that could be addressed through the use of semantic web approaches, specifically annotation with public ontologies.

12.5.2
Annotation with Public Sources and Ontologies

The Plaque Biology Study described below shows semantic integration of correlation networks with public domain knowledge on mechanisms, biological functions, and disease-related findings. This approach is designed to enable statistical findings to be placed into a biological context, allowing researchers and clinicians to make sense of these results from a biological viewpoint.

Fundamental challenges can be summarized in three basic questions:

i) How can statistical correlations and correlation networks be described with known mechanisms?
ii) How can we gain better biological understanding of multimodal responses involving multiple overlapping functions?
iii) Will understanding of differences between different disease stages (stable and ruptured plaque) at the molecular level provide decision support in drug discovery and healthcare?

In practice, meaningfully connecting data networks (including analytical results across experimental methods) with knowledge networks from qualified sources such as curated reference databases or via text mining from peer-reviewed publications has been extremely difficult and time-consuming. Recent maturation of experimental platforms, analytical methods, and semantic integration technologies has lowered these barriers for creation of data networks, and for annotation and visualization with knowledge resources. Relationships represented in correlations are not necessarily functionally linked within a given biochemical network, despite residing in the same context, such as arising from the same disease or drug perturbation. The observed cross-omics expression changes may represent very different biological processes and are in most cases the sum of effects from multiple pathway interactions. Biological systems understanding to predict phenotypic outcome on treatment responses is still very incomplete and in its infancy. For this reason, insights in biological systems from public domain knowledge are essential to predict phenotypic outcomes. Merging data networks with public knowledge networks using a semantic model has become effective and more easily achievable as the method of choice for such functional integration.

Building on these methods within a broader systems biology approach, researchers and clinicians can more easily reduce network complexity and can effectively identify and understand genetic, protein, and metabolic classifiers with multiple common pathway relationships, but very different responses. Furthermore, these methods make it possible to efficiently test and refine hypotheses on dynamically changing patient data and validation datasets, and to apply the resulting knowledge to new patients for practical personalized medicine applications.

In the following, examples of how such systems biology approaches lead to new insights, as well as to confirmation of previously described findings, are presented. Being able to use tools such as "Applied Semantic Knowledgebase (ASK)" (IO Informatics) for predictive biology and screening allows them to be used as confident decision support on disease state (diagnostics) and/or efficiency of treatment options in patient-centric personalized medicine – thus, better understanding of biology naturally extends into applying those insights for the benefit of the patient.

These examples will provide evidence of how similar approaches can be used in predictive biology as confident decision support on disease state alterations and to assess comparative effectiveness of treatment options in patient-centric personalized medicine.

12.5.3
Results and Benefits of the Approach

Results following from more efficient and flexible connection of data-driven correlation networks with canonical knowledge resources include: creation, visualization, and refinement of hypotheses; identification and qualification of biomarkers; and development of new insights and methods to support various research and personalized medicine applications [38].

Previous to the emergence and maturation of these combined experimental methods (genetic, proteomic, metabolic, and high content imaging, among others) and semantic integration approaches, these sorts of results were practically unachievable due to limitations following from traditionally expensive, time-consuming, and inflexible techniques. To make these results achievable, the creation of practically useful knowledge networks increasingly draws on combined applications of thesauri, ontologies, and data linking under semantic integration methods.

A key finding of the Plaque Biology Study was downregulation of integrin signaling, specifically the downstream actin cytoskeletal/focal adhesion signal transduction cascade when comparing ruptured plaque slices to stable plaque slices. This finding was observed at both the gene transcript and protein level (Figure 12.7).

The integrin receptor aggregation suggests a likely loss of cytoskeletal structures in ruptured plaque and a loss of internal focal adhesion. The observations are consistent in both proteomics and transcriptomics measurements. There was a striking agreement among both, proteins and transcripts, for actin, α-actinin, talin (all complexes), and vinculin. This is suggestive of a weakening of the actin cytoskeleton in ruptured plaques, possibly due to integrin signaling.

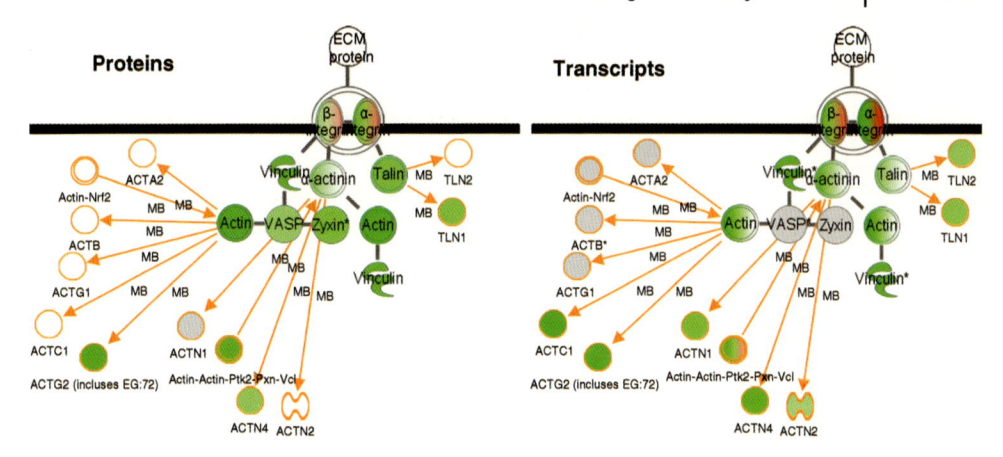

Figure 12.7 Protein (left) and transcript (right) regulation in integrin signaling in ruptured plaque. Green denotes downregulated analytes or complexes and red denotes upregulated analytes or complexes (a green/red mix represents a complex with both up and downregulated members). Gray shapes denote measured analytes that did not change in the ruptured/stable plaque comparison. (©Ingenuity Systems, 2009.) This pathway was generated through the use of Ingenuity Pathways Analysis (Ingenuity® Systems, www.ingenuity.com).

Other systems, such as peroxisome proliferator-activated receptor-α (PPARα)/retinoid X receptor-α (RXRα) signaling did not have strong correspondence between gene (Figure 12.8) and protein (Figure 12.9) expression.

In this example, marker analytes do not indicate a strong protein response. Many of the proteins measured may be of low abundance, and therefore difficult to measure with mass spectrometry, or the gene transcripts not made into proteins, or, perhaps, some combination of the two is possible. Overall, this observation does not provide a very conclusive picture on biological events and requires additional explanation.

12.5.3.1 Integral Informatics Approach

Correlation datasets from experiments included three cross-omics sets, containing 149 metabolites, 714 proteins and 6800 transcript markers. The semantically merged data network then was enriched via public reference data sources: Entrez Gene [39]; OMIM (Online Mendelian Inheritance in Man) [40], Protein [41–44], and PubChem [45] from the National Center for Biotechnology Information (NCBI); UniProt [46, 47]; HMDB (Human Metabolome Database) Compound [48, 49]; KEGG (Kyoto Encyclopedia of Genes and Genomes) [50]; and BioCarta (www.biocarta.com).

The semantic integration of data networks with knowledge network leads to mechanistically relevant functional biology insights which can be applied in form of arrays of models (expressed as SPARQL graph queries, containing ranking and weighing of relationships) to predictive decision support for new patient populations. (SPARQL is an RDF query language; its name is a recursive acronym that stands for SPARQL Protocol and RDF Query Language. It was standardized by the RDF Data Access Working Group of the World Wide Web Consortium and is considered a key semantic web technology. On 15 January 2008, SPARQL became an official W3C

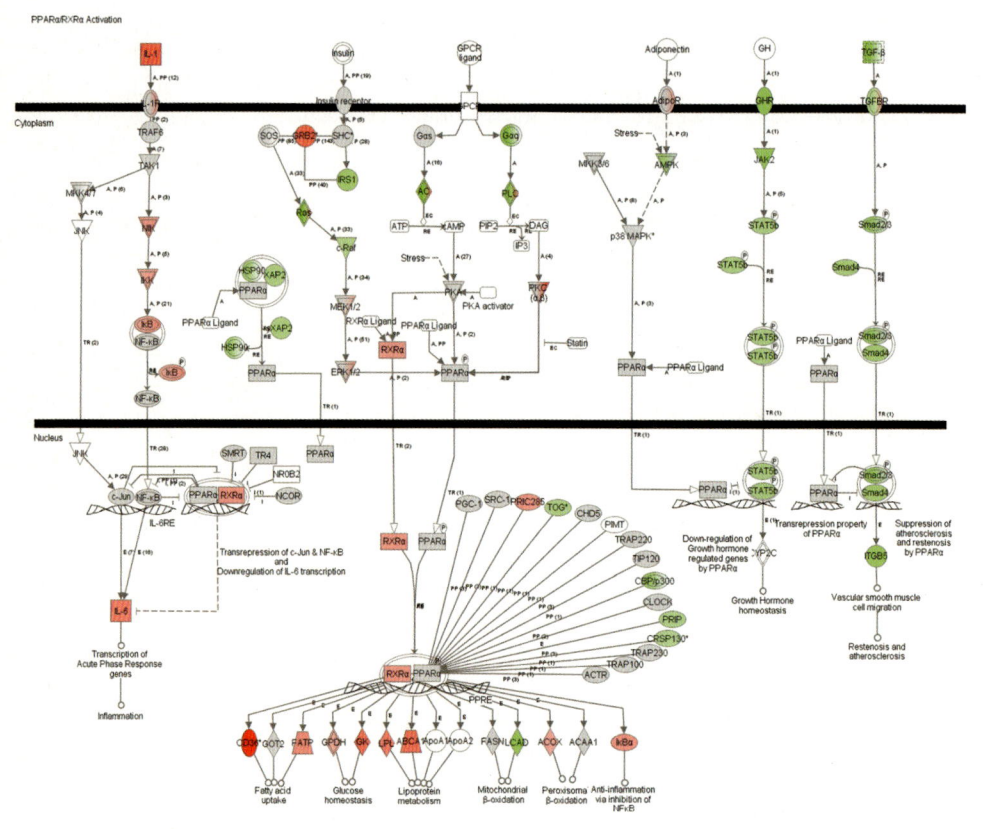

Figure 12.8 PPARα/RXR activation in gene transcripts, measured in ruptured plaque. Colors as described in Figure 12.7. (©Ingenuity Systems, 2009.) This pathway was generated through the use of Ingenuity Pathways Analysis (Ingenuity® Systems, www.ingenuity.com).

Recommendation. SPARQL allows for a query to consist of triple patterns, conjunctions, disjunctions, and optional patterns; http://www.w3.org/TR/rdf-sparql-query/.)

i) Unify and query data by statistic relevancy.
ii) Create semantic networks to visualize and explore markers.
iii) Enrich experimental correlation network with public knowledge.
iv) Focus on common, but differentially affected pathways.

An example of the merged ontology used is in Figure 12.10. This "local ontology" combines practical experimental class hierarchies with subsets from formal ontologies such as Gene Ontology [51, 52], Pathway Ontology (from the National Center for Biomedical Ontology (NCBO) [53]) and elements from literature reference source classifications (see below).

Table 12.2 summarizes the annotations applied to the Plaque Biology Study correlation network using semantic techniques within three biological processes of both statistical and biological significance in plaque rupture.

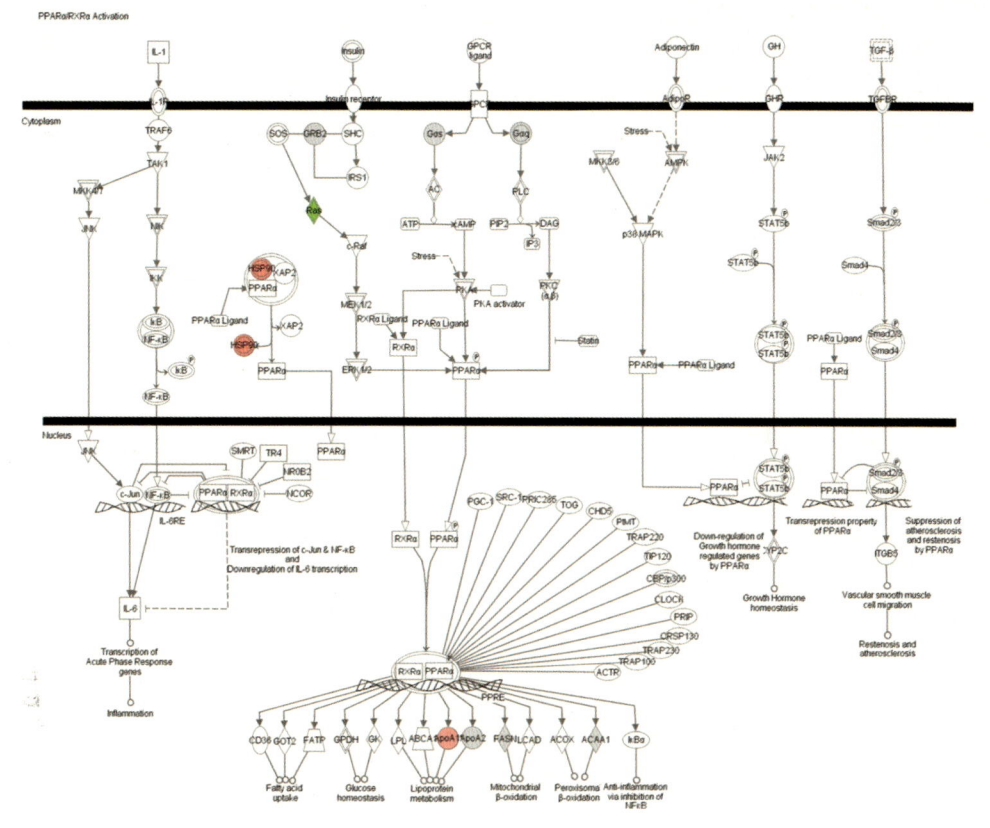

Figure 12.9 PPARα/RXR activation in proteins, measured in ruptured plaque. Colors as described in Figure 12.7. (©Ingenuity Systems, 2009.) This pathway was generated through the use of Ingenuity Pathways Analysis (Ingenuity® Systems, www.ingenuity.com).

The following example illustrates semantic knowledge annotation by visualizing the relevant subnetworks leading towards novel findings in plaque biology.

12.5.3.1.1 **Biological Networks in the Semantic Web Context** The entire covariance matrix containing the primary correlation networks in the Plaque Biology Study was translated from a "nodes" table and "edges" table into RDF triplets:

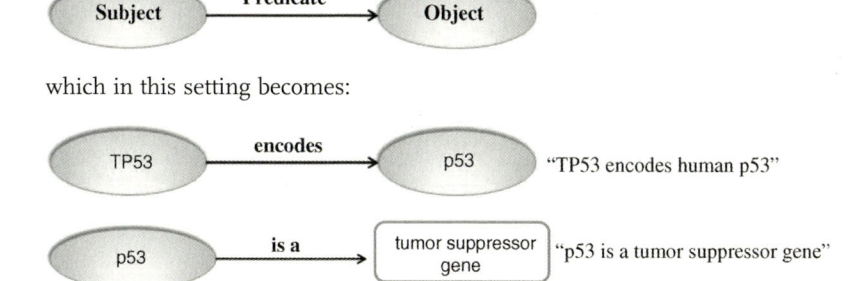

Subject → Predicate → Object

which in this setting becomes:

TP53 → encodes → p53   "TP53 encodes human p53"

p53 → is a → tumor suppressor gene   "p53 is a tumor suppressor gene"

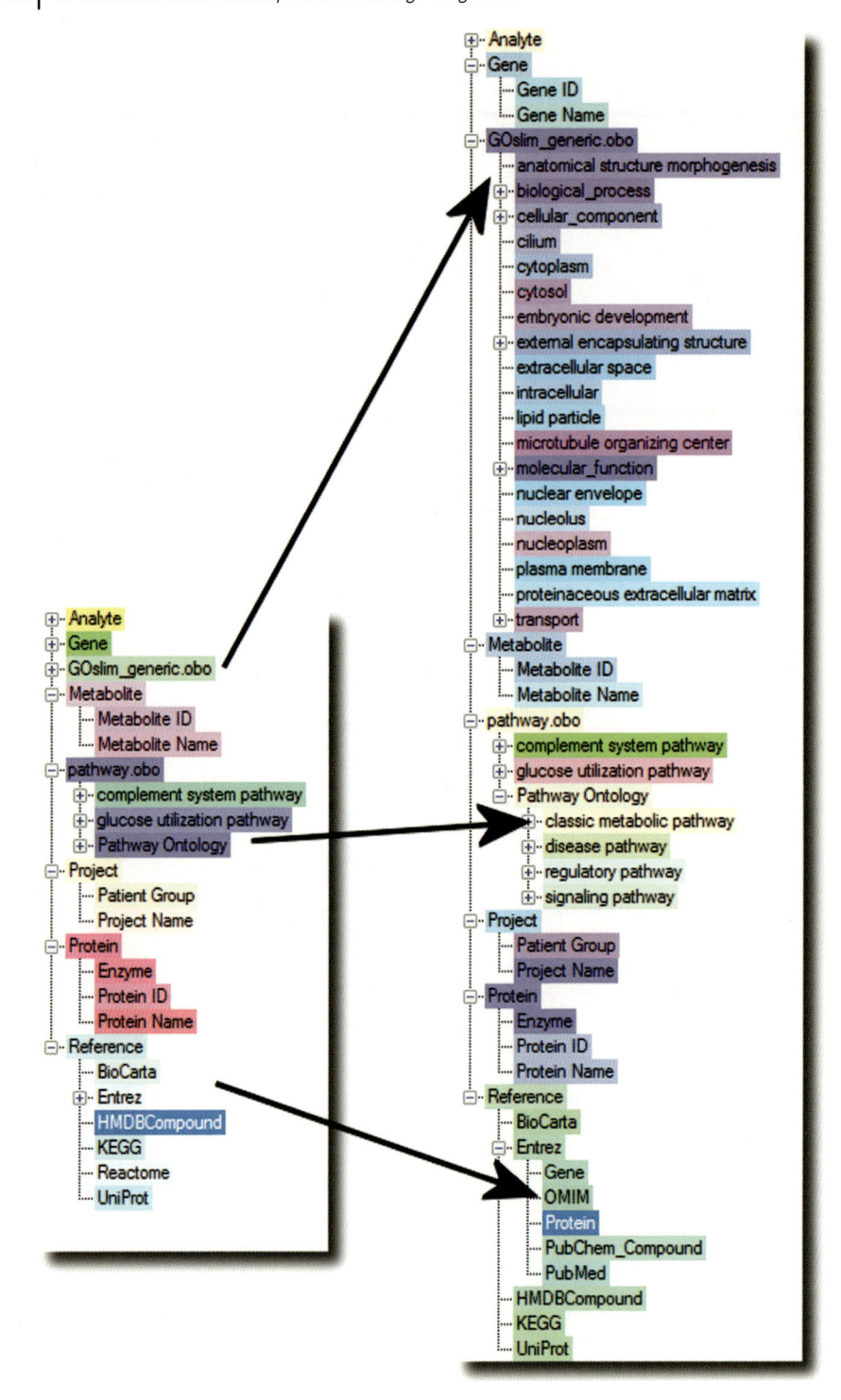

Figure 12.10 Example of merged and partially expanded "local ontology."

12.5 Knowledge Annotation for Networks

Table 12.2 Enrichment of pathways/processes in the ruptured/stable plaque comparison.

Source	Pathway/process		
	Integrin, plaque related	PPAR, plaque related	Histidine metabolism, plaque related
PubMed (citations and abstracts only)	428	115	4
PubMed Central	1206	203	721
Nucleotide	39	5	
Protein	18		1
Genome	9	4	
Gene	74	29	5
Biosystems (pathways and interactions)	18		
GEO profiles	56	2	

TP53 —located at→ chromosome 17   "TP53 is located on chromosome 17"

Triples can be connected to a graph, like in this example:

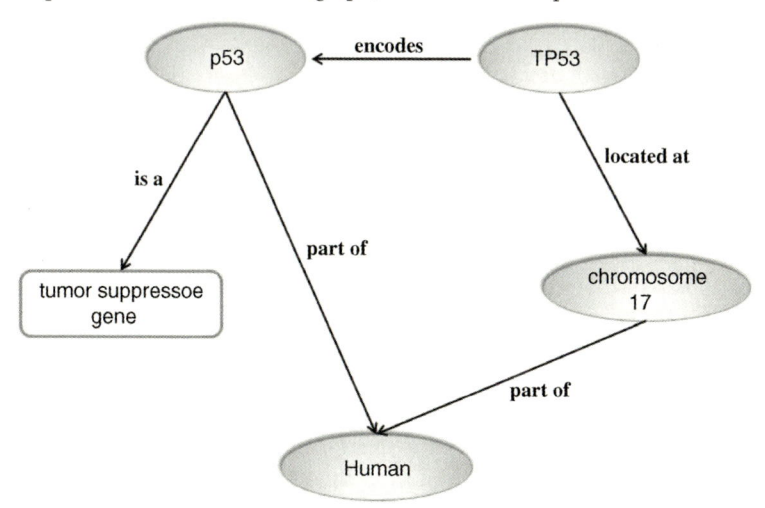

p53 ←encodes— TP53
p53 —is a→ tumor suppressoe gene
p53 —part of→ Human
TP53 —located at→ chromosome 17
chromosome 17 —part of→ Human

The big difference in using this data model is that the graph data structure makes merging data with shared identifiers trivial. Triples act as a least-common denominator for expressing data. Metadata used for naming removes ambiguity as the same identifier means the same thing no matter where the data came from. This provides the ability to infer and reason across connected data.

Semantic data represented in RDF can be visualized as a focused subnetwork, as shown in Figure 12.11.

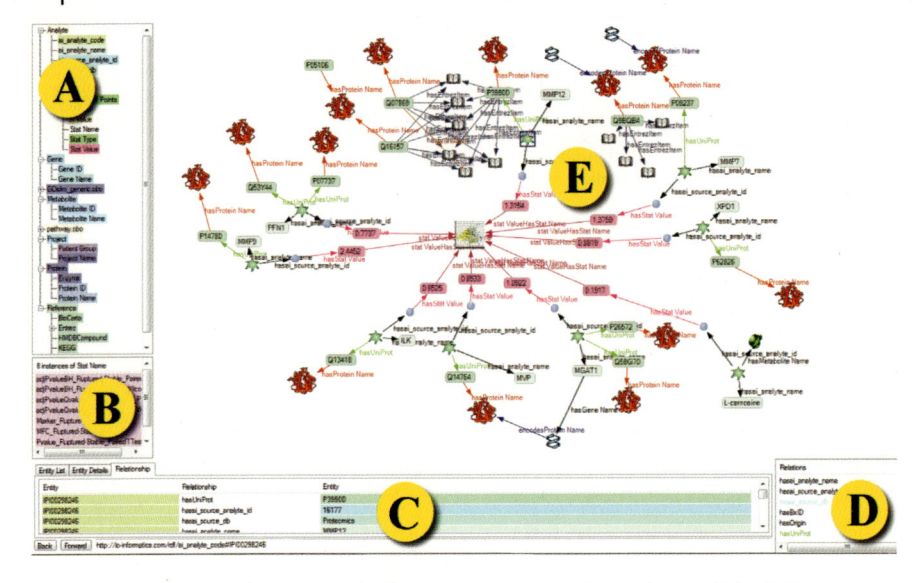

Figure 12.11 Correlation network of multimodal experimental data. Panel (A) shows the dynamically generated ontology, and the instances for each selected class are shown in panel (B). Panel (C) lists the entities and panel (D) their relationships to each other. Panel (E) displays the current cross-omics experimental network. Within panel (E) green stars represent transcript biomarkers and blue helices designate genes encoding for regulated proteins, which are displayed as red protein icons. Metabolites are represented via dark green small molecule icons. In this network representation, proteins, genes and metabolites have been reduced to icons for easier identification.

In this specific implementation, icons, relationships, and their corresponding labels can be turned on/off, and labels become visible when any entity or relationship is highlighted or "moused" over. Relationships between graph entities are color-coded based on their originating sources; they can be weighed and scaled for visualization according to numerical properties depending on the needs of the researcher.

This multimodal subnetwork was initially created with five types of correlation filters, based on node statistics, edge statistics, and edge relationships. It represents semantically merged correlations with data in the network assembled based on p-value and regression (r-value log-transformed Pearson) to reduce complexity and increase relevance. Node significance was defined as $p \leq 0.05$ and $q \leq 0.1$ for either paired t-test or Wilcoxon signed-rank test; with a q-value threshold of ≤ 0.001 for genes. Correlation edge significance was defined as $p \leq 0.05$ and $r \geq |0.8|$. Modalities were compared in a stepwise fashion, starting with metabolites versus proteins, then transcripts versus proteins, and then metabolites versus transcripts. The goal was to evaluate relationships across the different omics modalities.

Data from different sources were integrated under a common, dynamically generated ontology, supported by thesauri and semantically based reasoning to assign data elements to desired nomenclature and positions within the ontological framework. Ontologies and thesauri are merged, extended, edited, and managed with governance and source versioning. For instances, subsets of formal ontologies applicable to a scientific use case can be aggregated to a so-called "local ontology," and its classes and subclasses harmonized using a thesaurus for class names and relationships. In the same way more than one thesaurus may be applied to account for synonym harmonization across multiple data sources during import and/or merge as entities are mapped to preferred terms at that time. This is crucial to any merging as data coming from different domains (such as experimental biology, functional biology, clinical observations, molecular interactions, drug chemistry) have vastly diversified nomenclature for the same thing, and synonym and term harmonization becomes a must.

Being able to address this via automated tools that account for proper assignment of preferred terms for classes, instances, and relationships makes import of and merging with new data into a local ontology effortless and convenient (Figure 12.12).

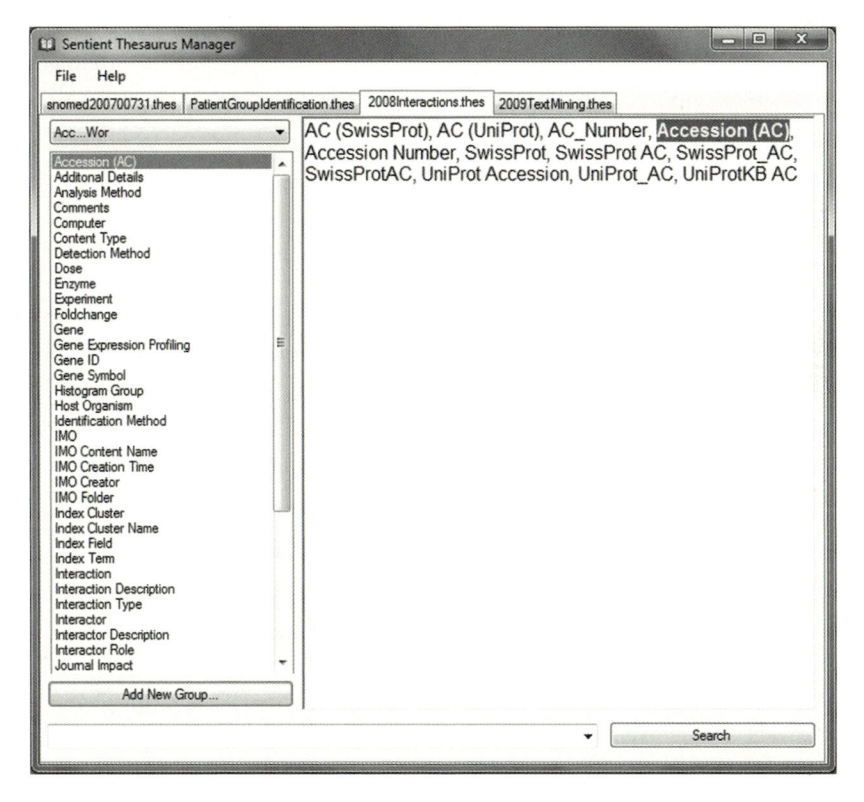

Figure 12.12 Managing synonyms during data import and merging. In this example implementation (Sentient Thesaurus Manager; IO Informatics), several thesauri are used to harmonize classes, instances, and relationships from multiple domains such as experimental biology data, clinical records, and molecular interactions.

12 Correlation Network Analysis and Knowledge Integration

12.5.3.1.2 Annotating Biological Networks with Public Knowledge Resources From the initial correlation network in Figure 12.11, knowledge network connections via node-selective drillout, query, and import from public resources were added as described in Table 12.2, generating an annotated network (Figure 12.13)

To annotate the experimental molecular correlation networks, curated public references were added to the network to facilitate mechanistic and functional biology understanding. These included 240 disease references, 112 functional references, 16 pathways, 28 proteins, and five key enzymes.

Each circle in Figure 12.13 denotes a pathway involved in the subnetwork; the yellow pathway labels are obtained from direct mapping to Pathway Ontology, and the protein identifiers (accession numbers) from UniProt.

Correlation edges were restored to the network and are presented in Figure 12.14.

When merged with the experimental correlation network (Figure 12.11) into one common network graph, the effectiveness of integrating public knowledge networks (Figure 12.14) for mechanistic insights and decoding of overlapping biological functions is now achievable as individual relationships between biomolecules and their involvement in different pathways become apparent. This way, regulative changes are interpretable in context of contributions to multiple pathway activities, each of which may contribute to observed fold changes. In a semantic environment no additional analytic process needs to be applied; by simply merging the data

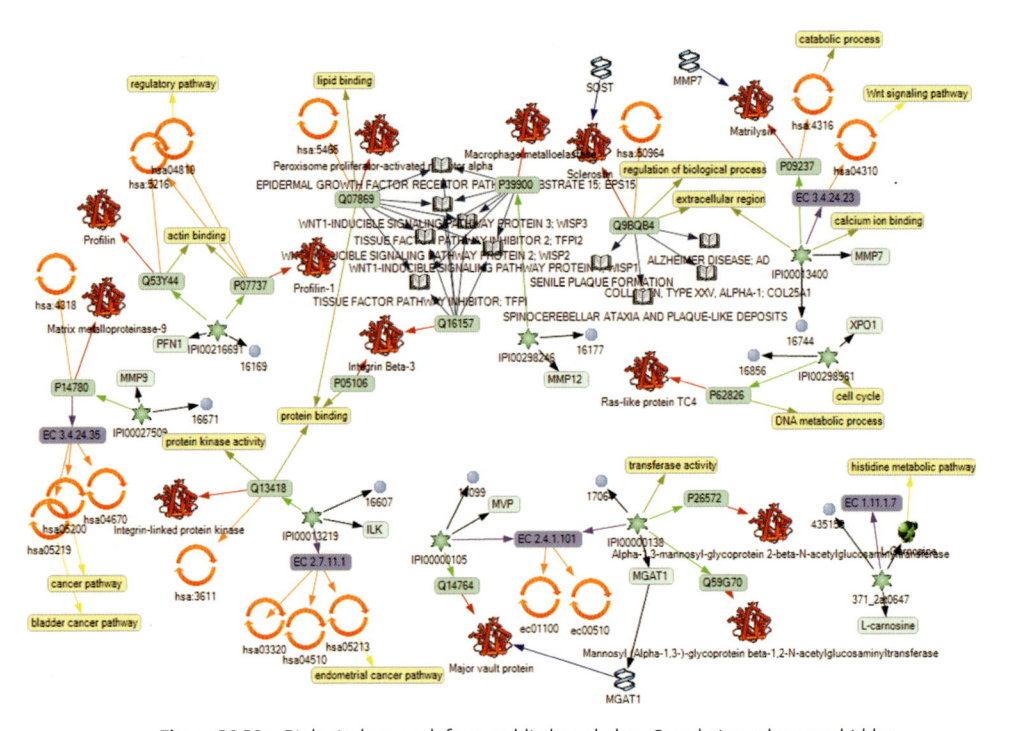

Figure 12.13 Biological network from public knowledge. Correlation edges are hidden.

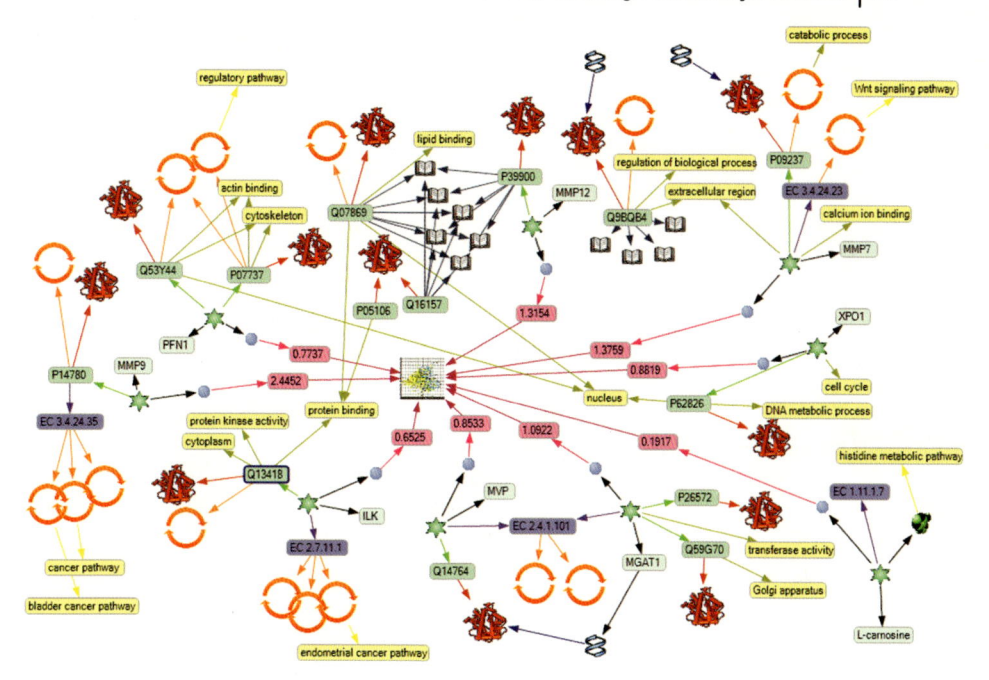

Figure 12.14 Functional multimodal correlation and knowledge network.

network with functional knowledge into a common network, biological mechanistic relevancy is revealed. This, of course, assumes that the knowledge source contributes a mechanistic model for which experimental data and correlation are an instance of that model.

The combined network can effectively be used to explore, test, and refine hypotheses. SPARQL queries can be generated by selecting subgraphs within the larger network, and automatically parsed and saved for refinements. Such refinements include weighing of selected nodes (all genes are not equally important in their contribution to the disease state) and their relationships to other biological entities as well as ranges for numeric parameters to define exclusions or intersections of subnetworks and include those in the model. The key benefit of "slicing-and-dicing" the network is being able to reduce its complexity without diminishing the relevancy of the model. Using arrays of SPARQL query-represented network graphs such as ASK with scoring provides a method for confident decision support based on mechanisms mapped to experimental results. Such models are used to qualify putative biomarkers, establish classifiers, and refine arrays of them through extension of training sets.

Arrays of SPARQL graph queries created in this fashion can effectively represent complex biological events, and are used in biomarker validation on blinded studies and as decision support for new patient populations to score risk for specific disease stages, such as ruptured plaque. An example of such a SPARQL query used to screen new or unknown datasets for match is depicted in Figure 12.15.

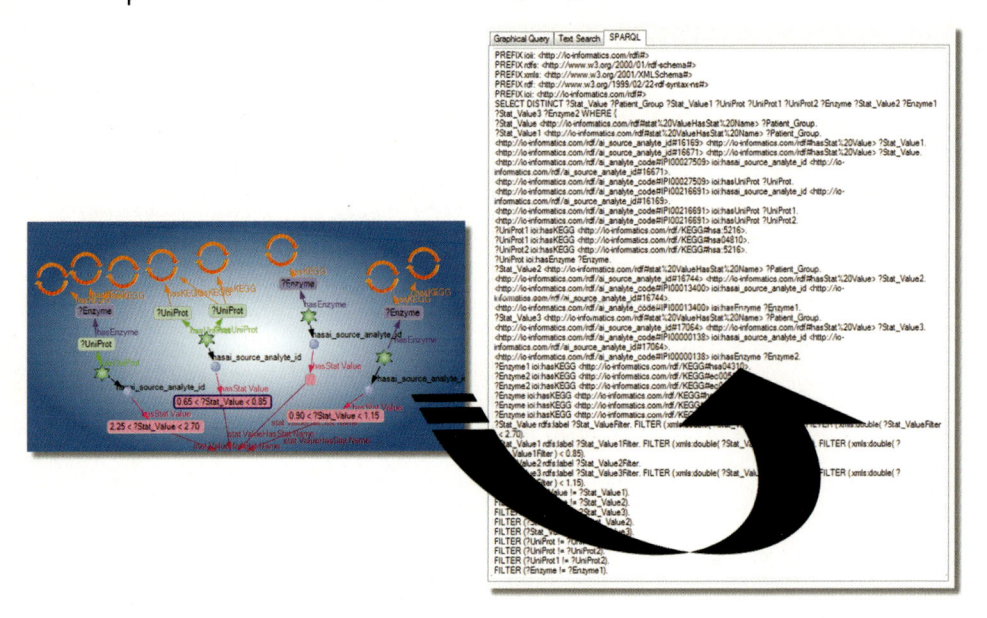

Figure 12.15 Graphical SPARQL query generation directly off the network: highlighting elements in the network graph (left) and setting confidence ranges creates automatically the SPARQL query (right) without any programming or knowledge of SPARQL syntax required.

12.5.3.1.3 Histidine Pathway Focus The observed L-carnosine regulation results from the combined effect of events in the histidine and alanine metabolic pathways. In the absence of this integration, the experimentally observed changes would not suggest contributions from either pathway (Figure 12.11) nor would it associate interleukin-6 (IL-6) or interleukin-8 (IL-8). In combination with the pathway and mechanistic information obtained from public reference resources on the example of ruptured plaque (Figure 12.13), the selection of L-carnosine (shown in the lower right portion of the integrated functional multimodal correlation and knowledge network in Figure 12.14 expanded in Figure 12.16). At this stage, the dependencies are clearly revealed in their functional context as the referenced pathway regulation coincides with observed IL-8 and histidine decarboxylate changes. The activity of enzyme complex for histidine decarboxylase (EC 4.1.1.22) and the observed IL-8 genomic and proteomic response (P10145) render inflammatory response and angiogenesis keying off the histidine pathway a strong contributing factor in the L-carnosine regulation while the alanine pathway involvement plays only a minor role in its regulation. The IL-6 change is overlapped effect of the observed interferon-β_2 (IFNB2) and IL-6 transcriptomic changes. The network graph in Figure 12.16 bridges between observations and mechanistic expectations, and as such helps decoding overlapping fold changes.

Inflammation is a widely observed response in many diseases; thus, historically it has been mostly downplayed in looking deeper into its relevancy in the characterization of a specific disease stage such as plaque building and rupture. However, this

12.5 Knowledge Annotation for Networks | 273

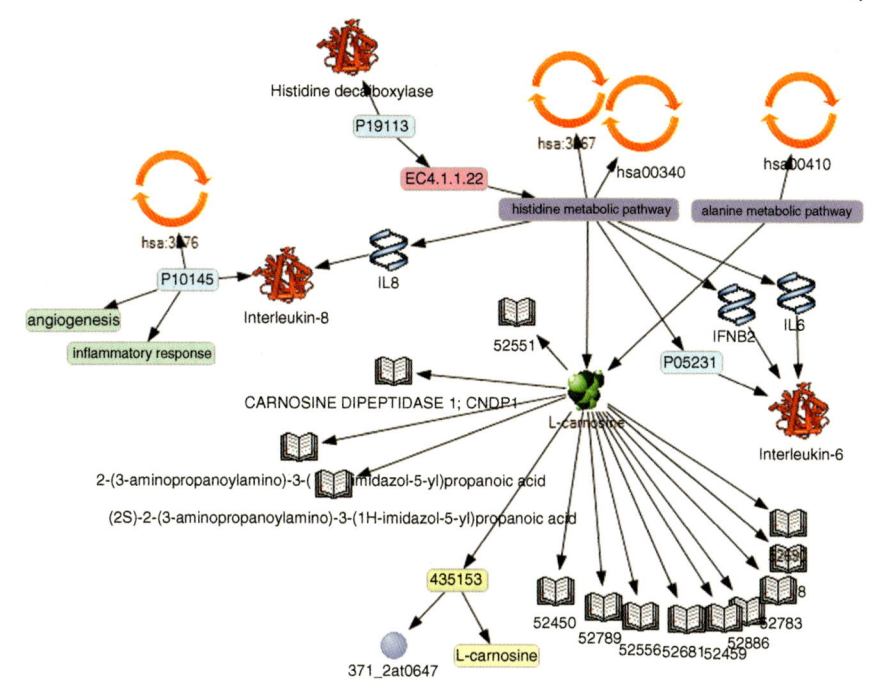

Figure 12.16 Inflammatory response subnetwork in ruptured plaque The connection depth exploration for L-carnosine is shown two levels down from Figure 12.14.

mechanistically enhanced correlation network can provide meaningful clues around the cause and impact of inflammation in this case. The exploration of histidine and alanine metabolic pathways, along with the likely inflammatory response, illustrate one route used to elucidate the contribution of these pathways to plaque rupture. Although exploration of integrin and actin signaling, along with PPARα/RXRα signaling, is beyond the scope of this work, similar approaches can be taken to show their contribution to plaque rupture.

One of the biggest benefits of semantic mining of integrated networks is the fact that this would not be possible from connecting data using canonical pathway analysis alone. By bringing in ontologies and mapping all findings – experimental and functional reference-related – to one common network graph makes it easy to not only to qualify biological relevance of statistically significant observations, but also to validate the overlapping mechanistic functions involved in complex disease stages at the same time. In this example, findings were consistent with the observed inflammatory response in ruptured plaque, and provided a deeper insight in the role of inflammatory processes and their pathway interactions in the transition between stable and ruptured plaque.

In conclusion, semantic data integration methods have been applied to integrate correlation networks with peer-reviewed public resources, and can easily be further reviewed and curated in a visual environment for effective, persistent reproducible

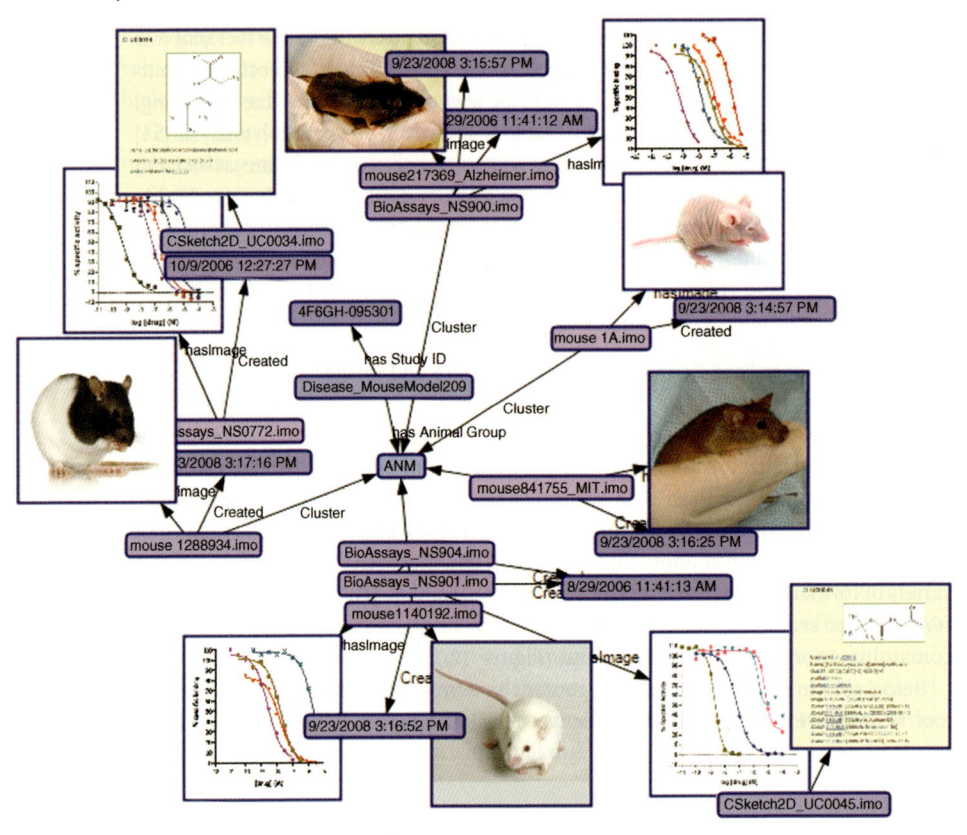

Figure 12.17 Example of a semantic network graph with images and chemical structures. In this example, conclusive animal assay results on different disease treatments can be visualized within the model for treatment responses at a defined stage of a disease.

12.6.5
Improved Integration of Text and Structured Data

While many resources provide access to specific classes of information, internal documents or scientific literature are not necessarily available in digested or semantically relevant formats. This is emphasized by the importance of natural language processing (NLP) or text mining according to harmonized relationships that are capable of capturing the meaning in a concise, cross-domain manner. Particularly within the chemistry and medical domains themselves, a need for consolidation of relationships beyond the language aspect has been historically challenging, let alone across all the multiple scientific disciplines that account for the "big picture" in the understanding of complex disease patterns. Lately, several approaches have been reported that can help minimize the human curation required in such cases, so the future of integrative semantically sound text and structured resources looks very bright.

12.6.6
New Classes of Knowledge-Based Applications Such as Network Pattern Based Screening and Prediction

As knowledge integration emerges, new classes of knowledge-based applications have been developed to address the needs for better and more efficient pharmaceutical drug development and clinical patient treatment. With the advent of the ability to save and apply relevant subnetworks from systems biology approaches in the form of SPARQL queries to new data populations and to refine those queries based upon result feedback from extended training sets of known responders, the possibility for

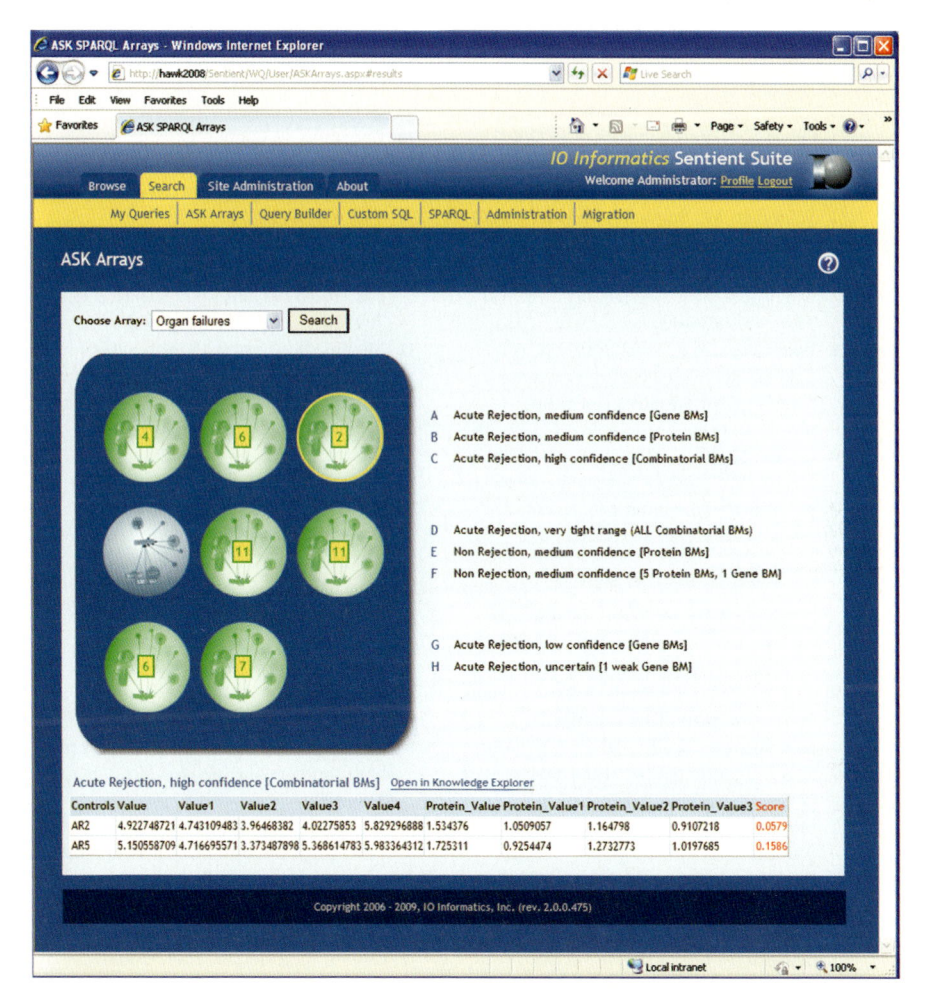

Figure 12.18 Example of patient screen for likelihood of organ failure. Each circle represents a profile for rejection/nonrejection; the number indicates the patients matching the profile. Selecting a profile will list the patients with their biomarker values and confidence score.

confident decision support based on predictive biology became a real opportunity. One example of such an implementation is ASK, which uses a web-based tool for screening (Figure 12.18).

The practical use of such applications has major implications in pharma for target validation, compound efficacy, toxicity profiling, and compound safety. It has led to disease signatures that allow for state classification based on phenotypical marker responses – all of which being used for patient stratification for clinical trials, companion diagnostics for certain therapies to ensure the treatment works on that patient, and, even more general, for patient-centric predictive screening (e.g., for presymptomatic vital organ failure in transplant patients) and for decision support on treatment options based on responder profiles in personalized medicine.

References

1 Oltvai, Z.N. and Barabasi, A.L. (2002) Systems biology. Life's complexity pyramid. *Science*, **298**, 763–764.
2 Schreiber, M., Sugihara, M. *et al.* (2006) Quantum mechanical studies on the crystallographic model of bathorhodopsin. *Angew. Chem. Int. Ed. Engl.*, **45**, 4274–4277.
3 Behrenfeld, M.J., Westberry, T.K. *et al.* (2009) Satellite-detected fluorescence reveals global physiology of ocean phytoplankton. *Biogeosciences*, **6**, 779–794.
4 van der Greef, J., Martin, S. *et al.* (2007) The art and practice of systems biology in medicine: mapping patterns of relationships. *J. Proteome Res.*, **6**, 1540–1559.
5 Ito, T., Chiba, T. *et al.* (2001) A comprehensive two-hybrid analysis to explore the yeast protein interactome. *Proc. Natl. Acad Sci. USA*, **98**, 4569–4574.
6 Giot, L., Bader, J.S. *et al.* (2003) A protein interaction map of Drosophila melanogaster. *Science*, **302**, 1727–1736.
7 Li, S., Armstrong, C.M. *et al.* (2004) A map of the interactome network of the metazoan C. elegans. *Science*, **303**, 540–543.
8 Rual, J.F., Venkatesan, K. *et al.* (2005) Towards a proteome-scale map of the human protein–protein interaction network. *Nature*, **437**, 1173–1178.
9 Stelzl, U., Worm, U. *et al.* (2005) A human protein–protein interaction network: a resource for annotating the proteome. *Cell*, **122**, 957–968.
10 Krogan, N.J., Cagney, G. *et al.* (2006) Global landscape of protein complexes in the yeast *Saccharomyces cerevisiae*. *Nature*, **440**, 637–643.
11 Lee, T.I., Rinaldi, N.J. *et al.* (2002) Transcriptional regulatory networks in *Saccharomyces cerevisiae*. *Science*, **298**, 799–804.
12 Medini, D., Covacci, A. *et al.* (2006) Protein homology network families reveal step-wise diversification of Type III and Type IV secretion systems. *PLoS Comput. Biol.*, **2**, e173.
13 Wingender, E., Chen, X. *et al.* (2001) The TRANSFAC system on gene expression regulation. *Nucleic Acids Res.*, **29**, 281–283.
14 Qiu, P. (2003) Recent advances in computational promoter analysis in understanding the transcriptional regulatory network. *Biochem. Biophys. Res. Commun.*, **309**, 495–501.
15 Ekins, S., Nikolsky, Y. *et al.* (2007) Pathway mapping tools for analysis of high content data. *Methods Mol. Biol.*, **356**, 319–350.
16 Ganter, B. and Giroux, C.N. (2008) Emerging applications of network and pathway analysis in drug discovery and development. *Curr. Opin. Drug Discov. Devel.*, **11**, 86–94.
17 Ganter, B., Zidek, N. *et al.* (2008) Pathway analysis tools and toxicogenomics reference databases for risk assessment. *Pharmacogenomics*, **9**, 35–54.

18 Oresic, M., Clish, C.B. et al. (2004) Phenotype characterisation using integrated gene transcript, protein and metabolite profiling. *Appl. Bioinformatics*, **3**, 205–217.

19 Adourian, A., Jennings, E. et al. (2008) Correlation network analysis for data integration and biomarker selection. *Mol. BioSyst.*, **4**, 249–259.

20 Adourian, A., Plasterer, T.N. et al. (2009) Systems pharmacology, biomarkers, and biomolecular networks, in *Drug Efficacy, Safety, and Biologics Discovery: Emerging Technologies and Tools*, vol. **1** (eds S. Ekins and J.J. Xu), John Wiley & Sons, Inc., Hoboken, NJ, pp. 75–113.

21 McBurney, R.N., Hines, W.M. et al. (2009) The liver toxicity biomarker study: phase I design and preliminary results. *Toxicol. Pathol.*, **37**, 52–64.

22 Tim, B.-L., Hendler, J., and Lassila, O. (2001) The Semantic Web. Scientific American Magazine.

23 W3C (2002) Feature Synopsis for OWL Lite and OWL: W3C Working Draft 29 July 2002. http://www.w3.org/TR/2002/WD-owl-features-20020729/.

24 W3C (2004) OWL Web Ontology Language Overview: W3C Recommendation 10 February 2004. http://www.w3.org/TR/owl-features/, http://www.w3.org/TR/2004/REC-owl-features-20040210/

25 Hollander, M. and Wolfe, D.A. (1973) *Nonparametric Statistical Inference*, John Wiley & Sons, Inc., New York.

26 Schnohr, P., Jensen, G. et al. (1977) The Copenhagen City Heart Study. A prospective cardiovascular population study of 20,000 men and women. *Ugeskr. Laeger*, **139**, 1921–1923.

27 Schnohr, P., Jensen, J.S. et al. (2002) Coronary heart disease risk factors ranked by importance for the individual and community. A 21 year follow-up of 12 000 men and women from The Copenhagen City Heart Study. *Eur. Heart. J.*, **23**, 620–626.

28 Best, D.J. and Roberts, D.E. (1975) Algorithm AS 89: the upper tail probabilities of Spearman's rho. *Appl. Stat.*, **24**, 377–379.

29 Butte, A.J. and Kohane, I.S. (2000) Mutual information relevance networks: functional genomic clustering using pairwise entropy measurements. *Pac. Symp. Biocomput.*, 418–429.

30 Gardner, T.S., di Bernardo, D. et al. (2003) Inferring genetic networks and identifying compound mode of action via expression profiling. *Science*, **301**, 102–105.

31 di Bernardo, D., Thompson, M.J. et al. (2005) Chemogenomic profiling on a genome-wide scale using reverse-engineered gene networks. *Nat. Biotechnol.*, **23**, 377–383.

32 Yan, A.T., Yan, R.T. et al. (2008) Relationships between plasma levels of matrix metalloproteinases and neurohormonal profile in patients with heart failure. *Eur. J. Heart Fail*, **10**, 125–128.

33 Shannon, P., Markiel, A. et al. (2003) Cytoscape: a software environment for integrated models of biomolecular interaction networks. *Genome Res.*, **13**, 2498–2504.

34 Christmas, R., Avila-Campillo, I. et al. (2005) Cytoscape: a software environment for integrated models of biomolecular interaction networks, *AACR Educ. Book*, 12–16.

35 Cline, M.S., Smoot, M. et al. (2007) Integration of biological networks and gene expression data using Cytoscape. *Nat. Protoc.*, **2**, 2366–2382.

36 Gao, J., Ade, A.S. et al. (2009) Integrating and annotating the interactome using the MiMI plugin for Cytoscape. *Bioinformatics*, **25**, 137–138.

37 Thiene, G. and Basso, C. (2010) Myocardial infarction: a paradigm of success in modern medicine. *Cardiovasc. Pathol.*, **19**, 1–5.

38 Racunas, S.A., Shah, N.H. et al. (2004) HyBrow: a prototype system for computer-aided hypothesis evaluation. *Bioinformatics*, **20** (Suppl. 1), i257–i264.

39 Maglott, D., Ostell, J. et al. (2007) Entrez Gene: gene-centered information at NCBI. *Nucleic Acids Res.*, **35**, D26–D31.

40 McKusick, V.A. (1998) *Mendelian Inheritance in Man. A Catalog of Human Genes and Genetic Disorders*, Johns Hopkins University Press, Baltimore, MD.

41 Bernstein, F.C., Koetzle, T.F. et al. (1977) The Protein Data Bank: a computer-based archival file for macromolecular structures. *J. Mol. Biol.*, **112**, 535–542.

42 Benson, D.A., Boguski, M.S. et al. (1999) GenBank. *Nucleic Acids Res.*, **27**, 12–17.

43 Wu, C.H., Yeh, L.S. et al. (2003) The protein information resource. *Nucleic Acids Res.*, **31**, 345–347.

44 Pruitt, K.D., Tatusova, T. et al. (2005) NCBI Reference Sequence (RefSeq): a curated non-redundant sequence database of genomes, transcripts and proteins. *Nucleic Acids Res.*, **33**, D501–D504.

45 Bolton, E.E., Wang, Y. et al. (2008) PubChem: integrated platform of small molecules and biological activities. *Annu. Rep. Comput. Chem.*, **4**, 217–241.

46 Consortium, U. (2010) The Universal Protein Resource (UniProt) in 2010. *Nucleic Acids Res.*, **38**, D142–D148.

47 Jain, E., Bairoch, A. et al. (2009) Infrastructure for the life sciences: design and implementation of the UniProt website. *BMC Bioinformatics*, **10**, 136.

48 Wishart, D.S., Tzur, D. et al. (2007) HMDB: the Human Metabolome Database. *Nucleic Acids Res.*, **35**, D521–D526.

49 Wishart, D.S., Knox, C. et al. (2009) HMDB: a knowledgebase for the human metabolome. *Nucleic Acids Res.*, **37**, D603–D610.

50 Kanehisa, M., Goto, S. et al. (2010) KEGG for representation and analysis of molecular networks involving diseases and drugs. *Nucleic Acids Res.*, **38**, D355–D360.

51 Ashburner, M., Ball, C.A. et al. (2000) Gene Ontology: tool for the unification of biology. The Gene Ontology Consortium. *Nat. Genet.*, **25**, 25–29.

52 Berardini, T., Li, D. et al. (2010) The Gene Ontology in 2010: Extensions and refinements. *Nucleic Acids Res.*, **38**, D331–D335.

53 Twigger, S.N., Shimoyama, M. et al. (2007) The Rat Genome Database, update 2007 – easing the path from disease to data and back again. *Nucleic Acids Res.*, **35**, D658–D662.

54 Storey, J.D. (2002) A direct approach to false discovery rates. *J.R. Statist. Soc.*, **64**, 479–498.

55 Royer, L.c., Reimann, M. et al. (2008) Unraveling protein networks with power graph analysis. *PLoS Comput. Biol.*, **4**, e1000108.

13
Network Screening: A New Method to Identify Active Networks from an Ensemble of Known Networks
Shigeru Saito and Katsuhisa Horimoto

13.1
Introduction

The global development of resources for "omics" data enables us to handle the thousands of activities of molecules simultaneously, providing a comprehensive overview of complicated biological systems. However, it is difficult to explain the systems only by each individual molecular function; it is natural to consider that these are due to interactions among molecules (i.e., a network). Therefore, elucidating networks responsible for the mechanisms of biological systems from measurements of molecules has been an important research area of systems biology. Accordingly, extensive studies have been undertaken in order to infer gene regulatory networks from gene expression data; several methods have been developed such as the Boolean network, the Bayesian network, graphical Gaussian modeling, and the mutual information base network [1–4]. Since inferred networks are expected to detect particular interactions regarding biological phenomena in theory, those methods have been widely used for practical problems [5–8]. Nevertheless, it is often difficult to interpret for biologists, because the reliability of inferred relationships in the networks is frequently uncertain. Two reasons can be listed: (i) the number of measured points is far less than the number of genes, and (ii) the biological and measurement noise strongly affect inference precision. Consequently, in order to obtain biologically meaningful results from the inferred interactions, careful validation by biological experimentation, requiring a lot of effort, is required.

Several methods to detect consistent networks responding to measured data from predefined known networks have been proposed [9, 10]. Known networks detected by those methods are easy to understand, since they are based on the accumulation of much biological knowledge. Therefore, recently, several databases of experimentally validated interactions have been developed extensively [11–13]. Such a vast amount of

Applied Statistics for Network Biology: Methods in Systems Biology, First Edition.
Edited by M. Dehmer, F. Emmert-Streib, A. Graber, and A. Salvador.
© 2011 Wiley-VCH Verlag GmbH & Co. KGaA. Published 2011 by Wiley-VCH Verlag GmbH & Co. KGaA.

collected interactions is related to various biological phenomena in various species, organs, and environments. Unfortunately, because of their size, those cannot be understood intuitively; it is necessary to focus on content regarding the research purpose, but this is not a trivial matter. An assumption that whole interactions in the database relate to a particular biological phenomena is unacceptable. Thus, developing a novel method for detecting specific networks responsible for biological phenomena from a vast amount of known networks is essential for a comprehensive understanding of biological machinery. Although the approach based on known networks is attractive, it has several pitfalls in practice. First, all interactions in the database should not be treated equally, because those are an ensemble of results discovered under different circumstances, by different researchers, and for different objects. Moreover, we cannot obtain new findings for structural information from identified known networks. In terms of these points, the known network approach and inference network approach are a tradeoff. One solution to this problem is the use of chromatin immunoprecipitation (ChIP) data. The ChIP technologies (e.g., ChIP-on-Chip, ChIP sequencing) allow us to measure the ability with which a particular protein binds to DNA [14, 15]. By using binding networks of a transcription factor based on ChIP data instead of already-known networks, we can evaluate a specific network relating to phenotype (this is neither knowledge-based nor statistical inference). However, no effective method has yet been developed using the combination of ChIP and gene expression data.

In this chapter, we first define a statistical method to evaluate the goodness-of-fit between a network and measured data. We then propose a concept, which we have named "network screening," that selects the specific network candidates that are consistent with the data measured under a particular condition from an ensemble of known networks. To demonstrate our method, we show an example using a toy model and artificially generated data. Furthermore, the possible application of our method to real data and network screening is shown by using the measurement data of *Escherichia coli*. Lastly, we show the effectiveness of the network screening for ChIP data by applying our method to transcription networks for Yamanaka factors based on ChIP data in induced pluripotent stem cells (iPSCs).

13.2
Methods

Here, we mainly describe a method to evaluate consistency of a given network structure and data according to our previous work [10]. In our method, a given network is described as a causal graph and the causal graph is parameterized on a graphical model; as a result, the log-likelihood of the graphical model is calculated. In order to evaluate the statistical significance for the log-likelihood, a new statistical measure, which we have named the graph consistency probability (*GCP*), is estimated on the log-likelihood distribution of randomly generated graphs. Figure 13.1 provides a summary overview of our method.

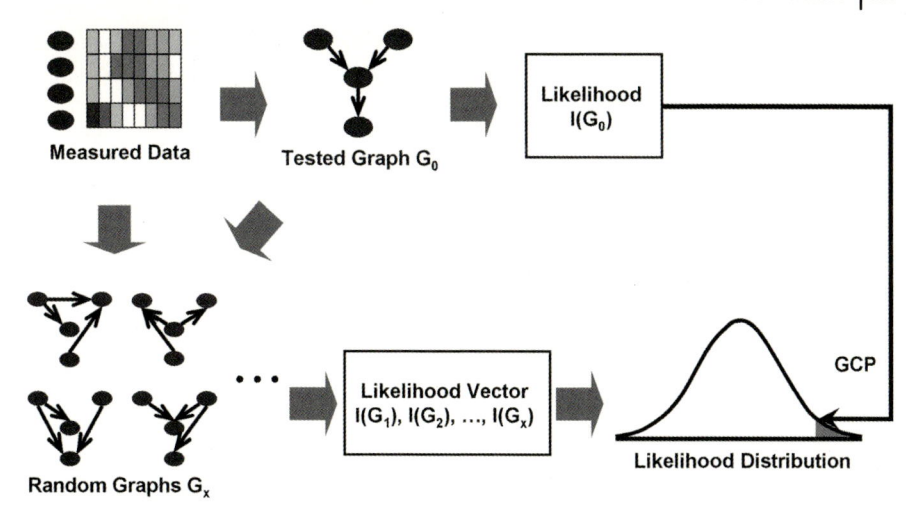

Figure 13.1 Overview of our method.

13.2.1
Causal Graph

A causal graph is known as a model that expresses causal relationships among variables [16].

Figure 13.2 shows a causal graph described as a DAG (directed acyclic graph), which denotes that measurements of variable X_3 depend on the condition of X_1 and X_2. As a graphical model corresponding the DAG, the probability density function of the graph G can be described by the product of the conditional probability density function with node X_i and its parent node $pa\{X_i\}$:

$$f(G) = f(X_1, \ldots, X_n) = \prod_{i=1}^{n_v} f(X_i | pa\{X_i\}) \tag{13.1}$$

where $pa\{X_i\}$ is the set of variables corresponding to the parents of X_i in graph G. On the assumption that the probability variable X_i is subjected to a multiple normal

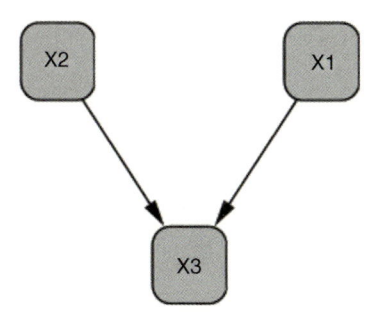

Figure 13.2 Causal graph.

distribution, named a "Gaussian network" (GN) [17], each conditional function in Equation 13.1 is obtained by linear regression of the constituent nodes (molecules) measured at m points as:

$$f(X_i|pa\{X_i\}) = \frac{1}{\sqrt{2\pi\sigma_i^2}} \exp\left[\frac{-1}{2\sigma_i^2}\sum_{k=1}^{m}\left(x_{ik} - \sum_{j=1}^{n_i}\beta_{ij}x_{jk}\right)^2\right] \tag{13.2}$$

where x_{ik} is the measured value of X_i at the kth point and n_i is the number of variables corresponding to the parents of X_i.

13.2.2
Quantification of Consistency Between a Model and Data

To quantify the goodness-of-fit of the model, the logarithm of the likelihood of Equation 13.1 is calculated for the measured data as:

$$l(G_0) = \ln\prod_{i=1}^{n_v} f(X_i|pa\{X_i\}) = -\frac{1}{2}\sum_{i=1}^{n_v}\sum_{j=1}^{n_i}\left\{\frac{1}{\sigma_i^2}\sum_{k=1}^{m}\left(x_{ik} - \sum_{j=1}^{n_i}\beta_{ij}x_{kj}\right)^2 + \ln(2\pi\sigma_i^2)\right\} \tag{13.3}$$

Thus, the GN allows us to quantify a given network into the corresponding numerical value from the measured data, according to the network structure. A relatively large log-likelihood indicates that the model is acceptable in the context of the GN. Note that calculation of the likelihood itself requires no assumptions on the relationships between variables. Indeed, the likelihood can be calculated in the case of nonlinear regressions, such as arbitrary spline regression.

13.2.3
Statistical Methods for Evaluation

Now that we can obtain a log-likelihood as a consistency measure, how can we evaluate whether the log-likelihood is statistically significantly high? In this section, we describe statistical methods to evaluate significance for the log-likelihood due to graph structure. Generally, the distribution for the log-likelihood of a graphical model that depends on graph structure is unknown. Therefore, we propose an affinity metric of a graph and data, GCP [10], which is calculated by sampling techniques. In the following sections, we will describe two methods: generalized extreme value (GEV) distribution and nonparametric statistics.

13.2.3.1 GEV Distribution
We can estimate the probability of $l(G_0)$ in Equation 13.3 by using the GEV distribution [18] on the assumption that $l(G_0)$ is extremely high against random structure graphs. First, the log-likelihoods of an ensemble of n random networks generated according to the GN are calculated and then the maximum log-likelihood is selected from them. The above procedure is iterated m times:

$$l^1_{max} = \text{Max}\{l(G^1_1), l(G^1_2), \ldots, l(G^1_n)\}$$
$$l^2_{max} = \text{Max}\{l(G^2_1), l(G^2_2), \ldots, l(G^2_n)\}$$
$$\vdots \tag{13.4}$$
$$l^m_{max} = \text{Max}\{l(G^m_1), l(G^m_2), \ldots, l(G^m_n)\}$$

The distribution of the maximum values by m iterations is expected asymptotically to be a GEV distribution:

$$G(l_{max}) = \exp\left\{-\left[1 + \xi\left(\frac{l_{max} - \mu}{\sigma}\right)\right]^{-1/\xi}\right\} \tag{13.5}$$

defined on the set:

$$\left\{l_{max} : 1 + \xi\left(\frac{l_{max} - \mu}{\sigma}\right) > 0\right\}$$

where the parameters satisfy $-\infty < \mu < \infty$, $\sigma > 0$, and $-\infty < \xi < \infty$. The model has three parameters: μ, σ, and ξ are a location parameter, a scale parameter, and a shape parameter, respectively. Maximization of the log-likelihood of Equation 13.5 with respect to the parameter vector (μ, σ, ξ) leads to the maximum likelihood estimate for any given dataset, using standard numerical optimization. We used the extRemes package [19] to fit the data to the GEV distribution. Note that the standard likelihood ratio test cannot be applied straightforwardly to the present case. This is because the density function of the population and the degrees of freedom in the likelihood ratio test are unclear when maximizing the likelihoods of the randomly generated graphs. In the present method, the GEV distribution of the maximum values of log-likelihoods in the blocks of generated graphs is adopted analogically, instead of the maximum likelihood in the likelihood ratio test. The utilization of the GEV distribution requires model fitting to the data, but allows us to set the significance probability arbitrarily, as usual in statistical tests. If the goodness-of-fit of the maximum values from the generated graphs is ascertained, then the occurrence probability of a given graph, named the *GCP*, can be directly estimated by corresponding the $l(G_0)$ in Equation 13.3 to the probability density function of GEV obtained in Equation 13.5:

$$P(l(G_0)) = \int_{l(G_0)}^{+\infty} G(l_{max}) \, dl_{max} \tag{13.6}$$

13.2.3.2 Nonparametric Statistic

There are two pitfalls in the calculation of the *GCP* by the GEV method. One problem relates to an empirical parameter – the parameter of sampling period n in the GEV method is only determined empirically. The other is the assumption about the distribution form of the log-likelihood. In the sampling step of randomized graphs, it is possible to skew a shape of the distribution due to topological restrictions of the graphs. Therefore, we propose a method that does not require empirical parameters (i.e., only sampling times required) and does not depend on the distribution form, as follows:

$$GCP = \frac{N_s}{N_r} \tag{13.7}$$

where N_r is total number of randomly generated graphs and N_s is the number of random graphs such that the log-likelihood is greater than that of the tested graph. Here, let $GCP = N_r^{-1}$, if $N_s = 0$.

13.2.4
Simulation Study

In this section, we perform a simulation study for the demonstration of our method. First, we generate the numerical data for 10 nodes with 50 sampling dimensions by using the following structural equations that correspond to the causal graph in Figure 13.3:

$$\begin{cases} X_{1l} = N(0, \sigma) \\ X_{2l} = N(0, \sigma) \\ X_{3l} = N(0, \sigma) \\ X_{4l} = \alpha_{1,4} X_{1l} + N(0, \sigma) \\ X_{5l} = \alpha_{1,5} X_{1l} + N(0, \sigma) \\ X_{6l} = \alpha_{2,6} X_{2l} + \alpha_{3,6} X_{3l} + N(0, \sigma) \\ X_{7l} = \alpha_{4,7} X_{4l} + N(0, \sigma) \\ X_{8l} = \alpha_{4,8} X_{4l} + N(0, \sigma) \\ X_{9l} = \alpha_{6,9} X_{6l} + \alpha_{7,9} X_{7l} + N(0, \sigma) \\ X_{10l} = \alpha_{9,10} X_{9l} + N(0, \sigma) \end{cases} \tag{13.8}$$

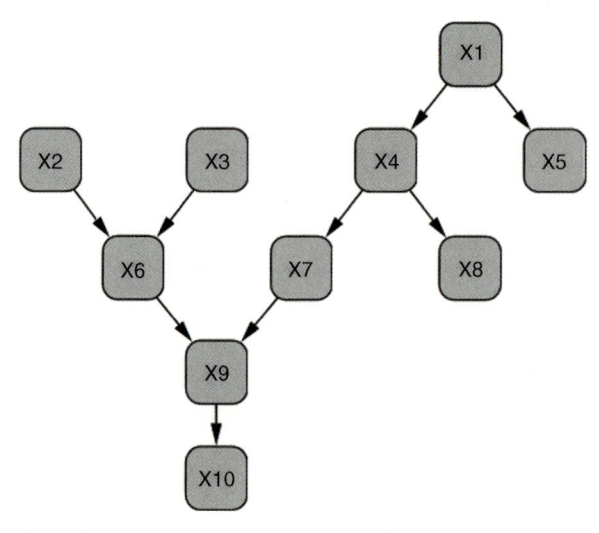

Figure 13.3 Toy model for simulation study.

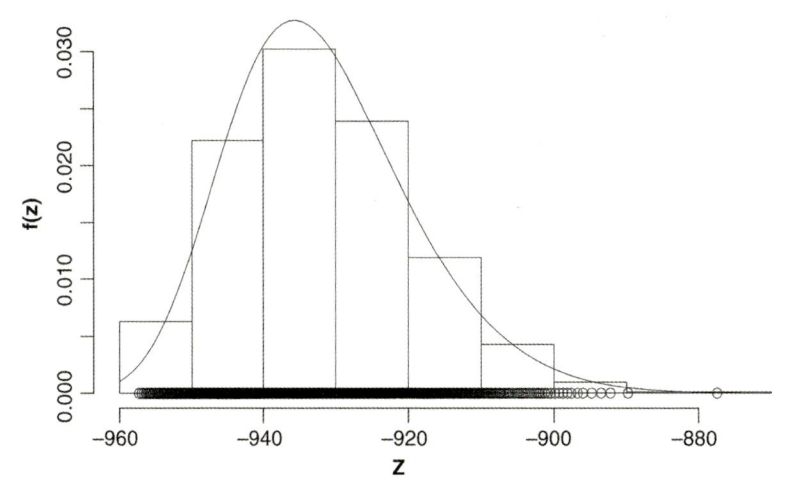

Figure 13.4 Estimated generalized extreme distribution of log-likelihood.

where $N(0, \sigma)$ means a normal distribution with a zero mean and a standard deviation of σ, and $\alpha_{i,j}$ is a path coefficient relating variables i and j. Here, we set $\sigma = 0.1$ and $\alpha_{i,j} = 0.5$.

Figure 13.4 shows the estimated generalized extreme distribution of log-likelihood, when sampling period $n = 50$ and sampling time $m = 1000$. The parameters for the distribution, μ, σ, and ξ, were estimated as -937.19, 11.33 and -0.13, respectively. Since, the log-likelihood of the tested graph is -862.39 and the resulting the GCP is 6.87e-7, the tested model is quite significant.

Figure 13.5 shows the log-likelihood distribution when $N_r = 50\,000$ with the nonparametric method. The log-likelihood of the tested graph is -862.39 and the GCP is 2.0e-5, which shows statistically significance. Thus, the nonparametric method also works successfully. Here, note that the proposed methods that have been demonstrated work robustly, and do not depend on the network topology, number of data points, noise level, and noise distribution [10]. In the following analysis, we use the nonparametric method to estimate the GCP because of its simplicity.

13.2.5
Network Screening

In the previous section we showed that the GCP reflects the consistency of a pair of graph and measured data by a simulation study. However, we frequently encounter cases where a tested graph structure cannot be determined or assumed. For example, in general, when a cell is stimulated by a particular signal, it is not obvious which network response relates to the signal in a vast amount of regulatory networks in the cell. Therefore, a technique to detect active networks in many predefined functional networks is important. We define this concept as "network screening" (Figure 13.6). Fortunately, large informative databases for metabolic pathways and signal trans-

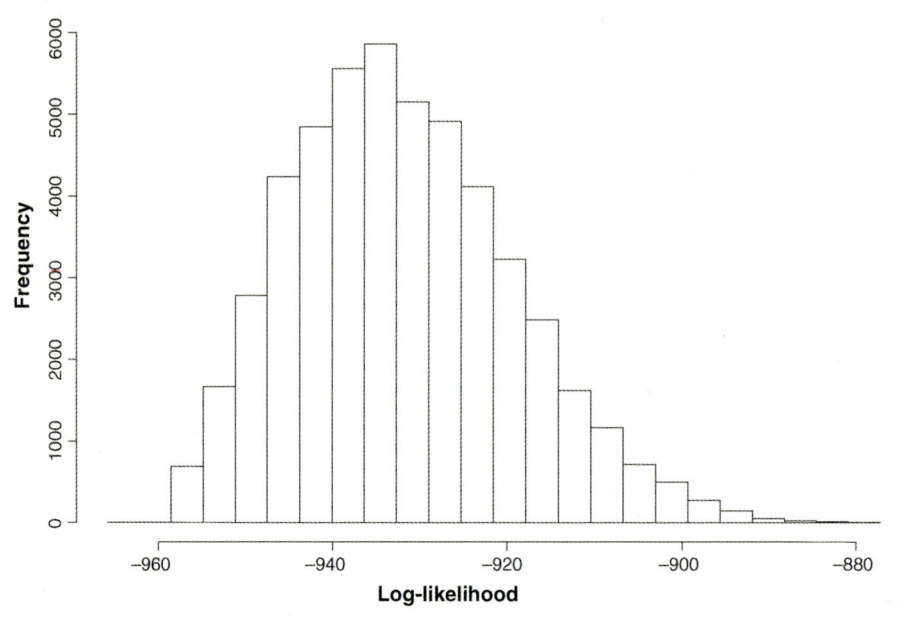

Figure 13.5 Null log-likelihood distribution in randomized graphs.

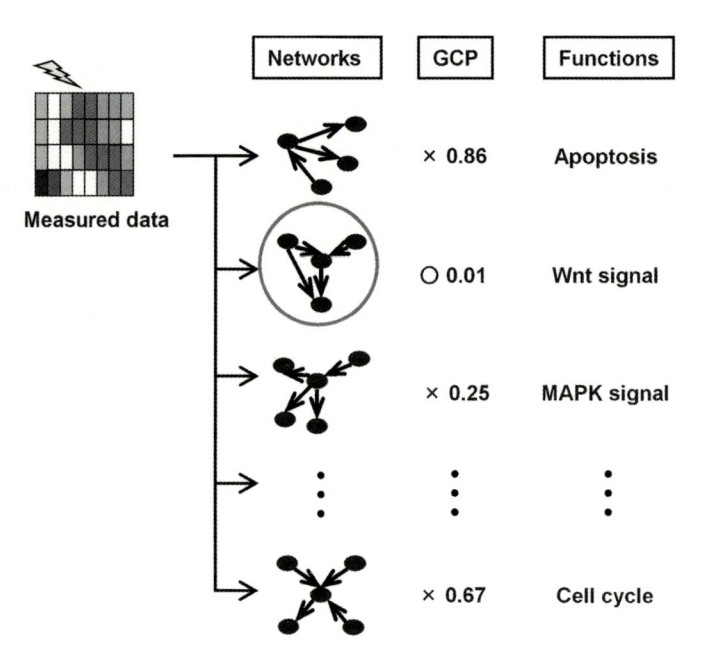

Figure 13.6 Schematic diagram of network screening.

duction networks are becoming increasingly available [11–13]. Thus, we can execute network screening by combination of the *GCP* and these databases.

13.3
Example Applications

In this section, we show several applications of our method. First, we examined the performance of the present method with two sets of actual networks in *E. coli* and the corresponding actual measured data. One set is a regulatory network for the SOS response system with the promoter activity of the constituent genes measured by Green Fluorescent Protein (GFP) [20] and the other is 30 networks classified by gene functions, with the gene expression under anaerobic conditions measured by microarray analysis [21]. The former examination is a verification of the *GCP* method for an actual network and the latter is a demonstration of network screening as a high-throughput search of network candidates, which is consistent with the data measured under particular conditions. Lastly, we show an example of the application of network screening to ChIP data that consists of four Yamanaka factor networks in several iPSCs.

13.3.1
Evaluation of the *E. coli* SOS Network

The gene network in the SOS DNA repair system is schematically shown in Figure 13.7 and is a well-characterized transcriptional network in *E. coli* [22, 23].

One of the SOS genes, *recA*, acts as a sensor of DNA damage and a master repressor. *lexA*. binds sites in the promoter regions of these operons. The corresponding data for the constituent genes in the network are measured with real-time monitoring by means of low-copy reporter plasmids, in which a promoter controls GFP [20]. By using our method, the *GCP* of the SOS network with the corresponding measured data was estimated as 0.049 and the network structure was estimated to be

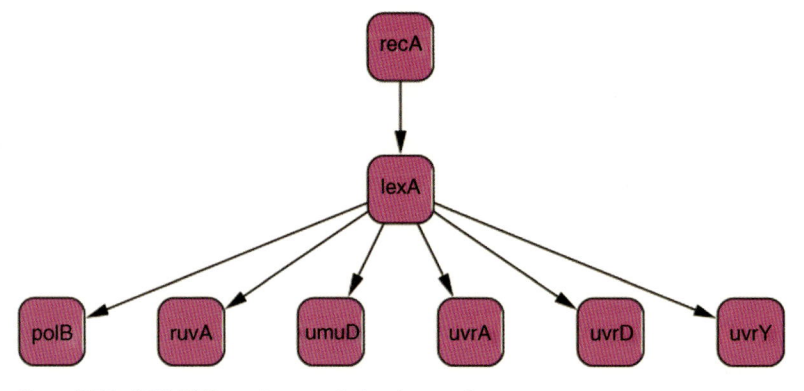

Figure 13.7 SOS DNA repair transcriptional network.

consistent with the data measured from the examined network. However, the GCP is slightly large in comparison with the GCPs in the simulation studies in Section 13.2.4. This is partly because the cyclic relationship of *recA* is neglected in the examined network. At any rate, the performance of the present method was verified by a well-known actual network, with a size similar to that in the simulation study and with the corresponding data measured by an experimental study.

13.3.2
Network Screening for *E. coli* Networks Under Anaerobic Conditions

Subsequently, we applied the network screening to practical data. We first classified transcription factors and regulated genes compiled in EcoCyc [24], according to the classification scheme of gene functions (http://biocyc.org/ECOLI/class-tree?object=Genes). Using the functional gene sets of the transcription factors and regulated genes, we next reconstructed the networks, so that the network would be a "connected graph" (i.e., it contains a directed path from v to u for every pair of nodes u,v). Thus, we obtained 130 regulatory networks characterized by biological functions. The nodes for which expression profiles were not found in expression data were excluded from the 130 networks. Since the network screening algorithm requires a certain number of nodes in the examined graph due to the variety of structures for randomize graphs, we removed the small networks with less than eight edges; as a result, 30 networks were used for evaluation. The consistency of each of the 30 networks was estimated with expression profiles measured under 22 different anaerobic conditions (GSE1107) [21] cited in the National Center for Biotechnology Information's Gene Expression Omnibus (GEO) [25]. The expression profiles were standardized by the average and standard deviation in each condition (i.e., z-score normalization), as preprocessing of the measured data.

Table 13.1 shows the network screening results with $N_r = 1000$; the analyzed networks and the corresponding GCPs and estimated false discovery rate (FDR) q-values of the 30 networks. Here, we set significance level to FDR $q < 0.25$ (bold lines).

The result show that a variety of different size networks, "Carbon compounds," "SOS response," "DNA repair," "Anaerobic respiration," and "Cytoplasm," were significantly identified. The most significant network for "Carbon compounds" was identical to the maltose regulon [26] (Figure 13.8).

This is an unexpected result, because glucose utilization as a carbon source for respiration is speculated from culture conditions. This suggests a possibility of pH response, indicating activation of the maltose regulon depending on the pH in the culture medium [27]. Three significant networks, "SOS response," "DNA repair," and "Cytoplasm," are based on *lexA* regulon; thus, several regulated genes are overlapped (blue nodes, *recA*, *umuC–D*, *uvrA–D* in Figure 13.9).

Indeed, it was reported that the SOS response network is activated under anaerobic conditions [28]; these *lexA* networks have the possibility of playing some role in respiration in anaerobic circumstances. In Figure 13.10, "Anaerobic respiration," as

Table 13.1 Network screening result for *E. coli* under anaerobic conditions.

ID	Description	Node	Edge	GCP	FDR q
C9449	Carbon compounds	10	9	3.0e-03	8.0e-02
C9337	SOS response	11	10	7.0e-3	8.0e-02
C9354	DNA repair	11	10	8.0e-03	8.0e-02
C9490	Anaerobic respiration	89	161	1.1e-02	8.3e-02
C9376	Cytoplasm	12	11	2.7e-02	1.6e-01
C9340	Flagella	9	8	5.5e-02	2.8e-01
C9474	Nucleotide and nucleotide conversion	11	15	6.7e-02	2.9e-01
C9372	Transcription related	91	143	1.5e-01	5.4e-01
C9449	Carbon compounds	10	11	1.6e-01	5.4e-01
C9362	Nucleoproteins, basic proteins	9	8	1.9e-01	5.6e-01
C9331	Motility, chemotaxis, energytaxis	9	8	2.7e-01	7.3e-01
C9333	Detoxification	6	8	3.0e-01	7.5e-01
C9523	Activator	58	92	5.3e-01	1.0e + 00
C9449	Carbon compounds	8	7	5.9e-01	1.0e + 00
C9528	Repressor	52	77	6.3e-01	1.0e + 00
C9426	Colanic acid (M-antigen)	6	9	6.4e-01	1.0e + 00
C9448	Amino acids	6	9	6.6e-01	1.0e + 00
C9449	Carbon compounds	9	8	6.9e-01	1.0e + 00
C9462	Amino acids, formyl-THF biosynthesis	7	10	6.9e-01	1.0e + 00
C9448	Amino acids, formyl-THF biosynthesis	7	10	7.0e-01	1.0e + 00
C9401	Tryptophan	9	8	7.8e-01	1.0e + 00
C9376	Cytoplasm	10	9	8.2e-01	1.0e + 00
C9509	Operon	6	9	8.7e-01	1.0e + 00
C9383	Arginine	11	10	1.0e + 00	1.0e + 00
C9393	Isoleucine/valine	13	12	1.0e + 00	1.0e + 00
C9394	Leucine	14	17	1.0e + 00	1.0e + 00
C9420	Purine biosynthesis	13	12	1.0e + 00	1.0e + 00
C9449	Carbon compounds	6	9	1.0e + 00	1.0e + 00
C9493	Fermentation	11	10	1.0e + 00	1.0e + 00
C9504	Phosphorous metabolism	23	22	1.0e + 00	1.0e + 00

THF = tetrahydrofolate.

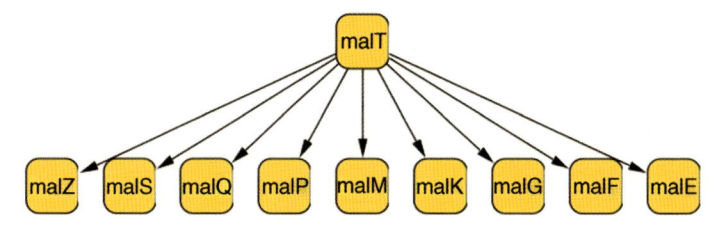

Figure 13.8 Significant network: "Carbon compounds."

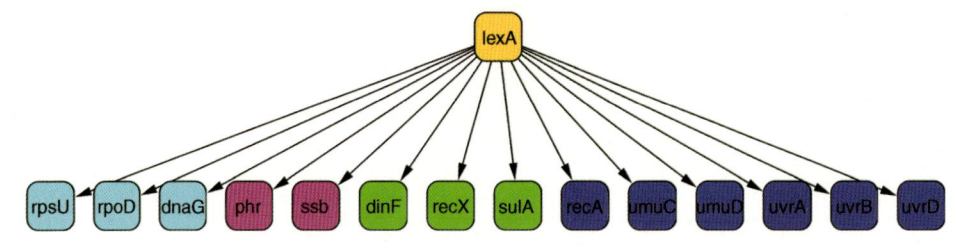

Figure 13.9 Merged network of "SOS response," "DNA repair," and "Cytoplasm."

the name suggests, is related to respiration under anaerobic conditions; even though the network consists of a lot of genes, our method detected its response. These results show that network screening is applicable for practical analysis with high confidence.

13.3.3
Network Screening for ChIP networks on iPSCs

Here, we show an example of the application of network screening using ChIP data, when transcription factors that relate to a specific biological phenomena are already known (or assumable). In recent years, iPSCs have has been studied extensively as a

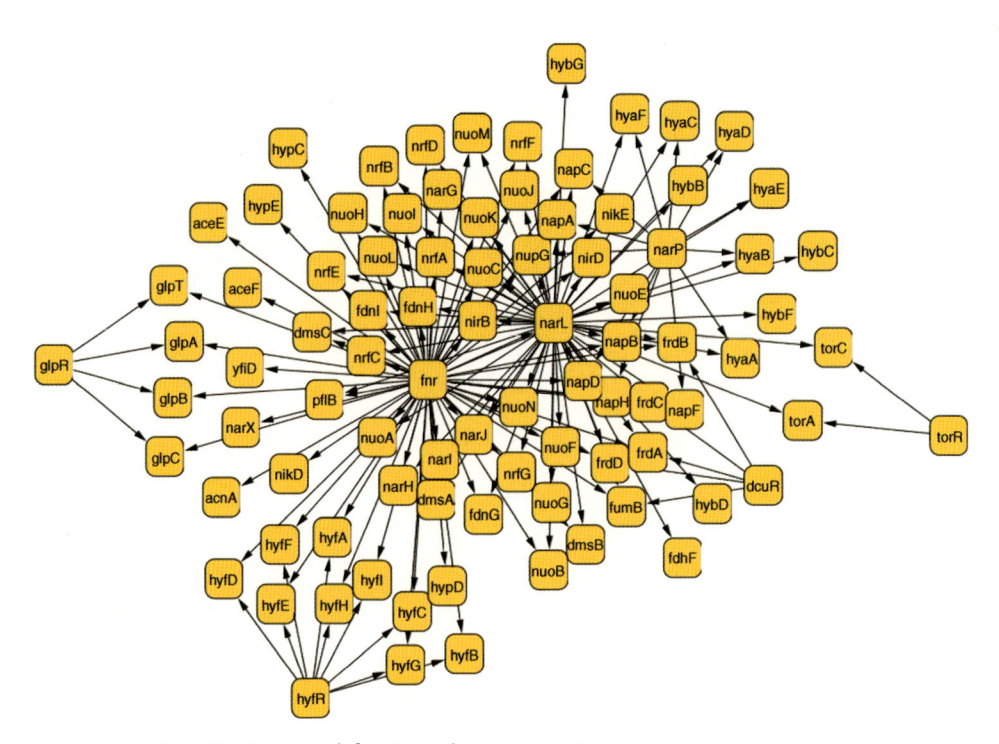

Figure 13.10 Network for "Anaerobic respiration."

differentiation and development model in terms of pluripotency and self-renewal, and as a resource for drug development and tissue engineering for regenerative medicine [29]. Genetic reprogramming to a pluripotent state of somatic cells was first achieved by ectopic expression of four transcription factors, also called Yamanaka factors (SOX2, OCT4, KLF4, and c-MYC) using retrovirus. One of the major challenges for iPSC research is to elucidate the mechanisms for pluripotency and self-renewal by the factors. In order to elucidate the mechanisms, ChIP data for the factors and several other master transcription factors have been measured; as a result, it was shown that the factors regulated various genes in mouse and human iPSCs [30–37]. However, the functional characteristics of those downstream genes remain largely unclear. Therefore, we attempted to elucidate those by network screening using previously reported ChIP data [36] and global gene expression profiles of four types of somatic cells (amniotic mesodermal, placental artery endothelial, uterine endometrium, and MRC-5 cells) and the corresponding iPSCs [38]. In this study, first, we identified iPSC-specific gene expression (expression signature) by differential analysis using t-test and fold-change with microarray data of 51 samples (12 somatic cells and 39 iPSCs) [38]; as a result, 2502 genes were differentially expressed. Secondly, we constructed networks of the four factors based on the ChIP data [36] and collected subnetworks such that regulated genes share the same functionality by means of the gene set database (MSigDB c2 curated gene set [39]). To obtain consistent networks, we applied the network screening to the dataset with $N_r = 1000$ and significance level FDR $< 5\%$. Finally, we evaluated enrichment of iPSC expression signature genes into each of the significant networks by a hyper-geometric distribution using [40]:

$$p = 1 - \sum_{i=0}^{k-1} \frac{\binom{M}{i}\binom{N-M}{n-i}}{\binom{N}{i}} \quad (13.9)$$

where N is the total number of genes in the whole networks, n is the number of iPSC specific genes, M is the number of genes in the significant network, and k is the number of genes existing in the iPSC-specific genes and the significant genes of the network screening. As a result, 28 networks were significantly detected (Figure 13.11).

Each network was classified manually into signal transduction pathways (yellow), cell–cell communications (blue), glycan biosynthesis pathways (red), cancer-related pathways (green), and others (black), in terms of gene annotations. Figure 13.11 shows that a combined network based on the 28 significant networks suggests that the four factors regulate directly many genes having a wide range of functionality.

Every detected signaling pathway, mitogen-activated protein kinase [41], p53 [42], WNT [43], transforming growth factor-β [44], Hedgehog, and JAK–STAT [37], has previously reported relationships for pluripotency and self-renewal of iPSCs. Detection of networks of cell–cell communication is also reasonable in terms of the importance of signaling among cells for cell developmental processes [45]. Interestingly, several networks of glycan structure biosynthesis were identified. This is because a change of

Pathway
HSA04010-MAPK_SIGNALING_PATHWAY
HSA04115-P53_SIGNALING_PATHWAY
HSA04310-WNT_SIGNALING_PATHWAY
HSA04340-HEDGEHOG_SIGNALING_PATHWAY
HSA04350-TGF_BETA_SIGNALING_PATHWAY
HSA04630-JAK_STAT_SIGNALING_PATHWAY
HSA01430-CELL_COMMUNICATION
HSA04520-ADHERENS_JUNCTION
HSA04530-TIGHT_JUNCTION
HSA04540-GAP_JUNCTION
HSA04810-REGULATION_OF_ACTIN_CYTOSKELETON
HSA01030-GLYCAN_STRUCTURES_BIOSYNTHESIS_1
HSA01031-GLYCAN_STRUCTURES_BIOSYNTHESIS_2
HSA04916-MELANOGENESIS
HSA05210-COLORECTAL_CANCER
HSA05213-ENDOMETRIAL_CANCER
HSA05215-PROSTATE_CANCER
HSA05216-THYROID_CANCER
HSA05217-BASAL_CELL_CARCINOMA
HSA05218-MELANOMA
APOPTOSIS
APOPTOSIS_GENMAPP
CALCIUM_REGULATION_IN_CARDIAC_CELLS
HSA04060-CYTOKINE_CYTOKINE_RECEPTOR_INTERACTION
HSA04110-CELL_CYCLE
HSA04670-LEUKOCYTE_TRANSENDOTHELIAL_MIGRATION
HSA05120-EPITHELIAL_CELL_SIGNALING_IN_HELICOBACTER_PYLORI_INFECTION
SMOOTH_MUSCLE_CONTRACTION

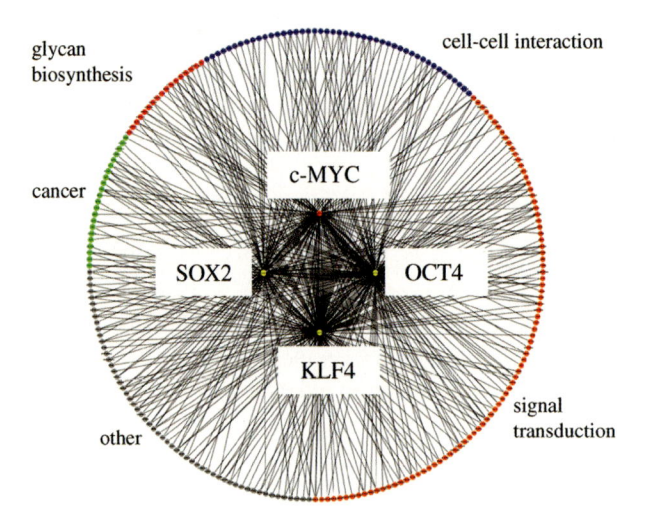

Figure 13.11 Activate networks in iPSCs.

surface glycan composition in iPSCs, which is linked to the corresponding level of synthesis and transfer enzyme expression, may affect cell–cell communications and stimulation for signal transduction pathways. Actually, the relation between the WNT pathway and heparan sulfate, which is one of the glycan structures, has reported in human embryonic stem cells [46]. Furthermore, detection of cancer-related networks is important from the viewpoint of carcinogenic risk; these networks may be a cue to the development of a method to safely establish iPSCs [47].

13.4
Discussion

We proposed a consistency measure, named the GCP, which indicates the goodness-of-fit between a given network and measured data. By using artificially generated data with a causal network, we show that the our method functioned successfully as we had expected. In addition, we applied the GCP and the network screening to *E. coli* data, demonstrating that our method is applicable even in real situations. Lastly, we used the human ChIP network for the network screening and we successfully identified networks having reasonable functionalities. These results suggest that our approach works not only for simple networks in prokaryotes, but also for complex regulatory networks in eukaryotes. Thus, the present method shows the possibility of bridging between static networks and the corresponding measurements, and will shed light on the network structure variations in terms of the changes in molecular interaction mechanisms that occur in response to the environment in a living cell. According to progress in measurement technologies, when ChIP data for a lot of transcription factors can be measured simultaneously, it is expected that network screening enable us to identify transcriptional regulatory networks controlling biological phenomena using the ChIP data and global gene expression profiles.

From the standpoint of methodology, although we used the Gaussian network as a probabilistic model in a present study, it does not depend on the type of kernel for the model – we can use a nonlinear kernel (e.g., β-spline) or categorical model (e.g., Bayesian network with categorical variables) according to data type. Note that several statistical methods to evaluate consistency between causal graphs and observed data have been proposed: SEM (structural equation model) [48] and d-sep [49]. Those methods enable us to check whether observed data come from a causal graph on the assumption of a certain probabilistic model in a strict sense (i.e., a data-centered approach). In contrast to that approach, our method focuses on structural specificity, indicating how frequently a pair of graph and data is accidentally observed (i.e., a structure-centered approach). Indeed, the SEM and the d-sep approach failed to identify event specific networks in actual analysis [10]. Of course, our approach also has some pitfalls. Although we use the DAG as causal model, it is well known that there are many feedback and undirected relations in actual biological systems. The limitation of our method due to structural restrictions may overlook important discoveries. Thus, what remains to be done is the development of an improved method applicable for various types of network structures.

References

1 Akutsu, T., Miyano, S., and Kuhara, S. (2000) Algorithms for inferring qualitative models of biological networks. *Pac. Symp. Biocomput.*, 293–304.
2 Friedman, N., Linial, M., Nachman, I., and Pe'er, D. (2000) Using Bayesian networks to analyze expression data. *J. Comput. Biol.*, 7, 601–620.
3 Toh, H. and Horimoto, K. (2002) Inference of a genetic network by a combined approach of cluster analysis and graphical

Gaussian modeling. *Bioinformatics*, **18**, 287–297.

4 Margolin, A.A., Nemenman, I., Basso, K., Wiggins, C., Stolovitzky, G., Dalla Favera, R., and Califano, A. (2006) ARACNE: an algorithm for the reconstruction of gene regulatory networks in a mammalian cellular context. *BMC Bioinformatics*, **7** (Suppl. 1), S7.

5 Genoud, T., Trevino Santa Cruz, M.B., and Métraux, J.P. (2001) Numeric simulation of plant signaling networks. *Plant. Physiol.*, **126**, 1430–1437.

6 Savoie, C.J., Aburatani, S., Watanabe, S., Eguchi, Y., Muta, S., Imoto, S., Miyano, S., Kuhara, S., and Tashiro, K. (2003) Use of gene networks from full genome microarray libraries to identify functionally relevant drug-affected genes and gene regulation cascades. *DNA Res.*, **10**, 19–25.

7 Ma, S., Gong, Q., and Bohnert, H.J. (2007) An *Arabidopsis* gene network based on the graphical Gaussian model. *Genome Res.*, **17**, 1614–1625.

8 Carro, M.S., Lim, W.K., Alvarez, M.J., Bollo, R.J., Zhao, X., Snyder, E.Y., Sulman, E.P., Anne, S.L., Doetsch, F., Colman, H., Lasorella, A., Aldape, K., Califano, A., and Iavarone, A. (2010) The transcriptional network for mesenchymal transformation of brain tumours. *Nature*, **463**, 318–325.

9 Herrgard, M.J., Covert, M.W., and Palsson, B.O. (2003) Reconciling gene expression data with known genome-scale regulatory network structures. *Genome Res.*, **13**, 2423–2434.

10 Saito, S., Aburatani, S., and Horimoto, K. (2008) Network evaluation from the consistency of the graph structure with the measured data. *BMC Syst. Biol.*, **2**, 84.

11 Kanehisa, M., and Goto, S. (2000) KEGG: Kyoto Encyclopedia of Genes and Genomes. *Nucleic Acids Res.*, **28**, 27–30.

12 Joshi-Tope, G., Gillespie, M., Vastrik, I., D'Eustachio, P., Schmidt, E., de Bono, B., Jassal, B., Gopinath, G.R., Wu, G.R., Matthews, L., Lewis, S., Birney, E., and Stein, L. (2005) Reactome: a knowledgebase of biological pathways. *Nucleic Acids Res.*, **33**, D428–D432.

13 Dahlquist, K.D., Salomonis, N., Vranizan, K., Lawlor, S.C., and Conklin, B.R. (2002) GenMAPP, a new tool for viewing and analyzing microarray data on biological pathways. *Nat. Genet.*, **31**, 19–20.

14 Lee, T.I., Rinaldi, N.J., Robert, F., Odom, D.T., Bar-Joseph, Z., Gerber, G.K., Hannett, N.M., Harbison, C.T., Thompson, C.M., Simon, I., Zeitlinger, J., Jennings, E.G., Murray, H.L., Gordon, D.B., Ren, B., Wyrick, J.J., Tagne, J.B., Volkert, T.L., Fraenkel, E., Gifford, D.K., and Young, R.A. (2002) Transcriptional regulatory networks in *Saccharomyces cerevisiae*. *Science*, **298**, 799–804.

15 Johnson, D.S., Mortazavi, A., Myers, R.M., and Wold, B. (2007) Genome-wide mapping of *in vivo* protein–DNA interactions. *Science*, **316**, 1497–1502.

16 Pearl, J. (1988) *Probabilistic Reasoning in Intelligent Systems*, Morgan Kaufman, San Mateo, CA.

17 Whittaker, J. (1990) *Graphical Models in Applied Multivariate Statistics*, John Wiley & Sons, Inc., New York.

18 Coles, S. (2001) *An Introduction to Statistical Modeling of Extreme Values*, Springer, London.

19 Gilleland, E., and Katz, R.W. (2006) Analyzing seasonal to interannual extreme weather and climate variability with the extremes toolkit (extRemes). 18th Conference on Climate Variability and Change/86th American Meteorological Society Annual Meeting, Atlanta, GA.

20 Ronen, M., Rosenberg, R., Shraiman, B.I., and Alon, U. (2002) Assigning numbers to the arrows: parameterizing a gene regulation network by using accurate expression kinetics. *Proc. Natl. Acad. Sci. USA*, **99**, 10555–10560.

21 Covert, M.W., Knight, E.M., Reed, J.L., Herrgard, M.J., and Palsson, B.O. (2004) Integrating high-throughput and computational data elucidates bacterial networks. *Nature*, **429**, 92–96.

22 Kenyon, C.J., and Walker, G.C. (1980) DNA-damaging agents stimulate gene expression at specific loci in *Escherichia*

coli. Proc. Natl. Acad. Sci. USA, **77**, 2819–2823.

23 Little, J.W. and Mount, D.W. (1982) The SOS regulatory system of *Escherichia coli*. *Cell*, **29**, 11–22.

24 Karp, P.D., Keseler, I.M., Shearer, A., Latendresse, M., Krummenacker, M., Paley, S.M., Paulsen, I., Collado-Vides, J., Gama-Castro, S., Peralta-Gil, M., Santos-Zavaleta, A., Peñaloza-Spinola, M.I., Bonavides-Martinez, C., and Ingraham, J. (2007) Multidimensional annotation of the *Escherichia coli* K-12 genome. *Nucleic Acids Res.*, **35**, 7577–7590.

25 Barrett, T., Troup, D.B., Wilhite, S.E., Ledoux, P., Rudnev, D., Evangelista, C., Kim, I.F., Soboleva, A., Tomashevsky, M., and Edgar, R. (2007) NCBI GEO: mining tens of millions of expression profiles – database and tools update. *Nucleic Acids Res.*, **35**, D760–D765.

26 Boos, W., and Shuman, H. (1998) Maltose/maltodextrin system of *Escherichia coli*: transport, metabolism, and regulation. *Microbiol. Mol. Biol. Rev.*, **62**, 204–229.

27 Alonzo, S., Heyde, M., Laloi, P., and Portalier, R. (1998) Analysis of the effect exerted by extracellular pH on the maltose regulon in *Escherichia coli* K-12. *Microbiology*, **144**, 3317–3325.

28 Eraso, J.M., and Weinstock, G.M. (1992) Anaerobic control of colicin E1 production. *J. Bacteriol.*, **174**, 5101–5109.

29 Takahashi, K., and Yamanaka, S. (2006) Induction of pluripotent stem cells from mouse embryonic and adult fibroblast cultures by defined factors. *Cell*, **126**, 663–676.

30 Boyer, L.A., Lee, T.I., Cole, M.F., Johnstone, S.E., Levine, S.S., Zucker, J.P., Guenther, M.G., Kumar, R.M., Murray, H.L., Jenner, R.G., Gifford, D.K., Melton, D.A., Jaenisch, R., and Young, R.A. (2005) Core transcriptional regulatory circuitry in human embryonic stem cells. *Cell*, **122**, 947–956.

31 Loh, Y.H., Wu, Q., Chew, J.L., Vega, V.B., Zhang, W., Chen, X., Bourque, G., George, J., Leong, B., Liu, J., Wong, K.Y., Sung, K.W., Lee, C.W., Zhao, X.D., Chiu, K.P., Lipovich, L., Kuznetsov, V.A., Robson, P., Stanton, L.W., Wei, C.L., Ruan, Y., Lim, B., and Ng, H.H. (2006) The Oct4 and Nanog transcription network regulates pluripotency in mouse embryonic stem cells. *Nat. Genet.*, **38**, 431–440.

32 Chen, X., Xu, H., Yuan, P., Fang, F., Huss, M., Vega, V.B., Wong, E., Orlov, Y.L., Zhang, W., Jiang, J., Loh, Y.H., Yeo, H.C., Yeo, Z.X., Narang, V., Govindarajan, K.R., Leong, B., Shahab, A., Ruan, Y., Bourque, G., Sung, W.K., Clarke, N.D., Wei, C.L., and Ng, H.H. (2008) Integration of external signaling pathways with the core transcriptional network in embryonic stem cells. *Cell*, **133**, 1106–1117.

33 Jiang, J., Chan, Y.S., Loh, Y.H., Cai, J., Tong, G.Q., Lim, C.A., Robson, P., Zhong, S., and Ng, H.H. (2008) A core Klf circuitry regulates self-renewal of embryonic stem cells. *Nat. Cell Biol.*, **10**, 353–360.

34 Kim, J., Chu, J., Shen, X., Wang, J., and Orkin, S.H. (2008) An extended transcriptional network for pluripotency of embryonic stem cells. *Cell*, **132**, 1049–1061.

35 Liu, X., Huang, J., Chen, T., Wang, Y., Xin, S., Li, J., Pei, G., and Kang, J. (2008) Yamanaka factors critically regulate the developmental signaling network in mouse embryonic stem cells. *Cell Res.*, **18**, 1177–1189.

36 Sridharan, R., Tchieu, J., Mason, M.J., Yachechko, R., Kuoy, E., Horvath, S., Zhou, Q., and Plath, K. (2009) Role of the murine reprogramming factors in the induction of pluripotency. *Cell*, **136**, 364–377.

37 Huang, J., Chen, T., Liu, X., Jiang, J., Li, J., Li, D., Liu, X.S., Li, W., Kang, J., and Pei, G. (2009) More synergetic cooperation of Yamanaka factors in induced pluripotent stem cells than in embryonic stem cells. *Cell Res.*, **19**, 1127–1138.

38 Saito, S., Onuma, Y., Ito, Y., Tateno, H., Toyoda, M., Akutsu, H., Nishino, K., Chikazawa, E., Fukawatase, Y., Miyagawa, Y., Okita, H., Kiyokawa, N., Shimma, Y., Umezawa, A., Hirabayashi, J., Horimoto, K., and Asashima, M. (2010) Potential linkages between the inner and outer cellular states of human

induced pluripotent stem cells. *BMC Syst. Biol.*, in press.

39 Subramanian, A., Tamayo, P., Mootha, V.K., Mukherjee, S., Ebert, B.L., Gillette, M.A., Paulovich, A., Pomeroy, S.L., Golub, T.R., Lander, E.S., and Mesirov, J.P. (2005) Gene set enrichment analysis: a knowledge-based approach for interpreting genome-wide expression profiles. *Proc. Natl. Acad. Sci. USA*, **102**, 15545–15550.

40 Boyle, E.I., Weng, S., Gollub, J., Jin, H., Botstein, D., Cherry, J.M., and Sherlock, G. (2004) GO: TermFinder – open source software for accessing Gene Ontology information and finding significantly enriched Gene Ontology terms associated with a list of genes. *Bioinformatics*, **20**, 3710–3715.

41 Eiselleova, L., Matulka, K., Kriz, V., Kunova, M., Schmidtova, Z., Neradil, J., Tichy, B., Dvorakova, D., Pospisilova, S., Hampl, A., and Dvorak, P. (2009) A complex role for FGF-2 in self-renewal, survival, and adhesion of human embryonic stem cells. *Stem. Cells*, **27**, 1847–1857.

42 Hong, H., Takahashi, K., Ichisaka, T., Aoi, T., Kanagawa, O., Nakagawa, M., Okita, K., and Yamanaka, S. (2009) Suppression of induced pluripotent stem cell generation by the p53–p21 pathway. *Nature*, **460**, 1132–1135.

43 Marson, A., Foreman, R., Chevalier, B., Bilodeau, S., Kahn, M., Young, R.A., and Jaenisch, R. (2008) Wnt signaling promotes reprogramming of somatic cells to pluripotency. *Cell Stem. Cell*, **3**, 132–135.

44 Maherali, N., and Hochedlinger, K. (2009) Tgfbeta signal inhibition cooperates in the induction of iPSCs and replaces Sox2 and cMyc. *Curr. Biol.*, **19**, 1718–1723.

45 Discher, D.E., Mooney, D.J., and Zandstra, P.W. (2009) Growth factors, matrices, and forces combine and control stem cells. *Science*, **324**, 1673–1677.

46 Sasaki, N., Okishio, K., Ui-Tei, K., Saigo, K., Kinoshita-Toyoda, A., Toyoda, H., Nishimura, T., Suda, Y., Hayasaka, M., Hanaoka, K., Hitoshi, S., Ikenaka, K., and Nishihara, S. (2008) Heparan sulfate regulates self-renewal and pluripotency of embryonic stem cells. *J. Biol. Chem.*, **283**, 3594–3606.

47 Miura, K., Okada, Y., Aoi, T., Okada, A., Takahashi, K., Okita, K., Nakagawa, M., Koyanagi, M., Tanabe, K., Ohnuki, M., Ogawa, D., Ikeda, E., Okano, H., and Yamanaka, S. (2009) Variation in the safety of induced pluripotent stem cell lines. *Nat. Biotechnol.*, **27**, 743–745.

48 Bollen, K.A. (1989) *Structural Equations with Latent Variables*, John Wiley & Sons, Inc., New York.

49 Shipley, B. (2000) A new inferential test for path models based on directed acyclic graphs. *Struct. Equation Model.*, **7**, 206–218.

14
Community Detection in Biological Networks
Gautam S. Thakur

14.1
Introduction

Networks are everywhere and evolving in various forms, following certain characteristic patterns like small-world and scale-free characteristics. The basis of these forms is not the random addition of nodes and vertices to the system under study, but a systematic process of mutual evolution. This makes many aspects of such systems worth studying. Researchers have invested many years interpreting such systems, for instance how biological networks evolve and why humans form societies to name a few. Thus, seminal to the behavior to these systems are the way internal components are connected to each other. These connections breed structures that exemplify the nature and the behavior of the overall system.

Advances in genomics have generated enormous amounts of data, currently gathered by biologists and now available to the scientific community at large. The categorization of this data naturally falls into various classes, ranging from protein sequences to complex molecules. Thus, challenges lie in understanding the properties and dynamics of such complex systems rather than gaining insight into the individual components. Fortunately, promises shown in statistical physics and network theory that offer a host of methods and tools to analyze large data can invariably be used in biology. Modeling interactions between proteins, cells, and other genomic data as network graphs and applying statistical techniques on them can greatly benefit us in scrutinizing the intricacies that exist in these datasets. These methods not only simplify the representations, but also aid in searching for general principles of organization and evolution of biological structures. For example, small-world analysis is used to explain protein–protein interaction networks and scale-free invariance properties exist for *in silico* evolution of metabolic networks, which encode an increasing amount of information about their environment. The questions often arises as to which properties of such systems are worth studying and how are the emergent behaviors actually affecting the overall functionality of such systems. Visualization is a potential tool that was used as a first step to display the connectivity patterns between various components and describe the structure as a whole. Another set of measurements are various centrality indices that quantify the connectivity of such systems. For example, the connectivity in the genetic interactions affects the pairing of genes and represented graph models of protein interactions can help predict a three-dimensional structure of a protein from its amino acid sequence. In the following, we discuss a basic measure that explains various types of such connectivity patterns in terms of centrality measures often used to study complex systems.

14.2
Centrality Measures

To quantify a network [1–3] it is important to standardize various measures that capture the internal structure of the system. In this section, we discuss some of the important measures that can be used to analyze complex systems. To adhere to the notion of generalizing systems, we consider them as a graph $G = (V, E)$, where V is a set of vertices and E is a set of edges that represents the interaction between the vertices. One advantage of using such notation is that we can compare different types of datasets on the same scale, and apply the same algorithm to discover features and patterns common to all of them. One such measure that is widely used in the literature is *centrality*. Centrality defines the importance of a vertex in a way such that it dominates the overall structure of the system. Some important centrality measures widely used in network theory [4, 5] include degree centrality, betweenness, eigenvector centrality, and closeness [6].

14.2.1
Degree Centrality

Degree centrality is the number of edges connected to a vertex. While an undirected graph has only degree centrality, a directed graph has both in- and out-degree. A vertex in a network with high degree centrality provides a measure of its reachability and quantitative depth to access information from various sources. For example, the railway connectivity of Berlin is the highest among other European cities, which shows it is easier to catch a train from Berlin and reach a significant part of European nations than from anywhere else. The degree centrality is measured as:

$$D_v = \frac{d(v)}{V}$$

where $d(v)$ is the degree of vertex v and V is the number of vertices in graph G.

14.2.2
Eigenvector Centrality

A relative measure of a vertex's importance is measured by its connections to other vertices, which are themselves important. Thus, the centrality score is assigned as a cumulative total of the measure of a vertex's neighbors. Initially, we calculate the centrality score of one vertex at a time and then calculate the centrality of each vertex as the weighted sum of centralities of all vertices in its neighborhood. We then divide the value of centrality by the largest eigenvalue (λ) present in the network and continue to follow this process until the relative measure of the centrality does not change in successive iterations. For a vertex v, its centrality score is proportional to the cumulative total of all the scores of its neighbors:

$$v'_j = \frac{1}{\lambda}\sum_{j=1}^{N} M_{ij}v_j$$

where M_{ij} is the vertex's adjacency matrix and N is the total number of vertices in the graph G.

14.2.3
Betweenness Centrality

This centrality measures the extent of connectivity a vertex provides in a graph. Basically, it is a measure that counts how many times a vertex is visited while traversing inside the graph [7]. Theoretically, for all possible sets of shortest paths that are present between all pairs of vertices, betweenness centrality of a vertex v is the number of paths that pass through v:

$$B_v = \sum \frac{\pi_{sp}(v)}{\pi_{sp}(G)}$$

where $\pi_{sp}(v)$ is the number of shortest paths that pass through vertex v and $\pi_{sp}(G)$ is the total number of shortest paths present in graph G. A downside of using betweenness centrality is the amount of resources it takes to compute the betweenness of each vertex. Invariably, it takes $O(n^2)$ space and $O(n^3)$ computation time, where n is the number of vertices in the graph G. A faster implementation is available in [8] for interested readers. An important study of betweenness centrality in large complex networks is done in [9].

14.2.4
Closeness Centrality

The concept of closeness centrality is based on the average distance between two vertices. A vertex with low average distance has a higher value of closeness centrality. Thus, a vertex with a large number of direct neighbors has high closeness with them and is assumed to have better access to these vertices as compared to others with a large number of hops. In social networks, this measure plays a key role in accessing the closeness between two entities. In [10], the authors approximate values of closeness centrality and present new rank algorithms based on this measure.

However, there are a few problems using this measure, which are described by Newman [11].

14.3
Study of Complex Systems

The interactions between entities spurs the development of emergent properties visually depicted in the form of connectivity networks. A mathematical study of such networks using network theory helps understand the internal structures of the Internet, world wide web, social networks, and various forms of biological structures. However, a common problem with these networks is their enormity and complexity, which makes them difficult to study. Recent development in statistical methods to quantify such large networks address these issues in the form of small-world analysis [12], scale-free networks [13, 14], percolation theory [15], and preferential attachments [16]. Thus, a natural way to study such systems is to break down large systems into smaller subsystems. From a theory point of view, the density distribution of vertices in a network subgraph is higher than expected in an equivalent network with edges placed at random. Here, we study this concept of a network that is widely used to understand the emergent properties in the investigation of groups or communities in the networks. The process by which a network evolves into components tells how the internal organization is built upon and helps us understand how the system is structured. Identification of interaction patterns in complex networks via community structures has received a lot of attention in recent research studies. Local community structures provide a better measure to understand and visualize the nature of interactions when the global knowledge of networks is not available. Recent research on local community structures [17–19], however, lacks the feature to adjust itself in the dynamic networks and depends heavily on the source vertex position. In this chapter, we discuss a novel approach to identify local communities in biological networks based on iterative agglomeration and local optimization. The algorithm strengthens the local community measure in each iteration by agglomerating the best possible set of vertices such that the proposed vertex and community rank criterion are suitable for dynamic networks where the interactions among vertices may change over time. An extensive set of experiments and benchmarking on computer-generated networks as well as real-world social and biological networks shows that the proposed algorithm can identify local communities, irrespective of the source vertex position, with more than 92% accuracy in the synthetic as well as in the real-world networks.

14.4
Overview

Identifying patterns of interaction has gained a lot of recent attention in finding community structures in complex networks [20–22]. As already mention in

Section 14.2, such networks are modeled as a graph $G = (V, E)$, where V is a set of vertices and E is a set of edges representing the interaction among the vertices. *A community in a network is defined as a group of vertices having denser edge connectivity than with vertices outside the group.* Thus, a relative degree distribution among the members of a community is greater than with noncommunity members. We identify community as a functional grouping of entities exhibiting certain generic characteristics. For instance, a community in a citation network can represent literature that belongs to the protein folding problem and a food web can represent feeding relationships between species within an ecosystem. Thus, widespread applicability has made the study of community structures a mainstay research topic in today's scenario.

Researchers from various disciplines analyze community structure using eigenvectors and sparse matrix formulations [23], algebraic connectivity [24], partitioning techniques [25], small-world effect [12], parameterized linear programming [26], and many other methods. These approaches provide better results, but are computationally expensive as they require global information of the network.

A recent work [18] has introduced the concept of *local community structure* that detects community given a source vertex and degree distribution of immediate neighbor vertices (*local information*). Subsequently, several methods have been proposed to reduce the complexity of this task; however, they suffer from one or more drawbacks. For instance, methods proposed in [13, 27] are specific to particular cases of either portal browsing or they require a minimal connected initial topology to promote the growth of a local community structure. The measure of modularity defined in [18] considers those vertices that reside on the boundary of a subgraph. The methods proposed in [28–30] expect some degree of global information to ascertain partitions that are not easily distinguishable. The shell-spreading method mentioned in [17] performs well only if the source vertex is in the middle of the enclosed community. An important issue in all approaches is the lack of ability to self-adjust with the evolving community structure when vertices are added and removed. Furthermore, it is not possible to capture changes that occur in highly dynamic networks where patterns of interaction change constantly.

Here, we explain a novel algorithm [19] to detect evolving local community structure using iterative agglomeration and local optimization. This algorithm is self-adjusting in nature and performs well in static as well as in dynamic networks. Based on the degree distribution of vertices it optimizes the community structure by selecting the vertices that are close enough to form a clique. As described in the later sections, the measure of "cliqueness" is calculated using the available local information of vertices degree and intercommunity rankings.

We evaluate the efficiency of the this algorithm on computer-generated dynamic and sparse networks with different source vertex positions. The analysis of the results shows that the algorithm is very effective in addressing a wide variety of social [31] and biological network problems [32, 33]. We also evaluate the performance of our algorithm on three real-world networks: the Zachary Karate

Club, N-methyl-D-aspartate receptor complex/MAGUK associated signaling complex (NRC/MASC) receptor complexes, and Massachusetts Institute of Technology (MIT) wireless mobile users. In the Zachary Karate Club, 94% of club members, regardless of source vertex position (including border vertices), are correctly identified within their original communities. These results are better than the results in [17, 22] that constrain the presence of the source vertex in the middle of an enclosed community. Another experiment to understand protein interactions mapping in NRC/MASC receptor complexes yields two big clusters enclosing the local community of motifs with 92% accuracy level in core MASC protein functional elements. In the study of wireless mobile users, we found 31 active communities consisting of 890 users based on their preferential attachment to locations.

Section 14.5 outlines this algorithm. The evaluation and benchmarking on computer generated and real-world networks is done in Section 14.6. Section 14.7 discusses some suggestions to further optimize the discovered communities. Finally, Section 14.8 concludes the chapter.

14.5
Proposed Algorithm

14.5.1
Overview

The local community structure can be formally defined as follows: given an undirected graph $G = (V, E)$ and a source vertex $s \in V$, in the absence of global knowledge, the graph G is explored one vertex at a time to form a community $G' = (V', E')$, where G' is a subgraph of G and s is more highly connected within V' than to any vertex in V/V'.

The quantitative measure of local community structure is independent of the global topology knowledge. Based on the available local information, the algorithm starts crawling the graph G from the source vertex s, first visiting immediate neighboring vertices. The method attempts to quantify the degree distribution of neighbor interactions that result in the formation of a clique. The optimization function ranks a subset of vertices from a large unknown size group that are more closely related to each other than to the remaining vertices of the group.

The proposed algorithm has three mains steps:

i) Initially, a series of random walks is performed to generate an initial set of communities C.
ii) The set of communities and their vertices obtained in step (i) are ranked using local optimization functions, which are defined later in this section.

iii) Based on these ranks, the vertices among the communities are exchanged to optimize the formation of the local community structure. The algorithm stops when the local optimization function cannot improve any further or if the size of the highest ranked community reaches a user-defined value k. Finally, the algorithm returns the highest ranked community as the solution.

14.5.2
Algorithm Description

The algorithm begins with a series of random walks from the source vertex s and explores unknown portions of the graph G to generate an initial set of communities. These communities are formed as a subset of vertices and edges of an unknown size graph G. Each vertex of a community is assigned a value γ, which is defined as its degree divided by the number of its immediate neighbors present in the same community. Within a community, the enclosed set of vertices are positioned with respect to γ. This implies that a vertex with larger γ has a higher probability to discover immediate neighbors in other communities; such vertices are called *active vertices*. It should be noted that, because of the inherent nature of the random walk algorithm, a vertex can be included in more than one community, but with different γ values.

Similarly, each community is assigned the value Γ, which is calculated as a cumulative average of γ values of its enclosed set of vertices. These communities are positioned with respect to their Γ value. This reflects two important aspects: (i) communities with large Γ are more active to form a local community structure and (ii) communities with small Γ are proclaimed local community structures or they are insufficient in terms of the basic requirement. Thus, a community C should either siphon off its active vertices or acquire the immediate neighbors of its active vertices to yield a local community structure.

In order to accomplish this, we perform iterative agglomeration on these communities in decreasing order of their Γ values. The agglomeration takes place through the exchange of active vertices among the higher ranked communities. It is now apparent that the set of vertices that belongs to these communities has a higher probability to discover their immediate neighbors in other communities. This exchange results in the change of a vertex's neighborhood and subsequently its γ value changes. This, in turn, changes the ranking of these communities. As a result of successive iterations, vertices come closer to satisfying their neighbor requirements. In doing so, their respective γ values will decrease, which further contributes towards the creation of a local community structure. This process continues until the algorithm has agglomerated k (given as input) number of vertices or the algorithm has discovered an entire local community structure. Fundamentally, the algorithm involves a process of building a community structure by clustering together a pattern of highly interactive vertices.

In the following subsections, we first discuss the random walk algorithm that generates an initial set of communities. Then, we discuss a method to find vertex and community positions, and present formulae to calculate their respective rankings. We also discuss the underlying rules to exchange active vertices among the higher ranked communities. Finally, dynamic optimization to achieve community formation in dynamic networks is presented.

14.5.3
Random Walk

There exists a variety of random walk algorithms based on different parameters, such as Euclidean time distance [34], reversible Markov chains [35], timing parameters [36], and available global information [37]. However, these methods do not meet the requirements to compute a local community structure due to two main reasons: (i) the source vertex position to initialize the local community structure and (ii) the minimum convergence time to form a community based on the available local information. Hence, we propose a random walk based on vertex degree distribution, which meets all the basic requirements to form a local community. The basic assumption of the proposed random walk is that the probability to explore a vertex is proportional to its degree distribution.

The random walk on the graph G is performed in order to generate an initial set of communities. At first, the given source vertex s is labeled and added to a community C_i. Then, one of its immediate neighboring vertices v is selected with a probability of $P_s = 1/d(s)$, where $d(s)$ is the degree of source vertex s. The selected vertex v is labeled and added to the community C_i. Continuing in this fashion, subsequent vertices are added to the same community until a previously labeled vertex encounters (hence, backward walk is not possible).

Iteratively, such random walks are performed to generate a set C of initial communities. Let $C = \{C_1, C_2, \ldots, C_\eta\}$ be the set of initial communities generated through a series of random walks, where $|C_i| \leq l$ for all $i = \{1, \ldots, \eta\}$. There are two different ways of deciding the total number of communities to generate: (i) user-provided input of value η and (ii) a random walk that is generated twice. The first option is useful when the user wants to restrict the number of generated communities, while with the second option, the algorithm maintains unique hashes (e.g., MD5 hashes) of performed random walks. A hash is generated from the unique set of nodes and edges encountered in that walk. If the same hash is generated twice the random walk algorithm is terminated and the number of communities that are generated till that iteration is termed η. Figure 14.1 shows a number of generated communities with user input option for the source vertex $s = 34$. Furthermore, this finite Markov chain random walk also has the capability to reach significantly obscure locations of a graph. Vertices that are not part of any community are considered to be *leftover vertices*. Later on, these leftover vertices form a community, which is checked against a classified local community for any possible exchange of vertices. Pseudo-code for user input option is given in Algorithm 14.1.

14.5 Proposed Algorithm

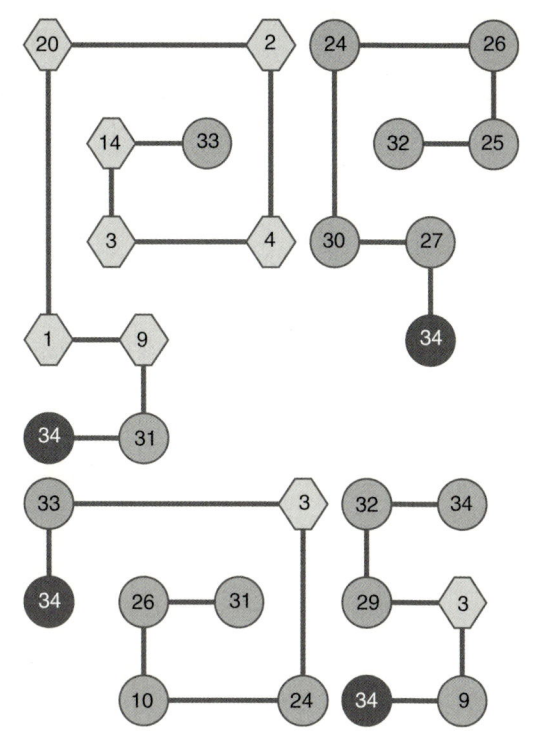

Figure 14.1 Four communities generated from the random walk. Vertex 34 is the source vertex s used in the Zachary Karate Club experiment.

Algorithm 14.1 Random Walk Algorithm

Input: η-number of communities to generate, s-source vertex
Output: Initial Solution of η communities

for $i \leftarrow 1$ **to** η **do**
    Initialize new community C_i
    add s to C_i
    set $v \leftarrow s$
    set $j \leftarrow 0$
    while $j \neq l$ **do**
        if $flag\_status\ (v) == true$ **then**
            break
        end
        srand(time(NULL))
        set $neib\_id \leftarrow rand()\%\ d(v)$
        add $neib\_id$ to C_i
        set flag$(v) \leftarrow true$
        set $v \leftarrow neib\_id$
        j++
    end
end

Algorithm 14.1 can be modified to implement the hash-based generation of initial communities. Instead of user input of value η, the algorithm will maintain the hash of random walks generated. After every walk the new hash is compared with existing hashes and if a match is found the algorithm terminates.

14.5.4
Local Optimization

The classification to partition a graph into local community structure can be interpreted as optimizing a quantitative notion of a community structure [18]. In the case of local community structures, the biggest challenge is the lack of global knowledge of the network. Therefore, any optimization must rely only on the available *local information*. Here, the local information of a community is the edge connectivity of its enclosed set of vertices and the knowledge of their immediate neighbors. Thus, from edge connectivity (adjacency matrix) it can be determined how many immediate neighbors of a vertex are within its own community and how many of them are in some different community.

Furthermore, if two or more such communities share their local information, it is inferential that a local community structure can be discovered. During this process, a vertex will look out for its immediate neighbors in other communities and then it will either join the community of one of its immediate neighbors or ask its immediate neighbors to join its own community. This results in the formation of a local community structure. Thus, *local optimization* is defined as a process to discover community structure based on the exchange of available local information between communities.

To further understand local optimization, we define two formulae: (i) vertex ranking and (ii) community ranking. In vertex ranking, the vertices of a community are ranked on the basis of their available local information related to their neighbor degree requirements. In community ranking, a community is ranked based on its state and capability to transform itself into a local community structure. Let us define these formulae.

14.5.4.1 Vertex Ranking

From the discussion it is clear that the vertex exchange between communities takes place on a one-on-one basis. Thus, we can think to define some criteria to optimize the processing time and reduce the vertex-to-vertex comparison. Note that the purpose of maximizing the local community structure is to identify a set of vertices that are close enough to form a clique. However, to identify a clique even if we know the network's global information is an NP-complete problem [38].

To solve this problem, we first define a vertex rank function to rank all the vertices explored during the random walk. The *rank of a vertex* decides its position within its community and is the basis on which the vertices in general are agglomerated with high probabilistic coefficient to form a clique.

Let $C = \{C_1, C_2 \ldots C_\eta\}$ be the set of communities generated from the random walk. Each community C_i contains a set of vertices $V_{C_i} = \{v_1, v_2 \ldots v_l\}$. The vertex rank is defined as:

$$\gamma_{ji} = \frac{(2d_{ji})}{(k_{ji}(k_{ji}-1))}$$

where j is a vertex in a community C_i, d_{ji} is the degree of vertex j, and k_{ji} is the possible number of its immediate neighbors that are present in the same community. An *active vertex* within a community is a vertex that has the greatest γ value. This implies that such an active vertex has the highest probability to find more neighbors if it is transferred to some other community. This further explains the exchange of vertices from one community to another community as a way to mutually decrease the value γ, which in turn brings all the neighboring vertices together to form a local community structure.

Pseudo-code of this process is given in Algorithms 14.2 and 14.3. The algorithm performs multiple iterations to agglomerate a set of vertices enclosed within a local community structure. Using multiple iterations, the algorithm self-adjusts the community by adding the neighboring vertices from several other communities and by siphoning off the vertices that may belong to some other community. Finally, the algorithm outputs an assorted set of vertices that are close enough to form a clique.

Algorithm 14.2 Pseudo-Code to Find Local Community Structure

Input: s, η, l, n, k
Output: Community with k vertices
Perform RandomWalk($\eta - 1, s, l,$)
Add one more solution of non-picked vertices
for $i \leftarrow 1$ **to** $\eta\text{-}1$ **do**
    for $j \leftarrow 1$ **to** l_i **do**
        | Compute (γ_{ji})
    end
    sortVertices(C_i)
    Compute (Γ_i)
end
sortCommunities()
set $cflag \leftarrow true$;
while $cflag \neq false$ **do**
    set $cflag \leftarrow false$;
    for $i \leftarrow 1$ **to** $\eta\text{-}1$ **do**
        set $exchange \leftarrow do\_exchange(C_i, C_{i+1})$
        if *(exchange == true)* **then**
            | set $cflag \leftarrow true$;
        end
    end
    for $i \leftarrow 1$ **to** η **do**
        | *remove_empty_communities*(C_i)
    end
    Compute η
    sortCommunities()
end
Let C_α be the fittest community.
do_crossover(C_α, C_η)
if $|C_\alpha| > k$ **then**
    | remove $|C_\alpha| - k$ vertices from C_α
end
return C_α

Algorithm 14.3 Pseudo-Code to Compute Possible Exchange of Vertices

Input: C_1, C_2
Output: Exchange Communities
$remove\_common\_vertices(C_1, C_2)$
set $flag \leftarrow false$
for $i \leftarrow 1$ to l_1 do
    for $j \leftarrow 1$ to l_2 do
        if $\gamma_i \geqslant \gamma_j$ then
            set $val1 \leftarrow \gamma_i$
            Compute new Γ_2 by adding v_i to its enclosing community. Let new value v_i be γ'_i
            if $val1 > \gamma'_i$ then
                add (C_2, v_i)
                remove (C_1, v_i)
                Compute (Γ_1)
                Compute (Γ_2)
                set $i \leftarrow 1$
                set $j \leftarrow i$
                set $flag \leftarrow true$
                continue;
            end
        else if $\gamma_j > \gamma_i$ then
            set $val1 \leftarrow \gamma_j$
            Compute new Γ_1 by adding v_j to its enclosing community. Let new value v_j be γ'_j
            if $val1 > \gamma'_j$ then
                add (C_1, v_j)
                remove (C_2, v_j)
                Compute (Γ_1)
                Compute (Γ_2)
                set $i \leftarrow 1$
                set $j \leftarrow i$
                set $flag \leftarrow true$
                continue;
            end
        else
            continue;
        end
    end
end
if (*flag* == *true*) then
    return *true*
end
return *false*

14.5.4.2 Community Ranking

A vertex's neighbor requirement is very much local to its community. For a vertex to look out for its potential neighbors, it is required to have knowledge of other communities. However, from the optimization point of view, to explore all the generated communities is not feasible. Hence, we should derive some quantitative notion of a community to selectively perform neighbor-lookup for its enclosed set of vertices.

The *rank of a community* is defined as an average γ value of its enclosed set of vertices. Thus, community rank Γ_i of a community C_i is defined as:

$$\Gamma_i = \frac{\sum_{v_{ji} \in C_i} \gamma_{ji}}{|C_i|}$$

From the above equation it is clear that the Γ value of a community is directly proportional to the sum of γ values of all the vertices enclosed in that community. From the discussion on vertex rank, we can conclude further that a community is relatively more active if it encloses more active vertices. Such *active communities* are preferred for neighbor-lookup and possible exchange of their vertices.

14.5.5
Mutual Exchange

The mutual exchange of vertices among communities is an act to mutually decrease the Γ value of these communities. At any point in time, a community with the lowest Γ value is the fittest community and is considered as a candidate to become the local community structure. The fundamental act of mutual exchange of the vertices between a pair of active communities is performed in the decreasing order of their Γ values.

Thus, when a vertex is added to a new community, it has more neighbors than it had in its previous community and so this vertex is closer to satisfying its immediate neighbor requirement. It is very likely that this vertex was originally a part of this new community. During the exchange, if common edges match, both vertex and community rankings decrease with the movement of vertices from one community to another. The vertices are thus grouped together and their new rankings are recalculated. Similarly, ranks of the communities are recalculated with respect to their new set of vertices.

The mutual exchange criteria within a pair of communities is defined as: given two top-ranked communities C_A and C_B such that $\Gamma_A \geq \Gamma_B$, $v_a \in C_A$ and value γ_a.

i) An exchange of v_a exists if $\gamma'_a < \gamma_a$, where γ'_a is the new γ value calculated for v_a considering its neighboring vertices are enclosed in community C_B.

ii) For C_A and C_B, a vertex exchanges occurs if $\text{Avg}(C'_A, C'_B) < \text{Avg}(C_A, C_B)$, where $\text{Avg}(C_A, C_B) = \frac{\Gamma_A + \Gamma_B}{2}$.

14.5.5.1 Dynamic Optimization

Previous published work on local community structure does not allow any change in the already formed intermediate communities [13, 18, 27–30, 39]. However, it is an inexpensive computational approach that might not produce an optimized community.

In the current approach, using the dynamic optimization, whenever a vertex is exchanged between two sets of communities, its ranking is recalculated with respect to its immediate neighbors (if any) in this new community. Thus, the iteration always concludes with the most recent value of γ for the enclosed set of vertices belonging to that community.

As illustrated in Algorithm 14.2, in each iteration the algorithm calculates the maximum average reduction for the individual community $C_A = \{v_{a1}, v_{a2}, \ldots v_{al_a}\}$ and $C_B = \{v_{b1}, v_{b2}, \ldots v_{bl_b}\}$, and the vertex to be exchanged. The algorithm compares their respective averages before the transfer of their vertices based on the conditions mentioned in the mutual exchange scheme. When a vertex $v \in C_A$ community is transferred to C_B community, its γ value changes, which in turn, causes a change in value Γ of both C_A and C_B communities. The changed values are calculated as:

i) For vertex v, the changed ranking value is calculated as:

$$\gamma'_v = \frac{2d_v}{(k'_v * (k'_v - 1))}$$

where k' is the number of neighbors enclosed in new community. This is true for all other vertices.

ii) Changed ranking value of C_A is calculated as:

$$\Gamma'_A = \frac{\sum_{v_{ai} \in C_A} \gamma_{ai}}{l'_A}$$

where l'_A is the number of vertices after v_a is removed.

iii) Changed ranking value of C_B is calculated as:

$$\Gamma'_B = \frac{\sum_{v_{bi} \in C_B} \gamma_{bi}}{l'_B}$$

where l'_B is the number of vertices after v_a is added.

14.6
Experiments

14.6.1
Benchmarking

In this section, we use the benchmarking criteria as illustrated in [40] to provide an objective comparison with the existing methods of finding local community structures. Initially, a Barabási–Albert graph $G = (V, E)$ of $V = 512$ and $e_0 = 8$ is created, which is then randomly partitioned into four *reference* communities to contain nearly equal numbers of vertices. Every vertex in these communities has an average degree $z = z_i + z_{out} = 16$. Clearly, a small z_{out} displays a strong community structure. Finally, vertices are rewired through a pair of edges within same communities without changing their degree distribution.

14.6.1.1 Evaluation
The algorithm generates a community C and a noncommunity of V/C of vertices. For the comparison purpose, the *reference* partition is termed $P_R = \{C_R, C'_R\}$ and the

found partition is termed $P_F = \{C_F, C'_F\}$. The evaluation criteria uses normalized mutual information (NMI) [41–43] to measure the likeliness in P_R and P_F. In NMI, a *confusion matrix* **N** is created, with its columns corresponds to the found communities and rows to the reference communities. The similarity measure is defined as:

$$I(P_R, P_F) = \frac{-2\sum_{i=1}^{C_R}\sum_{j=1}^{C_F} N_{ij}\log\left(\frac{N_{ij}N}{N_{i.}N_{.j}}\right)}{\sum_{i=1}^{C_R} N_{i.}\log\left(\frac{N_{i.}}{N}\right) + \sum_{j=1}^{C_F} N_{.j}\log\left(\frac{N_{.j}}{N}\right)}$$

where N_{ij} is the number of nodes in the reference community i that appear in the found community j. C_R is the number of reference communities and C_F is the number of found communities. $N_{i.}$ is the sum over row i and $N_{.j}$ is the sum over column j.

Thus, $I(P_R, P_F)$ gives a quantitative measure of how much the found community is similar to the reference community. A score of 1 shows both communities are identical and a score of 0 is when they are totally independent.

As shown in Figures 14.2 and 14.3, the algorithm performed very well and comparable to the illustrated results in [40]. The algorithm is able to correctly identify the communities with small values of z_{out}, with a decrease in accuracy as the values of z_{in} and z_{out} became equal. In Figure 14.3, the algorithm performs very well with a large number of rewirings and its accuracy decreases as less edges are rewired, which indicates the communities are not sharply separated with a lesser rewiring count.

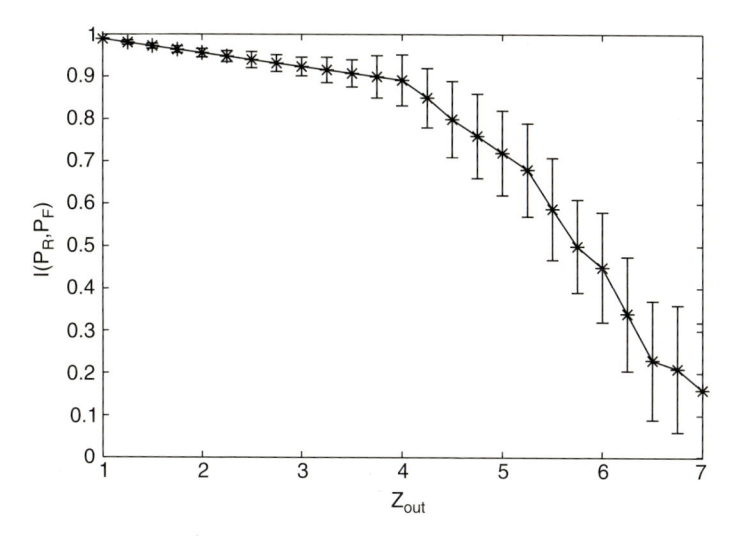

Figure 14.2 Performance analysis of the algorithm for a 128-node network, averaged over 1000 realizations. The algorithm performs well for low z_{out}, indicating strongly separated communities, and decreases in accuracy for higher values, meaning that community separation blurs. The error bars show the deviation from the mean value.

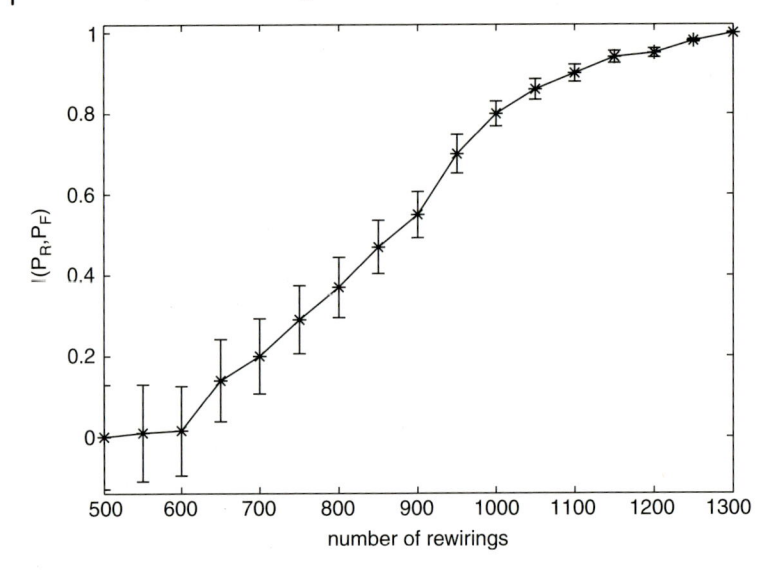

Figure 14.3 Performance analysis of the algorithm on a 512-node rewired network, averaged over 1000 realizations. For a large number of rewirings the algorithm performs well, but gradually decreases in accuracy as lower numbers of edges are exchanged. The error bars show the deviation from the mean value.

14.6.2
Computer-Generated Experiments

This section illustrates the experimental setup and the application of the proposed algorithm to problems related to socio-physics and computational biology. To verify that the algorithm can identify and recognize boundaries of a local community, we first apply the proposed algorithm on computer-generated graphs (the experimental random graphs are generated using VGJ and MATLAB Bioinformatics tool box) [44] that have well-known community structures. As previously mentioned, the proposed algorithm can also be used in dynamic networks. To verify this scenario, we artificially add and subtract vertices on the fly during the execution of the algorithm. A graph is constructed for $n = 120$ vertices, divided into four local community structures known already. Edges are placed independently at random between vertex pairs such that the expected average degree of each vertex is equal to 10. We can tune the sparseness of the graph by varying average degree of vertices.

Figure 14.4 shows a dendrogram generated from this algorithm. It contains a community of $n = 30$ vertices. As can be seen, all the vertices are identified with a major split at the top of the tree. Figure 14.5 shows the corresponding Γ value against number of iterations required to achieve it. Here, we cannot put a bound on the lower limiting value of Γ as it is related to an average degree of all vertices. However, our experiments have shown that any value $0 < \Gamma \leq 1$ is an indication of a good community.

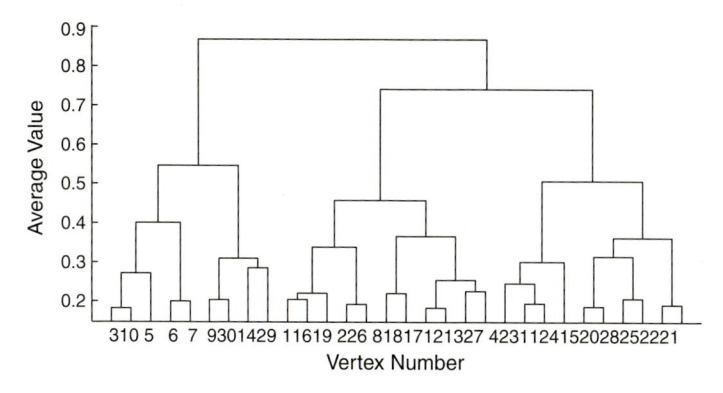

Figure 14.4 Plot of a dendrogram shown for a computer-generated local community structure. A community with a set of 30 vertices is identified correctly for a mean $\Gamma = 0.86$. Numbers at the bottom denote vertex identifiers belonging to that community. The top single line agrees with the formation of a complete community.

Figure 14.6 shows the result of an experiment that involves graph sparseness. A typical bar graph chart illustrates the percentage of vertices of a community identified correctly against a varying average vertex degree. The algorithm performed quite well for an average of $d = 6$ and onwards, identifying almost 92% of vertices correctly. For an average low degree ($d \leq 4$), the results deteriorate to a level of 75%. This result indeed clarifies an important aspect of the random walk. The initial solution generated from the random walk contains a significant number of vertices that contribute to other communities. As the sparseness level of a graph increases, the random walk traverses these nonlocalized parts of the community.

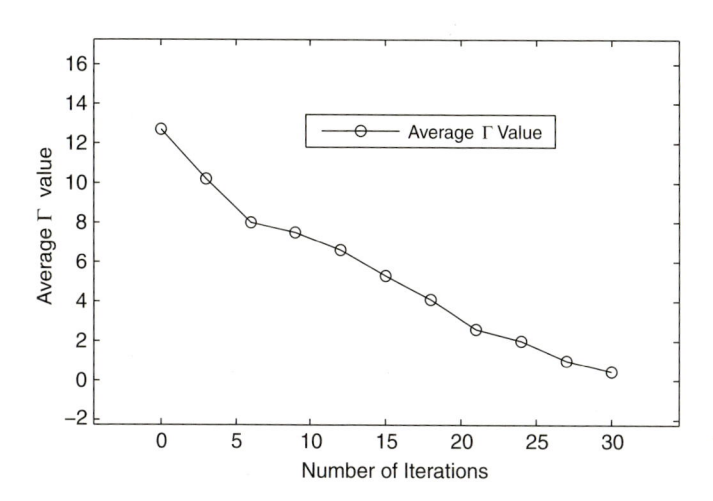

Figure 14.5 Plot against number of iterations and average Γ. When the number of iterations of the algorithm reaches 30, a value of $\Gamma = 0.86$ signifies that a local community structure is formed with all vertices classified correctly.

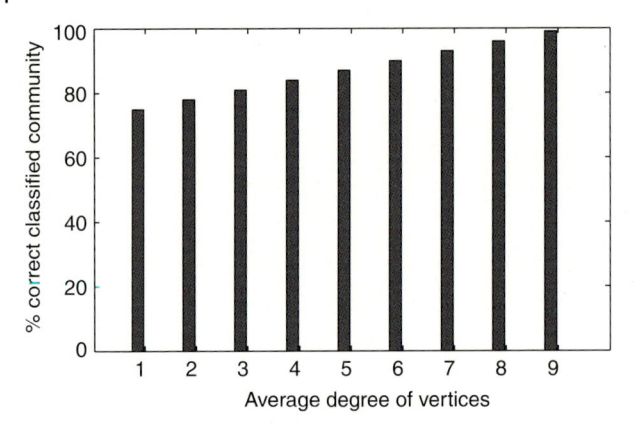

Figure 14.6 Bar graph showing the efficiency of the algorithm on various levels of graph sparseness. The algorithm performed quite well while the average degree of community is $d = 6$.

The dynamic property of our algorithm was also evaluated as shown in Figure 14.7. For a known community of $n = 43$ vertices, we feed the algorithm with vertices on the fly during its execution process. These externally added vertices are not part of any initial solution generated from the random walk. The spikes (squares) show a temporary increase in Γ when a set of vertices is fed to the network. A comparison is made against the static nature of the community (circles) with the same set of vertices. In a static community, all the vertices were part of an initial solution generated from the random walk. At the end of the execution, Γ merges to a common

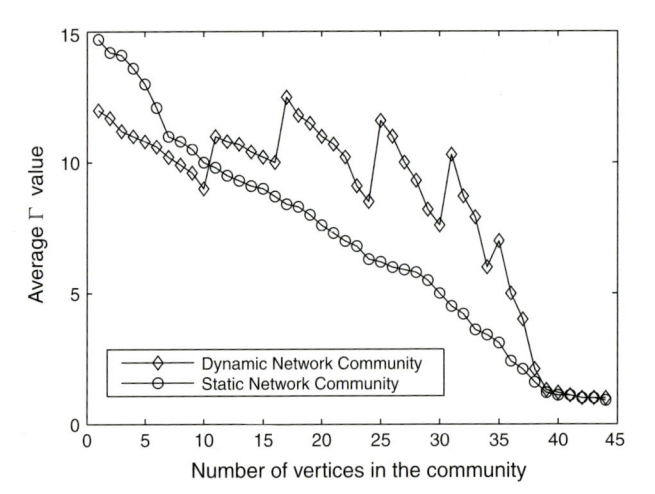

Figure 14.7 Relative comparison of static and dynamic community Γ values shown for the same set of vertices. The two curves show average Γ (y-axis) with number of vertices in communities (x-axis). As expected, dynamic communities (squares) show spikes in Γ when vertices are added on the fly. Also, static communities (circles), based on initial random solutions, show a steady decrease in Γ.

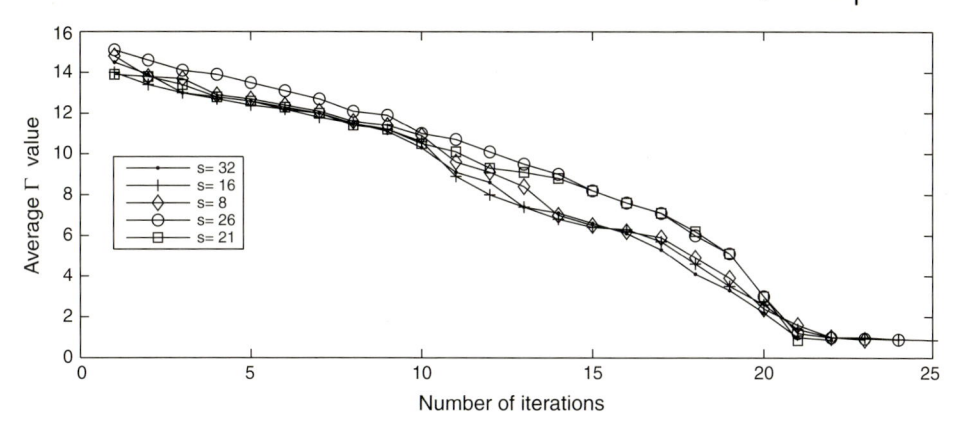

Figure 14.8 Plot showing an effect on Γ (y-axis) and number of iteration (x-axis) when s is chosen differently. The merge on the bottom right denotes convergence to local community structure with nearly the same Γ value. Almost 95% of vertices are correctly identified despite varying source vertices.

point for both communities. This is expected, given that the set of vertices eventually remains the same in both the communities.

Figure 14.8 presents a comparative study of local community convergence by varying the position of the source vertex. Several iterations of the proposed algorithm on a constant set of $n = 120$ vertices were carried out for a known community of $n = 43$ vertices. In each iteration, a different source vertex s was chosen to carry out the community formation. For the purpose of clarity, only five different source vertices are illustrated with their respective values of Γ against number of iteration, before they converge to a local community. As revealed in Figure 14.8, for all source vertices, final communities converge for an average $\Gamma = 0.82$, signifying good quality of local structures assuming all communities contain the same set of vertices.

Figure 14.9 presents execution time comparisons of various local and global community-based algorithms that are referenced in this chapter. As can be seen, local information-based algorithms have good execution time performance over global information-based algorithms. Also, within local community structures, the proposed method performs well over other local methods.

From these results, we conclude that the proposed algorithm effectively identifies the local community structure in static and dynamic networks.

14.6.3
Application to Real Networks

In this section, we evaluate the proposed algorithm on real-world social and biological networks. The first example is taken from the Zachary Karate Club in which a factional division led to a formal separation of the karate club into two organizations. It was earlier studied by [17, 45] with different classes of hierarchical clustering methods. We also consider a biological example to map protein interactions to create

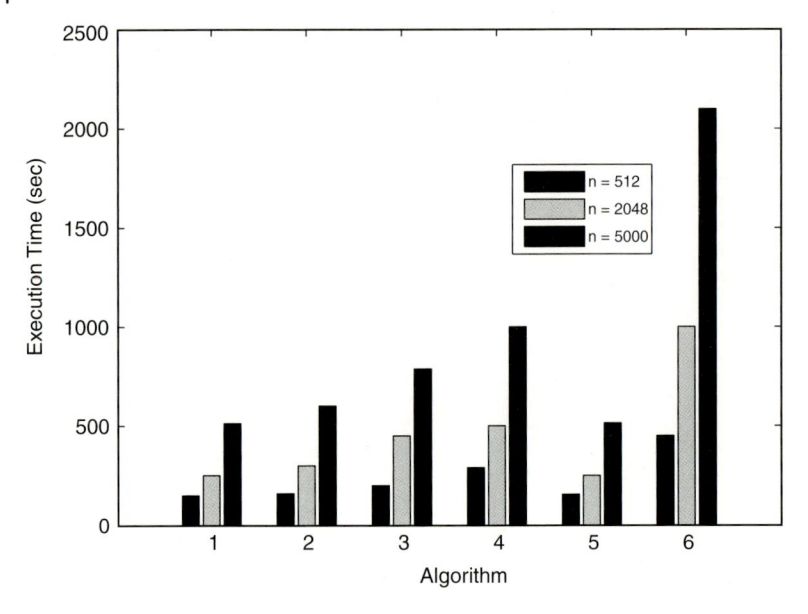

Figure 14.9 Comparison of the execution time with varying number of nodes in the network. The algorithms used are in the following order of references: [13, 27–29], the proposed algorithm, and [30]. As can be seen, local information-based community structure algorithms perform much better than global information-based algorithms. Also, among local algorithms, the proposed method performs better than other referenced local community structure algorithms.

a network representation of neurotransmitter receptor complexes. Finally, with the advancements in wireless technology we can continuously log mobile users' interactions and behavioral patterns in a social system. In one such obtained dataset, we apply our algorithm to discover communities based on mobile users' spatiotemporal preferences.

14.6.3.1 Zachary Karate Club

This example is built on the Zachary Karate Club [45], which presents data from a university-based karate club for $n = 34$ vertices, in which a factional division led to a formal separation of the club into two organizations.

The input consists of 34 vertices representing members of a big community. A disagreement developed between the administrator and instructor of the club, which resulted in the instructor leaving and starting a new club, taking about half of the original club members. We apply the proposed algorithm in an attempt to identify the factions and its members involved in the split. Figure 14.10 shows the original karate club, where circles represent members associated with the administrator's faction and squares represent the instructor's faction.

To justify the correctness of the proposed algorithm, we run the experiment for a distinct pair of source vertices from each local faction. Considering $s = 1$ as a border vertex and $s = 34$ as a center vertex with $\eta = 25$, we fed this network to the proposed

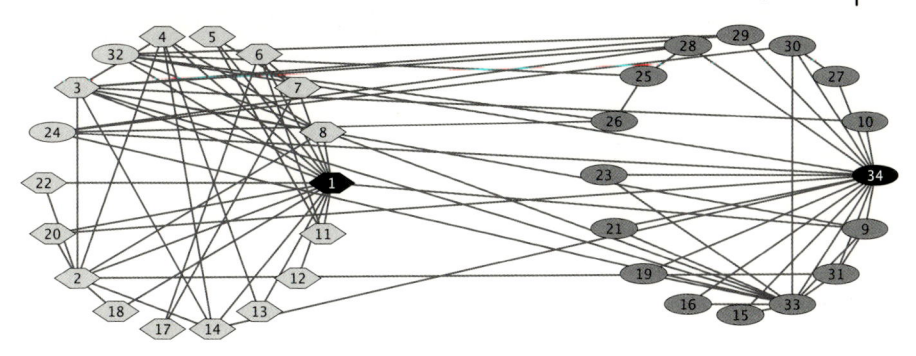

Figure 14.10 Actual breakdown of the Zachary Karate Club [45].

algorithm. Figure 14.11 shows the resultant generated communities. Except for vertices 9 and 10, the algorithm is able to correctly identify communities; however, it can be justified as both vertices are equally linked to both factions with the same number of edges outward to both communities. The graph in Figure 14.12 shows a steady convergence in Γ of the formed communities. It can also be interpreted that border vertices bring a sharp decrease in Γ of the community structure. With an average $\Gamma = 0.51$ for $s = 1$ and $\Gamma = 0.65$ for $s = 34$, an accuracy of 94% is achieved, which is superior to the previous results of [17, 22].

14.6.3.2 Neurotransmitter Receptor Complexes

Mapping protein interactions to create a network representation of the complexes helps reveal many important biological functions. Here, we explore the organization and function of NRC/MASC [46] using a local community approach. Using the information on the function, interactions patterns, and phylogeny of each protein, we are able to develop certain inferences on the structural and functional aspects of synapse complexes.

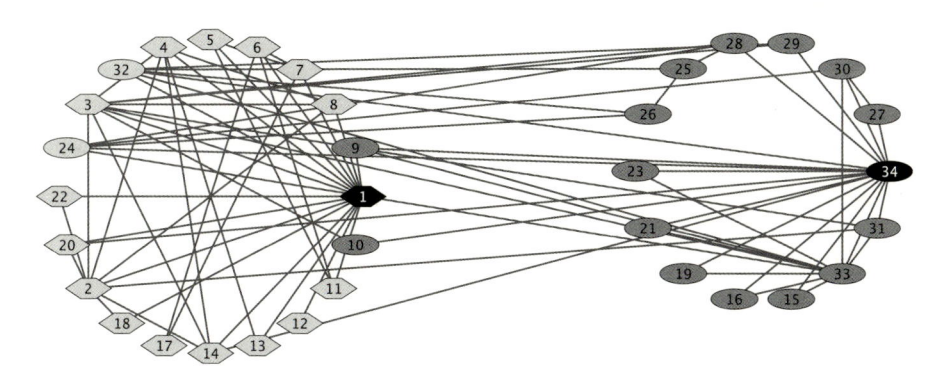

Figure 14.11 Two local factions in the Zachary Karate Club generated using the proposed algorithm: the community represented using circles with $s = 34$ and the community represented using squares with $s = 1$.

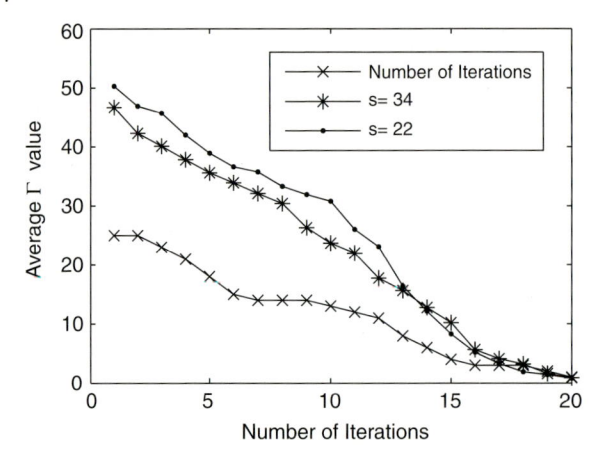

Figure 14.12 Zachary Karate Club convergence graph for the identification of communities. The graph shows the convergence of communities taking $s = 1$ and $s = 34$, for $\eta = 25$. The x-axis shows the number of times the iterative agglomeration is performed.

The annotation study carried before only considers the list of components, and does not take into consideration their interaction patterns and organization. When represented as an undirected graph, the input consisted of 101 protein vertices connected by 246 interaction edges. This constitutes the core functional elements of MASC proteins. It links together all glutamate receptors, and a high proportion of the signal transduction machinery responsible for the reception and integration of calcium and G-protein-coupled synaptic signaling. Given 13 different sparse and dense collections of network communities, we focused our work on the identification of ionotropic glutamate receptor proteins that also contain number of PDZ/DHR/GLGF scaffolding molecules and serine/threonine kinase protein kinase A, known as an integrator of signals in synaptic plasticity containing some traces of tyrosine kinases and SH2 motif proteins. Figure 14.13 show the results of the application forming two big clusters of local communities of motifs. The algorithm is able to identify more than 92% of them correctly for an average $\Gamma = 0.81$ as compared to the original community structure mentioned in [46]. The two clusters in Figure 14.13 show that the proteins perform fast information sharing and response coordination, first within themselves and then with other module members. As inferred, the clusters of proteins around ionotropic glutamate receptors and tyrosine protein kinase form the primary sites for signal reception regulating effector mechanisms and vesicular trafficking [46]. They are similar in nature, and influence a particular set of functional processes specific to their structure and individual behavior. Thus, the formed grouping shows a community of significant overlap in function and phenotypic annotations. This conveys an important result that, irrespective of the type of network, the proposed algorithm has performed consistently well across a wide range of implementations.

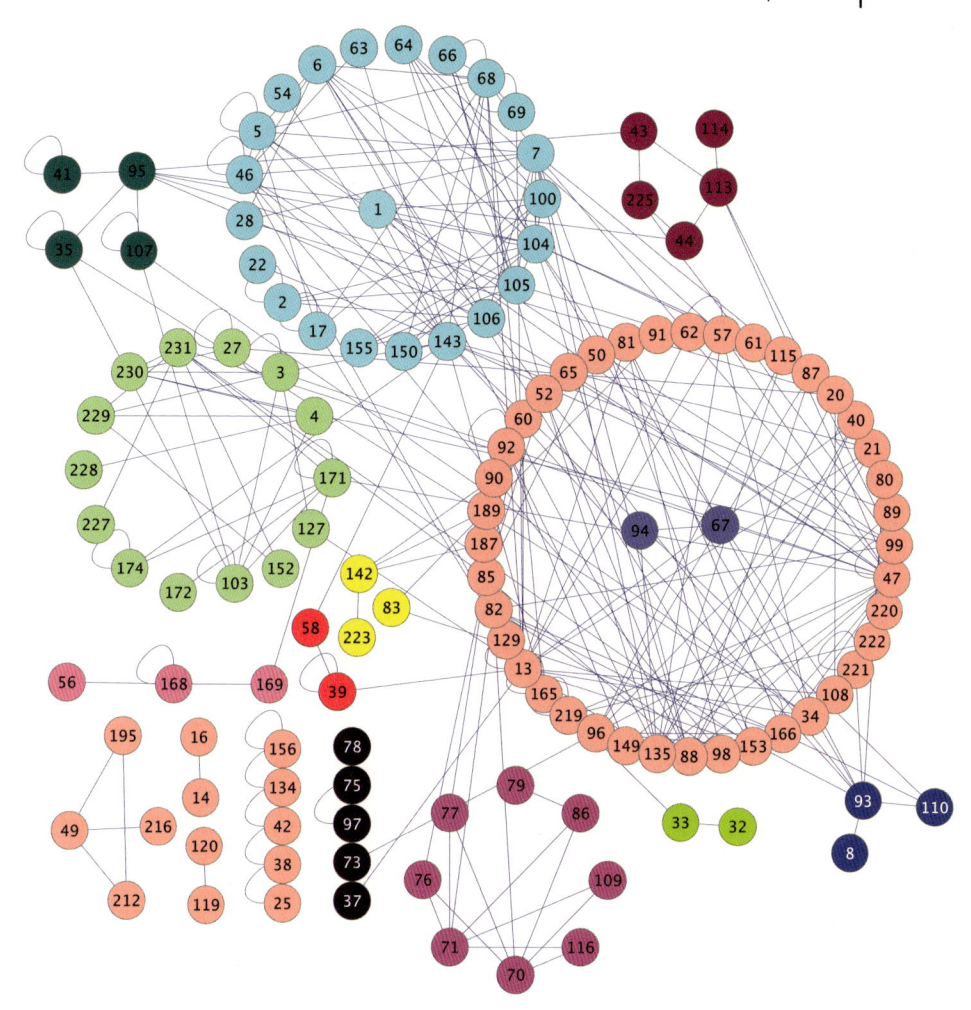

Figure 14.13 NRC/MASC analysis. Communities generated for neurotransmitter receptor complexes using [47].

14.6.4
Study of Wireless Mobile Users

Recently, the wide applicability of hand-held mobile devices (e.g., iPhones) equipped with multisensor (e.g., Bluetooth) capability has given rise to a whole new world of social analysis [48, 49]. The tight coupling [50] between user and device has opened a new avenue to sense their spatiotemporal activities. Continuous logging of such activities provides an insight into the behavioral patterns, likes/dislikes, and preferential attachment to locations of the mobile users. Traditionally, most network research was done without the specific assumption

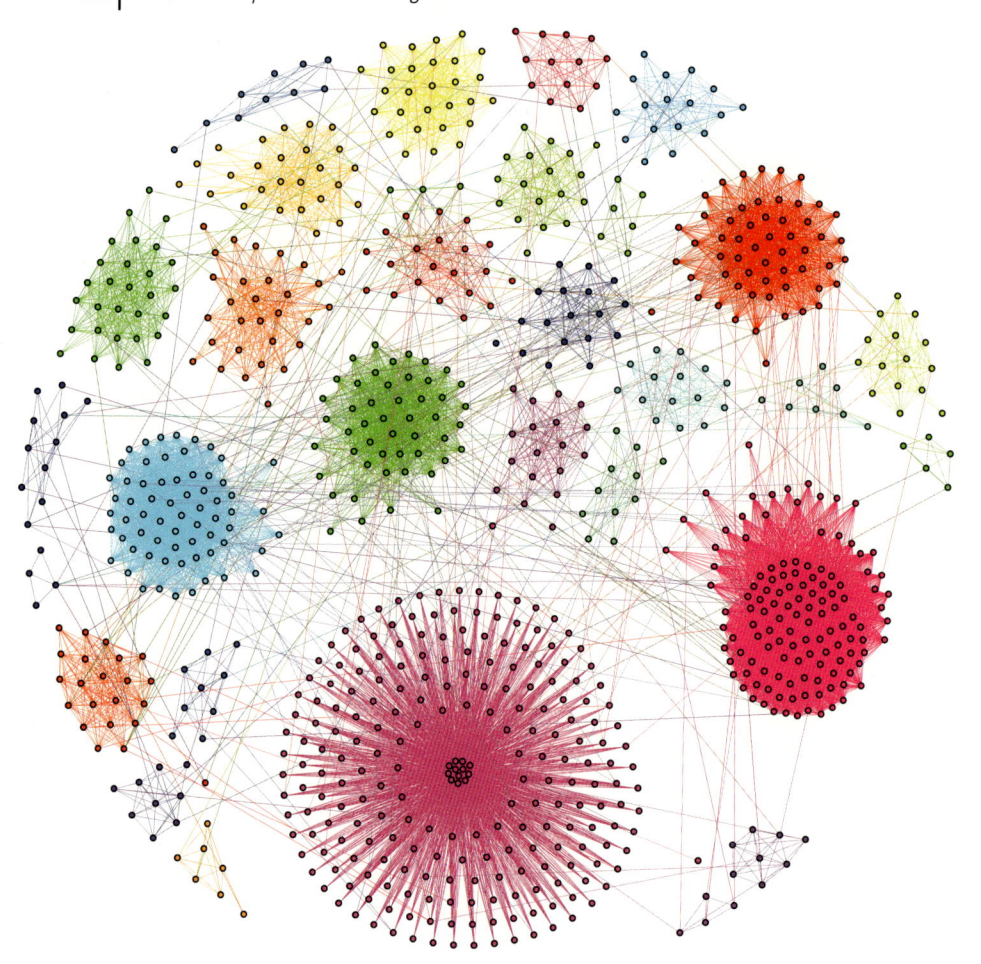

Figure 14.14 In the study of wireless local area network mobile users, 31 active communities are identified based on preferential attachments to on-campus locations. For clarity we removed interconnections among users of different communities. Community structure modeling is done using [53].

of users' presence, including other environmental factors. However, with the adoption of wireless systems there is now a need to develop protocols and networks infrastructures that consider user mobility, surroundings, and communication medium. Specifically, for a better utilization of available resources like bandwidth and limited battery power, we need protocols and services that are behavior-oriented, leveraging the underlying patterns of user activity and their mutual interaction. For example, a study of the density distribution of mobile users on a university campus can help investigate which locations contribute most network connectivity and during which time of the day. Thus, an analysis of such data collected from connecting wireless "access points" can tell us whether network administrators need to provide more bandwidth at these location as compared to other sparse locations.

To understand the macro mobility of mobile users, in this experiment we obtain a dataset consisting of 890 mobile users spread across MIT campus [51] (http://nile.cise.ufl.edu/MobiLib/). We analyze their month-long movements and approximated their preferential attachment to certain campus locations [52]. To understand the interaction patterns among these users we ran our algorithm to identify communities of users based on spatiotemporal similarity. Thus, two users are deemed closest if they visit similar locations most of the times during a 1-month period.

As shown in Figure 14.14, our algorithm finds 31 unique communities, illustrating the richness in user interaction. For more clarity we removed interconnections among users of different communities that are clustered based on an individual's location-visiting preferences. The most dominant clusters shows locations that are highly visited and hence contribute to a large set of mobile users. The largest cluster consists of 25.73% of users followed by 14.16 and 7.3% of mobile users, combining in all more than 50% of users located on university campuses. We also find 23 small clusters where the user population is less than 5% of the total. These results provide a very interesting insight into the distribution of mobile users, and an understanding and characterization of the structure of such community patterns, demonstrating similarity in their behavioral patterns. Invariably, concentrating only on three or four locations, we can identify more than 50% of the population in the dataset. Furthermore, such analysis is very important in developing group-aware services, user modeling, understanding network throughput from a macro mobility level, and discovering the social perspective of mobile users associated with certain locations.

14.7
Further Improvements

Once the agglomeration process is complete, additional activities can be performed to further optimize the discovered community. Successive merging of the communities is an important aspect that involves grouping small communities to form a big enclosed community (when $l < k$). By performing this process, we are also able to avoid possible local minima that are formed during community generation. Secondly, when agglomeration does not optimize top-ranked communities, an exchange operation between complementary ranked communities can instead be performed.

- **Community merging**: A set of small enclosed communities (with $l < k$) can represent a big community, although a single vertex exchange among them does not optimize the solution. Using the basis of vertex and community rank, the algorithm successively merges these communities together to form a bigger community.

- **Complementary exchange**: As discussed before, the top-ranked communities are considered for vertex exchange. Sometimes communities with a high value of Γ may not necessarily generate an optimize solution. They remain divided in themselves and look for those vertices that exist in comparably stable communities. During the course of iterations, if exchange does not generate a better solution, the algorithm marks such communities such that they will not take part in future iterations to save degenerative counts. To further optimize the solution, the algorithm performs an exchange operation among complementary ranked communities with respect to their adjacency and value of Γ.

14.8
Conclusions

In this chapter, we introduced a novel approach to find local community structure, when the complete information of a network is unknown. Using the random walk based on degree centrality, we explored each vertex at a time and added it to the initial set of communities. We then introduced the concept of vertex and community rankings. This helped the iterative agglomeration technique to improve the initial set of communities. During the course of iterative agglomeration, the vertices on the basis of rankings are transferred from one community to another in the search for immediate neighbors. This process brings the high-ranked communities closer to form a clique. Each iteration optimizes the current set of communities until k vertices are explored or a local community structure is discovered. An extension to this algorithm can be made for the identification of global community structure with different source vertex positions.

Computer-simulated experiments proved the effectiveness of the algorithm in both static as well as dynamic networks. Using NMI, we benchmark the proposed algorithm to quantify its accuracy to discover local communities. Experiments performed on the Zachary Karate Club, neurotransmitter receptor complexes, and wireless mobile user datasets show the applicability of the proposed algorithm in a wide range of applications.

Acknowledgments

The author would like to acknowledge his colleagues: Ahmed Helmy and My T. Thai, who helped formulate many of the ideas instrumental in this chapter, and Wei-Jen Hsu, who helped gather datasets for experimental purposes. The author also wishes to thank IET Publications for granting permission to reproduce figures and an opportunity to disseminate an extended version of an earlier work [19].

References

1 Cho, S., Park, S.G., Lee, D.O.H., and Park, B.C. (2004) Protein–protein interaction networks: from interactions to networks. *J. Biochem. Mol. Toxicol.*, **37**, 45–52.

2 Dunne, J.A., Williams, R.J., and Martinez, N.D. (2002) Food-web structure and network theory: the role of connectance and size. *Proc. Natl. Acad. Sci. USA*, **99**, 12917–12922.

3 Wasserman, S. and Faust, K. (1994) *Social Network Analysis*, Cambridge University Press, Cambridge.

4 Abraham, A., Hassanien, A.E., and Snášel, V.E. (2010) *Computational Social Network Analysis: Trends Tools and Research Advances*, 1st edn, Computer Communications and Networks.

5 Newman, M., Barabasi, A.-L., and Watts, D.J. (2006) *The Structure and Dynamics of Networks*, 1st edn, Princeton Studies in Complexity, Princeton University Press, Princeton, NJ.

6 Freeman, L. (1979) Centrality in social networks conceptual clarification. *Soc. Netw.*, **1**, 215–239.

7 Freeman, L.C. (2004) *The Development of Social Network Analysis: A Study in the Sociology of Science*, Empirical Press, Vancouver.

8 Brandes, U. (2001) A faster algorithm for betweenness centrality. *J. Math. Sociol.*, **25**, 163–177.

9 Barthelemy, M. (2004) Betweenness centrality in large complex networks. *Eur. Phys. J. B*, **38**, 163–1688.

10 Okamoto, K., Chen, W., and Li, X.-Y. (2008) Ranking of closeness centrality for large-scale social networks, in *Proceedings of the 2nd Annual International Workshop on Frontiers in Algorithmics*, Springer, Berlin, pp. 186–195.

11 Newman, M.E.J. (2010) *Networks: An Introduction*, Oxford University Press, Oxford.

12 Travers, J. and Milgram, S. (1969) An experimental study of the small world problem. *Sociometry*, **32**, 425–443.

13 Barbosa, V.C., Donangelo, R., and Souza, S.R. (2006) Emergence of scale-free networks from local connectivity and communication trade-offs. *Phys. Rev. E*, **74**, 016113.

14 Barabasi, A.-L. (2009) Scale-free networks: a decade and beyond. *Science*, **325**, 412–413.

15 Stauffer, D. and Aharony, A. (1992) *Introduction to Percolation Theory*, 2nd edn, Taylor & Francis, London.

16 Capocci, A., Servedio, V.D.P., Colaiori, F., Buriol, L.S., Donato, D., Leonardi, S., and Caldarelli, G. (2006) Preferential attachment in the growth of social networks: the internet encyclopedia Wikipedia. *Phys. Rev. E*, **74**, 036116.

17 Bagrow, J.P. and Bollt, E.M. (2005) Local method for detecting communities. *Phys. Rev. E*, **72**, 046108.

18 Clauset, A. (2005) Finding local community structure in networks. *Phys. Rev. E*, **72**, 026132.

19 Thakur, G., Tiwari, R., Thai, M., Chen, S.-S., and Dress, A. (2009) Detection of local community structures in complex dynamic networks with random walks. *IET Syst. Biol.*, **3**, 266–278.

20 Gloor, P.L., Laubacher, R., Dynes, S., and Zhao, Y. (2003) Visualization of interaction patterns in collaborative knowledge networks for medical applications. Human–Computer Interaction Conference, Crete.

21 Fortunato, S., Latora, V., and Marchiori, M. (2004) Method to find community structures based on information centrality. *Phys. Rev. E*, **70**, 056104.

22 Girvan, M. and Newman, M.E.J. (2002) Community structure in social and biological networks. *Proc. Natl. Acad. Sci. USA*, **99**, 7821.

23 Pothen, A., Simon, H.D., and Liou, K.-P. (1990) Partitioning sparse matrices with eigenvectors of graphs. *SIAM J. Matrix Anal. A*, **11**, 430–452.

24 Fiedler, M. (1973) Algebraic connectivity of graphs. *Czech. Math J.*, **23**, 298–305.

25 Kernighan, B.W. and Lin, S. (1970) An efficient heuristic procedure for partitioning graph. *Bell Syst. Tech. J.*, **49**, 291–307.

26 Chen, W.Y.C., Dress, A.W.M., and Yu, Q. (2007) Checking the reliability of a new approach towards detecting community structures in networks using linear programming. *IET Syst. Biol.*, **1**, 286–291.

27 Almeida, R.B. and Almeida, V.A.F. (2002) Local community identification through user access patterns. *Clin. Orthop. Relat. Res.*, cs.IR/0212045.

28 Farutin, V., Robison, K., Lightcap, E., Dancik, V., Ruttenberg, A., Letovsky, S., and Pradines, J. (2006) Edge-count probabilities for the identification of local protein communities and their organization. *Proteins*, **62**, 800–818.

29 Palla, G., Derenyi, I., Farkas, I., and Vicsek, T. (2005) Uncovering the overlapping community structure of complex networks in nature and society. *Nature*, **435**, 814.

30 Spirin, V. and Mirny, L.A. (2003) Protein complexes and functional modules in molecular networks. *Proc. Natl. Acad. Sci. USA*, **100**, 12123–12128.

31 de Solla Price, D.J. (1965) Networks of scientific papers. *Science*, **149**, 510–515.

32 Ito, T., Chiba, T., Ozawa, R., Yoshida, M., Hattori, M., and Sakaki, Y. (2001) A comprehensive two-hybrid analysis to explore the yeast protein interactome. *Proc. Natl. Acad. Sci. USA*, **98**, 4569–4574.

33 Kleinberg, J.M., Kumar, R., Raghavan, P., Rajagopalan, S., and Tomkins, A.S. (1999) The web as a graph: measurements, models and methods. *Lecture Notes Comput. Sci.*, **1627**, 1–17.

34 Fouss, F., Pirotte, A., and Saerens, M. (2005) A novel way of computing similarities between nodes of a graph, with application to collaborative recommendation. IEEE/WIC/ACM International Conference on Web Intelligence, Compiegne.

35 Aldous, D. and Fill, J.A. (2002) A second look at general Markov chains, in Reversible Markov Chains and Random Walks on Graphs, University of California, Berkeley.

36 Lovsz, L. (1993) *Combinatorics, Paul Erdos is Eighty, Bolyai Society Mathematical Studies*, János Bolyai Mathematical Society, Budapest.

37 Latapy, M. and Pons, P. (2005) Computing communities in large networks using random walks. *Lecture Notes Comput. Sci.*, **3733**, 284–293.

38 Garey, M.R. and Johnson, D.S. (1979) *Computers and Intractability: A Guide to the Theory of NP-Completeness*, Freeman, San Francisco, CA.

39 Newman, M.E.J. and Girvan, M. (2004) Finding and evaluating community structure in networks. *Phys. Rev. E*, **69**, 026113.

40 Bagrow, J.P. (2008) Evaluating local community methods in networks. *J. Stat. Mech. Theory. Exp.*, P05001.

41 Danon, L., Díaz-Guilera, A., Duch, J., and Arenas, A. (2005) Comparing community structure identification. *J. Stat. Mech. Theory. Exp.*, P09008.

42 Fred, A.L. and Jain, A.K. (2003) Robust data clustering. IEEE Computer Society Conference on Computer Vision and Pattern Recognition, Madison, WI.

43 Strehl, A. and Ghosh, J. (2002) Cluster ensembles – a knowledge reuse framework for combining multiple partitions. *J. Mach. Learn. Res.*, **3**, 583–617.

44 McCreary, C. and Barowski, L. (1998) VGJ: visualizing graphs through Java. *Lecture Notes Comput. Sci.*, **1547**, 454–455.

45 Zachary, W.W. (1977) An information flow model for conflict and fission in small groups. *J. Anthropol. Res.*, **33**, 452–473.

46 Pocklington, A.J., Armstrong, J.D., and Grant, S.G.N. (2006) Organization of brain complexity – synapse proteome form and function. *Brief. Funct. Genomics*, **5**, 66–73.

47 Shannon, P., Markiel, A., Ozier, O., Baliga, N.S., Wang, J.T., Ramage, D., Amin, N., Schwikowski, B., and Ideker, T. (2003) Cytoscape: a software environment for integrated models of biomolecular interaction networks. *Genome Res.*, **13**, 2498–2504.

48 Hui, P., Yoneki, E., Chan, S.Y., and Crowcroft, J. (2007) Distributed community detection in delay tolerant networks. Second ACM/IEEE International Workshop on Mobility in the Evolving Internet Architecture, New York.

49 Hsu, W.-J. and Helmy, A. (2006) On nodal encounter patterns in wireless LAN traces.

Second Workshop on Wireless Network Measurements, Boston, MA.

50 Hsu, W.-J., Dutta, D., and Helmy, A. (2008) Profile-cast: behavior-aware mobile networking. *SIGMOBILE Mob. Comput. Commun. Rev.*, **12**, 52–54.

51 Yeo, J., Kotz, D., and Henderson, T. (2006) CRAWDAD: a community resource for archiving wireless data at Dartmouth. *ACM SIGCOMM Comp. Commun. Rev.*, **36**, 21–22; Project overview.

52 Kotz, D. and Essien, K. (2005) Analysis of a campus-wide wireless network. *Wirel. Netw.*, **11**, 115–133.

53 Bastian, M., Heymann, S., and Jacomy, M. (2009) Gephi: an open source software for exploring and manipulating networks. International AAAI Conference on Weblogs and Social Media, San Jose, CA.

15
On Some Inverse Problems in Generating Probabilistic Boolean Networks
Xi Chen, Wai-Ki Ching, and Nam-Kiu Tsing

15.1
Introduction

The study of genetic regulatory networks and the development of efficient numerical algorithms for the regulatory interactions among DNA, RNA, proteins, and small molecules are important research topics in bioinformatics and the postgenome era [1, 2]. Understanding the complex patterns of behavior of the interactions among genes and such molecules creates a scientific challenge with potentially high payoffs. Thus, a lot of formalisms for modeling and simulation of gene regulation processes have been developed. The mathematical models include Bayesian networks [3, 4], Boolean networks (BNs) [5–7], correlation-based methods [8], multivariate Markov chain models [9], linear regression models [10], evolutionary models [11], and probabilistic Boolean networks (PBNs) [12–15]. Interested readers can find reviews on these models in [16, 17]. With such models and algorithms, the properties of the models and hypotheses can be tested via computer simulations, and this can save both industrial costs and experimental time.

Among all these mathematical models, BNs and their extension PBNs have received much attention as they are able to capture the switching behavior of a biological process [2]. In a BN model, each gene of a network is regarded as a vertex of a graph and the gene expression states are quantized to only two levels: on and off (represented as 1 and 0). We remark that more states can be assigned, although this will increase the computational difficulty. The target gene can be predicted by several genes called its input genes via a *Boolean function (predictor function)*. When all the input genes and Boolean functions are defined or given, we say that a BN is then defined. Actually, a BN is a deterministic model, the only randomness comes from its initial state. Considering the inherent deterministic directionality in BNs as well as only a finite number of possible states, it is straightforward to see that some states will be revisited infinitely, depending on the initial state. Such states are called *attractors* and the states that lead to them comprise their *basins of attraction*. The number of transitions needed to return to a given state in an attractor is called the *cycle length* [5, 18]. It is also well known that a BN will eventually enter into an attractor cycle and stay there forever. In fact, the cycles have biological significance [19], such as cell proliferation or cell apoptosis. For more details we refer the interested reader to [7, 15] and references therein.

Since the genetic regulation process exhibits uncertainty and the microarray datasets used to infer the model may have errors due to experimental noise in the complex measurement process, it is therefore more realistic to consider a probabilistic approach. We consider an extension of a BN to a class of stochastic models, PBNs. In a PBN, each gene can have more than one Boolean function (a set of Boolean functions with probabilities assigned to them). For the inference of a PBN, the dynamics (transitions) of a PBN can be studied using Markov chain theory [12, 15, 20]. The Boolean functions, the predictor sets, and the selection probabilities of the Boolean functions can be obtained using the methods proposed in [21–23]. Given a PBN, assuming the underlying Markov chain is irreducible, its long-run behavior is characterized by its steady-state probability distribution. It gives the first-order statistical information of a PBN. One can then understand a genetic network and identify the influence of different genes via such a network. An iterative method – a power method in conjunction with an efficient construction method for the transition probability matrix – has been used to get an approximation of the stationary distribution [24]. A matrix approximation method has also been proposed in [25] to compute an approximation of the stationary distribution when the network size is large. Furthermore, it is possible to control one or more genes in a network that can drive the whole network into a desirable state or a stationary distribution (a mixture of states). Therapeutic gene intervention or gene control policy can be developed and studied (e.g., [15, 18, 26]). Moreover, PBNs can also provide a logic model of pathway biology for potential drug discovery (e.g., [27]).

This chapter introduces some inverse problems arising from the construction of PBNs and they have been discussed in [28, 29]. The remainder of this chapter is structured as follows. In Section 15.2, we give a brief introduction to BNs and PBNs. In Section 15.3, we consider the problem of constructing a PBN from a given stationary probability distribution and a given set of BNs [29]. This is an important inverse problem of large size in network inference with steady-state data, as most microarray datasets are assumed to be obtained from sampling the steady state. This problem is ill-posed, which means that there can be many networks or even no network having the desirable properties. A modified conjugate gradient (CG) method is proposed to solve the problem. In fact, for the case of BNs, Pal *et al.* [30] have proposed two algorithms to find the attractors constituting a BN. However, to the best of our knowledge, the PBN case has not been addressed in the literature. In Section 15.4, we study the problem of constructing a PBN from a given transition probability matrix [28]. We propose some efficient heuristic algorithms for this problem. Finally, in Section 15.5, concluding remarks are given to address further research issues.

15.2 Reviews on BNs and PBNs

In this section, we briefly introduce the ideas of a BN and its extension PBN.

15.2.1 BNs

A BN $G(V, F)$ is represented by a set of vertices (genes):

$$V = \{v_1, v_2, \ldots, v_n\}$$

and a set of Boolean functions:

$$F = \{f_1, f_2, \ldots, f_n\}$$

where:

$$f_i : \{0, 1\}^n \to \{0, 1\}$$

We define $v_i(t)$ to be the state (0 or 1) of the gene i at time t. The rules of the regulatory interactions among the genes can then be represented by Boolean functions:

$$v_i(t+1) = f_i(\mathbf{v}(t)), \quad i = 1, 2, \ldots, n \tag{15.1}$$

where the Boolean vector:

$$\mathbf{v}(t) = (v_1(t), v_2(t), \ldots, v_n(t))^\mathrm{T}$$

is called the gene activity profile (GAP). The GAP can take any possible forms (states) from the set:

$$S = \{(v_1, v_2, \ldots, v_n)^\mathrm{T} : v_i \in \{0, 1\}\} \tag{15.2}$$

and thus totally there are 2^n possible states in the network.

The following is an example of a BN having two genes with the truth table given in Table 15.1. From the truth table, there are four states: (0,0), (0,1), (1,0), and (1,1). Let us label them by 1, 2, 3, and 4, respectively. We note that if the current state of the network is 1, the network will go to state 3 in the next step (with probability 1). Suppose the current state is 3, the network will go to state 1 in the next step (with

Table 15.1 Truth table.

States	$v_1(t)$	$v_2(t)$	$f^{(1)}$	$f^{(2)}$
1	0	0	1	0
2	0	1	1	0
3	1	0	0	0
4	1	1	1	1

probability 1). Similarly, if the current state is 2, the network will go to state 3 in the next step (with probability 1). Finally, if the current network state is 4, the network will stay with state 4 in the next step (with probability 1). The transition probability matrix (BN matrix) of the two-gene BN is then given by:

$$B = \begin{pmatrix} 0 & 0 & 1 & 0 \\ 0 & 0 & 0 & 0 \\ 1 & 1 & 0 & 0 \\ 0 & 0 & 0 & 1 \end{pmatrix} \quad (15.3)$$

The truth table gives the one-step transition probability (0 or 1 in the case of a BN) between any two states. Let column vectors **a** and **b** be any two states in the set S. By letting **a** and **b** take all the possible states in S, one can get the transition probability matrix of the two-gene BN. Since the network is a deterministic one, each column in B (the BN matrix) has only one nonzero element and the column sum is 1. We observe that there are two cycles (attractors) of period 1 and 2 given, respectively, as follows: (i) $(1,1) \leftrightarrow (1,1)$ and (ii) $(0,0) \leftrightarrow (1,0)$. Moreover, $(1,1)$ belongs to the basin of attraction of the one-period attractor cycle. Finally, we remark that there is a one-to-one relation between a BN and its corresponding BN matrix. This means if we are given a BN, there is a unique (up to ordering of states) BN matrix B to describe its dynamics and vice versa.

15.2.2
PBNs

Since a BN is a deterministic model, it is more realistic to extend it to a probabilistic setting. To extend the concepts of a BN to a stochastic model, for each vertex v_i in a PBN, instead of having only one Boolean function as in a BN, there are a number of Boolean functions (predictor functions) $f_j^{(i)}$ ($j = 1, 2, \ldots, l(i)$) to be chosen for determining the state of gene v_i and usually $l(i)$ is not very large. The probability of choosing $f_j^{(i)}$ as the predictor function is $c_j^{(i)}$:

$$0 \leq c_j^{(i)} \leq 1 \quad \text{and} \quad \sum_{j=1}^{l(i)} c_j^{(i)} = 1 \quad \text{for} \quad i = 1, 2, \ldots, n \quad (15.4)$$

In [21], the authors provide a method named coefficient of determination (COD), which can be used to estimate the probability $c_j^{(i)}$ with real gene expression datasets. We let f_j be the jth possible realization, where:

$$f_j = \left(f_{j_1}^{(1)}, f_{j_2}^{(2)}, \ldots, f_{j_n}^{(n)}\right), \quad 1 \leq j_i \leq l(i), \quad i = 1, 2, \ldots, n \quad (15.5)$$

Suppose that the selection of the Boolean function f_{j_i} for each gene i is an independent process, then the probability of choosing the corresponding BN with Boolean functions $(f_{j_1}, f_{j_2}, \ldots, f_{j_n})$ is given by:

$$q_{j_1 j_2 \ldots j_n} = \prod_{i=1}^{n} c_{j_i}^{(i)} \quad (15.6)$$

There are at most:

$$N = \prod_{i=1}^{n} l(i) \tag{15.7}$$

different possible realizations of BNs. We note that the transition process among the states in the set S forms a Markov chain process. Let **a** and **b** be any two column vectors in the set S. Then the transition probability:

$$\begin{aligned} P\{\mathbf{v}(t+1) = \mathbf{a} | \mathbf{v}(t) = \mathbf{b}\} = \\ \sum_{j=1}^{N} P\{\mathbf{v}(t+1) = \mathbf{a} | \mathbf{v}(t) = \mathbf{b}, \text{ the } j\text{th network is selected}\} \cdot q_j \end{aligned} \tag{15.8}$$

Here, we let:

$$q_j = q_{j_1 j_2, \dots, j_n} \quad \text{and} \quad j = j_1 + \sum_{i=2}^{n} \left((j_i - 1) \left(\prod_{k=1}^{i-1} l(k) \right) \right) \tag{15.9}$$

We can then use both of them when there is no confusion. By letting **a** and **b** take all the possible states in S, one can get the transition probability matrix of the Markov chain (or the PBN). The transition probability matrix can be written as [25]:

$$A = \sum_{j=1}^{N} q_j A_j \tag{15.10}$$

where A_j is the corresponding transition matrix of the jth BN (see Equation 15.3, for instance) and q_j is the probability of choosing the jth BN. We note that there are at most $N2^n$ nonzero entries in the transition probability matrix A and this means the matrix is very sparse (i.e., it has a lot of zero entries).

Note that there are several different kinds of PBNs. The instantaneously random PBN described above is the simplest one. To stabilize the network, one can add random gene perturbation to a PBN. The random gene perturbation is a description of the random inputs from the outside due to external stimuli. The effect of it is to make the genes flip from state 1 to state 0 or vice versa. This makes the underlying Markov chain of the PBN ergodic [31]. The instantaneously random PBN can also be extended to the context-sensitive PBN [32]. For simplicity of discussion, here we focus only on the class of instantaneously random PBNs.

15.3
Construction of PBNs from a Prescribed Stationary Distribution

In this section, we present the inverse problem of building a PBN from a prescribed steady-state distribution and we then present the modified CG method for solving the inverse problem.

Suppose that the possible BNs constituting the PBN are given and they are denoted by $(A_1, A_2, \ldots, A_N)$ and suppose steady-state behavior of the PBN, the stationary distribution $\mathbf{p}$ can also be observed. Then we have:

$$A\mathbf{p} = \mathbf{p} \quad \text{and} \quad A = \sum_{j=1}^{N} q_j A_j$$

Here, the inverse problem is to get the parameters q_j, $j = 1, 2, \ldots, N$. Now we let the matrix:

$$V = [\mathbf{v}_1, \mathbf{v}_2, \ldots, \mathbf{v}_N] \tag{15.11}$$

where:

$$\mathbf{v}_j = A_j \mathbf{p} \quad \text{and} \quad \mathbf{q} = (q_1, q_2, \ldots, q_N)^T \tag{15.12}$$

Then one possible way to get q_j is to consider the following minimization problem:

$$h(\mathbf{q}^*) = \min_{\mathbf{q}} \|V\mathbf{q} - \mathbf{p}\|_2^2 \tag{15.13}$$

subject to:

$$0 \le q_j \le 1 \quad \text{and} \quad \sum_{j=1}^{N} q_j = 1 \tag{15.14}$$

15.3.1
CG Method Approach

We note that in practice the matrix V is usually very large. It may not be possible to store the whole matrix and therefore one may seek an iterative method for solving the above minimization. One possible way is to consider the CG method (e.g., [33], p. 470). Given a symmetric positive definite $m \times m$ matrix H_m, a well-known and successful iterative method for solving the linear system $H_m \mathbf{x} = \mathbf{b}$ is the CG method. In general, the CG method will converge in at most m steps [33] (i.e., the number of iterations required for convergence is of $O(m)$).

We observe that:

$$\|V\mathbf{q} - \mathbf{p}\|_2^2 = (V\mathbf{q} - \mathbf{p})^T (V\mathbf{q} - \mathbf{p}) \tag{15.15}$$

and:

$$(V\mathbf{q} - \mathbf{p})^T (V\mathbf{q} - \mathbf{p}) = \mathbf{q}^T V^T V \mathbf{q} - 2\mathbf{q}^T V^T \mathbf{p} + \mathbf{p}^T \mathbf{p} \tag{15.16}$$

Thus, the minimization problem (15.13) without constraints is equivalent to:

$$\min_{\mathbf{q}} \{\mathbf{q}^T V^T V \mathbf{q} - 2\mathbf{q}^T V^T \mathbf{p}\} \tag{15.17}$$

15.3 Construction of PBNs from a Prescribed Stationary Distribution

If V is a full rank matrix then $V^T V$ is a symmetric positive definite matrix. The minimization problem without constraints is equivalent to solving:

$$V^T V \mathbf{q} = V^T \mathbf{p} \quad (15.18)$$

with the CG method (e.g., [34]) We note that if there is a probability distribution $\mathbf{q}$ satisfying the equation $V\mathbf{q} = \mathbf{p}$ with $\mathbf{1}^T \mathbf{q} = 1$ and $0 \leq q_i \leq 1$ for all i, then the CG method can yield the solution. Here, $\mathbf{1}$ is the vector of all 1s. To ensure that $\mathbf{1}^T \mathbf{q} = 1$, we add a row of $(w, w, \ldots, w)$ to the bottom of the matrix V to form a new matrix $\overline{V}$. At the same time, we add an entry w at the end of the vector $\mathbf{p}$ to get a new vector $\overline{\mathbf{p}}$. Here, w is a large positive number so as to make sure the constraint $\mathbf{1}^T \mathbf{q} = 1$ is active. Thus, we consider the revised equation:

$$\overline{V}^T \overline{V} \mathbf{q} = \overline{V}^T \overline{\mathbf{p}} \quad (15.19)$$

Since sometimes there may be no such vector $\mathbf{q}$, the CG algorithm has to be modified to ensure that the constraint $0 \leq q_i \leq 1$ is satisfied in each iteration step. We have to run the modified CG method for a number of times with different initial guesses to get the best solution in the sense of the smallest residual error in $\|\cdot\|_2^2$. We remark that the main computational cost of the CG method comes from the matrix-vector multiplication that takes $O(N2^n)$ operations. Since the number of iteration for convergence is $O(N)$, if we run the CG method for T times with T different initial guesses, then the total computational cost will be $O(TN^2 2^n)$. We remark that the PBN generated depends on the initial guess and is not unique.

15.3.2
Estimation of $C_j^{(i)}$

Once we get the estimates of the probabilities $q_{j_1 j_2, \ldots, j_n}$, we may have N equations of the form:

$$\prod_{i=1}^{n} c_{j_i}^{(i)} = q_{j_1 j_2, \ldots, j_n} \quad (15.20)$$

For ease of presentation of the analysis, in the following, we consider the special case that $l(i) = 2$ for $i = 1, 2, \ldots, n$. We remark that the techniques can be applied similarly to the cases $l(i) \geq 3$ for $i = 1, 2, \ldots, n$

Now we have $c_2^{(i)} = 1 - c_1^{(i)}$ and $N = 2^n$. To estimate $c_1^{(k)}$, we note that for $j_k = 1, 2$, we have, respectively:

$$c_{j_1}^{(1)}, \ldots, c_{j_{k-1}}^{(k-1)} c_1^{(k)} c_{j_{k+1}}^{(k+1)}, \ldots, c_{j_n}^{(n)} = q_{j_1, \ldots, j_{k-1} 1 j_{k+1}, \ldots, j_n} \quad (15.21)$$

and:

$$c_{j_1}^{(1)}, \ldots, c_{j_{k-1}}^{(k-1)} (1 - c_1^{(k)}) c_{j_{k+1}}^{(k+1)}, \ldots, c_{j_n}^{(n)} = q_{j_1, \ldots, j_{k-1} 2 j_{k+1}, \ldots, j_n} \quad (15.22)$$

Therefore, for any $j_1,\ldots,j_{k-1},j_{k+1},\ldots,j_n \in \{1,2\}$, we have:

$$\frac{c_1^{(k)}}{1-c_1^{(k)}} = \frac{q_{j_1,\ldots,j_{k-1}1j_{k+1},\ldots,j_n}}{q_{j_1,\ldots,j_{k-1}2j_{k+1},\ldots,j_n}} \tag{15.23}$$

Or equivalently:

$$c_1^{(k)} r_{j_1,\ldots,j_n}^{(k)} - q_{j_1,\ldots,j_{k-1}1j_{k+1},\ldots,j_n} = 0 \tag{15.24}$$

where:

$$r_{j_1,\ldots,j_n}^{(k)} = q_{j_1,\ldots,j_{k-1}1j_{k+1},\ldots,j_n} + q_{j_1,\ldots,j_{k-1}2j_{k+1},\ldots,j_n} \tag{15.25}$$

Since there may not exist $c_1^{(k)}$ satisfying all the equations, one possible way to estimate $c_1^{(k)}$ is to consider the minimizer of the following functional:

$$J\left(c_1^{(k)}\right) = \sum_{j_1,\ldots,j_{k-1}j_{k+1},\ldots,j_n \in \{1,2\}} \left(c_1^{(k)} r_{j_1,\ldots,j_n}^{(k)} - q_{j_1,\ldots,j_{k-1}1j_{k+1},\ldots,j_n}\right)^2 \tag{15.26}$$

and $0 \le c_1^{(k)} \le 1$. The minimizer of the problem (15.26) can be easily shown to be:

$$c_1^{(k)*} = \frac{\sum_{j_1,\ldots,j_{k-1}j_{k+1},\ldots,j_n \in \{1,2\}} r_{j_1,\ldots,j_n}^{(k)} \times q_{j_1,\ldots,j_{k-1}1j_{k+1},\ldots,j_n}}{\sum_{j_1,\ldots,j_{k-1}j_{k+1},\ldots,j_n \in \{1,2\}} \left(r_{j_1,\ldots,j_n}^{(k)}\right)^2} \tag{15.27}$$

It is straightforward to show that $c_1^{(k)*} \in [0,1]$ by using (15.25).

As there are many solutions, to evaluate the solution obtained, we define:

$$J\left(c_1^{(1)},\ldots,c_1^{(n)}\right) = \sum_{k=1}^{n} J\left(c_1^{(k)}\right)$$

as a measure of the fitness of the estimators. The smaller this value is, the better the estimators are. Thus, one may use:

$$J\left(c_1^{(1)*},\ldots,c_1^{(n)*}\right) = \sum_{k=1}^{n} J\left(c_1^{(k)*}\right) \tag{15.28}$$

together with the optimal value $h(\mathbf{q}^*)$ in (15.13) to rank different PBNs obtained.

15.3.3
Numerical Example

In this subsection, we give an example of a three-gene network taken from [29].

Example 15.1

In the first example, we consider a PBN having three genes $n = 3$. The truth table of the Boolean functions is given in Table 15.2.

15.3 Construction of PBNs from a Prescribed Stationary Distribution

Table 15.2 Truth table of the BNs (taken from [29]).

State	$(v_1(t), v_2(t), v_3(t))$	$f_1^{(1)}$	$f_2^{(1)}$	$f_1^{(2)}$	$f_1^{(3)}$	$f_2^{(3)}$
1	(0,0,0)	0	0	0	1	0
2	(0,0,1)	1	0	1	1	0
3	(0,1,0)	1	0	0	0	1
4	(0,1,1)	1	1	1	1	0
5	(1,0,0)	0	1	1	0	0
6	(1,0,1)	0	1	1	0	1
7	(1,1,0)	0	1	0	0	1
8	(1,1,1)	1	1	0	1	0

Since Genes 1 and 3 have two possible Boolean functions, there are four possible BNs. The transition probability matrices of the BNs A_1, A_2, A_3, A_4 are given as follows:

$$A_1 = \begin{pmatrix} 0 & 0 & 0 & 0 & 0 & 0 & 1 & 0 \\ 1 & 0 & 0 & 0 & 0 & 0 & 0 & 0 \\ 0 & 0 & 0 & 0 & 1 & 1 & 0 & 0 \\ 0 & 0 & 0 & 0 & 0 & 0 & 0 & 0 \\ 0 & 0 & 1 & 0 & 0 & 0 & 0 & 0 \\ 0 & 0 & 0 & 0 & 0 & 0 & 0 & 1 \\ 0 & 0 & 0 & 0 & 0 & 0 & 0 & 0 \\ 0 & 1 & 0 & 1 & 0 & 0 & 0 & 0 \end{pmatrix} \quad A_2 = \begin{pmatrix} 1 & 0 & 0 & 0 & 0 & 0 & 0 & 0 \\ 0 & 0 & 0 & 0 & 0 & 0 & 1 & 0 \\ 0 & 0 & 0 & 0 & 1 & 0 & 0 & 0 \\ 0 & 0 & 0 & 0 & 0 & 1 & 0 & 0 \\ 0 & 0 & 0 & 0 & 0 & 0 & 0 & 1 \\ 0 & 0 & 1 & 0 & 0 & 0 & 0 & 0 \\ 0 & 1 & 0 & 1 & 0 & 0 & 0 & 0 \\ 0 & 0 & 0 & 0 & 0 & 0 & 0 & 0 \end{pmatrix} \quad (15.29)$$

$$A_3 = \begin{pmatrix} 0 & 0 & 1 & 0 & 0 & 0 & 0 & 0 \\ 1 & 0 & 0 & 0 & 0 & 0 & 0 & 0 \\ 0 & 0 & 0 & 0 & 0 & 0 & 0 & 0 \\ 0 & 0 & 0 & 0 & 0 & 0 & 0 & 0 \\ 0 & 0 & 0 & 0 & 0 & 0 & 1 & 0 \\ 0 & 0 & 0 & 0 & 0 & 0 & 0 & 1 \\ 0 & 1 & 0 & 0 & 1 & 1 & 0 & 0 \\ 0 & 0 & 0 & 1 & 0 & 0 & 0 & 0 \end{pmatrix} \quad A_4 = \begin{pmatrix} 1 & 0 & 0 & 0 & 0 & 0 & 0 & 0 \\ 0 & 0 & 1 & 0 & 0 & 0 & 0 & 0 \\ 0 & 1 & 0 & 0 & 0 & 0 & 0 & 0 \\ 0 & 0 & 0 & 0 & 0 & 0 & 0 & 0 \\ 0 & 0 & 0 & 0 & 0 & 0 & 0 & 1 \\ 0 & 0 & 0 & 0 & 0 & 0 & 1 & 0 \\ 0 & 0 & 0 & 1 & 1 & 0 & 0 & 0 \\ 0 & 0 & 0 & 0 & 0 & 1 & 0 & 0 \end{pmatrix} \quad (15.30)$$

Suppose:

$$\left(c_1^{(1)}, c_2^{(1)}\right) = (0.25, 0.75) \quad \text{and} \quad \left(c_1^{(2)}, c_2^{(2)}\right) = (0.60, 0.40) \quad (15.31)$$

then we have:

$$(q_1, q_2, q_3, q_4) = (0.15, 0.10, 0.45, 0.30) \quad (15.32)$$

It is straightforward to check that the stationary distribution is given by:

$$\mathbf{p} = (0.1394, 0.1394, 0.1002, 0.0127, 0.1574, 0.1271, 0.2570, 0.0667)^\mathrm{T} \quad (15.33)$$

Now we may assume that **p** is the observed stationary distribution, and we wish to apply the modified CG method to find q_i and $c_j^{(i)}$ pretending that q_i and $c_j^{(i)}$ are not known. Using the modified CG method with $w = 100$, one can recover the solution in (15.32).

To obtain the probabilities $c_1^{(1)}, c_2^{(1)}, c_1^{(2)}$ and $c_2^{(2)}$ we have to make use of the following equations:

$$\begin{cases} c_1^{(1)} c_1^{(2)} = q_1 = 0.15 \\ c_1^{(1)} c_2^{(2)} = c_1^{(1)}(1-c_1^{(2)}) = q_2 = 0.10 \\ c_2^{(1)} c_1^{(2)} = q_3 = 0.45 \\ c_2^{(1)} c_2^{(2)} = c_2^{(1)}(1-c_1^{(2)}) = q_4 = 0.30 \end{cases} \quad (15.34)$$

Solving the above equations, we have the solutions as:

$$c_1^{(1)*} = \frac{1}{4}, \quad c_2^{(1)*} = \frac{3}{4}, \quad c_1^{(2)*} = \frac{3}{5}, \quad c_2^{(2)*} = \frac{2}{5} \quad (15.35)$$

same as (15.31). The sum of squares of errors $J(c_1^{(1)*}, c_1^{(2)*}) = 0$ in this example.

15.3.4
Computational Cost Analysis

In general, in the case of n genes and each gene has two possible Boolean functions there are $(2^{2^n})^{2n}$ truth tables and for each truth table there are $N = 2^n$ BNs. The computational cost for examining all the possible PBNs will be $O((2^{2^n})^{2n} TN^2 2^n) = O(T2^{n2^{n+1}+3n})$. Thus, to find the possible BNs (i.e., the matrices A_i) is still a challenging problem for future research.

15.4
Construction of PBNs from a Prescribed Transition Probability Matrix

In this section, we first present another inverse problem of constructing a PBN from a given transition probability matrix A and a set of BNs $\{A_i\}_{i=1}^M$. We then present some efficient heuristic algorithms for the problem discussed in [28].

We are interested in estimating the unknown parameters $q_i, i = 1, 2, \ldots, M$ when A is the given $2^n \times 2^n$ transition probability matrix. We suppose that each column of A has at most m nonzero entries. Thus, we have at most $m2^n$ equations for the unknown parameters when we consider the equality in (15.10). We remark that this is an inverse problem of huge size and the matrix A is usually very sparse. In [35], a maximum entropy rate method has been proposed to solve this problem. However, the computational cost is large.

15.4 Construction of PBNs from a Prescribed Transition Probability Matrix

We note that q_i and A_i are non-negative and there are at most $m2^n$ nonzero entries in A. In the following, we propose two simple and fast algorithms for constructing PBNs from a given transition probability matrix A. In particular, the complexity of the algorithm is $O(m2^n)$ and we have:

$$A = \sum_{i=1}^{M} q_i A_i$$

where:

$$M \leq m2^n$$

15.4.1
Heuristic Algorithms

Algorithm 15.1

Step 0: Set $R_1 = A$; $k = 0$
Step 1: $k \to k+1$
Step 2: Choose the smallest nonzero entry q_k from R_k. Then for each of the other columns, find the largest entry. All the entries are bigger than or equal to q_k. Suppose the concerned entries are given by

$$[R_k]_{k_1,1}, [R_k]_{k_2,2}, \ldots, [R_k]_{k_n,2^n}$$

Then we define the following BN matrix:

$$A_k = [e_{k_1,1}, \ldots, e_{k_{2^n},2^n}]$$

Here $e_{j,i}$ is the unit column vector whose jth entry is 1 for $i = 1, \ldots, 2^n$.
Step 3: $R_{k+1} = R_k - q_k A_k$
Step 4: If R_{k+1} is the zero matrix then go to Step 5; otherwise go to Step 1.
Step 5: $M = k$ and $A = \sum_{i=1}^{M} q_i A_i$.

Algorithm 15.2

Step 0: Set $R_1 = A$; $k = 0$
Step 1: $k \to k+1$
Step 2: For each of columns of A, find the largest entry. Suppose the concerned entries are given by:

$$[R_k]_{k_1,1}, [R_k]_{k_2,2}, \ldots, [R_k]_{k_{2^n},2^n}$$

and choose the smallest nonzero entry q_k from $[R_k]_{k_i,i}(i = 1, \ldots, 2^n)$. Then we define the following BN matrix:

$$A_k = [e_{k_1,1}, \ldots, e_{k_{2^n},2^n}]$$

Here, $e_{j,i}$ is the unit column vector whose jth entry is 1 for $i = 1, \ldots, 2^n$.

Step 3: $R_{k+1} = R_k - q_k A_k$
Step 4: If R_{k+1} is the zero matrix then go to Step 5; otherwise go to Step 1.
Step 5: $M = k$ and $A = \sum_{i=1}^{M} q_i A_i$.

15.4.2
Numerical Demonstration

We can demonstrate Algorithms 15.1 and 15.2 by a simple example. Suppose the transition probability matrix of a PBN is given by:

$$A = \begin{pmatrix} 0.1 & 0.2 \\ 0.9 & 0.8 \end{pmatrix}$$

Our work is to find a linear combination of BN matrices constituting the matrix A.

In Algorithm 15.1, first we choose the smallest nonzero element in A, which is 0.1 from column 1. We take this out from column 1. For the other columns, we are going to subtract this from the largest entry of each column. Then we have:

$$A = \begin{pmatrix} 0.1 & 0.2 \\ 0.9 & 0.8 \end{pmatrix} = 0.1 \begin{pmatrix} 1 & 0 \\ 0 & 1 \end{pmatrix} + \begin{pmatrix} 0.0 & 0.2 \\ 0.9 & 0.7 \end{pmatrix} \equiv 0.1 A_1 + R_2$$

We then do the same operations to R_2 and get:

$$R_2 = \begin{pmatrix} 0.0 & 0.2 \\ 0.9 & 0.7 \end{pmatrix} = 0.2 \begin{pmatrix} 0 & 1 \\ 1 & 0 \end{pmatrix} + \begin{pmatrix} 0.0 & 0.0 \\ 0.7 & 0.7 \end{pmatrix} \equiv 0.2 A_2 + R_3$$

Finally, we have:

$$R_3 = 0.7 \begin{pmatrix} 0 & 0 \\ 1 & 1 \end{pmatrix} \equiv 0.7 A_3$$

Therefore, we have the decomposition as:

$$A = 0.1 A_1 + 0.2 A_2 + 0.7 A_3$$

In Algorithm 15.2, we choose the largest entries from each of the columns of A, which are 0.9 from column 1 and 0.8 from column 2. Then we find the smaller one among these two entries and subtract this from the largest entry of each column. We then have:

$$A = \begin{pmatrix} 0.1 & 0.2 \\ 0.9 & 0.8 \end{pmatrix} = 0.8 \begin{pmatrix} 0 & 0 \\ 1 & 1 \end{pmatrix} + \begin{pmatrix} 0.1 & 0.2 \\ 0.1 & 0.0 \end{pmatrix} \equiv 0.8 A_1 + R_2$$

We then apply the same operations to R_2 and get:

$$R_2 = \begin{pmatrix} 0.1 & 0.2 \\ 0.1 & 0.0 \end{pmatrix} = 0.1 \begin{pmatrix} 1 & 1 \\ 0 & 0 \end{pmatrix} + \begin{pmatrix} 0.0 & 0.1 \\ 0.1 & 0.0 \end{pmatrix} \equiv 0.1 A_2 + R_3$$

Finally, we have:

$$R_3 = 0.1 \begin{pmatrix} 0 & 1 \\ 1 & 0 \end{pmatrix} \equiv 0.1 A_3$$

Therefore, we have:

$$A = 0.8A_1 + 0.1A_2 + 0.1A_3$$

15.4.3
Computational Cost Analysis

The following two theorems indicate that both Algorithms 15.1 and 15.2 are very efficient.

Theorem 15.1 *If each column of A has at most m nonzero entries, then Algorithm 15.1 will terminate in at most $m2^n$ iterations or $M \leq m2^n$.*

Proof We note that there are at most $m2^n$ nonzero entries in A. Each time from Step 2 to Step 3 (from R_k to R_{k+1}), the number of nonzero entries decreases by at least one and each of the column sum will decrease by the same amount q_k. Thus, we conclude that the algorithm will terminate in at most $m2^n$ iterations.

For Algorithm 15.2, we have a similar result. Since the proof is similar to the proof of Theorem 15.1, we omit it here.

Theorem 15.2 *If each column of A has at most m nonzero entries, then Algorithm 15.2 will terminate in at most $m2^n$ iterations or $M \leq m2^n$.*

15.4.4
Modifications of Algorithms 15.1 and 15.2

The disadvantage of using these two algorithms is that it may give only one solution, but there is no way to improve the solution even if there is some measure for the fitness of the solution obtained. The reason lies in Step 2; for each of the columns, we choose the position with the largest entry to form part of a BN matrix and deduct the value q_k. However, in fact, one can choose any one of the nonzero entries and proceed with the algorithm. Here we introduce a probabilistic approach to modify Step 2. In Algorithm 15.1, after the selection of q_k, instead of choosing the largest entry from each of the other columns, we assume in the ith column there are m nonzero entries $R_{1i}, R_{2i}, \ldots, R_{mi}$, then we define the probability of choosing R_{ji} to be:

$$\frac{R_{ji}}{R_{1i} + R_{2i} + , \ldots, + R_{mi}} \tag{15.36}$$

In Algorithm 15.2, similarly, instead of choosing the largest entry in Step 2, one can define the probability of choosing R_{ji} according to the probability given in (15.36). Then we choose q_k as the smallest one of the chosen entries and continue with the algorithm.

We may further define a measure of fitness of the solution q_i by its entropy as:

$$-\sum_{i=1}^{M} q_i \log q_i \tag{15.37}$$

The modified algorithm can then be run for a number of times and to get a better solution in the sense of (15.37). Entropy [36] is a measure of the uncertainty associated with a random variable. It actually measures, in the sense of an expected value, the information contained in a message. Entropy can also be regarded as a measure of the multiplicity associated with the possible realizations of BNs contained in a PBN.

Then one may consider two kinds of optimization problems. First, one may consider the maximization problem:

$$\max_{q_i} \left\{ -\sum_{i=1}^{M} q_i \log q_i \right\} \quad (15.38)$$

subject to:

$$\sum_{i=1}^{M} q_i = 1 \quad \text{and} \quad q_i \geq 0, \quad i = 1, 2, \ldots, M$$

The objective function is the entropy of the probability distribution **q** and it is maximized if **q** is the uniform distribution. On the other hand, one may consider the minimization problem:

$$\min_{q_i} \left\{ -\sum_{i=1}^{M} q_i \log q_i \right\} \quad (15.39)$$

subject to:

$$\sum_{i=1}^{M} q_i = 1 \quad \text{and} \quad q_i \geq 0, \quad i = 1, 2, \ldots, M$$

The optimal value is then equal to 0 and this occurs when one of the q_i is 1.

Since there are two algorithms and two objective functions, apart from Algorithms 15.1 and 15.2, we suppose four more modified algorithms. Algorithms 15.1.1 and 15.1.2 are Algorithm 15.1 with maximization problem (15.38) and minimization problem (15.39), respectively. Algorithms 15.2.1 and 15.2.2 are Algorithm 15.2 with maximization problem (15.38) and minimization problem (15.39), respectively. We note that Algorithm 15.1 tends to include more BNs with medium probability in the construction of the PBN and Algorithm 15.2 tends to include BNs with relatively large probability. Thus, we expect Algorithm 15.1.1 will give the highest entropy solution, whereas Algorithm 15.2.2 will give the lowest entropy solution.

15.4.5
Numerical Example

In [12], Shmulevich et al. proposed a PBN consisting of three genes that has been frequently used in the literature. The function sets are given by:

$$F_1 = \{f_1^{(1)}, f_2^{(1)}\}, \quad F_2 = \{f_1^{(2)}\}, \quad F_3 = \{f_1^{(3)}, f_2^{(3)}\}$$

The truth table is given in Table 15.3.

15.4 Construction of PBNs from a Prescribed Transition Probability Matrix

Table 15.3 Truth table (taken from [28]).

$x_1x_2x_3$	$f_1^{(1)}$	$f_2^{(1)}$	$f_1^{(2)}$	$f_1^{(3)}$	$f_2^{(3)}$
000	0	0	0	0	0
001	1	1	1	0	0
010	1	1	1	0	0
011	1	0	0	1	0
100	0	0	1	0	0
101	1	1	1	1	0
110	1	1	0	1	0
111	1	1	1	1	1
$c_j^{(i)}$	0.6	0.4	1	0.5	0.5

From the truth table (Table 15.3), one can get the transition probability matrix B as:

$$B = \begin{pmatrix} 1 & 0 & 0 & 0.2 & 0 & 0 & 0 & 0 \\ 0 & 0 & 0 & 0.3 & 0 & 0 & 0 & 0 \\ 0 & 0 & 0 & 0 & 1 & 0 & 0 & 0 \\ 0 & 0 & 0 & 0 & 0 & 0 & 0 & 0 \\ 0 & 0 & 0 & 0.2 & 0 & 0 & 0.4 & 0 \\ 0 & 0 & 0 & 0.3 & 0 & 0 & 0.6 & 0 \\ 0 & 1 & 1 & 0 & 0 & 0.4 & 0 & 0 \\ 0 & 0 & 0 & 0 & 0 & 0.6 & 0 & 1 \end{pmatrix}$$

There are four possible realizations with corresponding vector functions given by

$$f_1 = \left(f_1^{(1)}, f_1^{(2)}, f_1^{(3)}\right), \quad f_2 = \left(f_1^{(1)}, f_1^{(2)}, f_2^{(3)}\right),$$
$$f_3 = \left(f_2^{(1)}, f_1^{(2)}, f_1^{(3)}\right), \quad f_4 = \left(f_2^{(1)}, f_1^{(2)}, f_2^{(3)}\right)$$

respectively. Here, we indicate the corresponding BN matrices of these four possible realizations by B_1, B_2, B_3, and B_4 respectively. Then from the truth table (Table 15.3), one can get the following selection probabilities:

$$q_1 = c_1^{(1)} \cdot c_1^{(2)} \cdot c_1^{(3)} = 0.3, \quad q_2 = c_1^{(1)} \cdot c_1^{(2)} \cdot c_2^{(3)} = 0.3,$$
$$q_3 = c_2^{(1)} \cdot c_1^{(2)} \cdot c_1^{(3)} = 0.2, \quad q_4 = c_2^{(1)} \cdot c_1^{(2)} \cdot c_2^{(3)} = 0.2$$

Then we have:

$$B = 0.3B_1 + 0.3B_2 + 0.2B_3 + 0.2B_4$$

where the BN matrices B_i are given in Table 15.4. The rows represent the positions of the nonzero entries.

For the transition probability matrix B, we apply the proposed algorithms to find $\mathbf{q}$ and BNs constituting this PBN. Table 15.5 shows the results of different algorithms. We remark that, in this case, Algorithms 15.2.2 and 15.1.2 can recover exactly the

Table 15.4 BN matrices B_i (taken from [28]).

q	BN matrix	1	2	3	4	5	6	7	8
0.3	B_1	1	7	7	6	3	8	6	8
0.3	B_2	1	7	7	2	3	8	6	8
0.2	B_3	1	7	7	5	3	7	5	8
0.2	B_4	1	7	7	1	3	7	5	8

Table 15.5 Comparison of results of different algorithms (taken from [28]).

Algorithm	Entropy
Algorithm 15.1	1.5571
Algorithm 15.1.1	1.5571
Algorithm 15.1.2	1.3662
Algorithm 15.2	1.5048
Algorithm 15.2.1	1.7481
Algorithm 15.2.2	1.3662

original PBNs. The solution obtained by Algorithm 15.1 is the same as that obtained by Algorithm 15.1.1 and is given by:

$$B = 0.3B_{11} + 0.2B_{12} + 0.2B_{13} + 0.2B_{14} + 0.1B_{15}$$

where the BN matrices B_{1i} are given in Table 15.6. The solution obtained by Algorithm 15.2 is:

$$B = 0.3B_{21} + 0.3B_{22} + 0.2B_{23} + 0.1B_{24} + 0.1B_{25}$$

where the BN matrices B_{2i} are given in Table 15.7. The solution obtained by Algorithm 15.2.1 is:

$$B = 0.2B_{31} + 0.2B_{32} + 0.2B_{33} + 0.2B_{34} + 0.1B_{35} + 0.1B_{36}$$

where the BN matrices B_{3i} are given in Table 15.8.

Table 15.6 BN matrices B_{1i} (Algorithms 15.1 and 15.1.1) (taken from [28]).

q	BN matrix	1	2	3	4	5	6	7	8
0.3	B_{11}	1	7	7	2	3	8	6	8
0.2	B_{12}	1	7	7	5	3	8	6	8
0.2	B_{13}	1	7	7	1	3	7	5	8
0.2	B_{14}	1	7	7	6	3	7	5	8
0.1	B_{15}	1	7	7	6	3	8	6	8

Table 15.7 BN matrices B_{2i} (Algorithm 15.2) (taken from [28]).

q	BN matrix	1	2	3	4	5	6	7	8
0.3	B_{21}	1	7	7	6	3	8	6	8
0.3	B_{22}	1	7	7	2	3	7	5	8
0.2	B_{23}	1	7	7	5	3	8	6	8
0.1	B_{24}	1	7	7	1	3	7	5	8
0.1	B_{25}	1	7	7	1	3	8	6	8

Table 15.8 BN matrices B_{3i} (Algorithm 15.2.1) (taken from [28]).

q	BN matrix	1	2		4	5	6	7	8
0.2	B_{31}	1	7	7	6	3	7	6	8
0.2	B_{32}	1	7	7	2	3	8	6	8
0.2	B_{33}	1	7	7	5	3	7	6	8
0.2	B_{34}	1	7	7	1	3	8	5	8
0.1	B_{35}	1	7	7	6	3	8	5	8
0.1	B_{36}	1	7	7	2	3	8	5	8

15.5 Conclusions

In this chapter, we study two problems. First, we consider the problem of constructing PBNs from a given stationary distribution and a set of BNs. This is an inverse problem of large size. We have formulated the inverse problem as a constrained least-squares problem and proposed a heuristic method based on the CG method to solve the resulting least-squares problem. From the tested numerical examples, we conclude that the computational cost for examining all the possible PBNs using our proposed algorithms is huge, and heuristic methods such as genetic algorithms [37] and particle swarm optimization methods [38] will be developed to solve the problem efficiently.

Then, we study another inverse problem of generating PBNs from a given transition probability matrix. We propose some efficient algorithms for constructing a PBN when its transition probability matrix is given. The tested numerical results show that all the algorithms are able to recover the dominated BNs and therefore the major structure of the network. In particular, Algorithm 15.2 and its modifications are relatively more effective for this purpose.

One of the main disadvantages of the proposed CG method is that the PBN generated depends on the initial guess and is not unique. From the insight obtained from the entropy theory, in [35], an optimization formulation was proposed to construct a sparse transition probability matrix of a Markov chain from its observed steady-state distribution. In fact, they formulate the problem as an entropy maximizing problem:

$$\max_{\mathbf{q}} \left\{ \sum_{i=1}^{N}(-q_i \log q_i) \right\} \tag{15.40}$$

subject to:

$$\overline{V}\mathbf{q} = \overline{\mathbf{p}} \quad \text{and} \quad 0 \leq q_i, \quad i = 1, 2, \ldots, N$$

We remark that the constraint $q_i \leq 1$ can be discarded as we required that:

$$\sum_{i=1}^{N} q_i = 1 \quad \text{and} \quad 0 \leq q_i, \quad i = 1, 2, \ldots, N$$

By considering the dual problem of the maximization problem [39], a constraint-free minimization problem can be obtained. Newton's method [40] in conjunction with the CG method can then be applied to solving the minimizer. By taking advantage of the convexity of the objective function, a unique solution can be obtained in this case.

Acknowledgments

The authors would like to thank the publishers, Institution of Engineering and Technology and Elsevier Science, for their permissions to reproduce the material in [28, 29], respectively. Research supported in part by Hong Kong Research Grants Council grant 7017/07P, Hong Kong University Strategy Research Theme Fund on Computational Sciences, National Natural Science Foundation of China grant 10971075, and Guangdong Provincial Natural Science grant 9151063101000021.

References

1 Celis, J.E., Kruhøfferm, M., Gromova, I., Frederiksen, C., Østergaard, M., and Ørntoft, T.F. (2000) Gene expression profiling: monitoring transcription and translation products using DNA microarrays and proteomics. *FEBS Lett.*, **480**, 2–16.

2 Huang, S. (1999) Gene expression profiling, genetic networks, and cellular states: an integrating concept for tumorigenesis and drug discovery. *J. Mol. Med.*, **77**, 469–480.

3 Kim, S., Imoto, S., and Miyano, S. (2003) Dynamic Bayesian network and nonparametric regression for nonlinear modeling of gene networks from time series gene expression data. *Lecture Notes Comput. Sci.*, **2602**, 104–113.

4 Slezak, D. (2009) Degrees of conditional (in)dependence: a framework for approximate Bayesian networks and examples related to the rough set-based feature selection. *Inform. Sci.*, **179**, 197–209.

5 Kauffman, S. (1969) Metabolic stability and epigenesis in randomly constructed gene nets. *J. Theor. Biol.*, **22**, 437–467.

6 Kauffman, S. (1969) Homeostasis and differentiation in random genetic control networks. *Nature*, **224**, 177–178.

7 Kauffman, S. (1993) *The Origins of Order: Self-Organization and Selection in*

Evolution, Oxford University Press, New York.

8 Lindløf, A. and Olsson, B. (2002) Could correlation-based methods be used to derive genetic association networks? *Inform. Sci.*, **146**, 103–113.

9 Ching, W., Fung, E., Ng, M., and Akutsu, T. (2005) On construction of stochastic genetic networks based on gene expression sequences. *Int. J. Neural Syst.*, **15**, 297–310.

10 Zhang, S., Ching, W., Tsing, N., Leung, H., and Guo, D. (2009) A new multiple regression approach for the construction of genetic regulatory networks. *J. Artif. Intell. Med.*, **48**, 153–160.

11 Ando, S., Sakamoto, E., and Iba, H. (2002) Evolutionary modeling and inference of gene network. *Inform. Sci.*, **145**, 237–259.

12 Shmulevich, I., Dougherty, E., Kim, S., and Zhang, W. (2002) Probabilistic Boolean networks: a rule-based uncertainty model for gene regulatory networks. *Bioinformatics*, **18**, 261–274.

13 Shmulevich, I., Dougherty, E., Kim, S., and Zhang, W. (2002) Control of stationary behavior in probabilistic Boolean networks by means of structural intervention. *J. Biol. Syst.*, **10**, 431–445.

14 Shmulevich, I., Dougherty, E., Kim, S., and Zhang, W. (2002) From Boolean to probabilistic Boolean networks as models of genetic regulatory networks. *Proc. IEEE*, **90**, 1778–1792.

15 Shmulevich, I. and Dougherty, E. (2007) *Genomic Signal Processing*, Princeton University Press, Princeton, NJ.

16 de Jong, H. (2002) Modeling and simulation of genetic regulatory systems: a literature review. *J. Comput. Biol.*, **9**, 69–103.

17 Smolen, P., Baxter, D., and Byrne, J. (2000) Mathematical modeling of gene network. *Neuron*, **26**, 567–580.

18 Akutsu, T., Hayasida, M., Ching, W., and Ng, M. (2007) Control of Boolean networks: hardness results and algorithms for tree structured networks. *J. Theor. Biol.*, **244**, 670–679.

19 Huang, S. and Ingber, D.E. (2000) Shape-dependent control of cell growth, differentiation, and apoptosis: switching between attractors in cell regulatory networks. *Exp. Cell Res.*, **261**, 91–103.

20 Ching, W. and Ng, M. (2006) *Markov Chains: Models, Algorithms and Applications, International Series on Operations Research and Management Science*, Springer, New York.

21 Dougherty, E., Kim, S., and Chen, Y. (2000) Coefficient of determination in nonlinear signal processing. *Signal Process.*, **80**, 2219–2235.

22 Kim, S., Dougherty, E.R., Bittner, M.L., Chen, Y., Sivakumar, K., Meltzer, P., and Trent, J.M. (2000) General nonlinear framework for the analysis of gene interaction via multivariate expression arrays. *J. Biomed. Opt.*, **5**, 411–424.

23 Li, P., Zhang, C., Perkins, E.J., Gong, P., and Deng, Y. (2007) Comparison of probabilistic Boolean network and dynamic Bayesian network approaches for inferring gene regulatory networks. *BMC Bioinformatics*, **8** (Suppl. 7) S13.

24 Zhang, S., Ching, W., Ng, M., and Akutsu, T. (2007) Simulation study in probabilistic Boolean network models for genetic regulatory networks. *J. Data Mining Bioinform.*, **1**, 217–240.

25 Ching, W., Zhang, S., Ng, M., and Akutsu, T. (2007) An approximation method for solving the steady-state probability distribution of probabilistic Boolean networks. *Bioinformatics*, **23**, 1511–1518.

26 Ching, W., Zhang, S., Jiao, Y., Akutsu, T., and Wong, A. (2009) Optimal control policy for probabilistic Boolean networks with hard constraints. *IET Syst. Biol.*, **3**, 90–99.

27 Watterson, S., Marshall, S., and Ghazal, P. (2008) Logic models of pathway biology. *Drug Discov. Today*, **13**, 447–456.

28 Ching, W., Chen, X., and Tsing, N. (2009) Generating probabilistic Boolean networks from a prescribed transition probability matrix. *IET Syst. Biol.*, **6**, 453–464.

29 Zhang, S., Ching, W., Chen, X., and Tsing, N. (2010) Generating probabilistic Boolean networks from a prescribed stationary distribution. *Inform. Sci.*, **180**, 2560–2570.

30 Pal, R., Ivanov, I., Datta, A., Bittner, M., and Dougherty, E. (2005) Generating Boolean networks with a prescribed attractor structure. *Bioinformatics*, **21**, 4021–4025.

31 Shmulevich, I., Dougherty, E., and Zhang, W. (2002) Gene perturbation and intervention in probabilistic Boolean networks. *Bioinformatics*, **18**, 1319–1331.

32 Pal, R., Datta, A., Bittner, M., and Dougherty, E. (2005) Intervention in context-sensitive probabilistic Boolean networks. *Bioinformatics*, **21**, 1211–1218.

33 Axelsson, O. (1996) *Iterative Solution Methods*, Cambridge University Press, Cambridge.

34 Kincaid, D. and Cheney, W. (2001) *Numerical Analysis: Mathematics of Scientific Computing*, 3rd edn, Brooks Cole, Pacific Grove, CA.

35 Ching, W. and Cong, Y. (2009) A new optimization model for the construction of Markov chains. Second International Joint Conference on Computational Sciences and Optimization, Hainan.

36 Shannon, C.E. (1948) A mathematical theory of communication. *Bell Syst. Tech. J.*, **27**, 379–423.

37 Lozano, M., Herrera, F., and Cano, J. (2008) Replacement strategies to preserve useful diversity in steady-state genetic algorithms. *Inform. Sci.*, **178**, 4421–4433.

38 Wang, Y. and Yang, Y. (2009) Particle swarm optimization with preference order ranking for multi-objective optimization. *Inform. Sci.*, **179**, 1944–1959.

39 Ching, W., Scholtes, S., and Zhang, S. (2004) Numerical algorithms for dynamic traffic demand estimation between zones in a network. *Eng. Optimiz.*, **36**, 379–400.

40 Conn, A., Gould, B., and Toint, P. (2000) *Trust-Region Methods*, SIAM, Philadelphia, PA.

16
Boolean Analysis of Gene Expression Datasets
Debashis Sahoo

16.1
Introduction

The effect of the genetic constitution of a cell on its growth and function has been a fascinating source of knowledge for many decades. Parts of it directly influence the understanding of life in general; that once seemed unbreakable has now become practical and accessible. The fundamental genetic code (DNA) has been cracked and laid down. The genetic constitution and evolution of various proteins can be answered in tremendous detail. However, we still do not fully understand the genetic specifications that give rise to various disease processes. From direct observation it looks like a complex interactions of proteins, RNA, DNA, and other molecules, and their subtle modifications inside a cell control this activity. The quality of the measurement techniques for these molecules has been improving rapidly.

Advances in DNA microarray technology that enable the simultaneous measurement of thousands of genes in a single experiment have revolutionized current molecular biology. The twenty-first century is witnessing an explosion in the amount of biological information on normal and disease biological processes. Scalable approaches for the analysis and the organization of this vast amount of information into usable forms is needed. We introduce StepMiner – a tool for the analysis of time-course microarray data that discovers stepwise changes in the time course. StepMiner groups genes whose expression patterns change in the same direction and at the same time. In addition, we present the use of Boolean implication relationships for mining massive amounts of publicly available gene expression microarray datasets. The Boolean analysis results are easily understandable and directly interpretable from visual and computational inspection of the heatmap and the scatter plots. We demonstrate how Boolean analysis is used to understand gene regulation, gene function, and various human diseases. Some of the new biological hypothesis generated using our Boolean implication network have been validated experimentally.

16.2
Boolean Analysis

Boolean algebra (or Boolean logic) is a logical calculus of truth values, developed by George Boole in the late 1830s. It is an algebra of two values. These are usually taken to be 0 and 1, or false and true (in the context of gene expression, we use low and high). There are basic operations, namely conjunction $x \wedge y$ (AND), disjunction $x \vee y$ (OR), and negation $\neg x$ (NOT). Boolean operations can be represented in multiple ways (Figure 16.1).

Conjunction of two Boolean variables is true when both variables are true. The first column and first row of Figure 16.1 tabulates four possible valuations of $x \wedge y$. The second row of the first column shows the logic gate representation of $x \wedge y$ that is used in digital electronics. A conjunction is dual to disjunction via De Morgan's law: $\neg(x \wedge y) = \neg x \vee \neg y$ and $\neg(x \vee y) = \neg x \wedge \neg y$. De Morgan's equivalent of $x \wedge y$ is shown in the third row of the first column in the form of a logic gate. The circles represent negation operations. The conjunction (AND) operation can also be represented by a Venn diagram as shown in Figure 16.1 (the fourth row of the first column).

Similarly, other operations such as disjunction (second column), implication (third column), and exclusive-OR (fourth column) are shown in Figure 16.1. The Boolean implication is a binary operation that is false when x is true and y is false. This can also be expressed as $x \rightarrow y = \neg x \vee y$. According to the *contrapositive law* $x \rightarrow y$ is logically equivalent to $\neg y \rightarrow \neg x$. In this chapter we discuss Boolean implication relationships between two genes A and B. For example, A high $\Rightarrow$ B low. This means that if A is high then B is low. If A is low, then B can be high or low in this case.

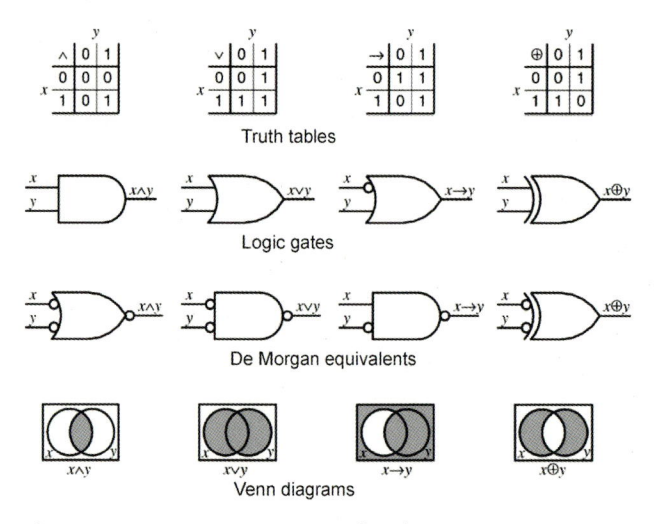

Figure 16.1 Various representations of Boolean operations.

16.3
Main Organization

Two main methods are discussed in this chapter:

- **StepMiner** is a new method for analyzing microarray time courses by identifying genes that undergo abrupt transitions in expression level and the time at which the transition occurs. StepMiner matches the sequence of expression levels for each gene against temporal patterns having one or two transitions between two expression levels. StepMiner reports a p-value for the matching pattern of each gene and a global false discovery rate (FDR) can also be computed. After matching, genes can be sorted by the direction and time of transitions. Genes can be partitioned into sets based on the direction and time of change for further analysis, such as comparison with Gene Ontology (GO) annotations or binding site motifs. StepMiner is evaluated on simulated and actual time-course data. Using microarray data for budding yeast, it is shown that the groups of genes that change in similar ways and at similar times have significant and relevant GO annotations.
- **BooleanNet** is a new kind of gene expression network consisting of Boolean implications between gene pairs obtained from very large amounts of gene expression microarray data, where the whole genome is analyzed with each gene that is assigned a binary expression state of "low" or "high." Our Boolean implication network identifies not only symmetric relationships between gene pairs, such as coexpression, but also a much larger number of previously unidentified asymmetric relationships. For example, the rule "if gene A is high, then B is high" is asymmetric if it holds, but the converse rule "if gene B is high, then A is high" does not. We refer to the set of symmetric and asymmetric "if–then" relationships between binary states of gene pairs as "Boolean implications." We identified all the Boolean implications in publicly available data from thousands of microarrays for humans, mice, and fruit flies. The resulting Boolean implication network consists of hundreds of millions of Boolean implications between genes, and identifies biologically meaningful information about gender differences, tissue differences, development, and differentiation. The Boolean implication network contains relationships that describe known biological phenomena; it also identifies many previously unidentified relationships that can serve to generate new biological hypotheses. Many Boolean implication relationships are conserved between humans, mice, and fruit flies.

16.4
StepMiner

An obvious approach to studying a biological process, such as the reaction of cells to a stimulus, is to measure the activity of the cell at a sequence of time points. However, when the measurements consist of high-throughput gene expression microarrays, it is not obvious how to extract biologically meaningful results. We describe a new computational method, called StepMiner, the primary goal of which is to assist biologists in understanding the temporal progression of genetic events and biological processes following a stimulus, based on gene expression microarray data.

At the most basic level, StepMiner identifies genes that undergo one or more binary transitions over short time courses. It directly addresses the more basic questions one can ask of time-course data: "Which genes transition are up- or downregulated as a result of the stimulus?" and "When does the gene transition to up- or downregulated?"

16.5
StepMiner Algorithm

StepMiner extracts three types of binary temporal patterns. The first type, shown in Figure 16.2(a and b), describes *one-step* transitions, where the expression level of a gene transitions from a high to a low value or from a low to a high value. The second type, shown in Figure 16.2(c and d), describes two-step transitions. Genes in this category appear that turn on then back off or vice versa. The third type is reserved for genes whose temporal expression pattern is neither of the first or second type. The expression levels for up- and down-regulated are chosen that best fit the data.

Fitting the patterns of one- and two-step transitions requires an algorithm that evaluates every possible placement of the transitions (or step) between time points, and chooses the one that gives the best fit. This process is called *adaptive regression*.

16.5.1
Fitting One- or Two-Step Functions

The objective of StepMiner is to find a one- or two-step function that best fits n time points, $X_1, X_2, \ldots X_n$. The algorithm evaluates all possible step positions. For each position, it finds the values of constant segments using linear regression. The fitted

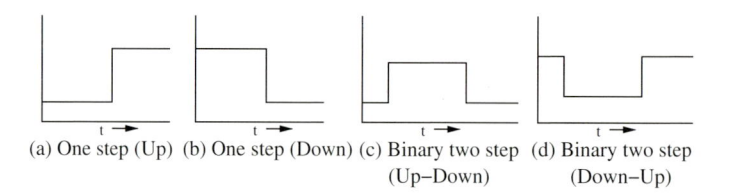

(a) One step (Up) (b) One step (Down) (c) Binary two step (Up–Down) (d) Binary two step (Down–Up)

Figure 16.2 Signals of interest. Different types of binary temporal patterns that need to be extracted from the time-course microarray data. (a) Gene expressions transition from a low value to a high value. (b) Gene expressions transition from a high value to a low value. (c) Gene expressions transition from low to high and return back to low again. (d) gene expressions transition from high to low and return back to high again.

values $\hat{X}_1, \hat{X}_2, \ldots, \hat{X}_n$ for each step give the square error (SSE_{step}). The adaptive regression scheme chooses the step positions that minimize the square error. For the two-step curve, the first and third constant segment are assumed to have the same value.

Let $\hat{X}_1, \hat{X}_2, \ldots, \hat{X}_n$ be the fitted value from the adaptive regression and $\bar{X}$ be the mean of the n original time points. The total sum of squares is defined to be:

$$SSTOT = \sum_{i=1}^{n} (X_i - \bar{X})^2$$

The degrees of freedom for $SSTOT$ is $n-1$ (the calculation of degrees of freedom appears below). The sum of squares error SSE is defined to be:

$$SSE = \sum_{i=1}^{n} (X_i - \hat{X}_i)^2$$

Let the degrees of freedom for SSE be $n-m$. The regression sum of squares SSR is:

$$SSR = \sum_{i=1}^{n} (\hat{X}_i - \bar{X})^2 = SSTOT - SSE$$

Therefore, the degrees of freedom for SSR is $(n-1)-(n-m) = m-1$. We define the regression mean square MSR as:

$$MSR = SSR/(m-1)$$

and the error mean square MSE as:

$$MSE = SSE/(n-m)$$

The regression test statistic is:

$$F = \frac{MSR}{MSE}$$

This F-statistic follows an F-distribution with $(m-1, n-m)$ degrees of freedom. Let $\mathcal{F}_{n-m}^{m-1}$ be a random variable that has this distribution. The p-value corresponding to the tail probability of this distribution is computed as:

$$p = \Pr[\mathcal{F}_{n-m}^{m-1} > F]$$

A low p-value represents a good fit of the curve to the data.

16.5.2
Selecting the Best Step Function

The p-values for the three different patterns can be computed, using the statistic mentioned in Section 16.5.1. Let F_1 and F_2 be the F-statistic described in

Section 16.5.1 for the one- and two-step patterns. The algorithm selects the best step positions adaptively for patterns. Let SSE_1 and SSE_2 be the sum of squares error for one- and two-step patterns, and let $n-m_1$ and $n-m_2$ be their corresponding degrees of freedom.

F_{12} represents the *relative* goodness of fit of a one- Versus a two-step pattern. This is an F-distribution whose p-value represents the probability of the same result on random data:

$$F_{12} = \frac{(SSE_1 - SSE_2)/(m_2 - m_1)}{SSE_2/(n - m_2)}$$

StepMiner uses the following algorithm to select the best patterns for each gene:

```
SelectBestModels(){
oneStep = F-Significant(F_1) && Not-F-Significant(F_12)
twoStep = F-Significant(F_2) && NotIn(oneStep)
other = NotIn(oneStep, twoStep)
}
```

This algorithm was found in simulation to be superior to the standard forward stepwise and backward stepwise algorithms [1]. It first selects the genes for which a one-step pattern fits well and a two-step pattern does not fit significantly better, based on whether the appropriate p-values fall under the specified threshold. Next, genes are selected from those remaining where two-step patterns fit very well according whether the p-value F_2 is under the threshold. Genes that do not fall into any of the above categories are added to the "other" category.

16.5.3
Degrees of Freedom

The construction of a regression test statistic involves estimating *degrees of freedom* for the fitted pattern, which adjusts the statistic to eliminate the advantage that a more complex curve has over a simpler curve in fitting a given set of points. The degrees of freedom is estimated using random simulation, since it is nontrivial to derive it analytically in an adaptive framework. Gaussian $\mathcal{N}(0, 1)$ data for 10 000 simulated genes with 15 time points for each gene was generated. The regression sum of squares (SSR) for both the one- and the two-step pattern was calculated and the tail probabilities ($\#\{SSR > \alpha\}/10\,000$ for different α) were plotted as shown in Figure 16.3. The χ^2-distribution with different degrees of freedom is also plotted in Figure 16.3. As can be easily seen, the degrees of freedom for SSR_1 and SSR_2 can be approximated as 3 and 4. The estimated degrees of freedom is in the range 2–3 for one-step and 3–4 for two-step patterns. (This is consistent with the results of Owen [2], who estimated that a broken line uses two to three degrees of freedom.)

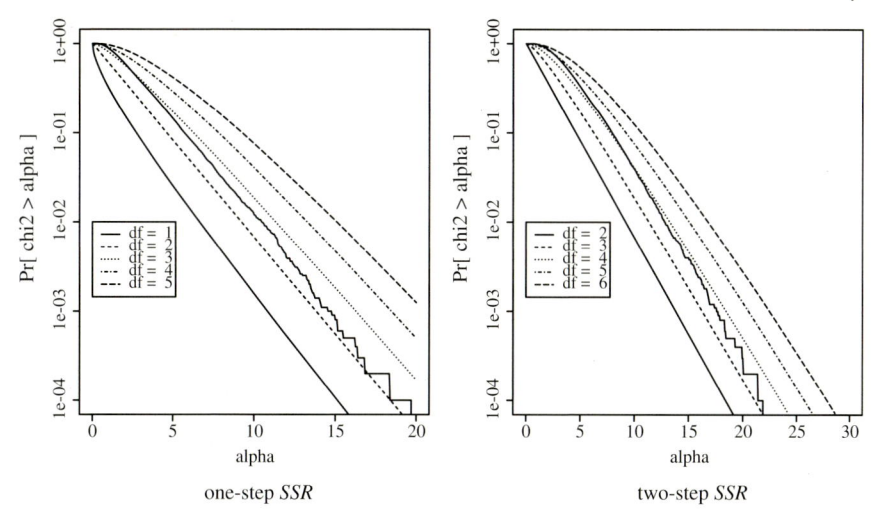

Figure 16.3 Estimating degrees of freedom for one-step SSR and two-step SSR.

16.5.4
FDR

A *false discovery* occurs when the algorithm finds a one- or two-step pattern, but the data contains no steps; a *true discovery* occurs when the algorithm finds a one- or two-step pattern, when the data contains a one- or two-step pattern (the algorithm does not have to find the correct number of steps or a step at the correct time to produce a true discovery as defined here.) The FDR in StepMiner is the ratio of false discoveries to true discoveries.

To estimate the FDR, StepMiner extracts binary transitions from many random permutations of the time points, and the results for each gene are assigned *p*-values from the F_1 or F_2 statistic, depending on whether the behavior is categorized as one- or two-step. Then, the *p*-values are sorted for each permutation in increasing order. The number of significant genes in the original order and the average number of significant genes in the random permutations are computed using a user-specified *p*-value threshold. The ratio of the average number of significant genes to the original number of significant genes is an estimate of the FDR [3]. The FDR can be adjusted by setting the *p*-value threshold used in the matching algorithm.

16.6
BooleanNet

A large and exponentially growing volume of gene expression data from microarrays is now available publicly. Since the quantity of data from around the world dwarfs the output of any individual laboratory, there are opportunities for mining this data that can yield insights that would not be apparent from smaller, less diverse datasets. Consequently, numerous approaches to extract large networks of relationships from large amounts of public-domain gene expression data have emerged. Almost all of this work constructs networks of pairwise relationships between genes, indicating that the genes are coexpressed [4–8]. Coexpression is a symmetric relationship between a gene pair, because if A is related to B, then B is related to A, such as correlation.

We present a new approach to identify a larger set of relationships between gene pairs across the whole genome from a vast amount of gene expression microarray data. We first classify the expression level of each gene on each array as "low" or "high" relative to an automatically determined threshold that is derived individually for each gene. We then identify all Boolean implications between pairs of genes. An implication is an if–then rule, such as "if gene A's expression level is high, then gene B's expression level is almost always low," or more concisely, "A high implies B low," written "A high $\Rightarrow$ B low."

In general, Boolean implications are asymmetric: "A high $\Rightarrow$ B high" may hold for the data without "B high $\Rightarrow$ A high" holding. However, it is also possible that both of these implications hold, in which case A and B are said to be Boolean equivalent. Boolean equivalence is a symmetric relation. Equivalent genes are usually strongly correlated as well. A second kind of symmetric relation occurs when A high $\Rightarrow$ B low and B high $\Rightarrow$ A low. In this case, the expression levels of A and B are usually strongly negatively correlated, and A and B are said to be opposite. In total, six possible Boolean relationships are identified: two symmetric (equivalent and opposite) and four asymmetric (A low $\Rightarrow$ B low, A low $\Rightarrow$ B high, A high $\Rightarrow$ B low, B high $\Rightarrow$ A high). Below, "symmetric relationship" means a Boolean equivalence or opposite relationship; "asymmetric relationship" means any of the four kinds of implications, when the converse relationship does not hold; and "relationship" means any of the four asymmetric or two symmetric relationships.

A network of Boolean implications can be constructed as a directed graph, which connects nodes with arrows, instead of the more common undirected graphs used for many biological networks. It is important to understand that a Boolean implication is an empirically observed invariant on the expression levels of two genes and does not necessarily imply any causality.

One way to understand the biological significance of a Boolean implication is to consider the sets of arrays where the two genes are expressed at a high level. The asymmetric Boolean implication A high $\Rightarrow$ B high means that "the set of arrays where A is high is a subset of the set of arrays where B is high." For example, this may occur when gene B is specific to a particular cell type and gene A is specific to a subclass of those cells. Alternatively, this implication can be the result of a regulatory relationship, so A high $\Rightarrow$ B high could hold because A is a transcription factor that increases expression of B or because B is a transcription factor that represses expression of A. On the other hand, the asymmetric Boolean implication A high $\Rightarrow$ B low means that A and B are rarely high on the same array – the genes are "mutually exclusive." A possible explanation for this is that A and B are specific to distinct cell types (e.g., brain versus prostate).

Boolean implications capture many more significant relationships between pairs of genes than existing methods that scale to large amounts of data, which generally find only symmetric relationships. There may be a highly significant Boolean implication between genes whose expression is only weakly correlated. The Boolean implication network described here is more comprehensive than such networks constructed previously from microarray data. The Boolean implication network identifies Boolean implications that describe known biological phenomena, as well as many new relationships that can serve to generate new hypotheses. Moreover, this approach discovers many previously unidentified relationships that are conserved between humans, mice, and fruit flies.

16.6.1
Boolean Implications in Gene Expression Microarray Data

Boolean implication networks are constructed by finding Boolean implications between pairs of probe sets in hundreds to thousands of microarrays belonging to the same platform (see Section 16.7.2). The logarithm (base 2) of each expression level is used. To find a Boolean implication between a pair of genes, each probe set is assigned an expression threshold t (see Figure 16.11 below). A scatter plot where each point represents gene A's expression versus gene B's expression for a single sample is divided, based on the thresholds, into four quadrants: (A low, B low), (A low, B high), (A high, B low), and (A high, B high). A Boolean implication exists when one or more quadrants is sparsely populated according to a statistical test, and there are enough high and low values for each gene (to prevent the discovery of implications that follow from an extreme skew in the distribution of one of the genes). The test produces a score and a cutoff is chosen for the presence or absence of an implication to obtain an acceptable FDR (see Section 16.7.3). To reduce sensitivity to small errors in the choice of t and noise in the data, points within an interval around the threshold are ignored (see Figure 16.11 below). A visual examination of the scatter plots is a straightforward way to understand the implications and to check the quality of the results (Figure 16.4).

There are four possible asymmetric Boolean relationships, each occurring when a particular quadrant is sparse. Figure 16.4(a) shows an example low $\Rightarrow$ low implication; here the quadrant for gene PTPRC low and gene CD19 high is sparse, so PTPRC low $\Rightarrow$ CD19 low. Figure 16.4(b) shows a high $\Rightarrow$ low implication; here XIST high $\Rightarrow$ RPS4Y1 low; this relationship was recently identified in study of the Celsius microarray database Day et al. [9], which annotated microarrays by gender. Figure 16.4 (d) shows a low $\Rightarrow$ high implication; here FAM60A low $\Rightarrow$ NUAK1 high. In this case, when FAM60A expression level is low, NUAK1 expression level is high, but when FAM60A expression level is high, NUAK1 expression level is evenly distributed between high and low. Finally, Figure 16.4(e) shows a high $\Rightarrow$ high implication; here, COL3A1 high $\Rightarrow$ SPARC high. This particular relationship may be viewed as complex, since it involves a combination of multiple types of relationships including linear and constant. However, from a Boolean perspective, this is a simple and clear logical implication.

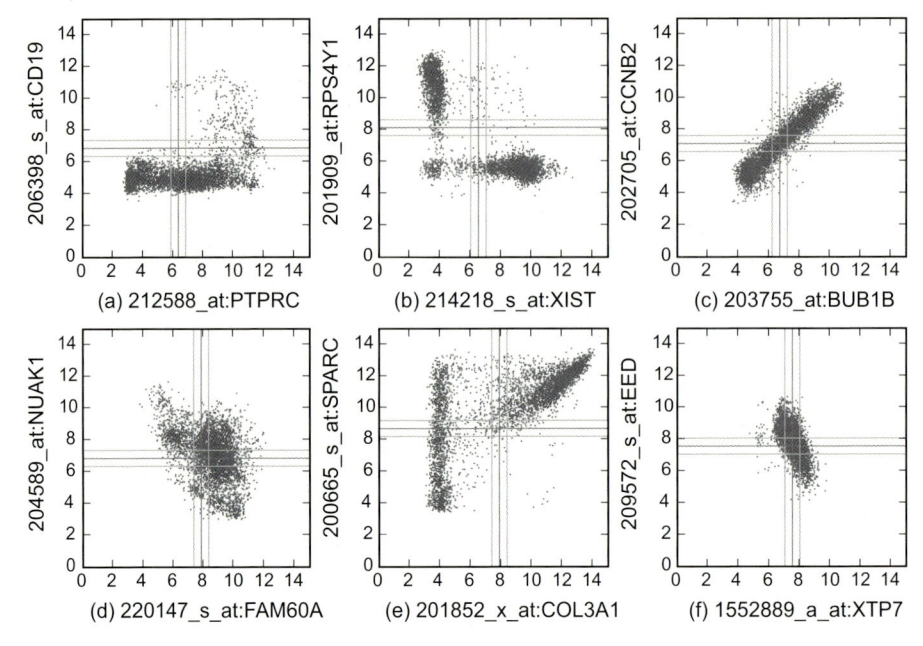

Figure 16.4 Boolean implications in gene expression microarray data. Six different types of Boolean relationships between pairs of genes taken from the Affymetrix U133 Plus 2.0 human dataset. Each point in the scatter plot corresponds to a microarray experiment, where the two axes correspond to the expression levels of two genes. There are 4787 points in each scatter plot. (a) PTPRC low ⇒ CD19 low. (b) XIST high ⇒ RPS4Y1 low. (c) Equivalent relationship between CCNB2 and BUB1B. (d) FAM60A low ⇒ NUAK1 high. (e) COL3A1 high ⇒ SPARC high. (f) Opposite relationship between EED and XTP7.

For each of the above asymmetric Boolean implications there is always a contrapositive Boolean relationship that holds. For example, PTPRC low ⇒ CD19 low, so CD19 high ⇒ PTPRC high. Similarly, XIST high ⇒ RPS4Y1 low, so RPS4Y1 high ⇒ XIST low; FAM60A low ⇒ NUAK1 high, so NUAK1 low ⇒ FAM60A high; and COL3A1 high ⇒ SPARC high, so SPARC low ⇒ COL3A1 low.

The two possible symmetric Boolean relationships correspond to two sparse diagonally opposed quadrants in a scatter plot. First, the low-high and high-low quadrant can be sparse as shown in Figure 16.4(c), which shows that CCNB2 and BUB1B are equivalent in the human network. Highly positively correlated genes are almost always equivalent. Alternatively, the low-low and high-high quadrants can be sparse, as shown in Figure 16.4(f), which shows that EED and XTP7 are opposite. Negatively correlated genes are often opposite. An important reason for ignoring points that are close to the low/high threshold is to enable discovery of equivalence and opposite relationships. As is clear in Figure 16.4(c), if points inside the intermediate region were considered, there would be a significant number of points in all four quadrants. Empirically, the interval width of 1 results in the discovery of many equivalent genes. Notice that it is not possible to have both the low-low and high-low quadrants be sparse because that would require the second gene to be always low; similarly, it is not possible for the low-high and low-low quadrants both to be sparse.

Boolean implication networks for human, mouse, and fruit fly were constructed from publicly available microarray data. A very large number of Boolean implications were found for each individual species. Approximately 3 billion probe set pairs are checked for possible Boolean implications in the human dataset. There are 208 million implications in the human dataset, even with a stringent requirement for significance (a permutation test yields a FDR of 10^{-4}). Similarly, 2 billion and 196 million probe set pairs are checked for possible Boolean implications in mouse and fruit fly datasets, respectively. The mouse dataset has 336 million implications (FDR $= 6 \times 10^{-5}$) and the fruit fly dataset has 17 million implications (FDR $= 6 \times 10^{-6}$). Of the 208 million implications in the human dataset, 128 million are high $\Rightarrow$ low, 38 million are low $\Rightarrow$ low, 38 million are high $\Rightarrow$ high, 2 million are low $\Rightarrow$ high, 1.6 million relationships are equivalences, and 0.4 million are opposite.

Table 16.1 summarizes the number of Boolean relationships found in each dataset. In all cases, Boolean implications of the type high $\Rightarrow$ low are most common and the opposite relationships are rare. As can be seen from Table 16.1, in the human dataset 1% of the total Boolean relationships are symmetric, while the remaining 99% are asymmetric. Similarly, in the mouse dataset 1.4% of the total Boolean relationships are symmetric and 98.6% are asymmetric. However, in the fruit fly dataset 12% of the Boolean relationships are symmetric. The number of low $\Rightarrow$ low relationships is the same as the number of high $\Rightarrow$ high relationships because of contrapositives. One reason for the large number of high $\Rightarrow$ low relationships is that there are many genes that are specific to particular cell and tissue types and n mutually exclusively expressed genes give rise to $n * (n - 1)$ high $\Rightarrow$ low relationships.

An interesting fact about array technology is that alternative probe sets for the same gene are not always equivalent in the network; instead, there is often a low $\Rightarrow$ low relationship between them. This is consistent with previous findings of low average correlation among them [10]. Boolean implications might be helpful in pointing out

Table 16.1 Size of Boolean network: number (in millions) of Boolean relationships in human, mouse, and fruit fly datasets.

Dataset	Total	Low $\Rightarrow$ high	High $\Rightarrow$ low	Low $\Rightarrow$ low	High $\Rightarrow$ high	Equivalent	Opposite
Human	208	2	128	38	38	1.6	0.4
Mouse	336	8	208	57.6	57.6	4.1	0.7
Fruit fly	17	0.3	7.3	3.7	3.7	1.9	0.1

In the human dataset, 1% of the all Boolean relationships are symmetric (equivalence + opposite) and 99% are asymmetric (low $\Rightarrow$ low + low $\Rightarrow$ high + high $\Rightarrow$ low + high $\Rightarrow$ high) relationships of the total Boolean relationships. The mouse dataset has 1.4% symmetric (equivalence + opposite) and 98.6% asymmetric (low $\Rightarrow$ low + low $\Rightarrow$ high + high $\Rightarrow$ low + high $\Rightarrow$ high) relationships. The fruit fly dataset has 12% symmetric (equivalence + opposite) and 88% asymmetric (low $\Rightarrow$ low + low $\Rightarrow$ high + high $\Rightarrow$ low + high $\Rightarrow$ high) relationships.

16.6.2
Biological Interpretations of Boolean Implications

Boolean implication captures a wide variety of currently known biological phenomena. The generated networks contain relationships that show gender differences, development, differentiation, tissue difference, and coexpression, suggesting that the Boolean implication network can potentially be used as a discovery tool to synthesize new biological hypotheses. The scatter plot between XIST and RPS4Y1 in Figure 16.5(a) is an example of an asymmetric Boolean relationship that shows gender difference. RPS4Y1 is expressed only in certain male tissues because it is present solely on the Y chromosome [11] and XIST is normally expressed only in female tissues [12, 13], so RPS4Y1 and XIST are rarely expressed together on the same array. Hence, the implications RPS4Y1 high ⇒ XIST low and XIST high ⇒ RPS4Y1 low hold. Moreover, in the network, RPS4Y1 is equivalent to four other genes, all of

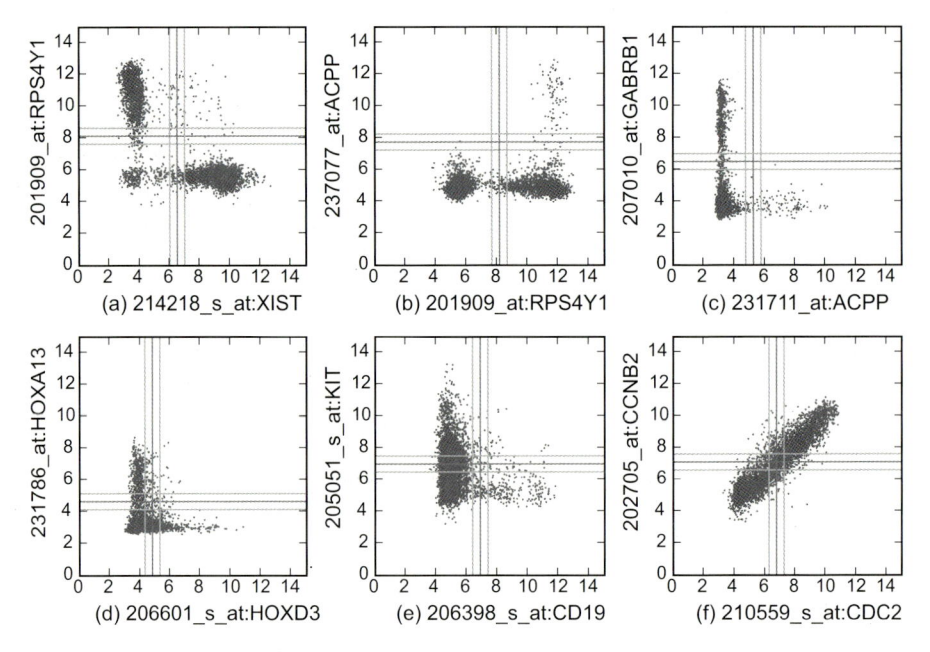

Figure 16.5 Boolean relationships follow known biology. (a) Gender difference: XIST high ⇒ RPS4Y1 low, male and female genes are not expressed in the same sample. (b) Gender tissue-specific: RPS4Y1 low ⇒ ACPP low, prostate cells are from males. (c) Tissue difference: ACPP high ⇒ GABRB1 low, prostate and brain genes are not expressed in the same samples. (d) Development: HOXD3 high ⇒ HOXA13 low, anterior is different from posterior. (e) Differentiation: KIT high ⇒ CD19 low, Differentiated B cell is different from hematopoietic stem cell. (f) Coexpression: CDC2 versus CCNB2.

which are Y-linked. Also, RPS4Y1 low ⇒ ACPP low (Figure 16.2b), KLK2 low, and KLK3 (PSA) low, and ACPP, KLK2, and KLK3 are all prostate-specific [14].

Boolean implications can capture the hierarchy of tissue types. Figure 16.5(c) shows ACPP high ⇒ GABRB1 low. GABRB1 is specific to the central nervous system [15] and ACPP is prostate-specific [14]; hence, ACPP high ⇒ GABRB1 low appears sensible because the prostate is distinct from the central nervous system. On the other hand, GABRA6 is primarily expressed in the cerebellum and we find that GABRB1 low ⇒ GABRA6 low, because the cerebellum is part of the central nervous system. This can be taken more literally to mean that if a sample is not part of the central nervous system, it is also not part of the cerebellum.

To show an example of a Boolean implication between two developmentally regulated genes, we identify HOXD3 and HOXA13 as shown in Figure 16.5(d). HOXD3 and HOXA13 have their evolutionary origin from fruit fly antennapedia (Antp) and ultrabithorax (UBX) respectively [16]. It was recently discovered that HOXD3 and HOXA13 are expressed in human proximal and distal sites, respectively [17] – a pattern of expression that is evolutionarily conserved from fruit flies. The human Boolean network indeed shows that high expressions of HOXD3 and HOXA13 are mutually exclusive (HOXD3 high ⇒ HOXA13 low), which is consistent with the above paper [16]. (Unlike the findings of that paper [16], this relationship is not highly conserved in our analysis because orthologous mouse and fruit fly probe sets for the desired genes did not have a good dynamic range in the dataset.)

Implications between genes expressed during differentiation of specific tissue types also appear in the network. For example, a Boolean implication between two key marker genes from B cell differentiation, KIT and CD19, is shown in Figure 16.5(e). KIT is a hematopoietic stem cell marker [18] and CD19 is a well-known B cell differentiation marker [19]. KIT and CD19 are rarely expressed together, in the form of Boolean implications CD19 high ⇒ KIT low and its contrapositive KIT high ⇒ CD19 low.

From inspecting the human network, it is clear that hundreds of genes are coexpressed that are related to the cell cycle. Two such genes, CDC2 and CCNB2, are shown in Figure 16.5(f).

16.6.3
Conserved Boolean Implications

A network consisting of the Boolean implication relationships that hold between orthologous genes in multiple species was constructed (see Section 16.7.5). The network of relationships that are conserved in humans and mice network has a total of 3.2 million Boolean implications consisting of 8000 low ⇒ high, 2 million high ⇒ low, 0.5 million low ⇒ low, 0.5 million high ⇒ high, 10 814 equivalent, and 94 opposite implications. Applying the same analysis to randomized human and mouse datasets yielded no conserved Boolean relationships, for an estimated FDR of less than 3.1×10^{-7}. An analogous network of implications conserved across human, mouse, and fruit fly has 41 260 Boolean relationships: 24 544 high ⇒ low, 8060 low ⇒ low, 8060 high ⇒ high, 596 equivalent, and 0 opposite. The FDR for the conserved

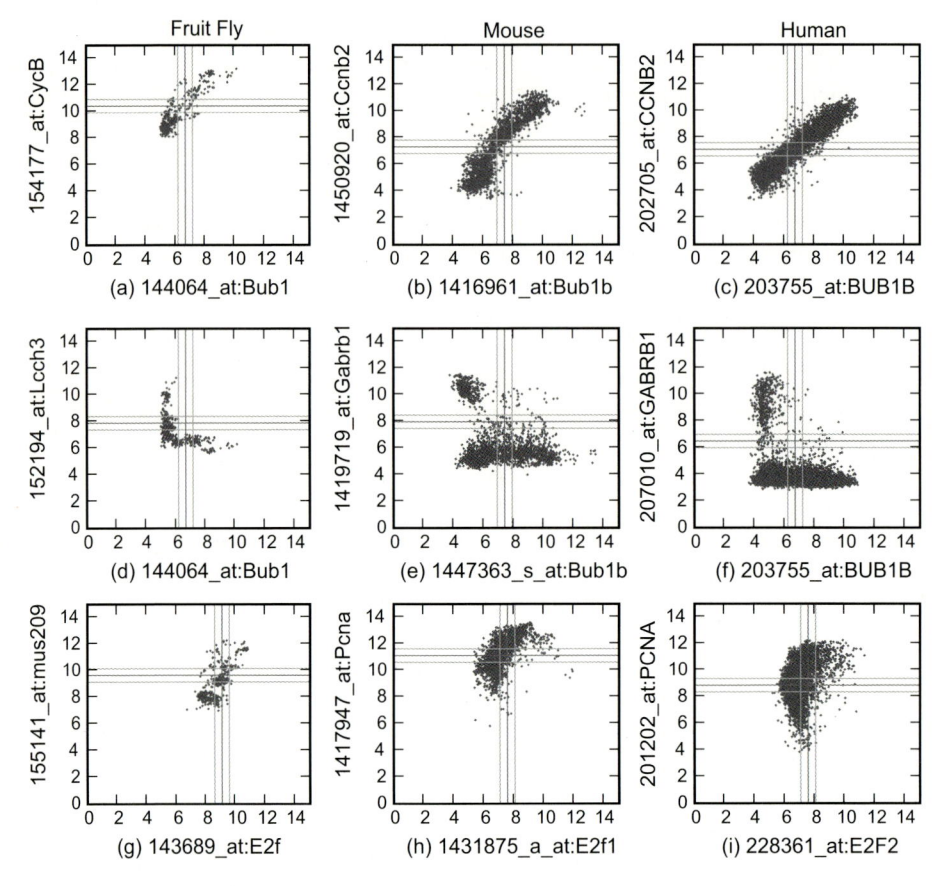

Figure 16.6 Highly conserved Boolean relationships. Orthologous CCNB2 and BUB1B equivalent relationships: (a) Bub1 versus CycB in fruit fly, (b) Bub1b versus Ccnb2 in mouse, and (c) BUB1B versus CCNB2 in human. Orthologous BUB1B high ⇒ GABRB1 low: (d) Bub1 versus Lcch3 in fruit fly, (e) Bub1b versus Gabrb1 in mouse, and (f) BUB1B versus GABRB1 in human. Orthologous E2F2 ⇒ PCNA high: (g) E2f versus mus209 in fruit fly, (h) E2f1 versus Pcna in mouse, and (i) E2F2 versus PCNA in human.

human, mouse, and fruit fly Boolean implication network is less than 2.4×10^{-5}. Figure 16.6 shows three examples of Boolean relationships that are conserved in human, mouse, and fruit fly. The first row in Figure 16.6 is an example of an equivalent relationship that is conserved in all three species, and the middle and bottom rows show highly conserved high ⇒ low and high ⇒ high relationships, respectively.

The top row in Figure 16.6 shows that CCNB2 (CycB in fruit fly) and BUB1B are equivalent in all three species. It is well known that both CCNB2 and BUB1B are related to the cell cycle [20, 21]. In this case, a network of correlated genes would also be able to find these conserved relationships because they are very well correlated in each species. The connected components of the network of equivalent relationships conserved in human, mouse, and fruit fly were examined (see Section 16.7.6). (A connected component of an undirected graph is a set of genes where there is a path between every pair of genes.) The algorithm found 13 different connected components, of which two are relatively large components. The largest component has 178 genes, including BUB1B, EZH2, CCNA2, CCNB2, and FEN1. The genes belong to this component were analyzed using DAVID functional annotation tools [22, 23], and found to be enriched for DNA replication (2.03×10^{-14}, 19 genes) and cell cycle process (1.06×10^{-13}, 30 genes) as significant gene ontology annotations. The functional annotation analysis also reported proteasome and cell cycle as significant KEGG (Kyoto Encyclopedia of Genes and Genomes) pathways for the largest component. The second largest component has 32 genes with transport (2.55×10^{-8}, 16 genes) and synaptic transmission (1.04×10^{-8}, eight genes) as significant gene ontology annotations. Further, this component is enriched for calcium signaling pathway in the KEGG database.

The connected components described above have biologically meaningful relationships. CCNB2 and BUB1B play roles in mitosis [21, 24], EZH2 is a histone methyltransferase [25], CCNA2 is required for G_1/S transition [26], and FEN1 has endonuclease activity during DNA repair [27]. Surprisingly, all these genes are highly correlated in all three species. Further, it is worthwhile to note that of the two human homologs of *Drosophila* polycomb group gene Enhancer-of-zeste (E(z)), EZH1 and EZH2, only EZH2 maintains the functional associations with other cell cycle genes. EZH1 might have evolved to acquire a different function than EZH2 in mammals. In addition, there are highly conserved equivalent genes that are part of the same protein complexes such as CDC2–CCNB2, EED–EZH2, RELB–NFKB2, RFC1–RFC2–RFC4, and MSH2–MSH6. Moreover, there is a conserved cluster of four genes: NDUFV1, IDH3B, CYC1, and UQCRC1, which are all related to generation of energy through oxidative phosphorylation and the electron transport chain.

The middle row in Figure 16.6 shows an asymmetric relationship that is conserved in all three species: BUB1B high $\Rightarrow$ GABRB1 low. GABRB1 is a receptor for an inhibitory neurotransmitter in vertebrate brains [28]. It is surprising to see that the Boolean implications between GABRB1 and BUB1B is conserved in vertebrates and arthropods (fruit fly). This relationship suggests that cells expressing the GABRB1 neurotransmitter are less likely to be proliferating. The bottom row in Figure 16.6 shows an asymmetric relationship between two well-known cell cycle regulators, E2F2 and PCNA [29, 31].

Furthermore, Figure 16.7 shows the Boolean implications between MYC and ribosomal genes that are present in the conserved human and mouse network. The implication is MYC high $\Rightarrow$ ribosomal genes high for both large and small ribosomal subunits. This implication is consistently observed for 19 genes for large subunits of the ribosome ($p < 3 \times 10^{-26}$) and 15 genes for small subunits of the ribosome ($p < 1 \times 10^{-22}$). MYC has been shown to regulate ribosomal genes in a recently comparative study between human and mouse [32]. In this study the high expression levels of MYC and ribosomal genes in human lymphoma were compared with the gene signature associated with MYC-induced tumorigenesis in mice.

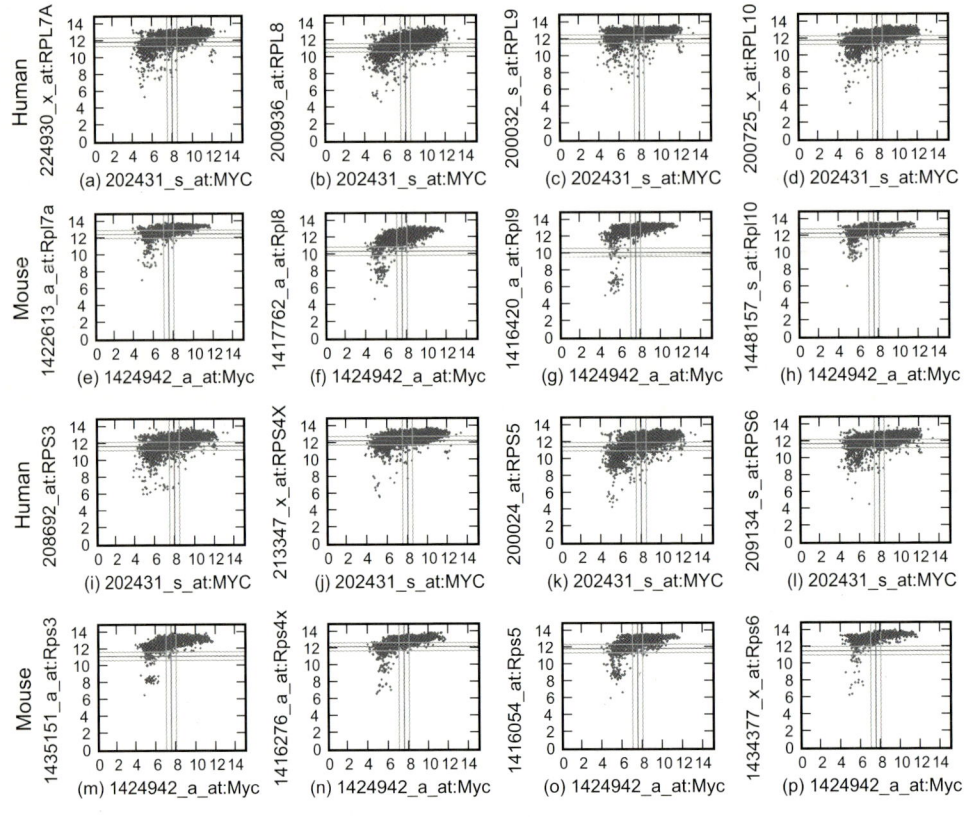

Figure 16.7 Conserved Boolean relationships between MYC and ribosomal genes. (a–h) The scatter plots show Boolean relationships between MYC and a few selected genes for large ribosomal subunits in both human and mouse datasets. (i–p) Boolean relationships between MYC and a few selected ribosomal small subunit genes in both human and mouse datasets. (a–d, i–l) human datasets. (e–h, m–p) mouse datasets.

(a) MYC high ⇒ RPL7a. (b) MYC high ⇒ RPL8 high. (c) MYC high ⇒ RPL9 high. (d) MYC high ⇒ RPL10 high. (e) Myc high ⇒ Rpl7a. (f) Myc high ⇒ Rpl8 high. (g) Myc high ⇒ Rpl9 high. (h) Myc high ⇒ Rpl10 high. (i) MYC high ⇒ RPS3. (j) MYC high ⇒ RPS4X high. (k) MYC high ⇒ RPS5 high. (l) MYC high ⇒ RPS6 high. (m) Myc high ⇒ Rps3. (n) Myc high ⇒ Rps4x high. (o) Myc high ⇒ Rps5 high. (p) Myc high ⇒ Rps6 high.

16.6.4
Comparison against Correlation Network

To compare the properties of Boolean implication networks to correlation-based networks, both types of networks were constructed based on human CD (cluster of differentiation) antigen genes. This set of genes were chosen because it is a relatively small and coherent subset of biologically interesting genes, and a correlation network can be constructed much more rapidly than if all the probe sets on the arrays were used. The correlation-based network on human CD genes was computed as described in Section 16.7.4.

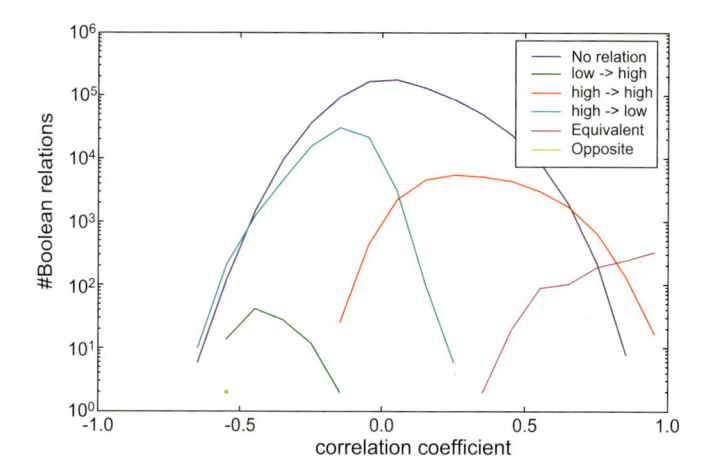

Figure 16.8 Comparison of Boolean implications with correlation on human CD (clusters of differentiation) genes, this plot shows the histogram of different types of Boolean relationships. (blue) No relationships. (green) Low $\Rightarrow$ High. (red) High $\Rightarrow$ High. (cyan) High $\Rightarrow$ Low. (magenta) Equivalent. (yellow) Opposite.

Figure 16.8 shows histograms of the various kinds of Boolean relationships with respect to the Pearson's correlation coefficients between expression levels of the same pairs of genes. As expected, highly correlated genes generally correspond to symmetric Boolean relationships, 80% of the symmetric Boolean relationships have correlation coefficients more than 0.65. Figure 16.8 shows that the number of Boolean equivalent pairs increases linearly with the correlation coefficient suggesting that most of the Boolean equivalence have good correlation coefficients. Therefore, gene pairs with high correlation coefficients are almost always found to be Boolean equivalent.

On the other hand, asymmetric Boolean relationships usually display poor correlation. 98.8% of the asymmetric Boolean relationships on the human CD genes have correlation coefficients ranging from -0.65 to 0.65 (correlation-based networks are often based on gene pairs having a threshold of 0.7 or greater for the correlation coefficient [6, 7, 33]). While the correlation coefficients of genes that have asymmetric relationships are distributed differently from the correlation coefficients of the gene pairs with no Boolean relationships, the histograms in Figure 16.8 suggest that it would be very difficult to find approximately the same asymmetric relationships using a filter based on correlation coefficients, because the number of nonrelationships in a given range of correlation coefficients almost always greatly exceeds the number of asymmetric relationships.

16.6.5
Boolean Implication Networks are Not Scale-Free

It has often been observed that other biological networks are scale-free [34–39]. To study the global properties of Boolean implication networks, we plotted the

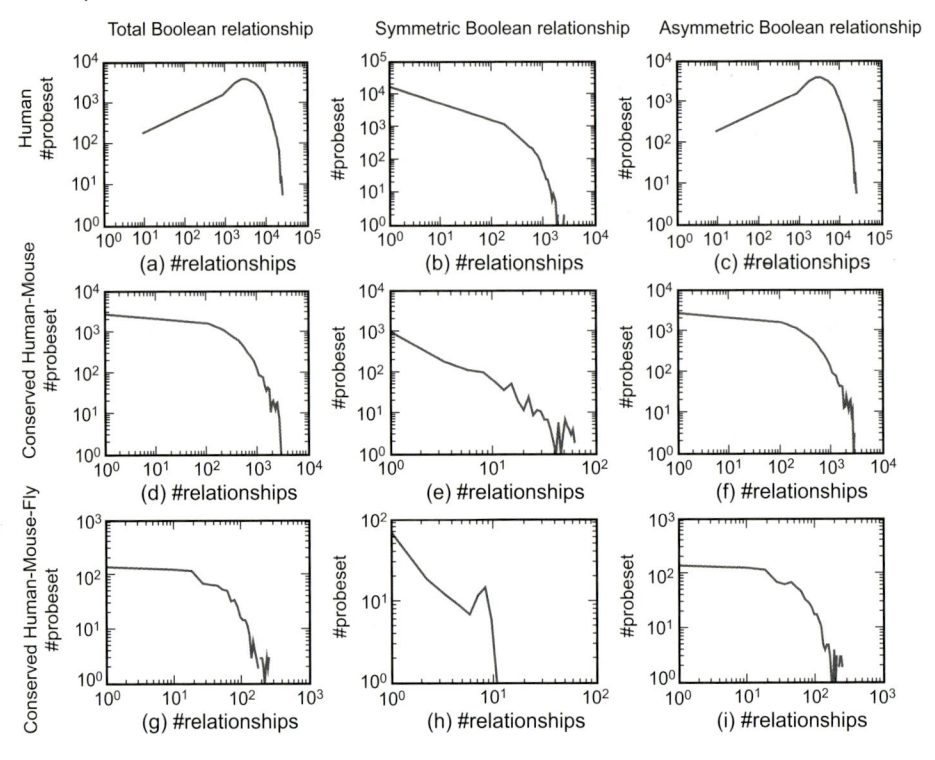

Figure 16.9 Properties of Boolean implication network. Log–log plot of the histogram of the probe sets with respect to their number of Boolean relationships. Human Boolean network: (a) total, (b) symmetric, and (c) asymmetric Boolean relationships. Conserved human and mouse Boolean network: (d) total, (e) symmetric, and (f) asymmetric Boolean relationships. Conserved human, mouse, and fruit fly Boolean network: (g) total, (h) symmetric, and (i) asymmetric Boolean relationships.

frequency of the probe sets against their degree (number of Boolean relationships) as shown in Figure 16.9. Each log–log plot shows on the horizontal axis the degree, while the vertical axis shows the number of probe sets that have the number of relationships to other probe sets. The top row in Figure 16.9 corresponds to the human Boolean implication network. From left to right are shown the total Boolean relationships, only symmetric Boolean relationships, and only asymmetric Boolean relationships. These plots are comparable to the Boolean implication networks for mouse and fruit fly (as shown in Figure 16.10). The middle row in Figure 16.9 corresponds to the conserved Boolean implication network between human and mouse, constructed of relationships that are present in both human and mouse. Finally, the bottom row in Figure 16.9 shows the conserved Boolean implication network between human, mouse, and fruit fly. As can be seen from these figures, the plots for symmetric Boolean relationships (second and third columns in Figure 16.9) are close to linear. However, the plots for total Boolean relationships (first column in Figure 16.9) are nonlinear. Therefore, the overall Boolean implication network is not scale-free.

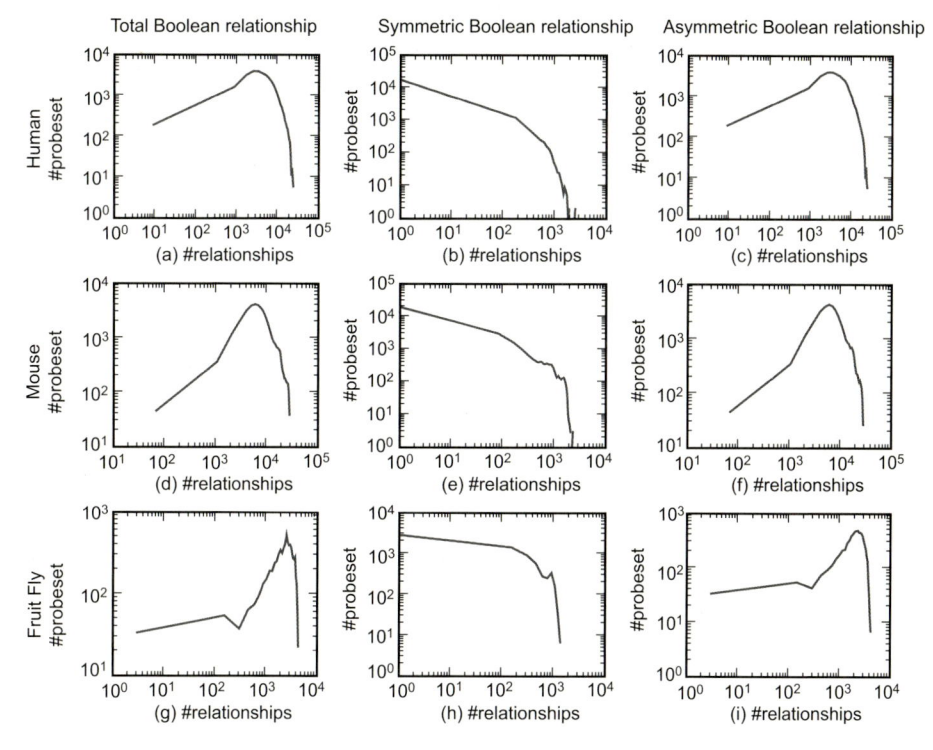

Figure 16.10 Properties of human, mouse, and fruit fly Boolean implication networks. Log–log plot of the histogram of the probe sets with respect to their number of Boolean relationships. Human Boolean network: (a) total, (b) symmetric, and (c) asymmetric Boolean relationships. Mouse Boolean network: (d) total, (e) symmetric, and (f) asymmetric Boolean relationships. Fruit fly Boolean network: (g) total, (h) symmetric, and (i) asymmetric Boolean relationships.

16.6.6
Computational Efficiency of BooleanNet

The total computation time to construct the network of implications for the human dataset was 2.5 h on a 2.4-GHz computer with 8 GB of memory. The human dataset consisted of a total of 54 677 distinct probe sets from 4787 microarrays. The computation time for the mouse dataset was 1.6 h. This dataset has 45 101 probe sets and 2154 microarrays. Finally, the computation time for fruit fly dataset, consisting of 14 010 probe sets and 450 microarrays, was 2 min.

Generating the Boolean implication network is conceptually a simple process. The relationships are immediately evident upon inspection of a scatter plot of the data points of expression levels for the two related genes, and are thus completely transparent and intuitive to biologists, unlike some approaches that find complex relationships, which can be more difficult for users to interpret.

16.7
BooleanNet Algorithm

16.7.1
Data Collection and Preprocessing

CEL files for 4787 Affymetrix U133Plus 2.0 human microarrays, 2154 Affymetrix 430 2.0 mouse arrays, and 450 Affymetrix Genome 1.0 *Drosophila* were downloaded from the National Center for Biotechnology Information's Gene Expression Omnibus (GEO) [40]. These array types were chosen because they are widely used and because results from different arrays can be compared more easily than results from two-channel arrays. The datasets were normalized using the standard RMA algorithm [41]; however, the available version of RMA uses excessive amounts of primary memory when normalizing thousands of arrays, so the program was rewritten to increase memory efficiency. Boolean expression levels were assigned for each gene in each array, using the log (base 2) of the expression values (Figure 16.11 illustrates this process). First, a threshold was assigned to each gene using the StepMiner algorithm [42], which was originally designed to fit step functions to time-course data. For this application, the expression values for each gene were ordered from low-to-high, and StepMiner was used to fit a rising step function to the data that minimizes the differences between the fitted and measured values. This approach places the step at the largest jump from low values to high values (but only if there are sufficiently many expression values on each side of the jump to provide evidence that the jump is not due to noise) and sets the threshold at the point where the step crosses that original data (as shown in Figure 16.11). In the case where the gene expression levels are evenly distributed from low to high, the threshold tends to be near the mean expression level. If the assigned threshold for a gene is t, expression levels above $t + 0.5$ are classified as "high," expression levels below $t - 0.5$ are classified as "low," and values between $t - 0.5$ and $t + 0.5$ are classified as "intermediate." In other words, the width of the "intermediate" region is set to 1, which is 2-fold change in terms of gene expression. Therefore, a minimum of 2-fold change is observed between the "low" and the "high" expression levels, which is a significant difference. Whenever more than two-thirds of the expression values of a gene were at an intermediate level of expression, the gene was excluded from further analysis, due to insufficient dynamic range in the expression values.

16.7.2
Discovery of Boolean Relationships

All pairs of features with sufficient dynamic range were analyzed to discover potential Boolean relationships. There are six possible Boolean relationships between genes A and B that are constructed from four possible Boolean implications: A low $\Rightarrow$ B low, A low $\Rightarrow$ B high, A high $\Rightarrow$ B low, and A high $\Rightarrow$ B high. Each of the above implications is detected by checking whether one of the four quadrants in the scatter plot of Figure 16.11 is significantly sparsely populated with points compared with the other

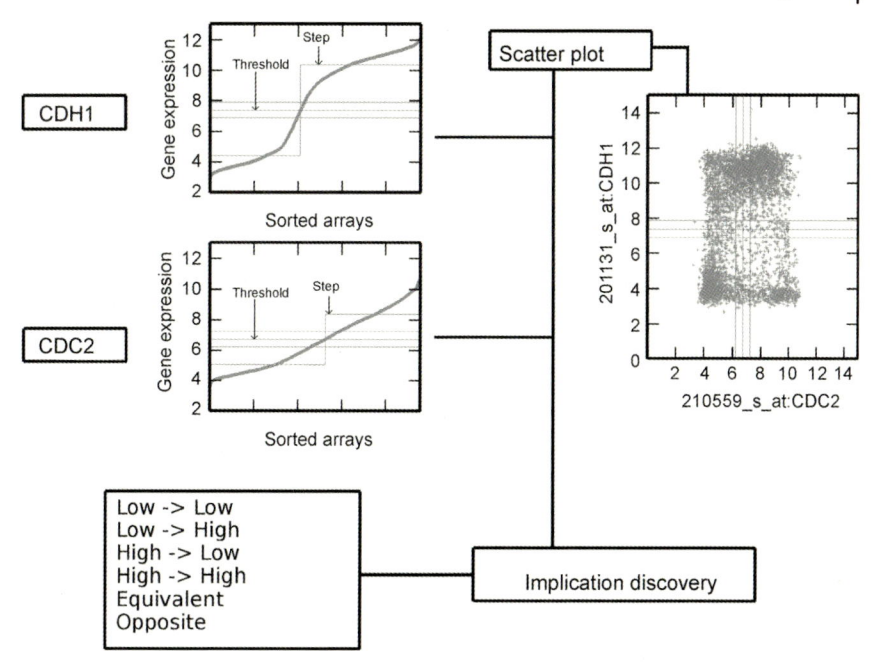

Figure 16.11 Boolean implication extraction process. The expression levels of each probe set are sorted and a step function is fitted (using StepMiner) to the sorted expression level that minimizes the square error between the original and the fitted values. A threshold t is chosen, where the step crosses the original data. The region between $t - 0.5$ and $t + 0.5$ is classified as "intermediate," the region below $t - 0.5$ is classified as "low," and the region above $t + 0.5$ is classified as "high." The examples show probe sets for two genes CDH1 and CDC2. As can be seen, CDH1 has a sharp rise between 6 and 9, and the StepMiner algorithm was able to assign a threshold in this region. CDC2, however, is very linear and the StepMiner algorithm assigns the threshold approximately in the middle of the line. A scatter plot is shown to illustrate the analysis. Each point in the scatter plot corresponds to a microarray experiment, where the value for the x-axis is CDC2 expression and the value for the y-axis is CDH1 expression. Boolean implication discovery analysis is performed on a pair of probe sets, which ignores all the points that lie in the intermediate region and analyzes the four quadrants of the scatter plot. Four asymmetric relationships (low $\Rightarrow$ low, low $\Rightarrow$ high, high $\Rightarrow$ low, and high $\Rightarrow$ high) are discovered, each corresponds to exactly one sparse quadrant in the scatter plot, and two symmetric relationships (equivalent and opposite) are discovered, each corresponds to two diagonally opposite sparse quadrants.

quadrants (intermediate values for A and B are ignored in this analysis). There are at most two possible sparse quadrants because the thresholds always separate a reasonable number of low and high expression levels for each gene. Each sparse quadrant corresponds to an implication. If A high $\Rightarrow$ B high and A low $\Rightarrow$ B low, A and B are considered to have equivalent levels of Boolean expression. When A high $\Rightarrow$ B low and A low $\Rightarrow$ B high, A and B are considered to have an opposite Boolean relationship. In both of these cases, two diagonally opposite quadrants are significantly sparse. In other cases, where there is only one sparse quadrant, the Boolean relationships between A and B have the same name as Boolean implications: A low ⇒ B low, A low ⇒ B high, A high ⇒ B low, and A high ⇒ B high. There are two tests that must succeed for the relationship between A and B to be considered an implication. For concreteness, let us consider whether the low-low quadrant is sparse, yielding an implication A low ⇒ B high. First, the number of expression values in the sparse quadrant must be significantly less than the number that would be expected under an independence model, given the relative distribution of low and high values for A and B. Specifically, if a_{00}, a_{01}, a_{10}, and a_{11} are the number of expression values where A and B are low and low, low and high, high and low, and high and high, respectively, a threshold on the following *statistic* is performed to test whether the low-low quadrant is sparse:

$$total = a_{00} + a_{01} + a_{10} + a_{11} \tag{16.1}$$

$$\text{number of A low expression values} = nAlow = (a_{00} + a_{01}) \tag{16.2}$$

$$\text{number of B low expression values} = nBlow = (a_{00} + a_{10}) \tag{16.3}$$

$$expected = \left(\frac{nAlow}{total} * \frac{nBlow}{total}\right) * total$$
$$= \frac{(a_{00} + a_{01}) * (a_{00} + a_{10})}{total} \tag{16.4}$$

$$observed = a_{00} \tag{16.5}$$

$$statistic = \frac{(expected - observed)}{\sqrt{expected}} \tag{16.6}$$

Second, the observed values in the sparse quadrant are considered erroneous points and a sparse quadrant must have a small number of erroneous points. A maximum likelihood estimate of the *error rate* is computed as:

$$error\ rate = \frac{1}{2}\left(\frac{a_{00}}{(a_{00} + a_{01})} + \frac{a_{00}}{(a_{00} + a_{10})}\right) \tag{16.7}$$

A second threshold on this *errorrate* is performed to ensure that the quadrant is really sparse. If the above tests succeed, the low-low quadrant is considered sparse and therefore, A low ⇒ B high is inferred. A threshold of 3 for the first *statistic* and a threshold of 0.1 for the *errorrate* are used here to discover the Boolean relationships. A FDR is computed (as described below) using these thresholds. A Boolean implication network (directed graph) is built from the implications, where each probe set A has two nodes, representing its low and high values, and edges are Boolean implications. For example, there is a directed edge from A low to B high if there is a Boolean implication A low ⇒ B high.

16.7.3
Computation of FDR

To compute the FDR [43], we permute randomly the expression values for each gene independently. Then, build a complete Boolean implication network as above. This analysis is repeated twenty times to compute the average number of Boolean relationships in the randomized data. The ratio of the average number of Boolean relationships in the randomized data to the original data is considered the FDR.

16.7.4
Correlation Network for Human CD Genes

Human CD genes were selected for comparison against a correlation-based network. The set of genes includes 966 Affymetrix U133 Plus 2.0 human probe sets. Pearson's correlation coefficients for all 466 095 pairs of genes were computed. Boolean implications are extracted from this data to compare the Boolean implication network with the correlation-based network.

16.7.5
Discovery of Conserved Boolean Relationships

Mouse and fruit fly orthologs for human genes were selected from the EUGene database [44]. For each Boolean relationship in the human dataset, a conserved relationship is detected if any of the mouse orthologs of the first human gene has a significant Boolean relationship with another mouse ortholog of the second human gene. To find conserved Boolean relationships in all three species, we check if any of the fruit fly orthologs of the first mouse gene has a significant Boolean relationship with another fruit fly orthologs of the second mouse gene for each conserved relationships in human and mouse.

16.7.6
Connected Component Analysis

Human genes for the highly conserved relationships in all three species were selected for the connected component analysis. An undirected graph was built with the gene names as nodes and the edges are from Boolean equivalent relationships. Connected component analysis was performed using a standard union-find algorithm [45] on the undirected graph to find clusters of genes that are connected together.

16.8
Conclusions

In this chapter, we presented two methods of analyzing gene expression datasets: StepMiner and BooleanNet. Both these tools provide a framework for Boolean analysis of large gene expression datasets.

StepMiner is a tool for the discovery of simple stepwise changes in time-course microarray data. It includes a new statistical test for fitting step function to the time-course data. We showed that it automatically retrieves biologically relevant gene expression changes. StepMiner can create a heatmap with genes arranged by the number, direction, and time of transitions, for easy visualization and exploration. The genes can be grouped into gene sets by these attributes for use with other tools to check whether genes with similar behavior have other shared characteristics. It provides a very simple overview of the massive amount of gene expression data. This simplification allows the biologists to focus on the genes relevant to the underlying biology. Further, it directly answers the most obvious question a biologist might ask: "When does gene expression transition between being up- and down-regulated?" None of the many other tools for analyzing microarray time courses answer this question directly. The method is statistically rigorous, producing p-values for individual matches and a global FDR. StepMiner is implemented in Java with a simple graphical user interface. StepMiner tool has been downloaded multiple times by researchers around the world.

BooleanNet discovers Boolean implication relationships between genes in a massive collection of microarray data. We developed a new statistical test for discovering Boolean implication relationships. It identifies a completely new set of huge number of asymmetric relationships that have not been well studied. We showed that these Boolean implication relationships are relevant to biology. Many Boolean implication relationships are shown to be conserved across multiple species. Further, BooleanNet is shown to be a platform for generating new biological hypotheses. We have validated one of the hypothesis generated in B cell development experimentally. We developed a web-based interface to better navigate around the BooleanNet. The web-based interface supports multiple features for retrieving Boolean implications for genes of interest. We also developed many tools to display scatter plots between two genes with an intelligent search interface. BooleanNet is attracting a lot of users from multiple laboratories at Stanford.

BooleanNet opens up a completely new field of research that could achieve a broader impact in new biological discoveries. It can serve as a general platform to start multiple lines of research. As explained here, BooleanNet can reveal fundamental biological processes such as gender difference, tissue difference, development, and differentiation. Therefore, it has tremendous potential to discover new biological relationships in such domains. BooleanNet can be used to model gene regulations in hematopoiesis. This could lead to new discoveries in less well-characterized developmental pathways. It is also possible to study cancer progression and metastasis as they are tightly related to developmental and differentiation-related abnormalities. A natural extension of our work is to include independent datasets from multiple sources such as protein–protein interaction, protein expression, DNA copy number variation, single nucleotide polymorphism data, and promoter occupancy measured by chromatin immunoprecipitation with unbiased genome-wide tiling microarrays and ultra-high-throughput DNA sequencing. BooleanNet could be used to integrate available genome-wide association studies, regulatory interactions, and model organism genetic data into a testable, system-level model of a specific human cancer.

Acknowledgements

This chapter includes the summary of two publications, Sahoo *et al.* [42] and Sahoo *et al.* [46], and the author's PhD thesis [1]. The content of this work also contains contributions by Professor David Dill, Professor Sylvia Plevritis, Professor Rob Tibshirani, and Andrew Gentles. The authors thank Mathew J. Brauer and Howard Chang for providing the necessary data for the analysis, and Professor Trevor Hastie for his useful advice on the StepMiner algorithm. This work was supported by the National Institutes of Health as part of the Integrative Cancer Biology Program, under grant NIH 5U56CA112 973-02. The contents of this chapter are solely the responsibility of the authors and do not necessarily represent the official views of the National Institutes of Health. For the StepMiner section, Figures, Tables and text from "Extracting binary signals from microarray time-course data" first published in Nucleic Acids Research on May 21, 2007 by Sahoo *et al.* 2007 were reproduced with the permission from Oxford University Press.

References

1. Sahoo, D. (2008) Boolean analysis of high-throughput biological datasets. PhD Thesis, Stanford University.
2. Owen, A. (1991) Discussion: multivariate adaptive regression splines. *Ann. Stat.*, **19**, 102–112.
3. Storey, J.D. and Tibshirani, R. (2003) Statistical significance for genomewide studies. *Proc. Natl. Acad. Sci. USA*, **100**, 9440–9445.
4. Allocco, D.J., Kohane, I.S., and Butte, A.J. (2004) Quantifying the relationship between co-expression, co-regulation and gene function. *BMC Bioinformatics*, **5**, 18.
5. Arkin, A. and Ross, J. (1995) Statistical construction of chemical reaction mechanisms from measured time-series. *J. Phys. Chem.*, **99**, 970–979.
6. Jordan, I.K., Marino-Ramirez, L., Wolf, Y.I., and Koonin, E.V. (2004) Conservation and coevolution in the scale-free human gene coexpression network. *Mol. Biol. Evol.*, **21**, 2058–2070.
7. Lee, H.K., Hsu, A.K., Sajdak, J., Qin, J., and Pavlidis, P. (2004) Coexpression analysis of human genes across many microarray data sets. *Genome Res.*, **14**, 1085–1094.
8. Tavazoie, S., Hughes, J.D., Campbell, M.J., Cho, R.J., and Church, G.M. (1999) Systematic determination of genetic network architecture. *Nat. Genet.*, **22**, 281–285.
9. Day, A., Carlson, M.R.J., Dong, J., O Connor, B.D., and Nelson, S.F. (2007) Celsius: a community resource for Affymetrix microarray data. *Genome Biol.*, **8**, R11.
10. Liao, B.Y. and Zhang, J. (2006) Evolutionary conservation of expression profiles between human and mouse orthologous genes. *Mol. Biol. Evol.*, **23**, 530–540.
11. Weller, P.A., Critcher, R., Goodfellow, P.N., German, J., and Ellis, N.A. (1995) The human Y chromosome homologue of XG: transcription of a naturally truncated gene. *Hum. Mol. Genet.*, **4**, 859–868.
12. Brockdorff, N., Ashworth, A., Kay, G.F., Cooper, P., Smith, S., McCabe, V.M., Norris, D.P., Penny, G.D., Patel, D., and Rastan, S. (1991) Conservation of position and exclusive expression of mouse Xist from the inactive X chromosome. *Nature*, **351**, 329–331.
13. Brown, C.J., Ballabio, A., Rupert, J.L., Lafreniere, R.G., Grompe, M., Tonlorenzi, R., and Willard, H.F. (1991) A gene from the region of the human X inactivation centre is expressed exclusively

from the inactive X chromosome. *Nature*, **349**, 38–44.

14 Sharief, F.S., Mohler, J.L., Sharief, Y., and Li, S.S. (1994) Expression of human prostatic acid phosphatase and prostate specific antigen genes in neoplastic and benign tissues. *Biochem. Mol. Biol. Int.*, **33**, 567–574.

15 Roth, R.B., Hevezi, P., Lee, J., Willhite, D., Lechner, S.M., Foster, A.C., and Zlotnik, A. (2006) Gene expression analyses reveal molecular relationships among 20 regions of the human CNS. *Neurogenetics*, **7**, 67–80.

16 Carroll, S.B. (1995) Homeotic genes and the evolution of arthropods and chordates. *Nature*, **376**, 479–485.

17 Rinn, J.L., Kertesz, M., Wang, J.K., Squazzo, S.L., Xu, X., Brugmann, S.A., Goodnough, L.H., Helms, J.A., Farnham, P.J., Segal, E., and Chang, H.Y. (2007) Functional demarcation of active and silent chromatin domains in human HOX loci by noncoding RNAs. *Cell*, **129**, 1311–1323.

18 Ikuta, K., Ingolia, D.E., Friedman, J., Heimfeld, S., and Weissman, I.L. (1991) Mouse hematopoietic stem cells and the interaction of c-kit receptor and steel factor. *Int. J. Cell Cloning*, **9**, 451–460.

19 Stamenkovic, I. and Seed, B. (1988) CD19, the earliest differentiation antigen of the B cell lineage, bears three extracellular immunoglobulin-like domains and an Epstein–Barr virus-related cytoplasmic tail. *J. Exp. Med.*, **168**, 1205–1210.

20 Bolognese, F., Wasner, M., Dohna, C.L., Gurtner, A., Ronchi, A., Muller, H., Manni, I., Mossner, J., Piaggio, G., Mantovani, R., and Engeland, K. (1999) The cyclin B2 promoter depends on NF-Y, a trimer whose CCAAT-binding activity is cell-cycle regulated. *Oncogene*, **18**, 1845–1853.

21 Davenport, J.W., Fernandes, E.R., Harris, L.D., Neale, G.A., and Goorha, R. (1999) The mouse mitotic checkpoint gene Bub1b, a novel Bub1 family member, is expressed in a cell cycle-dependent manner. *Genomics*, **55**, 113–117.

22 Dennis, G. Jr, Sherman, B.T., Hosack, D.A., Yang, J., Gao, W., Lane, H.C., and Lempicki, R.A. (2003) DAVID: database for annotation, visualization, and integrated discovery. *Genome Biol.*, **4**, P3.

23 Hosack, D.A., Dennis, G. Jr,G., Sherman, B.T., Lane, H.C., and Lempicki, R.A. (2003) Identifying biological themes within lists of genes with ease. *Genome Biol.*, **4**, R70.

24 Nurse, P. (1990) Universal control mechanism regulating onset of M-phase. *Nature*, **344**, 503–508.

25 Cao, R., Wang, L., Wang, H., Xia, L., Erdjument-Bromage, H., Tempst, P., Jones, R.S., and Zhang, Y. (2002) Role of histone H3 lysine 27 methylation in polycomb-group silencing. *Science*, **298**, 1039–1043.

26 Pagano, M., Pepperkok, R., Verde, F., Ansorge, W., and Draetta, G. (1992) Cyclin A is required at two points in the human cell cycle. *EMBO J.*, **11**, 961–971.

27 Hiraoka, L.R., Harrington, J.J., Gerhard, D.S., Lieber, M.R., and Hsieh, C.-L. (1995) Sequence of human FEN-1, a structure-specific endonuclease, and chromosomal localization of the gene (FEN1) in mouse and human. *Genomics*, **25**, 220–225.

28 Kirkness, E.F., Kusiak, J.W., Fleming, J.T., Menninger, J., Gocayne, J.D., Ward, D.C., and Venter, J.C. (1991) Isolation, characterization, and localization of human genomic DNA encoding the beta 1 subunit of the GABAA receptor (GABRB1). *Genomics*, **10**, 985–995.

29 Ivey-Hoyle, M., Conroy, R., Huber, H.E., Goodhart, P.J., Oliff, A., and Heimbrook, D.C. (1993) Cloning and characterization of E2F-2, a novel protein with the biochemical properties of transcription factor E2f. *Mol. Cell Biol.*, **13**, 7802–7812.

30 Mathews, M.B., Bernstein, R.M., Franza, B.R. Jr, and Garrels, J.I. (1984) Identity of the proliferating cell nuclear antigen and cyclin. *Nature*, **309**, 374–376.

31 Miyachi, K., Fritzler, M.J., and Tan, E.M. (1978) Autoantibody to a nuclear antigen in proliferating cells. *J. Immunol.*, **121**, 2228–2234.

32 Wu, C.H., Sahoo, D., Arvanitis, C., Bradon, N., Dill, D.L., and Felsher, D.W. (2008) Combined analysis of murine and human microarrays and chip analysis reveals genes associated with the ability of MYC to maintain tumorigenesis. *PLoS Genet.*, **4**, e1000090.

33 Tsaparas, P., Marino-Ramirez, L., Bodenreider, O., Koonin, E.V., and Jordan, I.K. (2006) Global similarity and local divergence in human and mouse gene co-expression networks. *BMC Evol. Biol.*, **6**, 70.

34 Barabasi, A.L. and Albert, R. (1999) Emergence of scaling in random networks. *Science*, **286**, 509–512.

35 Barabasi, A.L. and Oltvai, Z.N. (2004) Network biology: understanding the cell's functional organization. *Nat. Rev. Genet.*, **5**, 101–113.

36 Bhan, A., Galas, D.J., and Dewey, T.G. (2002) A duplication growth model of gene expression networks. *Bioinformatics*, **18**, 1486–1493.

37 Featherstone, D.E. and Broadie, K. (2002) Wrestling with pleiotropy: genomic and topological analysis of the yeast gene expression network. *BioEssays*, **24**, 267–274.

38 Jeong, H., Mason, S.P., Barabasi, A.L., and Oltvai, Z.N. (2001) Lethality and centrality in protein networks. *Nature*, **411**, 41–42.

39 Jeong, H., Tombor, B., Albert, R., Oltvai, Z.N., and Barabasi, A.L. (2000) The large-scale organization of metabolic networks. *Nature*, **407**, 651–654.

40 Edgar, R., Domrachev, M., and Lash, A.E. (2002) Gene Expression Omnibus: NCBI gene expression and hybridization array data repository. *Nucleic Acids Res.*, **30**, 207–210.

41 Irizarry, R.A., Bolstad, B.M., Collin, F., Cope, L.M., Hobbs, B., and Speed, T.P. (2003) Summaries of Affymetrix GeneChip probe level data. *Nucleic Acids Res.*, **31**, e15.

42 Sahoo, D., Dill, D.L., Tibshirani, R., and Plevritis, S.K. (2007) Extracting binary signals from microarray time-course data. *Nucleic Acids Res.*, **35**, 3705–3712.

43 Storey, J.D. and Tibshirani, R. (2003) Statistical significance for genomewide studies. *Proc. Natl. Acad. Sci. USA*, **100**, 9440–9445.

44 Gilbert, D.G. (2002) euGenes: a eukaryote genome information system. *Nucleic Acids Res.*, **30**, 145–148.

45 Galler, B.A. and Fisher, M.J. (1964) An improved equivalence algorithm. *Commun. ACM*, **7**, 301–303.

46 Sahoo, D., Dill, D.L., Gentles, A.J., Tibshirani, R., and Plevritis, S.K. (2008) Boolean implication networks derived from large scale, whole genome microarray datasets. *Genome Biol.*, **9**, R157.

Part Four
Systems Approach to Diseases

17
Representing Cancer Cell Trajectories in a Phase-Space Diagram: Switching Cellular States by Biological Phase Transitions

Mariano Bizzarri and Alessandro Giuliani

17.1
Introduction

Cell and tissue morphogenesis are complex processes driven by a set of both intracellular and environmental cues, which are coordinated in both time and space [1]. It is remarkable how cells are able to integrate simultaneous physical as well as biochemical inputs so as to produce just one of a few possible phenotypes.

How can we understand such a complex and fine-tuned regulated system? The classical, reductionistic approach focuses on single gene expression and how distinct functional pathways are regulated. However, the widely accepted molecular paradigm did not allow for a full understanding of the developmental processes, as it is increasing recognized [2, 3]. Indeed, molecular composition does not determine form; to explain how macroscopic form is generated, we need a "field-level" approach where general gene expression patterns as well as morphological changes at both tissue and cell levels are integrated so to generate a global picture of the relational order of the system [4].

Indeed, while we tend to assign specific, unambiguous functions to signaling molecules, the reality is that the information conveyed by the signal transduction machinery often cannot be localized to an individual cascade, rather it is distributed among several pathways. Moreover, the same signaling molecule can trigger paradoxical effects, depending on the microenvironment surrounding cells or on the cell activity state. These data outline that biological pathways are subjected to a nonlinear dynamics and suggest that signal specificity should be ascribed to the overall system configuration, rather than to single, specific components [5, 6].

A further complication comes from the existence of physical interacting cues. As a matter of fact, diffusible factors alone are not sufficient to fully explain cell fate regulation and even gradients of morphogenetic molecules ("morphogens") cannot entirely explain morphogenesis, as first proposed by the seminal paper of Turing [7]. Physical forces (electromagnetic and gravitational fields) and structural cues both affect cell behavior and phenotypic traits (shape and differentiation). Indeed, normal cells need to adhere to extracellular matrix (ECM) in order to survive, differentiate, and proliferate; ECM molecules bind to specific cell surface receptors (integrins) and therefore activate intracellular signaling pathways that drive a cell toward proliferation, differentiation, apoptosis, or even neoplastic transformation [8].

It is noteworthy that cells can be switched between entirely different gene expression patterns through alteration in ECM or cell shape, independently from growth factor binding or integrin binding [9]. Even though growth factors and ECM activated early signaling events, cyclin D1 and p27 levels remained, respectively, low and high in cells that were prevented from spreading; this activation pattern in round cells is nearly identical to the cell arrest induced by growth factor withdrawal or by inducing cell rounding from within by disrupting the actin cytoskeleton [10]. Moreover, compelling evidence about the relevance of cell shape in driving cell behavior has been provided by studies performed in microgravity. Cells exposed to low g fields underwent significant changes in both membrane and nuclear shape; in turn, these modifications were *per se* sufficient to trigger a dramatic and complex remodeling of cell phenotypes with subsequent metabolic as well as genomic alterations involving thousand of genes [11, 12].

The molecular mechanism by which a cell shape change could produce such a dramatic influence on complex biological functions is still poorly understood. Several lines of evidence suggest that tension-dependent integrity of the cytoskeleton network is essential in ensuring cell cycle control [13, 14]. Disrupting the cytoskeleton network by using cytochalasins or inhibiting cytoskeleton network tension generation using pharmacological modulators of actomyosin-based contractility both induce cell cycle arrest in G_1, similar to that produced by cell rounding. Moreover, focal adhesion formation is itself dependent on the form the cell acquires and on the associated tension generated by the interaction of the cytoskeleton network with the surrounding ECM [15]. Given the pluripotent role of force in tissue function, it is not surprising that several diseases, including cancer, are characterized by compromised tensional homeostasis [16]. On the contrary, normalizing the tissue tensional state of tumor cells can revert them towards a nonmalignant phenotype [17]. So far, changes in ECM and/or in the tensional balance of forces transmitted across focal adhesions, together with a change in cell shape, might account for the complex phenotypic and functional transformations occurring during tissue development or neoplastic transformation.

17.2
Beyond Reductionism

According to the reductionistic approach, the cell regulatory network is generally studied by breaking it down into individual signaling pathways, characterized by simple, linear dynamics. It is widely assumed that the "instructions" are encoded in the molecular structure of a ligand and its specific recognition by the receptor. In turn, receptors transmit a proper signal to the nucleus, hence activating the requested gene "program."

This simplistic framework has been cast on doubt by an impressive body of evidence that has accumulated over recent decades [18, 19].

The implicit assumption underlying the above-mentioned paradigm considers that the overall information is embedded into the gene, and the gene is able to govern cellular functions through a linear and deterministic process. However, we can no longer define a "gene" as a unitary component of the genome: every genomic element interacts either directly or indirectly with many other genomic and nongenomic components (proteins and RNA). Thus, the idea that any cellular or organismal character is "determined" by a single region of the genome has no logical connection with our knowledge of biogenesis. Genomic functions are inherently interactive in that isolated DNA is virtually inert: "DNA cannot replicate or segregate properly to daughter cells or template synthesis of RNA by itself. This fundamental biochemical reality alone would invalidate the central dogma" [20]. In addition, a large amount of data suggests gene expression is fundamentally a stochastic process [21]. Cells expressing the same phenotype and placed in homogenous environments should always express the same genes if they are controlled by a tight determinist mechanism. However, a large variability in gene activity has been observed both *in vivo* and in cultured cells [22]. In contrast to the classical view of development as a preprogrammed and deterministic process, these studies have demonstrated that stochastic perturbations of highly nonlinear systems may underlie the emergence and stability of biological patterns [23]. As a conclusion, it is now increasingly clear that the genome neither contains the future of the organism, in some preformative version of van Leeuwenhoek's homunculi, nor is it to be regarded, as in modern metaphors, as a blueprint or an information theorists' code-bearer.

17.3
Cell Shape as a Diagram of Forces

Across the animal kingdom there is no one-to-one correspondence between homologous genes and morphological structures [24]. The behavior of a cell expressing a given gene or subset of genes depends entirely on where that particular cell is in the body and at what point of development time. Thus, morphogenesis and phenotypic differentiation are time- and space-dependent processes [25]: morphological plasticity, rather than being the result of genetic "adaptation," reflects the influence of external physicochemical parameters on any material system and is therefore an inherent, inevitable property of organisms [26]. The physical milieu integrating the different chemical as well physical signals that drive cells and tissues towards differentiation it is known as the "morphogenetic field" [27, 28]. Within this field, morphogenetic cues exert short- and long-range influences by affecting gradients of morphogens and mechanical stresses. This process is strongly dependent on the geometry of the morphogenetic field, governing the topology of signaling cues. Within this framework, the geometric form a cell acquires (i.e., its shape) represents the integrated endpoint of the morphogenetic cues acting on the living system: morphogenesis is indeed the process through which a population of cells rearranges into a distinctive shape. According to D'Arcy Thompson, the form acquired by any portion of matter, "and the changes of form which are apparent in its movement and growth, may in all cases alike be described as due to the action of force. In short, the shape of an object is a 'diagram of forces' in this sense, at least, that from it we can judge or deduce the forces that are acting or have acted upon it" [29]. In turn, cellular and tissue geometries are acting as both a template and instructive cue for further morphogenesis [30].

Indeed, studies on cell shape control have revealed that a generalized extracellular mechanical stimulus leading to cell shape distortion is sufficient to modulate cell responsiveness to regulatory cues and to govern whether cells will switch between different gene expression patterns [31, 32]. In brief, this means that the switch between different cell fates could be considered dependent on cell distortion [33]. Such a "nonspecific," physical signal can direct cells towards the same fates as molecular factors that bind with high specificity to their cognate receptors. These results suggest that the regulation of cell functions can be thought of as *physical and topological structured networks* of molecular interactions and physical forces, behaving according to a nonlinear dynamics and located at different hierarchical levels.

Thus, the specific phenotypes cells exhibit are emergent functional and morphological configurations arising as a specific feature of a complex system. As such, biological entities and their morphofunctional transitions cannot be "reduced" to a linear transduction of the "genetic program:" the properties of living, complex organisms cannot be predicted from the individual study of the elements of the system. Moreover, morphologic processes arise in living as well as in nonliving systems operating far from the equilibrium and therefore the emergence of highly organized structures does not seem to stem from specific genetic information, but it is likely to be governed by general physical rules [34]: cellular form is generated by a web of interacting chemical and physical processes, whose every strand is woven of multiple gene products. So far, the relationship between genes and form is indirect and cumulative: morphogenesis must then be addressed as a problem not of molecular genetics, but of cellular physiology [35, 36]. That is to say, we need to know the organization principles that are involved, the dynamic relationships between components, the way they are arranged in space, and the patterns of change they undergo in time in order to explain what forms the system can take (Figure 17.1) [37, 38]. Morphologic and phenotypic traits are not random assemblages of working parts and their emergence reflects a deep pattern of ordered relationships. DNA plays an important role by stabilizing some aspects of this spatial and temporal order. However, it does not generate the order. So far, where does this order come from?

17.4
Morphologic Phenotypes and Phase Transitions

A crucial element to be taken into consideration is the finding that gradual variation in a single control parameter (e.g., cell shape) can switch cells between distinct phenotypes. This behavior is reminiscent of phase transitions – abrupt macroscopic changes between qualitatively discrete stable states – observed in physical systems.

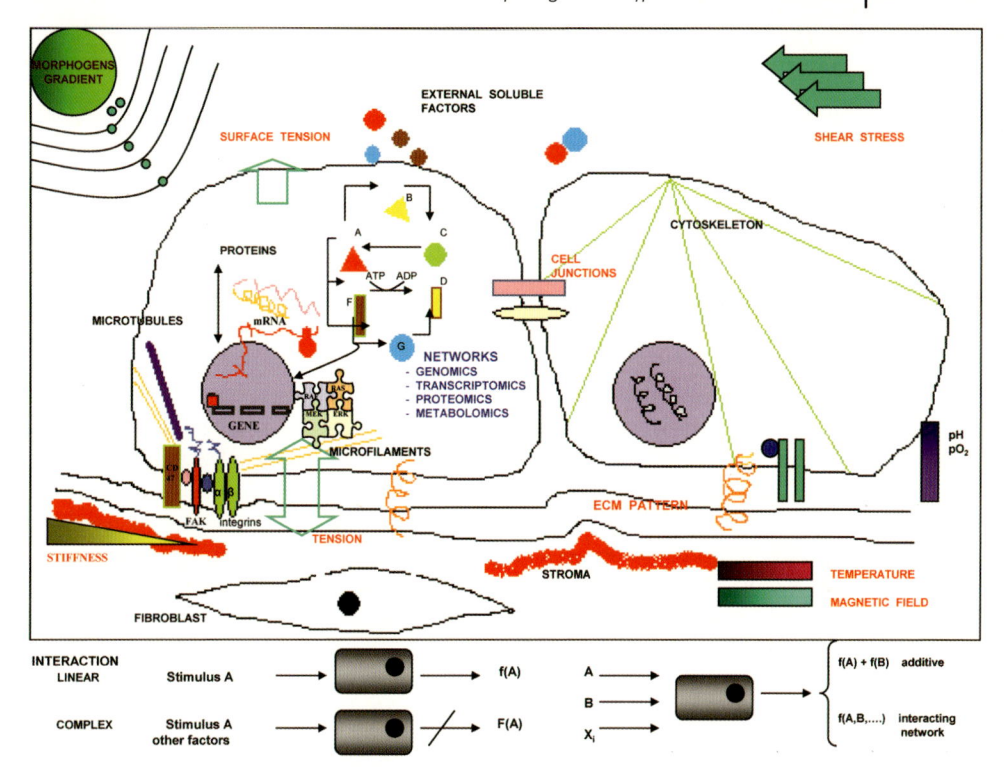

Figure 17.1 Complex biochemical and mechanical interplay between cells and their microenvironment (ECM, stromal cells, laminin, and collagen fibers). Cells are subjected to several physicochemical (surface and tissue tension, shear stress, pH and pO_2 gradients) and biochemical cues (morphogen gradient, and genomic and proteomics interactions). DNA (genomics) is transcribed to mRNA (transcriptomics) and translated into protein (proteomics), which can catalyze reactions that act on and give rise to metabolites (metabolomics), glycoproteins and oligosaccharides (glycomics), and various lipids (lipidomics). These processes are widely influenced and dictated by both molecular interactions, intracellular processing (alternative splicing), and microenvironmental physical and chemical stimuli, provided by the tissue geometrical architecture (mediated by focal adhesion kinases (FAK), integrins, microfilaments, and microtubules representing the cytoskeleton). Multiple stimulations can reveal new features in cell biology by analyzing their superimposed ($f(A) + f(B) + \ldots$), synergistic, and diverse interacting ($f(A, B, \ldots)$) effects.

However, in order to give an answer to this question we need to adopt a new approach – a systems biology view – by which changes in shape and functions may be ascribed to the overall system and not to a single component [39].

The different phenotypes – physiological as well pathological – a cell may experience can be considered as "cellular states" and the switches between them as "biological phase transitions." This framework allows describing the collective ordered behavior of a cell population and the underlying control network, without focusing on the individual molecular components. Namely, the fact that a wide set of different physical and biological stimuli can lead a cell to the same phenotype clearly suggests that cell fates should be considered as "common end programs" or attractors.

The mathematical formalism of the phase-space diagram is described by the theory of dynamical nonequilibrium systems [40]. The phase-space describes an energy landscape characterized by attractors ("valleys") – surrounded by basins – separated by "hills:" the difference in the behavioral potential between cells lies in their *position* on the landscape and the associated accessibility to attractors (Figure 17.2). In this view, cell fate regulation is based on the selection between pre-existing, intrinsically robust fates.

The phenotypic state of a cell at any time is indicated by the position of the state vector (represented by a marble in the Waddington diagram [41]) on that landscape. The marble will spontaneously roll down the valleys (stable developmental paths) leading to a distinct phenotype, corresponding to an attractor. In some instances, the marble may experience anomalous trajectories leading to unexpected positions (unstable or *metastable* states), characterized by reduced stability; they are therefore highly sensitive to minimal fluctuations and could lead to different even unexpected phenotypes, like that characterizing a pathological state. In this respect it is worth noting that in the great majority of cases we are dealing with population data so that the position in the phase-space is defined as an average over millions of cells; however, in analogy with statistical mechanics, the between-cells variance in the population (playing a role analogous to that played by temperature in nonliving systems) is crucial in allowing the entire population to skip to other phenotypes (subattractors).

The topology of the attractor, according to Huang's definition [42], is the "invisible hand" driving the system functions into coherent behavioral states: they are *self-organizing* structures and can "capture" the gene expression profiles associated with cell fates [43]. In other words, the phenotypic traits of the organism are embedded into the dynamical attractors of its underlying regulatory network [44].

A discrete finite number of attractor classes exist, corresponding to configurations allowed by thermodynamics constraints [45]. This implies a particular link between the multiplicity of microscopic states and the relative paucity of the corresponding macroscopic states that is both at the basis of the impossibility of a one-to-one correspondence between molecular and tissue-level representations, and of the substantial resilience of biological systems that allows for a wide variety of microscopic variability by keeping substantially invariant higher-level features.

Functional states depicted as attractors are conceived in Huang's perspective as mainly specified by the gene regulatory network. However, "the stability of functional states clearly also depends on external cues" [46]. So far, the system's dynamics in the phase-space cannot be "reduced" only to the genetic wiring diagram or, at best, to the integrated functioning of the genome–proteome–metabolome network. Additional influences (i.e., products of cell–stroma interaction) must be taken into consideration in order to give a more reliable definition of the attractor. This revised conceptualization of the "attractor" could capture many observations on tissue dynamics,

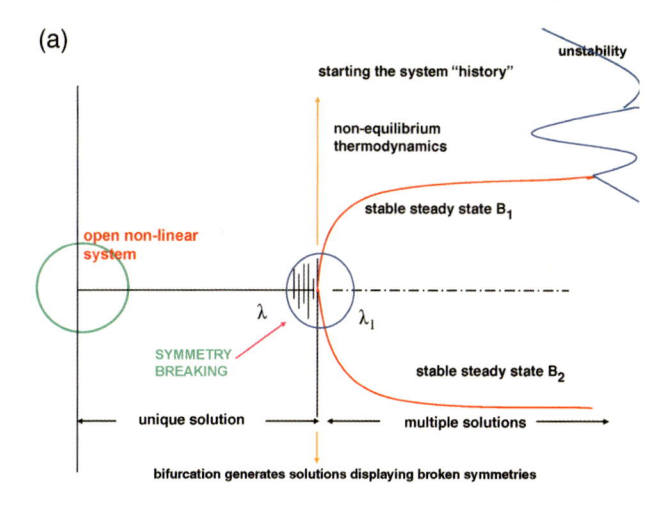

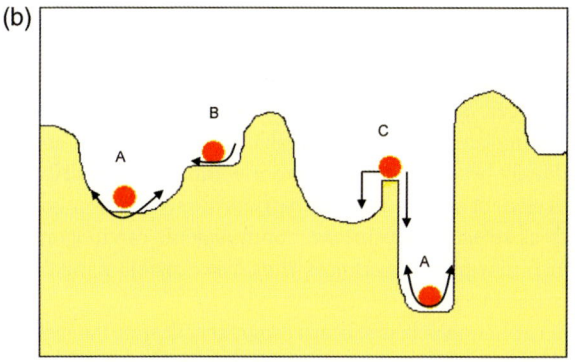

Figure 17.2 Emergence of complexity in living systems. (a) The system is started with an initial value of parameter λ. When λ increases to λ_1, the system enters a nonequilibrium state that leads to symmetry breaking and to the emergence of hysteresis. Thereafter, a new ordered behavior emerges, characterized by multiple choices and different stable states. (b) A two-dimensional landscape in which stable (A), metastable (B), and unstable (C) states are depicted. Stable states represent attractor states robust to many perturbations. Attractors are self-organizing structures and can "capture" the stability of gene expression profiles associated with cell fates. Transitions from one phenotype to another can be triggered by both internal and external regulatory signals that essentially "reset" the state by displacing the network state to another place in the state space. Minimal external perturbations will move the marble away from point C, while the marble will generally return to its attractor within its basin of attraction (A) even under the action of more important perturbations. The metastable state (B) is a week attractor from which the marble may relatively easily reach a neighboring stable attractor.

the existence of different dynamical regimes, as well as the transitions between them [47]. Indeed, such transitions are triggered by external regulatory signals by displacing the network state to another location in the phase-space. Chemical as well as physical external signals do not need to actually instruct cells which particular genes to up- or downregulate in order to "produce" a specific response. Instead, they merely have to "destabilize" a network state by displacing it and thus driving the system into a different basin of attraction. It is important to stress these attractors are implicit in the structure of the relation of the system that has a fixed repertoire of possible attractors. Such perturbations carry no specific information encoded by a molecular structure and, hence, do not operate at the level of single stimulus-specific receptors, but instead impose a *distributed set of constraints* on structural elements, such as the cytoskeleton, that are able to simultaneously influence numerous signaling pathways inside the cell. A process based on selection of pre-existing states is *robust*, and tolerates random fluctuations and errors [48].

17.5
Cancer as an Anomalous Attractor

The previously described regulation model explains how several hundreds of genes can be collectively and coherently involved in phase transitions from one attractor to another [49]. In some instances, such transitions are likely to rewire the gene regulatory network such that unoccupied and inaccessible *archaic* (typical of embryonic stages) attractors suddenly become accessible from within an adult cell phenotype. This issue becomes evident once the normal barriers are somehow modified by epigenetic stimuli, such as structural distortion, nonmutagenic carcinogens, or chronic inflammation, which may promote *attractor transitions*, enabling the cell to fall within "pathological" attractors.

A limited number of cancer phenotypes have been recognized, despite the fact that gene variability might lead, in principle, to a transfinite number of configurations. So far, this means that cancer phenotypes cannot be described as pertaining to a continuous distribution of abnormal tissue phenotypes despite the chaotic variety of cancer-associated random mutations.

Indeed, cancer cells are characterized by unstable karyotypes, despite their clonal origin and notwithstanding the fact that they share the same malignant phenotype [50, 51]. It is worth noting that tumors of the same type typically show no congruently overlapping but quite differing sets of mutations. Only a few mutations are found in the majority of tumors, whereas the majority of mutations are observed in less than 5% of tumors [52]. These controversial results have been tentatively explained by arguing that driver mutations affect genes that belong to the same "functional" class and, hence, need not to be identical although the functional classes are rather broadly defined [53]. On the other hand, if we move from genomic alterations to the cancer transcriptome, considered as "a whole" a stunning degree of organization and order of global expression patterns is observed [54]. Cluster analysis of tumor transcriptomes readily classifies tumors into a small number of discretely distinct groups [55], reminiscent of the organization of transcriptomes of normal cell types into groups of related tissues [56]. Noteworthy, such a clustering is well correlated with the traditional classification of tumor types derived from morphology.

The discreteness of types and the robustness of cancer phenotypes to environmental perturbations, strongly argues in favor of the existence of pre-existing "pathological" attractor states.

On the other hand, the fact that the transition from a normal toward a neoplastic attractor may be triggered by a wide set of both chemical and physical stimuli without any molecular specificity casts doubt on the classical oncogenic paradigm.

Until now cancer has been considered as a gene-dependent disease [57, 58]. Nevertheless, is increasingly clear that "cancer is more than a disease of specific genes" [59] and advanced knowledge of the carcinogenic process claims to be acquired by means of supragenomic strategies [59, 60]. Despite the wide, aggressive effort by many laboratories around to support the somatic mutation theory [61] of cancer, and its ancient and modern variations, several contradictions are waiting to be explained and, more importantly, the promised objectives have been not attained [62–64]. Furthermore, there is an increasing accumulation of data emphasizing the role of tissue interactions in carcinogenesis [65–67], suggesting that cancer is an emergent, supracellular phenomena [68]. This almost imperceptible switch of targets from a cell- to a tissue-based etiopathogenesis does not fit easily into the theoretical framework – the gene-centered paradigm – into which the somatic mutation theory has been built [69, 70].

Compelling evidence suggests cancer can arise as a consequence of the disruption of the reciprocal interactions between cells and the microenvironment, leading to modifications in cell morphology, signaling pathways, and genomic functions [71, 72]. In nonlinear systems, switching between stable states or phenotypes requires that activities of signaling molecules in multiple pathways change in concert; this is the case with microenvironmental stimuli that produce pleiotropic effects contributing to both normal and neoplastic development [73]. A wide range of cellular and molecular processes are tightly controlled by mechanical pleiotropic factors depending on cytoskeleton organization and ECM mechanical properties [74–77]. On the other hand, modification of the cancer microenvironment, through soluble mitogens or by means of physical cues, has been demonstrated to induce dramatic effects on cancer cell behavior, eventually leading to reversion of the malignant phenotype [78, 79]. Needham [80] and Waddington [35] speculated that cancers represent an *escape* from a morphogenetic field like those that guide embryonic development; a similar statement has been recently put forward again by Simon *et al.*, who suggest that "any change that leads stem cells to escape from the niche [i.e., the morphogenetic field] would result in tumor formation" [81]. Indeed, cancer cells exposed to embryonic or maternal microenvironments are committed to die or to differentiate and eventually they could be rewired into a nonmalignant phenotype [82–84]. So far, disruption of cellular interactions could be considered as the "initial source" of abnormal gene expression in cancer cells [60, 85]. It is worth considering that such a change seems to be a direct consequence of the modifications produced by the biophysical constraints of the microenvironment on the cell shape.

Thus, cancer can be considered an emergent property of living *tissues* under attractor-destabilizing stimulation, involving not merely DNA or single somatic cell functions, but instead *systems features*.

17.6
Shapes as System Descriptors

Shapes and structures of cells and tissues are as diverse as the functions ascribed to them. A long time ago, a clear-cut link between cell geometry and cell function was established by means of simple morphological observations. Numerous cellular behaviors in culture, including proliferation [86], apoptosis, lamellipodial extension [87], migration [88], glucose metabolism [89], RNA processing [90, 91], differentiation [92, 93], epithelial–mesenchymal transition [94], and stem cell fate [95], have been found to be determined by cellular geometry. As an example, connective tissue cells differ greatly in phenotype. Although they descend from a common mesenchymal stem cell precursor, differentiated adipocytes are round and fat-laden, while osteoblasts vary from elongated to cuboidal, depending on their matrix deposition activity. The shape of these cells serves their specialized functions, while simultaneously driving their multicellular organization. These different cell morphologies are thought to arise from changes in the cytoskeleton network organization, and in both cell–ECM and cell–cell interactions during stem cell commitment (i.e., the process by which a cell chooses its fate) and differentiation (i.e., the following development of lineage-specific characteristics).

Biological entities are thought to potentially adopt an undefined possibility of configurations ("forms") "inside the realm" of a common general frame. However, only a limited number is actually observed.

In the case of protein structures, the number of folds is much lower than that expected when referring to the transfinite number of possible dispositions of N residues in space; different sequences may give rise to the same fold. This implies some sort of "energy minimization" drastically constraining the number of allowable stable states [96], with the consequent onset of preferred stable states (attractors, in dynamical terms). Given that protein folding results from the nonlinear interactions between internal (amino acid sequence) as well as microenvironmental constraints, the resulting configuration can be considered as the integrated output of such complex and dynamical interplay taking place at the mesoscopic level.

Despite the amazing growth in our understanding of the molecular mechanisms of cancer, as a matter of fact, most diagnosis is still done by visual examination of images and by the morphological examination of radiological pictures, microscopy of cell and tissues, and so forth [97]. Indeed, cancer diagnosis as well as that of other diseases has been routinely made by looking at the cell and tissue shapes elicited by such pathological entities [98].Therefore, shape descriptors are reliably used as overall indicators of macrostates and they should be considered in representing the system's evolution into a phase-space diagram.

However, a compelling theory explaining the link between shape and biochemical activity is still lacking. This is in partly due, on the one hand, to the limited knowledge about how biochemical reactions are associated to the cytoskeleton (i.e., the internal topology of structure-linked reactions) and, on the other hand, to a lack of a standardized and wide-accepted *measure* of cell shape complexity. The ability to correctly characterize shapes has become particularly important in biological and biomedical sciences, where morphological information about the specimen of interest can be used in a number of different ways, such as for taxonomic classification and research on morphology–function relationships. A quantitative method that lends itself particularly useful for characterizing complex irregular structures is fractal analysis.

17.7 Fractals of Living Organisms

Although classical Euclidean geometry works well for describing properties of regular smooth-shaped objects such as circles or squares by using measures such as the length of the object's perimeter, these Euclidean descriptions are not adequate for complex irregular-shaped objects that occur in nature (e.g., clouds, coastlines, and biological structures). These "non-Euclidean" objects are better described by fractal geometry, which has the ability to quantify the irregularity and complexity of objects with a measurable value called the fractal dimension.

Mandelbrot [99] introduced the term "fractal" (from the Latin *fractus*, meaning "broken") to characterize spatial or temporal phenomena that are continuous, but not differentiable. Fractal dimension differs from our intuitive notion of dimension in that it can be a noninteger value, and the more irregular and complex an object is, the higher its fractal dimension [100]. In fractal analysis, the Euclidean concept of "length" is viewed as a *process*. This process is characterized by a constant parameter D known as the fractal (or fractional) dimension. The fractal dimension has a thermodynamic meaning and can be viewed as an intensive measure [101] of the "overall" (morphologic) complexity [102].

Like many summary statistics (e.g., mean), the fractal dimension is obtained by "averaging" variation in data structure [103]. In doing so, information is necessarily lost. The estimated fractal dimension of a lakeshore, for example, tells us nothing about the actual size or overall shape of the lake, nor can we reproduce a map of the lake from the fractal dimension alone. However, the fractal dimension does tell us a great deal about the relative complexity of the lakeshore and as such is an important descriptor when used in conjunction with other measures. It should be emphasized that the fractal dimension is a descriptive, quantitative measure; it is a statistic, representing an attempt to estimate a single-valued number for a property (complexity) of an object with a sample of data from the object. One can view the fractal dimension "in much the same way that thermodynamics might view intensive measures as temperature" [101]. In other words, fractal dimension can be considered a systems property and, together with one or more independent variables, could enable constructing a diagram of phases, like that relying on temperature, pressure, and volume for gas/liquid/solid phase transitions.

Moreover, it is known from the Bendixon–Poincaré theorem [104] that if a dynamic process possesses a limit cycle, (i.e., an attractor), then that attractor has fractal dimension. And vice versa, the existence of fractal dimension for a given dynamic process denotes that the process has been measured at its attractor [105]. In nonequilibrium systems, the fractal attractor is a common feature because of the dissipative character of these systems. The information dimension can then be used to determine the number of undamped dynamical variables that are active in the motion of the system; this means that dimension is something related to the number of degrees of freedom of the system.

Although there may be many nominal degrees of freedom available, the physics of the system may organize the motion into *only a few* effective degrees of freedom. This collective behavior is often termed self-organization and it arises in dissipative dynamic systems whose post-transient behavior involves fewer degrees of freedom than are nominally available. The system is *attracted* to a lower-dimensional phase-space and the dimension of this reduced phase-space represents the number of active degrees of freedom in the self-organized system. A similar trend can be observed during the shift from one morphotype to another in the course of the specialization/differentiation of a cell lineage: a cell type proceeds through a discrete number of morphotypes along its specializing/differentiating pathway and every morphotype could be considered as a *stable steady state* [106]. In a similar way, morphologic characterization of a cell population by means of fractal analysis could provide at least an independent variable to be used to construct a (measurable) space phase of the evolving system, in order to characterize the attractors and the location of bifurcations.

17.8
Fractals and Cancer

Fractal analysis can lead to a remarkable improvement in both cytohistological and radiographic diagnostic accuracy [107, 108]. Applications of fractal measures to pathology and oncology [109] suggest that fractal analysis provides reliable and unsuspected information [110, 111]. For instance, fractal analysis has helped in discriminating benign from malignant tissues [112] and low- from high-grade tumors [113].

Fractal studies elucidated some aspects of the complex interplay between cancer cells and stroma by suggesting that tumor vascular architecture is determined by heterogeneity in the cellular interaction with the ECM rather than by simple gradients of diffusible angiogenic factors [114]. Moreover, fractal analysis of the interface between cancer and normal cells helps in understanding how cell detachment from the primary mass and infiltration into adjacent tissue occurs through a nonmutational mechanism [115]. A correlation between the fractal dimensions of the epithelia/stroma interface in the oral cavity with the evolving lesions (from dysplasia to invasive carcinoma) was linked to increases in the irregularity of the surface and of the fractal number (from 1.0 of normal epithelium to 1.62 for invasive tumor) [116]. Also, both the global and local fractal dimension of the epithelium–stroma interface increased from normal through premalignant to malignant oral epithelium [117] implying that the involvement of the epithelium–stroma interface is not merely a consequence of tumor development, but instead is an intrinsic feature of the carcinogenic process. "Thus, the landmark of tumor aggressiveness would not be solely a localized tumor cell proliferation [...] *but a global switch or bifurcation between 'smooth margin' and 'fingering protrusions' surface patterns* [...] tumor growth occurs if surface tension or cell–cell adhesion is strong enough, but irregular, infiltrative structures resembling irregular outer boundaries of tumors emerge if these parameters are smaller" ([118], italics added). On the contrary, carcinoma cells P19 undergoing differentiation when treated with all-*trans* retinoic acid show a significant reduction in their fractal dimension [119].

Studies focusing on nuclear shape revealed strong correlations between shape change and modifications in cellular phenotype [120]. Moreover, microenvironmental-induced shape changes in chondrocyte nuclei correlate with changes in collagen synthesis [121] or in cartilage composition and density [122]. Changes in nuclear stiffness could be considered as a prerequisite of the increased motility observed in metastatic cancer cells [123]. In turn, these observed changes in nuclear shape may interfere with chromatin structure, and could modulate gene accessibility and nuclear elasticity required for translocation, leading to a large-scale reorganization of genes within the nucleus [124].

Collectively, these results highlight the relevance of shape–phenotype relationships that over two decades ago motivated Folkman and Moscona [125] to ask, "How important is shape?" [60]. An answer to this question largely remains unknown. However, Ingber claimed that "the importance of cell shape appears to be that it represents a *visual manifestation of an underlying balance of mechanical forces that in turn convey critical regulatory information to the cell*" ([126], italics added). Thus, cell distortion influences cytoskeleton function and a cell's adhesion to the ECM. Cell shape and cytoskeletal structure appear to be tightly coupled to cell proliferation. In this way, tissue structure limits the constitutive ability of cells to proliferate and to undergo neoplastic transformation [127]. Within this framework it seems that "function follows form, and not the other way around" [128].

17.9
Modifications in Cell Shape Precede Tumor Metabolome Reversion

Breast cancer cells (MCF-7 and MDA-MB-231) growing in an experimental maternal morphogenetic field (EMF) progressively undergo dramatic changes in cell shape and metabolism [129]. After 48 h, in breast cancer cells growing in an EMF, membrane profiles change, evolving into a more rounded shape, loosing spindle and invasive protrusions (Figure 17.3).

Those changes were quantified by means of fractal analysis, namely through the calculation of the normalized bending energy (BE) of the cell membrane (Figure 17.3b).

BE is an interesting global feature for shape characterization corresponding to the amount of energy needed to transform the specific shape under analysis into its lowest energy state (i.e., a circle) [130]. The BE concept was initially introduced by Young *et al.* [131] as a robust resource for global shape analysis and classification. BE has been used for analysis of the contour of the left ventricle of the heart [132], as

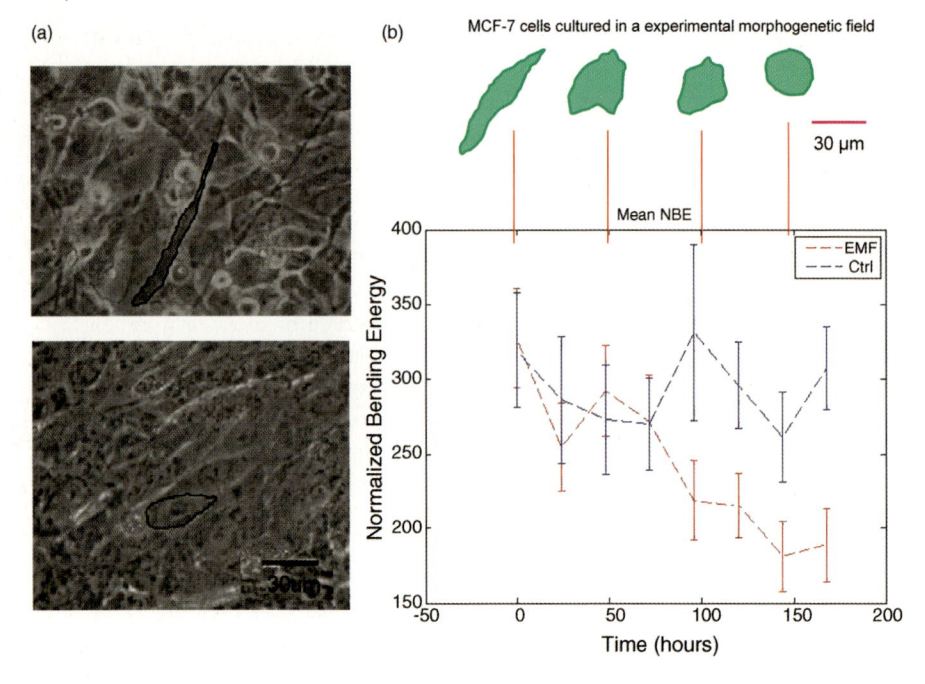

Figure 17.3 Cancer cell shape transitions after exposure to a morphogenetic field. (a) Examples of breast cancer cell membrane segmentation using the GVF-Snake method. MCF-7 cells in control culture conditions (top) and growing in an experimental morphogenetic field (bottom), after 96 h. (b). Transition of cancer cell shape from a spindle versus a round profile and their relationships with the corresponding normalized BE (NBE) values. At the bottom, mean normalized BE values in MCF-7 cell line computed at different experimental time in controls and treated conditions. Tumor cells exposed to an EMF experienced a morphophenotypic reversion, loosing malignant features whilst their fractal dimension is progressively reduced.

a shape descriptor for object recognition [133], and it was generalized to grayscale images [134]. The BE is deconvolved into a "curvegram" representing the spectral decomposition of BE at different spatial scales.

The "curvegram" can be accurately obtained by using digital signal processing techniques (more specifically through the Fourier transform) and it provides multi-scale representation of the curvature. BE is particularly suited for biological objects, such as membranes, nucleus, mitochondria and organelles [135]. As such, BE represents an interesting resource for translation and rotation-invariant shape classification, as well as a means of deriving quantitative information about the complexity of the shapes [136].

As previously reported [120], control cancer cells exhibit high membrane BE values, whereas EMT-treated cells showed a 2-fold reduction of cell membrane BE levels. This trend already becomes evident from the first 48 h of treatment. BE is inversely correlated with the surface tensions; in turn, surface tensions are reflective of intercellular adhesive intensities [137]. Thus, it is likely that EMF treatment increasing intercellular adhesion forces might influence cell behavior and metastatic spreading of cancer cells, as previously described [138]. Indeed, high BE values are associated to a "diffusive" shape, the form the cell acquires when it displays an invasive pattern, the stage that precedes the metastatic spreading [139]. On the contrary, breast cancer cells under treatment with EMF showed a significant reduction in their shape-related BE values. Reversion into a more "physiological" fractal dimension implies reduced morphologic instability and increased between-cell connectivity. On the other hand, it is troublesome that current cytotoxic anticancer treatments increase fractal dimension and thus they "may unwittingly contribute to tumor morphologic instability and consequent tissue invasion" [140]. Mild chemotherapeutic regimens did not modify tumor fractal dimension, whereas intensive cytotoxic chemotherapy increases fractal dimension values, and therefore enhances tissue disorder and chaotic tumor behavior, and finally promotes selection of more malignant phenotypes [141].

Alterations in cell shape lead individual cells to rearrange with respect to each other, so representing a driving force for tissue morphogenesis. Indeed, as a consequence of shape "normalization," breast cancer cells exposed to EMF display several differentiated structures (like ducts and mammary acini), reactivate signaling pathways (casein production, nodal activity), and, notably, they recover both tight and gap junctions (absent in control cells) as a consequence of increased expression of differentiating molecules, such as E-cadherin. It is well-known that gap junctional intercellular structures integrate and modulate both cellular and microenvironmental signals in order to inhibit proliferation and enhance differentiating processes [142]. So far, EMF-treated cells showed a significant proliferative inhibition, without experiencing a relevant apoptosis.

In addition, β-catenin distribution behaves in a significantly different manner in MDA-MB-231 cells exposed to EMF. β-Catenin has critical roles in morphogenesis and human cancer, through its dual function in adhesive complexes and as a transducer/transcriptional regulator in numerous signal transduction pathways, including the canonical WNT pathway [143], critical for both embryonic and adult mammary development [144, 145]. β-Catenin appears to be involved in organization of tissue architecture and polarity, whereas alternatively spliced isoforms of p120 catenin bind to the juxtamembrane domain of E-cadherin and influence cell adhesion. Membrane localization of p120 is reduced in many ductal carcinomas and becomes predominantly cytoplasmic in the majority of invasive lobular carcinomas, correlating with loss of E-cadherin [146]. β-Catenin in MCF-7 cells is normally distributed in the cytosol and does not translocate around the nucleus. On the contrary, in MDA-MB-231 cells β-catenin is prevalently represented into and around the nucleus. In MDA-MB-231 cells exposed to EMF, after 192 h, β-catenin is progressively displayed behind the cell membrane. Data collected by means of confocal microscopy were analyzed correlating spatial β-catenin distribution versus the inner cell membrane, by means of the Moran Index (MI) [147]. The MI (ranging from $+1$ and -1) quantifies the spatial correlation between a distribution and a given structure: 0 corresponding to absence of spatial correlation (random distribution), $+1$ indicating maximal overlapping with the given structure, and -1 corresponding to maximal avoidance of the reference structure. β-Catenin distribution is progressively modified in EMF-treated cells, evolving from a mean MI value of -0.7 to $+0.5$, thus evidencing how, after cell shape normalization, β-catenin moves form the nucleus to the membrane. On the contrary, in control cells, β-catenin presents an almost totally disorganized spatial distribution. Concomitantly, one can observe a significant increase in E-cadherin release. Adherens junctions are formed by cadherin molecules that mediate calcium-dependent cell–cell adhesion. Inhibition of cadherin function leads to a loss of the epithelial phenotype [148]; on the other hand, enhancing E-cadherin-based adhesion junctions triggers the development of epithelial morphology [149]. These results collectively suggest that in the experimental arm, treated cells were committed towards differentiating processes.

Shape modifications are associated with impressive changes in energy metabolism, when evaluating it through exometabolome analysis, performed with nuclear magnetic resonance spectroscopy. In EMF-treated breast cancer cells glycolytic fluxes were significantly reduced, with a parallel decrease in lactate, glutathione, and glutamine values. Citrate and *de novo* lipogenesis are, in turn, inhibited. When the cell shape reaches a new stable configuration at approximately 72 h, a complete metabolic reversion is observed.

In order to obtain a concomitant representation in the metabolomic space, principal component analysis (PCA) was carried out on a dataset constituted by the differences between each spectrum obtained after 48, 72, and 96 h of culture for treated and nontreated samples, and the corresponding average spectrum from the 0 h measurement. The obtained values are representative of net balances, with the positive ones being considered an estimate of net fluxes of production and the negative an estimate of the utilization of metabolites. Five principal components (PCs) were calculated and the corresponding model explained 80% of the total variance. A *t*-test, applied to the component scores to compare control and treated cells highlighted significant differences between the two groups on the first four PCs at each experimental time. Analysis of the PC1/PC2 score showed that PC1 is by far the major order parameter present in the data (42% of variation explained), and corresponds to the core energy metabolism as evident from its positive loading (correlation coefficient between original variable and component) with glucose utilization and its negative loadings with lactate. After 72 h, PC2 scores obtained from EMF-treated cells evidenced a meaningful metabolomic reversion, characterized by increased β-oxidation fluxes and reduced fatty acid synthesis. Therefore, the two principal metabolomic features of cancer metabolism (i.e., high glycolytic flux and lipogenesis) have been abolished under EMF treatment.

It is of utmost importance that PC1 mirrors the same divergence in time behavior of the control/treated differences observed as for the shape analysis, so pointing to an empirical correlation between the shape and metabolomic descriptions. What is worth noting is that the differentiation in shape between the control and treated groups seems to happen between 48 and 72 h, while in the case of the metabolic description the two experimental groups diverge between 72 and 96 h. This seems to indicate shape change precedes metabolic reversion. It should be emphasized that EMF-treated cells present increased glutamine consumption with respect to control cells. This increase in glutamine utilization does not correlate with a simultaneous increase in lactate (as expected if the difference between control and treated cell metabolism should be due to a mere diversification of energy sources) nor to an increase in fatty acid synthesis (as expected when *de novo* cell membrane production is required to sustain cell proliferation). Indeed, EMF-treated cells showed a statistically significant growth inhibition, confirming that glutaminolysis cannot be explained by energetic or proliferation needs: this implies the treated cells devote a higher portion of chemical energy to other anabolic work (construction of cellular structures) than control cells. Thus, excess of glutamine is preferentially transformed into proteins and does not appear as lactate. These preliminary data suggest that the structural reorganization fostered by EMF through shape reorganization induces an adaptive metabolomic reversion: EMF-treated cells loose both the glycolytic and lipogenic malignant phenotype, whereas differentiating processes took place.

These data highlight the neglected link between cell morphology and thermodynamics. According to the Prigogine–Wiame theory of development [150], during carcinogenesis, a living system constitutively deviates from a steady-state trajectory; this deviation is accompanied by an *increase* in the system dissipation function (Ψ) at the expense of coupled processes in other parts of the organism [151], where $\Psi = q_o + q_{gl}$ (where q_o is oxygen consumption and q_{gl} is glycolysis intensity). Keeping in mind that BE represents a "dissipative" form of energy, metabolomic data evidenced a significant reduction in glycolysis activity (in the presence of unchanged values of oxygen consumption) and it follows that in our experimental conditions Ψ decreased significantly until a stable state was attained. This stable state is characterized by a minimum in the rate of energy dissipation (principle of minimum energy dissipation) [152]. This behavior is exactly the opposite of what is expected in growing cancer cells and experimentally observed in tumor control cells.

17.10
Conclusions

Cell phenotypes can be considered as emergent modes that arise through collective interactions among different cellular (genes and signaling pathways) and microenvironmental components (cell adhesion structures, ECM, stroma) [153]. These pathways are driven by a nonlinear dynamics and belong to a complex network. How do distinct cell fates emerge? Several approaches have been proposed to give a reliable answer, involving gene regulatory networks and proteomics or metabolomics data provided by "omic" sciences. However, in order to provide an overall and suitable description of cell phenotype evolution it should be emphasized that "gradual variations in a single control parameter, such as cell shape, can switch living cells between distinct gene programs, including growth, differentiation and apoptosis" [154]. Such a parameter is likely to reflect a system property and then could be taken as a dimension of a phase-space diagram. Thereby, transition from one phenotype to another would be undoubtedly reminiscent of phase transitions observed in physical systems. Within this framework, cell phenotypes should be considered *stable cellular states* and the switches between them viewed as *biological phase transitions*. It is of utmost importance to outline that the level of organization that must be taken into consideration comprises both cells and their microenvironment. So far, influences from the surrounding milieu might be integrated into the whole network. Indeed, detailed knowledge of the wiring diagram of the genetic regulatory network and transcriptome is insufficient to predict the dynamic landscape of the biological system, even if data provided by "omics" high-throughput techniques (proteomics, genomics, metabolomics) are candidates to integrate, in the near future, the whole representation [155].

At least, a phase-space description of cell transition between different attractor states could be obtained by a set of morphological, quantitative, fractal parameters together with the dissipative function parameter Ψ. Hence, different cell phenotypes are mainly represented by a set of parameters describing their transition from a more towards a less dissipative configuration.

Viewing cancer cell populations as trapped in attractors defined by fractal dimensions raises an obvious question previously addressed by Huang and Kaufmann: "how can one perturb the malignant cells back to the trajectory that leads to the non-malignant, more differentiated cell? This will place the old, underexplored, idea of differentiation therapy in a new light" [54]. Relevant, preliminary data suggest that such a result could be obtained by both physical as well chemical morphogenetic factors, generally belonging to different embryonic morphogenetic fields [156].

References

1 Gilbert, S.F. and Sarkar, S. (2000) Embracing complexity: organicism for the twenty-first century. *Dev. Dyn.*, **219**, 1–9.

2 Goodwin, B. (1994) *How the Leopard Changed its Spots*, Princeton University Press, Princeton, NJ.

3 Van Speybroeck, L., De Waele, D., and van De Vijver, G. (2002) Theories in early embryology. Close connections between epigenesis, preformationism and self-organization. *Ann. NY Acad. Sci.*, **981**, 7–49.

4 Cumming, F.W. (1994) Aspects of growth and form. *Physica D*, **79**, 146–163.

5 Ricard, J. (1999) *Biological Complexity and the Dynamics of Life Processes*, Elsevier, New York.

6 Salazar-Ciudad, I., Newman, S.A., and Solè, R.V. (2001) Phenotypic and dynamical transitions in model genetic networks. Emergence of patterns and genotype–phenotype relationships. *Evol. Dev.*, **3**, 84–94.

7 Turing, A.M. (1952) The chemical basis of morphogenesis. *Trans. R. Soc. Lond. B Biol. Sci.*, **237**, 37–72.

8 Wang, N. and Ingber, D.E. (1994) Control of cytoskeletal mechanics by extracellular matrix, cell shape and mechanical tension. *Biophys. J.*, **66**, 2181–2189.

9 Ben-Ze'ev, A. (1991) Animal cell shape changes and gene expression. *BioEssays*, **13**, 207–212.

10 Huang, S., Chen, S.C., and Ingber, D.E. (1998) Cell shape-dependent control of $p27^{Kip}$ and cell cycle progression in human capillary endothelial cells. *Mol. Biol. Cell*, **9**, 3179–3193.

11 Hughes-Fulford, M., Rodenacker, K., and Jütting, U. (2006) Reduction of anabolic signals and alteration of osteoblast nuclear morphology in microgravity. *J. Cell Biochem.*, **99**, 435–449.

12 Hammond, T.G., Benes, E., O'Reilly, K.C. et al. (2000) Mechanical culture conditions effect gene expression: gravity-induced changes on the space shuttle. *Physiol. Genomics*, **3**, 163–173.

13 Ingber, D.E. (1997) Tensegrity: the architectural basis of cellular mechanotrasduction. *Annu. Rev. Physiol.*, **59**, 575–599.

14 Pourati, J., Maniotis, A., Speigel, D., Schaffer, J.L., Butler, J.P., Fredberg, J.J., Ingber, D.E., Stamenovic, D., and Wang, N. (1998) Is cytoskeletal tension a major determinant of cell deformability in adherent endothelial cells? *Am. J. Physiol.*, **274**, C1283–C1289.

15 Chen, C.S., Alonso, J.L., Ostuni, E., Whitesides, G.M., and Ingber, D.E. (2003) Cell shape provides global control of focal adhesion assembly. *Biochem. Biophs. Res. Commun.*, **307**, 355–361.

16 Davies, P.F., Spaan, J.A., and Krams, R. (2005) Shear stress biology of the endothelium. *Ann. Biomed. Eng.*, **33**, 1714–1718.

17 Paszek, M.J., Zahir, N., Johnson, K.R., Lakins, J.N., Rozenberg, G.I., Gefen, A., Reinhart-King, C.A., Margulies, S.S., Dembo, M., Boettiger, D., Hammer, D.A., and Weaver, V.M. (2005) Tensional homeostasis and the malignant phenotype. *Cancer Cell*, **8**, 241–254.

18 Rothman, S. (2002) *Lessons from the Living Cell. The Limits of Reductionism*, McGraw-Hill, New York.

19 Noble, D. (2006) *The Music of Life*, Oxford University Press, London.

20 Shapiro, J.A. (2009) Revisiting the central dogma in the 21st century. *Ann. NY Acad Sci.*, **1178**, 6–28.

21 Elowitz, M.B., Levine, A.J., Siggia, E.D., and Swain, P.S. (2002) Stochastic gene expression in a single cell. *Science*, **297**, 1183–1186.

22 Laforge, B., Guez, D., Martinez, M., and Kupiec, J.-J. (2005) Modeling embryogenesis and cancer: an approach based on an equilibrium between the autostabilization of stochastic gene expression and the interdependence of cells for proliferation. *Prog. Biophys. Mol. Biol.*, **89**, 93–120.

23 McAdams, H.H. and Arkin, A. (1997) Stochastic mechanisms in gene expression. *Proc. Natl. Acad. Sci. USA*, **94**, 814–819.

24 Wray, G.A. and Abouheif, E. (1998) When is homology not homology? *Curr. Opin. Genet. Dev.*, **8**, 675–680.

25 Nelson, C.M. and Bissell, M.J. (2006) Of extracellular matrix, scaffolds, and signaling: tissue architecture regulates development, homeostasis, and cancer. *Annu. Rev. Cell Dev. Biol.*, **22**, 287–309.

26 Newman, S.A., Forgas, G., and Muller, G.B. (2006) Before programs: the physical origination of multicellular forms. *Int. J. Dev. Biol.*, **50**, 289–299.

27 Belousov, L.V., Opitz, J.M., and Gilbert, S.F. (1997) Contributions to field theory and life of Alexander G. Gurwitsch. *Int. J. Dev. Biol.*, **41**, 771–779.

28 Gilbert, S.F., Opitz, J.M., and Raff, R.A. (1996) Resynthesizing evolutionary and developmental biology. *Dev. Biol.*, **173**, 357–372.

29 D'Arcy, W.T. (1992) *On Growth and Form*, Cambridge University Press, Cambridge.

30 Nelson, C.M. (2009) Geometric control of tissue morphogenesis. *Biochem. Biophys. Acta*, **1793**, 903–910.

31 Watson, P.A. (1991) Function follows form: generation of intracellular signals by cell deformation. *FASEB J.*, **5**, 2013–2019.

32 Watt, F.M., Jordan, P.W., and O'Neill, C.H. (1988) Cell shape controls terminal differentiation of human epidermal keratinocytes. *Proc. Natl. Acad. Sci. USA*, **85**, 5576–5580.

33 Chen, C.S., Mrksich, M., Huang, S., Withesides, G.M., and Ingber, D.E. (1997) Geometric control of cell life and death. *Science*, **276**, 1425–1428.

34 Newman, S.A. and Comper, W.D. (1990) Generic physical mechanisms of morphogenesis and pattern formation. *Development*, **49**, 1–18.

35 Harold, F.M. (1990) To shape a cell: an inquiry into the causes of morphogenesis of microorganisms. *Microbiol. Rev.*, **54**, 381–431.

36 Patwari, P. and Lee, R.T. (2008) Mechanical control of tissue morphogenesis. *Circ. Res.*, **103**, 234–243.

37 Joyce, A.R. and Palsson, B. (2006) The model organism as a system: integrating "omics" data sets. *Nat. Rev.*, **7**, 198–210.

38 Knox, S.S. (2010) From "omics" to complex disease: a systems biology approach to gene–environment interactions in cancer. *Cancer Cell Int.*, **10**, 1–13.

39 Conti, F., Valerio, M.C., Zbilut, J.P., and Giuliani, A. (2007) Will systems biology offer new holistic paradigms to life sciences? *Syst. Synth. Biol.*, **1**, 161–165.

40 Prigogine, I. and Nicolis, G. (1977) *Self-Organization in Non-Equilibrium Systems*, John Wiley & Sons, Inc., New York.

41 Waddington, C.H. (1957) *The Strategy of the Genes: A Discussion of Some Aspects of Theoretical Biology*, Allen & Unwin, London.

42 Huang, S. and Ingber, D.E. (2007) A non-genetic basis for cancer progression and metastasis: self-organizing attractors in cell regulatory networks. *Breast Dis.*, **26**, 27–54.

43 Chang, H.H., Hemberg, M., Barahona, M., Ingber, D.E., and Huang, S. (2008) Transcriptome wide noise controls lineage choice in mammalian progenitor cells. *Nature*, **453**, 544–547.

44 Albert, R. and Othmer, H.G. (2003) The topology of the regulatory interactions predicts the expression pattern of the segment polarity genes in *Drosophila melanogaster*. *J. Theor. Biol.*, **223**, 1–18.

45 Kauffman, S.A. (1993) *Origins of Order: Self-Organization and Selection in Evolution*, Oxford University Press, New York.

46 Huang, S. (1999) Gene expression profiling, genetic networks, and cellular states: an integrating concept for tumorigenesis and drug discovery. *J. Mol. Med.*, **77**, 469–480.

47 Aldana-Gonzalez, M., Coppersmith, S., and Kadanoff, L.P. (2003) Boolean dynamics with random couplings, in *Perspectives and Problems in Non-Linear Science* (eds E. Kaplan, J.E. Marsden, and K.R. Sreenivasan), Springer, New York, pp. 23–89.

48 Li, F., Long, T., Lu, Y., Ouyang, Q., and Tang, C. (2004) The yeast cell-cycle network is robustly designed. *Proc. Natl. Acad. Sci. USA*, **101**, 4781–4786.

49 Tsuchiya, M., Piras, V., Choi, S., Akira, S., Tomita, M., Giuliani, A., and Selvarajoo, K. (2009) Emergent genome-wide control in wildtype and genetically mutated lipopolysaccharides-stimulated macrophages. *PLoS ONE*, **4**, 1–13.

50 Sjoblom, T., Jones, S., Wood, L.D., Parsons, D.W., Lin, J., Barber, T.D. *et al.* (2006) The consensus coding sequences of human breast and colorectal cancers. *Science*, **314**, 268–274.

51 Greenman, C., Stephens, P., Smith, R., Dalgliesh, G.L., Hunter, C., Bignell, G. *et al.* (2007) Patterns of somatic mutation in human cancer genomes. *Nature*, **446**, 153–158.

52 Wood, L.D., Parsons, D.W., Jones, S., Lin, J., Sjoblom, T., Leary, R.J. *et al.* (2007) The genomic landscapes of human breast and colorectal cancers. *Science*, **318**, 1108–1113.

53 Rhee, S.Y., Wood, V., Dolinski, K., and Draghici, S. (2008) Use and misuse of the gene ontology 710 annotations. *Nat. Rev. Genet.*, **9**, 509–515.

54 Huang, S., Ernberg, I., and Kauffman, S. (2009) Cancer attractors: A systems view of tumours from a gene network dynamics and developmental perspective. *Semin. Cell Dev. Biol.*, **20**, 869–876.

55 Guo, Y., Eichler, G.S., Feng, Y., Ingber, D.E., and Huang, S. (2006) Towards a holistic, yet gene-centered analysis of gene expression profiles: a case study of human lung cancers. *J. Biomed. Biotechnol.*, **2006**, 1–11.

56 Su, AI., Wiltshire, T., Batalov, S., Lapp, H., Ching, KA., Block, D. *et al.* (2004) A gene atlas of the mouse and human protein-encoding transcriptomes. *Proc. Natl. Acad. Sci. USA*, **101**, 6062–6067.

57 Fearon, ER. and Vogelstein, B. (1990) A genetic model for colorectal tumorigenesis. *Cell*, **61**, 759–767.

58 Hahn, W.C., Counter, C.M., Lundberg, A.S., Beijersbergen, R.L., Brooks, M.W., and Weinberg, R. (1999) Creation of human tumor cells with defined genetic elements. *Nature*, **400**, 464–648.

59 Folkman, J., Hahnfeldt, P., and Hlatky, L. (2000) Cancer: looking outside of the genome. *Nat. Rev. Mol. Cell Biol.*, **1**, 76–79.

60 Capp, J.P. (2005) Stochastic gene expression, disruption of tissue averaging effects and cancer as a disease of development. *BioEssays*, **27**, 1277–1285.

61 Sonnenschein, C. and Soto, A.M. (1999) *The Society of Cells: Cancer and Control of Cell Populations*, Springer, New York.

62 Baker, S.G. and Kramer, B.S. (2007) Paradoxes in carcinogenesis: new opportunities for research directions. *BMC Cancer*, **7**, 151–157.

63 Bizzarri, M., Cucina, A., Conti, F., and D'Anselmi, F. (2008) Beyond the oncogenic paradigm: understanding complexity in cancerogenesis. *Acta Biotheor.*, **56**, 173–196.

64 Miklos, G.L.G. (2005) The Human Cancer Genome Project – one more misstep in the war on cancer. *Nat. Biotechnol.*, **3**, 535–537.

65 Olumi, A.F., Grossfeld, G.D., Hayward, S.W., Carroll, P.R., Tlsty, T.D., and Cunha, G.R. (1999) Carcinoma-associated fibroblasts direct tumor progression of initiated human prostatic epithelium. *Cancer Res.*, **59**, 5002–5011.

66 Barcellos-Hoff, M.H. and Ravani, S.A. (2000) Irradiated mammary gland stroma promotes the expression of tumorigenic potential by unirradiated epithelial cells. *Cancer Res.*, **60**, 1254–1260.

67 Maffini, M.V., Soto, A.M., Calabro, J.M., Ucci, A.A., and Sonnenschein, C. (2004) The stroma as a crucial target in rat mammary gland carcinogenesis. *J. Cell Sci.*, **117**, 1495–1502.

68 Weaver, V.M. and Gilbert, P. (2004) Watch the neighbour: cancer is a communal affair. *J. Cell Sci.*, **117**, 1495–1502.

69 Sonnenscheim, C. and Soto, A.M. (2000) Somatic mutation theory of carcinogenesis: why it should be dropped and replaced. *Mol. Carcinogen.*, **29**, 205–211.

70 Suresh, S. (2007) Biomechanics and biophysics of cancer cells. *Acta Biomater.*, **3**, 413–438.

71 Soto, A.M. and Sonnenschein, C. (2005) Emergentism as a default: cancer as a problem of tissue organization. *J. Biosci.*, **30**, 103–118.

72 Ronnov-Jessen, L. and Bissell, M.J. (2008) Breast cancer by proxy: can the microenvironment be both the cause and consequence? *Trends Mol. Med.*, **15**, 5–13.

73 Anderson, A.R.A., Weaver, A.M., Cummings, P.T., and Quaranta, V. (2006) Tumor morphology and phenotypic evolution driven by selective pressure from the microenvironment. *Cell*, **127**, 905–915.

74 Huang, S. and Ingber, D.E. (2005) Cell tension, matrix mechanics, and cancer development. *Cancer Cell*, **2**, 175–176.

75 Butcher, D.T., Alliston, T., and Weaver, V.M. (2009) A tense situation: forcing tumour progression. *Nat. Rev. Cancer*, **9**, 108–122.

76 Tilghman, R.W. and Parsons, J.T. (2008) Focal adhesion kinase as a regulator of cell tension in the progression of cancer. *Semin. Cancer Biol.*, **18**, 45–52.

77 Zaman, M.H., Trapani, L.M., Sieminski, A.L., Mackellar, D., Gong, H., Kamm, R.D., Wells, A., Lauffenburger, D.A., and Matsudaira, P. (2006) Migration of tumor cells in 3D matrices is governed by matrix stiffness along with cell–matrix adhesion and proteolysis. *Proc. Natl Acad. Sci. USA*, **103**, 10889–10894.

78 Weaver, V.M., Petersen, O.W., Wang, F., Larabell, C.A., Briand, P., Damsky, C., and Bissell, M.J. (1997) Reversion of the malignant phenotype of human breast cells in three dimensional culture and *in vivo* by integrin blocking antibodies. *J. Cell Biol.*, **137**, 231–245.

79 Krause, S., Maffini, M.V., Soto, A.M., and Sonnenschein, C. (2010) The microenvironment determines the breast cancer cells' phenotype: organization of MCF7 cells in 3D cultures. *BMC Cancer*, **10**, 263–270.

80 Needham, J. (1936) New advances in the chemistry and biology of organized growth. *Proc. R. Soc. London B. Biol. Sci.*, **29**, 1577–1626.

81 Ruiz-Vela, A., Aguilar-Gallardo, C., and Simón, C. (2009) Building a framework

for embryonic microenvironments and cancer stem cells. *Stem Cell Rev.*, **5**, 319–327.
82 Pierce, G.B. (1983) The cancer cell and its control by the embryo. *Am. J. Pathol.*, **113**, 116–124.
83 Cooper, M. (2009) Regenerative pathologies: stem cells, teratomas and theories of cancer. *Med. Stud.*, **1**, 55–66.
84 Ingber, D.E. (2008) Can cancer be reversed by engineering the tumour microenvironment? *Semin. Cancer Biol.*, **18**, 356–364.
85 Baker, S.G., Soto, A.M., Sonnenschein, C., Cappuccio, A., Potter, J.D., and Kramer, B.S. (2009) Plausibility of stromal initiation of epithelial cancers without a mutation in the epithelium: a computer simulation of morphostats. *BMC Cancer*, **9**, 89–95.
86 Singhvi, R., Kumar, A., Lopez, G.P., Stephanopoulos, G.N., Wang, D.I., Whitesides, G.M., and Ingber, D.E. (1994) Engineering cell shape and function. *Science*, **264**, 696–698.
87 Parker, K.K., Brock, A.L., Brangwynne, C., Mannix, R.J., Wang, N., Ostuni, E., Geisse, N.A., Adams, J.C., Whitesides, G.M., and Ingber, D.E. (2002) Directional control of lamellipodia extension by constraining cell shape and orienting cell tractional forces. *FASEB J.*, **16**, 1195–1204.
88 Jiang, X., Bruzewicz, D.A., Wong, A.P., Piel, M., and Whitesides, G.M. (2005) Directing cell migration with asymmetric micropatterns. *Proc. Natl. Acad. Sci. USA*, **102**, 975–978.
89 Bissell, M.J., Farson, D., and Tung, A.S. (1977) Cell shape and hexose transport in normal and virus-transformed cells in culture. *J. Supramol. Struct.*, **6**, 1–12.
90 Wittelsberger, S.C., Kleene, K., and Penman, S. (1981) Progressive loss of shape-responsive metabolic controls in cells with increasingly transformed phenotype. *Cell*, **24**, 859–866.
91 Chicurel, M.E., Singer, R.H., Meyer, C.J., and Ingber, D.E. (1998) Integrin binding and mechanical tension induce movement of mRNA and ribosomes to focal adhesions. *Nature*, **392**, 730–733.
92 Muschler, J., Lochter, A., Roskelley, C.D., Yurchenco, P., and Bissell, M.J. (1999) Division of labor among the alpha6beta4 integrin, beta1 integrins, and an E3 laminin receptor to signal morphogenesis and beta-casein expression in mammary epithelial cells. *Mol. Biol. Cell*, **10**, 2817–2828.
93 Shannon, J.M. and Pitelka, D.R. (1981) The influence of cell shape on the induction of functional differentiation in mouse mammary cells *in vitro*. *In Vitro*, **17**, 1016–1028.
94 Nelson, C.M., Khauv, D., Bissell, M.J., and Radisky, D.C. (2008) Change in cell shape is required for matrix metalloproteinase-induced epithelial–mesenchymal transition of mammary epithelial cells. *J. Cell. Biochem.*, **105**, 25–33.
95 McBeath, R., Pirone, D.M., Nelson, C.M., Bhadriraju, K., and Chen, C.S. (2004) Cell shape, cytoskeletal tension, and RhoA regulate stem cell lineage commitment. *Dev. Cell*, **6**, 483–495.
96 Denton, M. and Marshall, C. (2001) Laws of form revisited. *Nature*, **410**, 417.
97 Rosai, J. (2001) The continuing role of morphology in the molecular age. *Mod. Pathol.*, **14**, 258–260.
98 Virchow, R.L.K. (1978) *Cellular Pathology 1859 (Special Edition)*. Churchill, London, pp. 204–207.
99 Mandelbrot, B.B. (1975) Stochastic models for the Earth's relief, the shape and the fractal dimension of the coastlines, and the number-area rule for islands. *Proc. Natl. Acad. Sci. USA*, **72**, 3825–3828.
100 Mandelbrot, B.B. (1982) *The Fractal Geometry of Nature*, Freeman, New York.
101 Smith, T.G., Lange, G.D., and Marks, W.B. (1996) Fractal methods and results in cellular morphology – dimensions, lacunarity and multifractals. *J. Neurosci. Methods*, **69**, 123–136.
102 Cutting, J.E. and Garvin, J.J. (1987) Fractal curves and complexity. *Percept. Psychophys.*, **42**, 365–370.
103 Normant, F. and Tricot, C. (1991) Methods for evaluating the fractal dimension of curves using convex hulls. *Phys. Rev. A*, **43**, 6518–6525.

104 Poincaré, H. and Bendixson, I. (1901) Sur les courbes définies par des équations différentielles. *Acta Math.*, **24**, 1–88.
105 Scheck, F. (1990) *Mechanics*, Springer, Heidelberg.
106 Toussaint, O. and Schneider, E.D. (1998) The thermodynamics and evolution of complexity in biological systems. *Comp. Biochem. Physiol.*, **120**, 3–9.
107 Rangayyan, R.M. and Nguyen, T.M. (2007) Fractal analysis of contours of breast masses in mammograms. *J. Dig. Imag.*, **20**, 223–237.
108 Rangayyan, R.M., El-Faramawy, N.M., Desautels, J.E.L., and Alim, O.A. (1997) Measures of acutance and shape for classification of breast tumors. *IEEE Trans. Med. Imag.*, **16**, 799–810.
109 Losa, G.A., Merlini, D., Nonnenmacher, T.F., and Weibel, E.R. (eds) (2002) *Fractals in Biology and Medicine*, Birkhauser, Basel.
110 Baish, J.W. and Jain, R.K. (2000) Fractals and cancer. *Cancer Res.*, **60**, 3683–3688.
111 Cross, S.S. (1997) Fractals in pathology. *J. Pathol.*, **182**, 1–8.
112 Cross, S.S., McDonagh, A.J.G., and Stephenson, T.J. (1995) Fractal and integer-dimensional analysis of pigmented skin lesions. *Am. J. Dermatol.*, **17**, 374–378.
113 Claridge, E., Hall, P.N., and Keefe, M. (1992) Shape analysis for classification of malignant melanoma. *J. Biomed. Eng.*, **14**, 229–324.
114 Gazit, Y., Berk, D.A., Leunig, M., Baxter, L.T., and Jain, R.K. (1995) Scale-invariant behavior and vascular network formation in normal and tumor tissue. *Phys. Rev. Lett.*, **75**, 2428–2431.
115 Michaelson, J.S., Cheongsiatmoy, J.A., Dewey, F. *et al.* (2005) Spread of human cancer cells occurs with probabilities indicative of a nongenetic mechanism. *Br. J. Cancer*, **93**, 1244–1249.
116 Landini, G. and Rippin, J.W. (1996) How important is tumour shape? Quantification of the epithelial connective tissue interface in oral lesions using local connected fractal dimension analysis. *J. Pathol.*, **179**, 210–217.
117 Eid, R.A. and Landini, G. (2003) Quantification of the global and local complexity of the epithelial–connective tissue interface of normal, dysplastic, and neoplastic oral mucosa using digital imaging. *Pathol. Res. Pract.*, **199**, 475–482.
118 Tracqui, P. (2009) Biophysical model of tumor growth. *Rep. Prog. Phys.*, **72**, 1–30.
119 Waliszewski, P. and Konarski, J. (2002) Fractal structure of space and time is necessary for the emergence of self-organization, connectivity and collectivity in cellular system, in *Fractals in Biology and Medicine* (eds G.A. Losa, D. Merlini, T.F. Nonnenmacher, and E.R. Weibel), Birkhauser, Basel, pp. 15–24.
120 Lelièvre, S.A., Weaver, V.M., Nickerson, J.A., Larabell, C.A., Bhaumik, A., Petersen, O.W., and Bissell, M.J. (1998) Tissue phenotype depends on reciprocal interactions between the extracellular matrix and the structural organization of the nucleus. *Proc. Natl. Acad. Sci. USA*, **95**, 14711–14716.
121 Thomas, C.H., Collier, J.H., Sfeir, C.S., and Healy, K.E. (2002) Engineering gene expression and protein synthesis by modulation of nuclear shape. *Proc. Natl. Acad. Sci. USA*, **99**, 1972–1977.
122 Guilak, F. (1995) Compression-induced changes in the shape and volume of the chondrocyte nucleus. *J. Biomech.*, **28**, 1529–1541.
123 Wolf, K. and Friedl, P. (2006) Molecular mechanisms of cancer cell invasion and plasticity. *Br. J. Dermatol.*, **154**, 11–15.
124 Dahl, K.N., Ribeiro, A.J.S., and Lammerding, J. (2008) Nuclear shape, mechanics, and mechanotransduction. *Circ Res.*, **102**, 1307–1318.
125 Folkman, J. and Moscona, A. (1978) Role of cell shape in growth control. *Nature*, **273**, 345–349.
126 Ingber, D.E. (1999) How cells (might) sense microgravity. *FASEB J.*, **13**, S3–S15.
127 Sonnenschein, C. and Soto, A.M. (2008) Theories of carcinogenesis: an emerging perspective. *Semin. Cancer Biol.*, **18**, 372–377.
128 Ingber, D.E. (2005) Mechanical control of tissue growth: function follows form. *Proc. Natl. Acad. Sci. USA*, **102**, 11571–11572.
129 D'Anselmi, F., Valerio, M., Cucina, A., Galli, L., Proietti, S., Dinicola, S.,

Pasqualato, A., Manetti, C., Ricci, G., Giuliani, A., and Bizzarri, M. (2010) Metabolism and cell shape in cancer: a fractal analysis. *Int. J. Biochem. Cell Biol.*, Epub ahead of print.

130 Bowie, J.E. and Young, I.T. (1977) An analysis technique for biological shape. *Acta Cytol.*, **21**, 739–746.

131 Young, I.T., Walker, J.E., and Bowie, J.E. (1974) An analysis technique for biological shape. *Inform. Control*, **25**, 357–381.

132 Duncan, J.S., Lee, F.A., Smeulders, A.W.M., and Zaret, B.L. (1991) A bending energy model for measurement of cardiac shape deformity. *IEEE Trans. Med. Imag.*, **10**, 307–310.

133 Pernuš, F., Leonardis, A., and Kovačič, S. (1994) Two-dimensional object recognition using multiresolution non-information-preserving shape features. *Pattern Recognit. Lett.*, **15**, 1071–1076.

134 van Vliet, L.J. and Verbeeck, P.W. (1993) Curvature and bending energy in digitized 2D and 3D images. Eighth Scandinavian Conference on Image Analysis, Tromsø.

135 Castleman, K.R. (1996) *Digital Image Processing*, Prentice-Hall, Englewood Cliffs, NJ.

136 Cesar, R.M. Jr and Costa, L. daF. (1997) The application and assessment of multiscale bending energy for morphometric characterization of neural cells. *Rev. Sci. Instrum.*, **68**, 2177–2186.

137 Foty, R., Forgacs, G., Pfleger, C., and Steimberg, M. (1994) Liquid properties of embryonic tissues: measurement of interfacial tensions. *Phys. Rev. Lett.*, **72**, 2298–2301.

138 Foty, R., Pfleger, C., Forcas, G., and Steimberg, M. (1996) Surface tensions of embryonic tissues predict their mutual envelopment behaviour. *Development*, **122**, 1611–1620.

139 Rohrschneider, M., Scheuermann, G., Hoehme, S., and Drasdo, D. (2007) Shape characterization of extracted and simulated tumor samples using topological and geometric measures. 29th Annual International Conference of the IEEE EMBS, Lyon.

140 Cristini, V., Frieboes, H.B., Gatenby, R., Caserta, S., Ferrari, M., and Sinek, J. (2005) Morphologic instability and cancer invasion. *Clin. Cancer Res.*, **11**, 6772–6779.

141 Ferriera, S.C., Martins, M.L., and Villa, M.J. (2003) Morphology transitions induced by chemotherapy in carcinomas in situ. *Phys. Rev. E*, **67**, 1–9.

142 Trosko, J.E., Chang, C.-C., Upham, B.L., and Tai, M.-H. (2004) Ignored hallmarks of carcinogenesis: stem cells and cell–cell communication. *Ann. NY Acad. Sci.*, **1028**, 192–201.

143 Phillips, B.T. and Kimble, J. (2009) A new look at TCF and beta-catenin through the lens of a divergent *C. elegans* Wnt pathway. *Dev. Cell*, **17**, 27–34.

144 Tepera, S.B., McCrea, P.D., and Rosen, J.M. (2003) A β-catenin survival signal is required for normal lobular development in the mammary gland. *J. Cell Sci.*, **116**, 1137–1149.

145 Rowlands, T.R., Pechenkina, I., Hatsell, S.J., Pestell, R.G., and Cowin, P. (2003) Dissecting the roles of β-catenin and cyclin D_1 during mammary development and neoplasia. *Proc. Natl. Acad. Sci. USA*, **100**, 11400–11405.

146 Sarrio, D., Perez-Mies, B., Hardisson, D., Moreno-Bueno, G., Suarez, A., Cano, A., Martin-Perez, J., Gamallo, C., and Palacios, J. (2004) Cytoplasmic localization of $p120^{ctn}$ and E-cadherin loss characterize lobular breast carcinoma from preinvasive to metastatic lesions. *Oncogene*, **23**, 3272–3283.

147 da Silva, E.C., Silva, A.C., de Paiva, A.C., and Nunes, R.A. (2008) Diagnosis of lung nodule using Moran's index and Geary's coefficient in computerized tomography images. *Pattern Anal. Appl.*, **11**, 89–99.

148 Christofori, G. and Semb, H. (1999) The role of cell-adhesion molecule E-cadherin as a tumour-suppressor gene. *Trends Biochem. Sci.*, **24**, 73–76.

149 Gumbiner, B.M. (1996) Cell adhesion: the molecular basis of tissue architecture and morphogenesis. *Cell*, **84**, 345–357.

150 Prigogine, I. and Wiame, J.M. (1946) Biologie et Thermodynamique des phenomenes irreversibles. *Experientia*, **2**, 451–453.

151 Zotin, A.I. (1990) *Thermodynamic Bases of Biological Processes: Physiological Reactions and Adaptations*, de Gruyter, Berlin.

152 Zotin, A.A. and Zotin, A.I. (1997) Phenomenological theory of ontogenesis. *Int. J. Dev. Biol.*, **41**, 917–921.

153 Palumbo, M.C., Farina, L., De Santis, A., Giuliani, A., Colosimo, A., Morelli, G., and Ruberti, I. (2008) Collective behavior in gene regulation: post-transcriptional regulation and the temporal compartmentalization of cellular cycles. *FEBS J.*, **275**, 2364–2371.

154 Huang, S. and Ingber, D.E. (2000) Shape-dependent control of cell growth, differentiation, and apoptosis: switching between attractors in cell regulatory networks. *Exp. Cell Res.*, **261**, 91–103.

155 Griffin, J.L. (2004) Metabolic profiles to define the genome: can we hear the phenotypes? *Trans. R. Soc. Lond. B Biol. Sci.*, **359**, 857–871.

156 Ingber, D.E. (2008) Can cancer be reversed by engineering the tumour microenvironment? *Semin. Cancer Biol.*, **18**, 356–364.

18
Protein Network Analysis for Disease Gene Identification and Prioritization
Jing Chen and Anil G. Jegga

18.1
Introduction

Identifying disease causal genes is a problem of primary importance in the postgenomic era and in translational biomedical research. Traditional methods such as positional cloning, in spite of being successful in identifying monogenetic disease causal genes, are not effective for complex disease-associated gene discovery [1]. Recent technological advances, while providing novel methods and opportunities to understand the genetic basis of complex diseases, are replete with several limitations. To overcome some of these limitations, researchers, based on their prior knowledge, typically adopt a "candidate gene approach." A recent trend that mimics the researcher-based candidate gene approach is to use computational approaches to integrate multiple sources and features of human diseases and the genome (e.g., gene functions, sequence features, protein interactions, etc.), and use them as a "seed" or guideline to screen the whole genome for the most likely novel disease candidate genes. Broadly, these approaches can be classified into two types: functional annotation-based and network or topological analysis-based approaches. In this chapter, we review protein interaction network analysis-based methods for disease gene identification and prioritization.

18.2
Protein Networks and Human Disease

Following the findings in yeast that essential genes have higher connectivity (or "degree") [2], several research groups started applying similar analytical approaches to understand the molecular basis of human disease [3, 4]. Based on these findings, it can be concluded that genes associated with a particular phenotype or function are not randomly positioned in the network, and that they tend to exhibit high connectivity, cluster together, and occur in central network locations [5]. However, it should be noted that there could be a potential bias in the protein interaction datasets because typical disease-causing proteins tend to be more researched or better studied.

Applied Statistics for Network Biology: Methods in Systems Biology, First Edition.
Edited by M. Dehmer, F. Emmert-Streib, A. Graber, and A. Salvador.
© 2011 Wiley-VCH Verlag GmbH & Co. KGaA. Published 2011 by Wiley-VCH Verlag GmbH & Co. KGaA.

Another area in which protein interaction networks have played a role in understanding the molecular basis of human disease is in the prediction and prioritization of novel candidate genes. These studies are based on a "guilt-by-association" hypothesis. In other words, the likelihood of a network neighbor of a disease gene to cause either the same or a similar disease is higher [4, 6–9]. In one of the earliest efforts [10], Xu et al., built a K-nearest neighbors-based classifier using all disease genes from the OMIM (Online Mendelian Inheritance in Man) database and concluded that disease genes from OMIM are characterized by a larger degree, a tendency to interact with other disease genes, more common neighbors, and quick communication with each other [10]. In another study, Chen et al. [11], starting with a known gene list for Alzheimer's from the OMIM, expanded it based on protein interactions and proposed a graph connectedness-based scoring system to identify other Alzheimer's disease causal genes. Recently, Kohler et al. [12] and Chen et al. [9] used social and web network analysis algorithms to prioritize candidate disease genes. Vanunu and Sharan [13] developed a global, propagation-based method that integrates information on known causal disease genes and protein interaction confidence scores. Thus, the idea that proteins close to one another in a network cause similar diseases when mutated is becoming a central dogma for disease gene discovery. Although different approaches and different kinds of integrated protein interactions data are used to tackle disease gene prediction and prioritization problem, the central theme remains the same – all of the approaches involve superimposing a set of candidate genes alongside a set of known disease genes on a physical or functional network and analyzing the connectivity between them. "De novo" approaches that do not depend on prior knowledge of disease genes are yet to be developed [5].

While there is a wealth of protein–disease relationships in the published literature and a number of protein interaction resources are now available, relatively few studies have used protein interactions for prioritizing disease genes. Thus, making use of protein networks in the context of disease analyses is a relatively new challenge in the field [14]. As mentioned earlier, most of the current disease candidate gene prioritization methods [15–21] rely on functional annotations. However, the coverage of the gene functional annotations is a limiting factor. Although more than 1600 human disease genes have been documented, most of them are yet to be functionally characterized. Currently, only a fraction of the genome is annotated with pathways and phenotypes.

18.2.1
Ranking Algorithms for Network Analyses

Network-based analyses have been successfully used for protein function prediction [22], identification of functional modules [23], interaction prediction [24, 25], identification of disease candidate genes [14, 26] and drug targets [27, 28], and evolution [29–33]. When analyzing social networks, web graphs, and telecommunication networks, one of the principal objectives is to identify nodes that are "important" in terms of the extension of their relation with other nodes in the network. Visualization-centered approaches such as graph drawing, although useful to gain qualitative intuition about the structure, work primarily for small graphs, and it is difficult to use these approaches for large and complex networks. Therefore, several other approaches have been developed. For instance, to determine the "centrality" or "importance" of a node in a social network, sociologists have proposed using degree centrality [34], closeness centrality [35], and betweenness centrality [36]. The degree centrality of a node is a measure of the number of contacts or edges a node has. The betweenness centrality of a node is an estimate of the probability that the shortest path between any pair of nodes of the network passes through the original node. The closeness centrality of a node is an estimate of how closely connected a node is to all other nodes of the network. Similarly, in the area of web graphs, computer scientists have proposed a number of algorithms such as HITS [37] and PageRank [38] for automatically determining the "importance" of web pages. Biological networks are comparable to communication and social networks [39]. For instance, protein interaction networks and communication networks share several common features, such as scale-freeness and small-world properties, suggesting that the algorithms used for social and web networks should be equally applicable to biological networks. In the following sections, we describe the application of these ranking algorithms to prioritize disease candidate genes. The importance of the other genes relative to the root (or "seed") is calculated and ranked using a framework and algorithms proposed by White and Smyth [40].

18.2.2
Prioritization Methods

In our method, we represent the protein interaction network as an unweighted, undirected graph G, where proteins (genes) are nodes and interactions are edges. The set of all the proteins in the network is denoted as V, all the interactions as E. The set of known disease-genes (also referred as "seeds" or "training set") is denoted as R. The prioritization approaches are based on White and Smyth's methods [40], whose general framework, consisting of four successive problem formulations, each building on the next, defines the approach to ranking nodes in an unweighted digraph $G(V, E)$:

i) **Relative Importance of a node t with respect to a root node r.** Given G and r and t, where r and t are both nodes in G and r is the root or seed, compute the "importance" of t with respect to r. This importance is denoted as $I(t|r)$, a non-negative quantity.

ii) **Rank of importance of a set of nodes T with respect to a root node r.** Given G and a root node r in G, rank all vertices in T, a subset of vertices in G. For each node t in T, the value of $I(t|r)$ can be computed. Then the nodes can be ranked so that the largest values correspond to the highest importance.

iii) **Rank of importance of a set of nodes T with respect to a set of root nodes R.** Given G and a set of root node R in G, rank all vertices in T, a subset of vertices in G. The importance of node t to R is defined as the average sum of importance of t to each node in R:

$$I(t|R) = (1/|R|)\left(\sum(I(t|r))\right) \qquad (18.1)$$

iv) **Given G, rank all nodes**. This is a special case where $R = T = V$.

Based on White and Smyth's framework, the solution to problem (iii) described above is what is needed for prioritizing candidate genes. In other words, the problem is to prioritize a set of genes in the network based on their importance to a set of root genes (e.g., genes known to be associated to a disease). The importance of a gene to the set of root genes is the average sum of the importance of it to each of the individual root gene. It is to be noted that this framework was proposed for directed networks. However, we applied it to undirected networks also because undirected networks are nothing but a special case of directed networks. Thus, the goal is to find $I(t|r)$, the importance of node t with respect to a root node r. We used three White and Smyth [40] algorithms: PageRank with Priors, HITS with Priors, and K-step Markov. The following sections briefly describe the equations used for these three approaches. For additional details of the methods, the reader is referred to the original paper by White and Smyth [40].

PageRank with Priors is an extension to the original White and Smyth's PageRank algorithm. The iterative stationary probability equation is:

$$\pi(v)^{(i+1)} = (1-\beta)\left(\sum_{u=1}^{d_{in}(v)} p(v|u)\pi^{(i)}(u)\right) + \beta p_v \qquad (18.2)$$

In this equation, p_v represents the "prior bias" and $p_v = 1/|R|$ for v in R, the root node set; $p_v = 0$ otherwise. β, empirically defined on [0,1], represents a "back-probability." The other parameters are constants, where $d_{in}(v)$ is the in-degree of node v and $p(v|u)$ is the probability of arriving at v from u.

HITS with Priors is an extension of the original HITS algorithm. The iterative equations are defined as follows:

$$a^{(i+1)}(v) = (1-\beta)\left(\sum_{u=1}^{d_{in}(v)} \frac{h^{(t)}(u)}{H^{(i)}}\right) + \beta p_v$$

$$h^{(i+1)}(v) = (1-\beta)\left(\sum_{u=1}^{d_{out}(v)} \frac{a^{(t)}(u)}{A^{(i)}}\right) + \beta p_v \qquad (18.3)$$

where $d_{in}(v)$ and $d_{out}(v)$ are the in-degree and out-degree of v, respectively, and $H^{(i)}$ and $A^{(i)}$ are defined as:

$$H^{(i)} = \sum_{v=1}^{|V|} \sum_{u=1}^{d_{in}(v)} h^{(i)}(u)$$
$$A^{(i)} = \sum_{v=1}^{|V|} \sum_{u=1}^{d_{out}(v)} a^{(i)}(u) \qquad (18.4)$$

The definition of prior bias p_v and back-probability β is similar to PageRank with Priors. The authority score is set as the importance of the node.

The *K*-Step Markov approach computes the relative probability that the system will spend time at any particular node given that it starts in a set of roots *R* and ends after *K* steps. According to White and Smyth [40], the value of *K* controls the relative tradeoff between a distribution "biased" towards *R* and when *K* gets larger the steady-state distribution will converge to PageRank result. The equation to compute the *K*-Step Markov importance is:

$$I(t|R) = [\mathbf{A}\mathbf{p}_R + \mathbf{A}^2\mathbf{p}_R \ldots \mathbf{A}^K\mathbf{p}_R]_t \qquad (18.5)$$

where **A** is the transition probability matrix of size $n \times n$, $\mathbf{p}_R$ is an $n \times 1$ vector of initial probabilities for the root set *R*, and $I(t|R)$ is the *t*th entry in this sum vector.

The implementation of the three prioritization methods, namely, PageRank with Priors, HITS with Priors, and *K*-Step Markov approach, is available in the JUNG (Java Universal Network/Graph; jung.sourceforge.net) framework. It is a Java package that provides a common and extendible language for the modeling, analysis, and visualization of data that can be represented as a graph or network. We used version 2.0 and integrated it with other in-house programs to perform all the required functions.

18.2.3
Evaluation of Protein Interaction Network Topological Features

To evaluate our methodology, we used 19 diseases with 693 associated genes. This training data was identical to the one we used in an earlier published study [20]. The only difference was in the number of genes. Instead of 693 genes, 589 were used in the cross-validation because the remaining 104 genes do not have any known protein–protein interactions. The random training dataset, used as a control, was built with 19 random gene lists (each list comprising 31–38 genes). We used the three methods (i.e., *K*-Step Markov, PageRank with Priors, and HITS with Priors) to prioritize the disease gene with different parameter values. The random genes were prioritized using PageRank with Priors with back-probability set to 0.3. For additional details, the reader is referred to our published study [9]. To summarize our findings briefly, we observed that in terms of performance, HITS with Priors was similar to PageRank with Priors under different back-probability values. Therefore, we tested PageRank with Priors only for extreme back-probability values (0.01 and 0.05). We also observed that there was no significant difference among the best performance of the three methods [9].

18.3
ToppGene Suite of Applications

To facilitate easy access to these algorithms, we developed ToppNet, a web-accessible server, and made it available as part of our ToppGene Suite (toppgene.cchmc.org) of applications for candidate gene identification and prioritization [21]. The ToppGene Suite is a one-stop portal for (i) gene list functional enrichment ("ToppFun"),

(ii) candidate gene prioritization using functional annotations ("ToppGene"), (iii) candidate gene prioritization using network analysis ("ToppNet"), and (iii) identification and prioritization of novel disease candidate genes in the interactome ("ToppGenet"). The reader is referred to the ToppGene Suite web site for instructions and "help" for each of these modules. The back-end knowledgebase is updated periodically and the current status of the data (versions and coverage) can be accessed from the ToppGene Suite homepage (see "Database details" section from the ToppGene Suite homepage). Additionally, several worked out examples with stepwise instructions to demonstrate the utility of each of these modules are also provided (see "Supplementary" section from the ToppGene Suite homepage). Functional annotation-based disease candidate gene prioritization uses a fuzzy-based similarity measure to compute the similarity between any two genes based on semantic annotations. The similarity scores from individual features are combined into an overall score using statistical meta-analysis. A *p*-value of each annotation of a test gene is derived by random sampling of the whole genome [20, 21]. The protein interaction network-based disease candidate gene prioritization uses social and web network analysis algorithms (as described earlier) [9, 21]. For additional details about the protein interaction datasets used, algorithmic details, and validation, see our recently published study [9]. The ToppGenet module differs from ToppGene and ToppNet in that the test set is derived from the protein interactome. For a given training set of known disease genes, the test set is generated by mining the protein interactome and compiling the genes either directly or indirectly interacting (based on user input) with the training set. After any overlapping or common genes between test and training sets are removed, interactome-based test set genes can be prioritized using either a functional annotation-based method (ToppGene) or protein network analysis-based method (ToppNet). The human protein interactome dataset (file "interactions.gz"), a compilation of protein interactions from BIND [41], BioGRID [42], and HPRD [43], is downloaded from the National Center for Biotechnology Information's Entrez Gene FTP site (ftp://ftp.ncbi.nih.gov/gene/).

18.4
Conclusions

In the current chapter, we have described how successful graph analysis-based algorithms from computer science research can be used to address the disease gene prioritization problem. Our approach is based on the observation that biological networks share many properties with web and social networks. Using literature-based and manually curated protein interactions to form the base network, extended versions of the PageRank algorithm and HITS algorithm, as well as the *K*-Step Markov method are applied to prioritize disease candidate genes. Based on the results from large-scale cross-validations using a list of known disease-related genes, the following conclusions were drawn. First, under appropriate settings, such as a back-probability of 0.3 for PageRank with Priors and HITS with Priors, and walk length 4 for the *K*-Step Markov method, the three methods achieved the same area-under-the-curve value and, hence, similar performance. This suggests that based on the current knowledge of protein interaction networks, other similar methods (e.g., ranking of nodes in an unweighted graph) under the same framework may produce similar results. Second, the value of back-probability in PageRank with Priors and HITS with Priors could be from a fairly large range (e.g., 0.1–0.9), and has relatively little impact on the performance. However, when the back-probability is set to very low (e.g., 0.01), the performance drops significantly. This is expected because in both the methods (see Equations 18.3 and 18.4), as the back-probability reaches 0, the bias towards the "seeds" is eliminated and PageRank/HITS with Priors is the same as the original PageRank/HITS algorithm, and therefore the prioritization to the selected "seeds" fails. The performance of the K-Step Markov method, on the other hand, decreases significantly when the length of the random walk K is small (e.g., $K=1$). Under this condition, the method calculates the probability to spend time on each protein from the seeds with a random walk of length 1. The proteins that are not directly interacting with "seeds" will never be arrived at and scored 0. This suggests that if a true disease candidate is not directly interacting with the "seeds," it will be ignored when K is 1. The method converges to the best performance when K is set to 4. Any further increase in random walk length does not improve the performance. This could be due to the fact that the average shortest path length in the interaction network was only about 4.5.

Protein network-based disease candidate gene prioritization methods have certain limitations, however. Just like functional annotation-based methods, the performance depends on the quality of interaction data. Additionally, most of these algorithms were originally developed to identify "important" nodes in networks. Although we use extended versions of these algorithms to prioritize nodes to selected "seeds," they could still be biased towards hubs. Lastly, these approaches were designed for web and general networks; therefore, additional modifications may be required to make them fit better with biological networks, such as using weights on nodes (proteins) or edges (interactions).

References

1 McCarthy, M.I., Smedley, D., and Hide, W. (2003) New methods for finding disease-susceptibility genes: impact and potential. *Genome Biol.*, **4**, 119
2 Jeong, H., Mason, S.P., Barabasi, A.L., and Oltvai, Z.N. (2001) Lethality and centrality in protein networks. *Nature*, **411**, 41–42.
3 Wachi, S., Yoneda, K., and Wu, R. (2005) Interactome–transcriptome analysis reveals the high centrality of genes differentially expressed in lung cancer tissues. *Bioinformatics*, **21**, 4205–4208.
4 Goh, K., Cusick, M., Valle, D., Childs, B., Vidal, M. *et al.* (2007) The human disease network. *Proc. Natl. Acad. Sci. USA*, **104**, 8685–8690.
5 Ideker, T. and Sharan, R. (2008) Protein networks in disease. *Genome Res.*, **18**, 644–652.
6 Oti, M. and Brunner, H.G. (2007) The modular nature of genetic diseases. *Clin. Genet.*, **71**, 1–11.
7 Lage, K., Karlberg, E., Storling, Z., Olason, P., Pedersen, A. *et al.* (2007) A human phenome–interactome network of protein complexes implicated in genetic disorders. *Nat. Biotechnol.*, **25**, 309–316.

8 Franke, L., Bakel, H., Fokkens, L., de surJong, E., Egmont-Petersen, M. et al. (2006) Reconstruction of a functional human gene network, with an application for prioritizing positional candidate genes. *Am. J. Hum. Genet.*, **78**, 1011–1025.

9 Chen, J., Aronow, B.J., and Jegga, A.G. (2009) Disease candidate gene identification and prioritization using protein interaction networks. *BMC Bioinformatics*, **10**, 73

10 Xu, J. and Li, Y. (2006) Discovering disease-genes by topological features in human protein–protein interaction network. *Bioinformatics*, **22**, 2800–2805.

11 Chen, J.Y., Shen, C., and Sivachenko, A.Y. (2006) Mining Alzheimer disease relevant proteins from integrated protein interactome data. *Pac. Symp. Biocomput.*, 367–378.

12 Kohler, S., Bauer, S., Horn, D., and Robinson, P. (2008) Walking the interactome for prioritization of candidate disease genes. *Am. J. Hum. Genet.*, **82**, 949–958.

13 Vanunu, O. and Sharan, R. (2008) A propagation based algorithm for inferring gene-disease associations. German Conference on Bioinformatics, Dresden

14 Sam, L., Liu, Y., Li, J., Friedman, C., and Lussier, Y.A. (2007) Discovery of protein interaction networks shared by diseases. *Pac. Symp. Biocomput.*, 76–87.

15 Adie, E.A., Adams, R.R., Evans, K.L., Porteous, D.J., and Pickard, B.S. (2005) Speeding disease gene discovery by sequence based candidate prioritization. *BMC Bioinformatics*, **6**, 55

16 Adie, E.A., Adams, R.R., Evans, K.L., Porteous, D.J., and Pickard, B.S. (2006) SUSPECTS: enabling fast and effective prioritization of positional candidates. *Bioinformatics*, **22**, 773–774.

17 Aerts, S., Lambrechts, D., Maity, S., Van Loo, P., Coessens, B. et al. (2006) Gene prioritization through genomic data fusion. *Nat. Biotechnol.*, **24**, 537–544.

18 Tiffin, N., Adie, E., Turner, F., Brunner, H.G., van Driel, M.A. et al. (2006) Computational disease gene identification: a concert of methods prioritizes type 2 diabetes and obesity candidate genes. *Nucleic Acids Res.*, **34**, 3067–3081.

19 Turner, F.S., Clutterbuck, D.R., and Semple, C.A. (2003) POCUS: mining genomic sequence annotation to predict disease genes. *Genome Biol.*, **4**, R75

20 Chen, J., Xu, H., Aronow, B.J., and Jegga, A.G. (2007) Improved human disease candidate gene prioritization using mouse phenotype. *BMC Bioinformatics*, **8**, 392

21 Chen, J., Bardes, E.E., Aronow, B.J., and Jegga, A.G. (2009) ToppGene Suite for gene list enrichment analysis and candidate gene prioritization. *Nucleic Acids Res.*, **37**, W305–W311.

22 Nabieva, E., Jim, K., Agarwal, A., Chazelle, B., and Singh, M. (2005) Whole-proteome prediction of protein function via graph-theoretic analysis of interaction maps. *Bioinformatics*, **21** (Suppl. 1), i302–i310.

23 Lubovac, Z., Gamalielsson, J., and Olsson, B. (2006) Combining functional and topological properties to identify core modules in protein interaction networks. *Proteins*, **64**, 948–959.

24 Jansen, R., Yu, H., Greenbaum, D., Kluger, Y., Krogan, N.J. et al. (2003) A Bayesian networks approach for predicting protein–protein interactions from genomic data. *Science*, **302**, 449–453.

25 Wong, S.L., Zhang, L.V., Tong, A.H., Li, Z., Goldberg, D.S. et al. (2004) Combining biological networks to predict genetic interactions. *Proc. Natl. Acad. Sci. USA*, **101**, 15682–15687.

26 Goehler, H., Lalowski, M., Stelzl, U., Waelter, S., Stroedicke, M. et al. (2004) A protein interaction network links GIT1, an enhancer of huntingtin aggregation, to Huntington's disease. *Mol. Cell*, **15**, 853–865.

27 Ruffner, H., Bauer, A., and Bouwmeester, T. (2007) Human protein–protein interaction networks and the value for drug discovery. *Drug Discov. Today*, **12**, 709–716.

28 Neduva, V., Linding, R., Su-Angrand, I., Stark, A., de Masi, F. et al. (2005) Systematic discovery of new recognition peptides mediating protein interaction networks. *PLoS Biol.*, **3**, e405

29 Barabasi, A.L. and Albert, R. (1999) Emergence of scaling in random networks. *Science*, **286**, 509–512.

30 Berg, J., Lassig, M., and Wagner, A. (2004) Structure and evolution of protein interaction networks: a statistical model for link dynamics and gene duplications. *BMC Evol. Biol.*, **4**, 51

31 Eisenberg, E. and Levanon, E.Y. (2003) Preferential attachment in the protein network evolution. *Phys. Rev. Lett.*, **91**, 138701

32 Rzhetsky, A. and Gomez, S.M. (2001) Birth of scale-free molecular networks and the number of distinct DNA and protein domains per genome. *Bioinformatics*, **17**, 988–996.

33 Wagner, A. and Fell, D.A. (2001) The small world inside large metabolic networks. *Proc. Biol. Sci.*, **268**, 1803–1810.

34 Freeman, L.C. (1978) Centrality in social networks conceptual clarification. *Soc. Netw.*, **1**, 215–239.

35 Sabidussi, G. (1966) The centrality index of a graph. *Psychometrika*, **31**, 581–603.

36 Freeman, L.C. (1977) A set of measures of centrality based on betweenness. *Sociometry*, **40**, 35–41.

37 Jon, M.K. (1999) Authoritative sources in a hyperlinked environment. *J. ACM*, **46**, 604–632.

38 Page, L., Brin, S., Motwani, R., and Winograd, T. (1998) The PageRank citation ranking: bringing order to the web. Technical Report, Stanford InfoLab, Stanford, CA.

39 Junker, B.H., Koschutzki, D., and Schreiber, F. (2006) Exploration of biological network centralities with CentiBiN. *BMC Bioinformatics*, **7**, 219

40 White, S. and Smyth, P. (2003) Algorithms for estimating relative importance in networks. Ninth ACM SIGKDD International Conference on Knowledge Discovery and Data Mining, New York

41 Bader, G.D., Betel, D., and Hogue, C.W. (2003) BIND: the biomolecular interaction network database. *Nucleic Acids Res.*, **31**, 248–250.

42 Breitkreutz, B.J., Stark, C., Reguly, T., Boucher, L., Breitkreutz, A. *et al.* (2008) The BioGRID interaction database: 2008 update. *Nucleic Acids Res.*, **36**, D637–D640.

43 Peri, S., Navarro, J.D., Kristiansen, T.Z., Amanchy, R., Surendranath, V. *et al.* (2004) Human protein reference database as a discovery resource for proteomics. *Nucleic Acids Res.*, **32**, D497–D501.

19
Pathways and Networks as Functional Descriptors for Human Disease and Drug Response Endpoints

Yuri Nikolsky, Marina Bessarabova, Eugene Kirillov, Zoltan Dezso, Weiwei Shi, and Tatiana Nikolskaya

19.1
Introduction

Databases of pathways and cell processes such as KEGG (Kyoto Encyclopedia of Genes and Genomes) [1] and GO (Gene Ontology) [2] have been widely known and accessible for over a decade. However, until recently, they were mostly used as an interesting yet nonessential addition to statistical analysis of "omics" data (primarily microarray expression data), culminating in deducing gene-based classifiers for clinically important endpoints such as cancer metastases [3, 4] or drug-induced toxicity [5]. In parallel, algorithms of network and interactome analysis have been developed by a promising but niche area of bioinformatics known as "systems biology" [6–8]. The tools of calculation-intensive network biology were poorly understood and rarely used in mainstream omics experimentation, and the applications were mostly limited to "one-click" visualization of favorite genes connected by protein interactions in user-friendly tools such as MetaCore (Thomson Reuters; www.genego.com, [9]) or IPA (Ingenuity Biosystems; www.ingenuity.com).

However, we have recently seen a gradual "paradigm shift" in the applicability of functional analysis in the life sciences. The relative importance and awareness of functional methods has been growing quickly. Thus, "pathways" are announced as the strategic R & D focus at some major drug companies, such as Novartis [10, 11]. In several recent "visionary" publications, pathway and network analysis is introduced as the key engine for data interpretation in personalized and translational medicine as well as drug repositioning [12, 13] in the coming years. Arguably, the current rise of quantitative functional methods is fueled by an explosion of large-scale genome-wide association studies (GWAS) and sequencing data, where standard statistical procedures for gene-based classifiers are barely applicable [14]. In any event, quantitative methods of pathway and network analysis are becoming increasingly thought of in both fundamental life sciences and applications such as drug discovery, personalized, and translational medicine. Although the field is very young, there are a number of promising tools available, some of which are presented in this book and reviewed elsewhere [15, 16].

Applied Statistics for Network Biology: Methods in Systems Biology, First Edition.
Edited by M. Dehmer, F. Emmert-Streib, A. Graber, and A. Salvador.
© 2011 Wiley-VCH Verlag GmbH & Co. KGaA. Published 2011 by Wiley-VCH Verlag GmbH & Co. KGaA.

Here, we describe some methods of quantitative pathway and network analysis, which we have developed over the last few years and applied for the analysis of different types of omics data in collaborative disease and drug response studies. We also offer our opinion on the future of the field from the perspective of a 10-year veteran vendor of tools and databases for systems analysis.

19.2
Gene Content Classifiers and Functional Classifiers

Statistical analysis reduces thousands of data points (e.g., genome-wide gene expression values or, lately, single nucleotide polymorphism (SNP) significance values in GWAS studies) to a relatively short list of genes or SNP IDs. These genes represent a multivariant descriptor of the studied condition (endpoint) and may be further refined to distinguish between or predict phenotypic outcomes (a "gene signature"). Over the last decade, a number of gene signatures have been reported to predict metastases in breast cancer [3, 4], classify cancer subtypes [17, 18], or predict drug response and toxicity [5, 19–22]. Gene signatures became a mainstream source of multivariant biomarkers for disease states, including the US Food and Drug Administration-approved commercial products Onco*type* DX (Genomics Health; www.genomichealth.com) and MammaPrint (Agendia; www.agendia.com).

Although gene signatures have proven useful, several issues with statistical methods of feature selection for these signatures have arisen. For one, expression signatures perform well in a relatively small study, but lose performance (particularly, specificity) with an increasing number of repeated validation studies. Thus, the specificity of the prognostic MammaPrint's "Amsterdam" signature in a validation study of 307 patients [23] was reported at only 42% compared to 76% in the original study [2]. Eventually, it was concluded that MammaPrint "is not a sufficiently accurate predictor" for metastasis development [24]. Molecular signatures are also redundant, as variable selection methods (different statistical models and machine learning algorithms) lead to many quantitative solutions of equal reliability in terms of prediction performance [25–27]. This may occur because the various probe sets representing different genes capture information from the same complex molecular pathways that determine a particular clinical outcome (Figure 19.1) [28]. This is a logical "common sense" statement in accordance with the principle of biological modularity [29], which suggests that cellular functions are performed by groups of proteins that temporarily work in concert. It was proven that very large, multi-thousand sample studies are needed for a 70-gene signature to be 50% consistent (by gene content) between two studies [24]. Recently, the largest and most comprehensive study on gene signature generation models, MAQC-II run on six different datasets and 13 "endpoints," concluded that: (i) there is no "magic bullet" (i.e., the best statistical model that generates the best multivariant descriptors) and (ii) the single dominant parameter in model performance is the biology of the endpoint, with relatively poor performance of all models with complex phenotypes, such as disease survival [30].

19.2 Gene Content Classifiers and Functional Classifiers

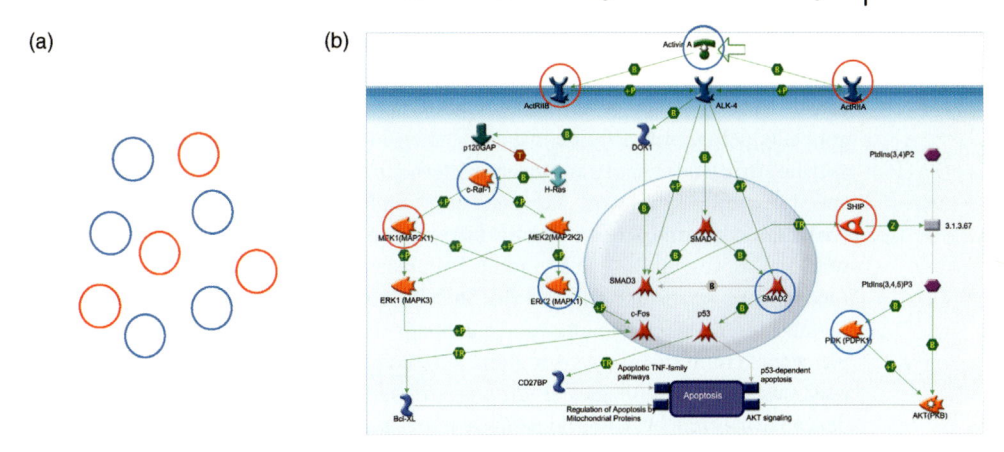

Figure 19.1 Gene content classifiers (gene signatures) and pathway classifiers. (a) Hypothetical gene signature classifiers of similar performance are usually different in content if calculated by different models or by the same model in different studies. "Blue" and "red" signatures for the same endpoint do not overlap. (b) A pathway classifier. The very same genes sets that do not overlap in signatures match to the same pathway, making it valid as a classifier.

Statistical methods of identifying gene signatures are almost inapplicable for the fastest growing body of omics data – genotyping in disease predisposition GWAS. GWAS experiments typically feature 2–3 orders of magnitude more data points than microarray expression studies, as hundreds of thousands of markers get tested in hundreds to tens of thousands of samples [14, 31, 32]. Microarray expression studies almost always produce hundreds or thousands of "statistically significantly" differentially expressed genes (DEGs) for deducing a classifier "gene signature." By contrast, in GWAS studies false discovery testing and other statistical validation procedures based on an assumption of independent variables (reviewed in [33]) typically eliminate all but a few SNPs as false-positives.

Functional analysis of high-content molecular ("omics") data aims to reveal the underlying biology driving observed changes, through identification of key pathways and mechanisms associated with the studied phenotype (reviewed in [34]). Early functional analysis consisted of "descriptive" mapping of expressed genes onto pathway maps such as KEGG [1] or functional ontologies such as GO [2]. More recently, functional analysis has recently expanded into three main "quantitative" approaches: enrichment in biological ontologies [35], biological network reconstruction, and interactome analysis [36, 37]. For example, we proposed using different types of "knowledge-based" functional groups such as networks [38] and signaling pathways as the statistical descriptors for a condition [39]. More recently, the power of "quantitative" functional analysis has been demonstrated in studies of common diseases [40–44], and in toxicity and drug response analyses [45, 46].

In "quantitative" analysis, the whole gene set of an entity in a certain functional category (pathways, biological processes, networks, or subnetworks) can be used as a classifier, or descriptor, to characterize an endpoint, such as a disease or toxic state.

It is essential that a functional descriptor makes biological sense as a whole – its gene content should constitute (be annotated as) a folder (term) in a certain "knowledge-based" ontology, such as the well-known GO of cellular processes [2]. GO terms generally define only a functional association between the category and the genes ascribed to it. The genes in any given GO term do not necessarily interact directly in a specific pathway or network, Furthermore, a high level of gene overlap between GO terms can confound the use of some statistical methods. Another type of functional ontology represents a set of graphical representations of biological pathways or mechanisms (maps or networks) that include not only molecular entities (proteins, metabolites, drugs, xenobiotics, and enzymatic reactions) but also, most importantly, functional connections between them, such as physical interactions and enzymatic reactions. Several ontologies of this type are assembled in the MetaCore/MetaDrug data analysis suite [9]: GeneGo pathway maps, process networks, disease biomarker networks, drug target networks, toxicity networks, and metabolic networks. The main advantage of these graphical ontologies is an improved mechanistic representation of fundamental biology, as they cover not only the participants in a particular biological process, but also indicate a chain of signaling interactions involved in the process.

As mentioned above, functional descriptors are essential in GWAS studies where standard statistical methods often eliminate all SNPs as "false-positives." Functional grouping must be applied for assembly of below threshold "weak hits" into functionally cohesive units such as pathways and for using p-value "group statistics" as a validation metrics. A number of pathway and network-based methods were offered for GWAS analysis recently [14, 47–51], reviewed in [16, 49].

19.3
Biological Pathways and Networks Have Different Properties as Functional Descriptors

Biological functions are carried out mostly by proteins that are interconnected by a series of direct physical interactions of different mechanisms and stringencies or mediated by biological transformations such as metabolic and transport reactions [52]. Moreover, over a third of endogenous human metabolites are physically associated with human proteins via ligand–receptor interactions. These interactions are mimicked by small-molecule drugs and biologically active compounds that perturb biological systems [53]. Therefore, we can consider binary protein–protein, protein–compound, protein–RNA interactions, and metabolic reactions as basic "building blocks" of cellular functionality. Protein interactions can be identified by a variety of wet lab, computational, and "knowledge-based" (literature annotation) methods (reviewed in [34]).

In an attempt to represent the complexity of cellular protein machinery, binary protein interactions can be assembled into more complex structures, namely pathways and networks. We see a clear distinction between these two concepts that defines different applications for "pathways" and "networks" in functional analysis. The "pathways" can be defined as directed chains of one-step metabolic

19.3 Biological Pathways and Networks Have Different Properties as Functional Descriptors

reactions, with a step defined by a single transformation of a biological molecule carried out by an enzyme. Metabolic pathways have been described since the first studies of biochemistry began, and were first assembled in pathway databases such as EMP [54], MPW [55], and BRENDA [56], and propagated by KEGG [1]. Later, the notion of a "pathway" was extended to signaling (signal transduction pathways). It is generally believed (a flux balance theory [57]) that a pathway is defined as the most energy-efficient way to organize multistep metabolic reactions [58] or signaling interactions [59]. This assumption makes pathways inherently stable and quite conserved (e.g., either this is a metabolic "TCA cycle" or a signaling "WNT signaling" pathway). Importantly, in a pathway not only just every binary step but the whole chain of interactions or reactions is confirmed experimentally. These pathways are usually displayed in the form of static maps (wire diagrams), typically with several dozen objects – a representation that is very convenient for ontology enrichment analysis, either by gene set enrichment analysis-like methods [35] or p-value distributions [9].

Unlike pathways, networks are usually assembled for a set of "seed nodes" (proteins, DEGs, drugs, and metabolites) out of building blocks (interactions) using certain rules (algorithms), which can vary greatly (some are summarized in the current book and elsewhere [60–64]). The difference between pathways and networks is illustrated by Figure 19.2. Protein kinases ATM and ATR have 126 and 66 interactions with human proteins, correspondingly (MetaBase; GeneGo). An "autoexpand" network that connects the two proteins in the shortest possible way and shows the immediate neighbors has 45 and 56 interactions for ATM and ATR (Figure 19.2a). The canonical pathway map for DNA damage, however, has only three and five interactions and 30 proteins relevant for this cellular function (Figure 19.2b). The main advantage of networks as a tool for functional analysis is in their dynamic

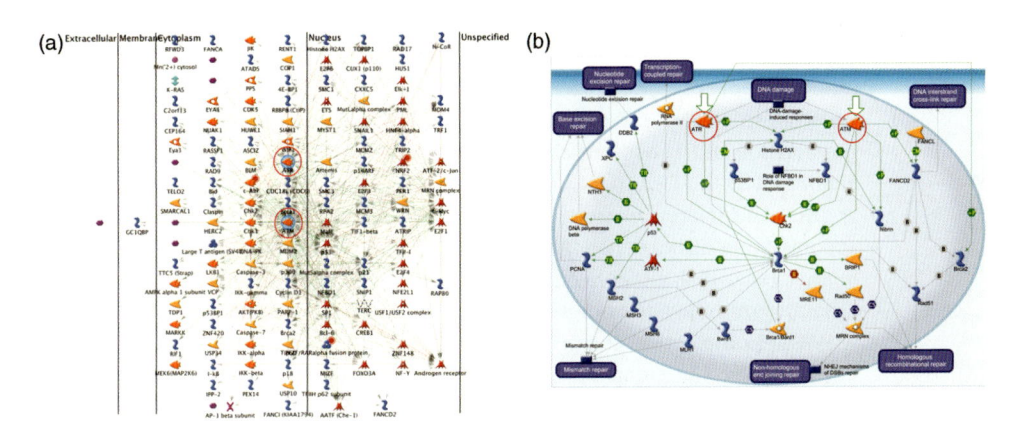

Figure 19.2 The difference between "canonical pathways" and networks. (a) An "autoexpand" close neighbor network built for the protein kinases ATR and ATM (red circles). Note dozens of interactions for these proteins. (b) A canonical pathway map for DNA damage. Only a small subset of interactions and protein neighbors connected with ATM and ATR are relevant.

nature and uniqueness for the analyzed datasets, which makes them a true hypothesis generation instrument. Also, networks are very flexible and can be adjusted for a specific analytical purpose by choosing specific types of objects (e.g., proteins of different functions), effects, and interaction mechanisms, as well as "filters" such as tissue expression specificity or species orthologs. For instance, in MetaCore, a network toolbox allows one to choose between over 30 protein functions, 20 mechanisms of interactions, 10 algorithms, and six filters, which results in thousands of possible versions of networks for a given seed dataset. However, the probabilistic nature of networks constitutes their main drawback. Even with experimentally proven binary interactions as building blocks, their combinations (networks and modules) are not validated and one cannot rely on them as a final biological result. Wet-lab network validation is usually a tricky, lengthy, and iterative process. Also, the unproven nature of networks somewhat limits their applicability as functional descriptors for clinical phenotypes.

19.4
Applications of Pathways as Functional Classifiers

A few years ago, we developed a novel method for using "knowledge-based" functional groupings (canonical pathways) as functional descriptors to distinguish between gene expression responses to endpoints such as drug response in rat livers [39]. The data consisted of Affymetrix profiling of phenobarbital, tamoxifen, and mestranol treatments. We hypothesized that three agents should have differential effects on the pathways, as they cause distinct "endpoint" phenotypes, but the compounds may be "paired" by expression profiles. Phenobarbital, an anticonvulsant used for treating neurological disorders such as epilepsy, is also a potent nongenotoxic carcinogen and can promote hepatocarcinogenesis in the mouse model system. On the contrary, tamoxifen is a well-known hormonal antineoplastic agent that inhibits cell growth via estrogen receptor (ER)-dependent and independent mechanisms. Mestranol directly activates the ER. ER-β may suppress breast cancer cell proliferation and tumor formation, and it has been reported to inhibit cyclin D1 and attenuate P53-induced apoptosis effects. Although the precise roles of ER-α and ER-β in liver cancer are unknown, it was shown that ER-β induces liver cancer cell apoptosis in a ligand-dependent manner. In view of this information, we expected to see similar biological effects between tamoxifen and mestranol, both essentially different from the effects of phenobarbital.

We precomputed a comprehensive set of overlapping "canonical pathways" derived from 450 signaling and metabolic pathway maps manually created at GeneGo. These pathways were then used as a framework for mapping experimental datasets. The pathways mostly represented portions of signaling cascades between secondary messengers and transcription factors. Due to branching (variability between the overlapping pathways) in many signaling cascades, the total number of such chains was quite large (about 12 000). We limited this number by selecting the pathways present on at least two different maps. The 2200 conserved pathway fragments with about 3000 proteins were preassembled and stored in a file with the following format:

Pathway ID. Network ID1 → Network ID2 → Network ID3 → ...

The network IDs were disassembled into lists of corresponding gene IDs and the files were parsed into MetaCore as gene lists. Next, we calculated relative distances between samples based on gene expression level of individual pathways. This idea is illustrated in Figure 19.3a. A pathway with three genes is used as an example (representation of a larger pathway would require a projection of multidimensional space). Samples are represented by points (vectors) in this space according to their expression values. After all the data was imported, we used a metric to compute the distances between samples in each pathway. For example, if we had a pathway defined as:

Pathway 1. [gene 1, gene 2, gene 3, gene 4] and each gene had data associated with it from the gene expression dataset:

	Sample 1	Sample 2	Sample 3	Sample 4
Gene 1	1	4	3	2
Gene 2	4	2	7	6
Gene 3	2	9	3	8
Gene 4	2	5	4	2

We then computed the distances for all combinations of samples $\binom{n}{k}$, where $k = 2$ and n is the total number of samples. This means the order in sample pairs is not important (i.e., Sample 1/Sample 4 are the same as Sample 4/Sample 1.) The possible

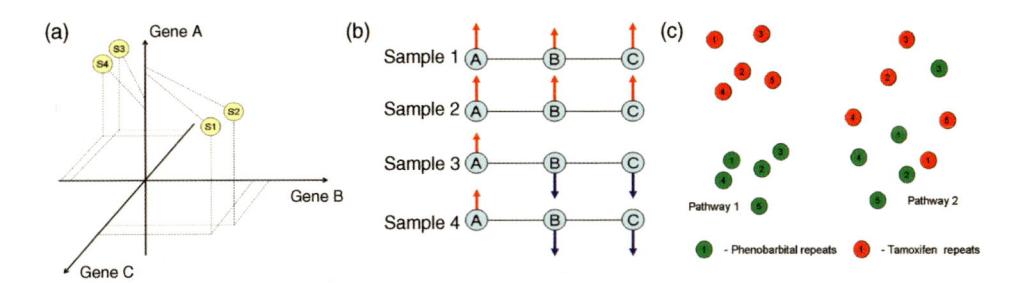

Figure 19.3 Calculation of distances between samples in the pathway's gene expression space. A hypothetical pathway consists of three proteins A, B and C. (a) Samples are represented as points in three-dimensional space of gene expression. Note the grouping of the samples. (b) Representation of the pathway: gene expression (fold change compared to control) is represented by arrows. (c) Clustering of samples in the pathway's gene expression space. In Pathway 1, the distance between samples corresponding to the same treatment is statistically smaller than the intertreatment distance. For Pathway 2, no such distinction can be made.

pathway combinations: Sample 1/Sample 2; Sample 1/Sample 3; Sample 1/Sample 4; Sample 2/Sample 3; Sample 2/Sample 4; Sample 3/Sample 4. Only the distances between the unique samples are considered, as the distance function is symmetric: the distance between Sample 1 and Sample 3 is the same as between Sample 3 and Sample 1. We applied Euclidean and Pearson distance functions for calculating the distances between samples.

$$\text{Euclidean distance } d_{x,y} = \frac{1}{n}\sum_{i=1}^{n}(x_i - y_i)^2$$

$$\text{Pearson distance } r_{x,y} = \frac{1}{n}\sum_{i=1}^{n}\left(\frac{x_i - \bar{x}}{\sigma_x}\right)\left(\frac{y_i - \bar{y}}{\sigma_y}\right)$$

Where $d_{x,y} = 1 - r_{x,y}$, $d_{x,y}$ is the distance between Sample x and Sample y, $r_{x,y}$ is the correlation coefficient for Sample x/Sample y, n is the number of genes in the pathway, x_i is the expression value for gene i in Sample x, y_i is the expression value for gene i in Sample y, $\bar{x}$ is the mean expression for Sample x, $\bar{y}$ is the mean expression for Sample y, σ_x is the standard deviation of expression values in Sample x, and σ_y is the standard deviation of expression values in Sample x.

For every distance function, we computed the distances between samples for all pathways:

	Sample 1/ Sample 2	Sample 1/ Sample 3	Sample 1/ Sample 4	Sample 2/ Sample 3	Sample 2/ Sample 4	Sample 3/ Sample 4
Pathway 1						
...						
Pathway n						

This procedure results in a matrix of distances between samples in the space of gene expression patterns of individual pathway modules. The distances can be grouped according to arbitrarily defined types of samples. We compared the distances between repeat samples for the same treatment versus samples from different treatments. For every pathway, we tested the statistical hypothesis that the average distance between repeats of the same treatment is significantly different than the average distance between different treatments. If this hypothesis is confirmed for a pathway, we can then calculate for which particular pair of treatments the repeat samples for the same treatment cluster together, while intertreatment distances are larger (Figure 19.3c). The criterion for success was identification of distinctive pathways or groups of pathways that distinguish between treatment responses (e.g., mestranol from phenobarbital). The selected pathways were grouped into "clusters" based on the t-test. In the next step, we selected clusters that clearly distinguish between two treatments. For instance, the pathways that distinguish mestranol from tamoxifen treatment had to satisfy the condition that both the distances between mestranol repeats and the distances between phenobarbital repeats were statistically smaller than mestranol–phenobarbital distances.

Upon selecting the groups, the genes from the pathways distinguishing between two particular treatments were imported into MetaCore for further analysis.

Based on the analysis described above, we found:

- Phenobarbital versus tamoxifen: 163 pathways containing a total of 131 unique genes.
- Mestranol versus tamoxifen: 3 pathways containing a total of 15 genes.
- Mestranol versus phenobarbital: 54 pathways containing a total of 151 genes.

No pathways were found that distinguished between responses to different tamoxifen concentrations. Another important observation is that the number of pathways that distinguish mestranol from tamoxifen was substantially smaller than the number distinguishing phenobarbital from both mestranol and tamoxifen treatments. No pathways were found that distinguish between responses to different tamoxifen concentrations. Mestranol and tamoxifen have similar structures and biological targets (both are ligands for the estrogen receptor); phenobarbital is distinct from this pair structurally as well as by mode of action.

Next, the lists of genes from the generated groups (the "unions" of genes in differential pathways) were imported into the current version of MetaCore to explore network connectivity. The "Direct interactions" network-building algorithm was applied. This algorithm retrieves all the interactions connecting genes of interest to each other (Figure 19.4a). The networks were overlaid with expression data that had been converted into log-ratios of gene expression levels between treated and untreated animals. The upregulated genes are marked with red circles, the downregulated ones with blue circles. The networks contained a substantial number of genes that are negatively correlated between treatments. Interestingly, the fold change values for most of these genes were relatively low (within 10–30%). However, these genes assembled into an interconnected network with synchronous differential response to treatments. The drug response networks produced by this method can be characterized by different parameters. For instance, their gene/protein content can be parsed onto such functional categories as canonical pathway maps, GO processes, preset process networks (GeneGo processes), as well as categories for diseases and toxicities. Smaller subgraphs of signature networks for the genes related to individual processes (e.g., a "mitosis" subgraph, Figure 19.4b) could be also considered as a functional descriptor between conditions. The p-value distribution of functional categories can be considered as the network parameters, and the functional distance between networks can be evaluated based on the number of matching functional categories and more subtle derivative parameters. According to these distributions (data not shown), the networks are essentially different, although one of the original datasets used for generating the networks is the same (phenobarbital expression profile). We would like to point out two important advantages of the network comparison method compared to conventional statistical procedures. These are illustrated in Figure 19.4c, which displays the highest scored pathway map, the CREB pathway, for the gene content of the tamoxifen/phenobarbital network. The expression data for phenobarbital and

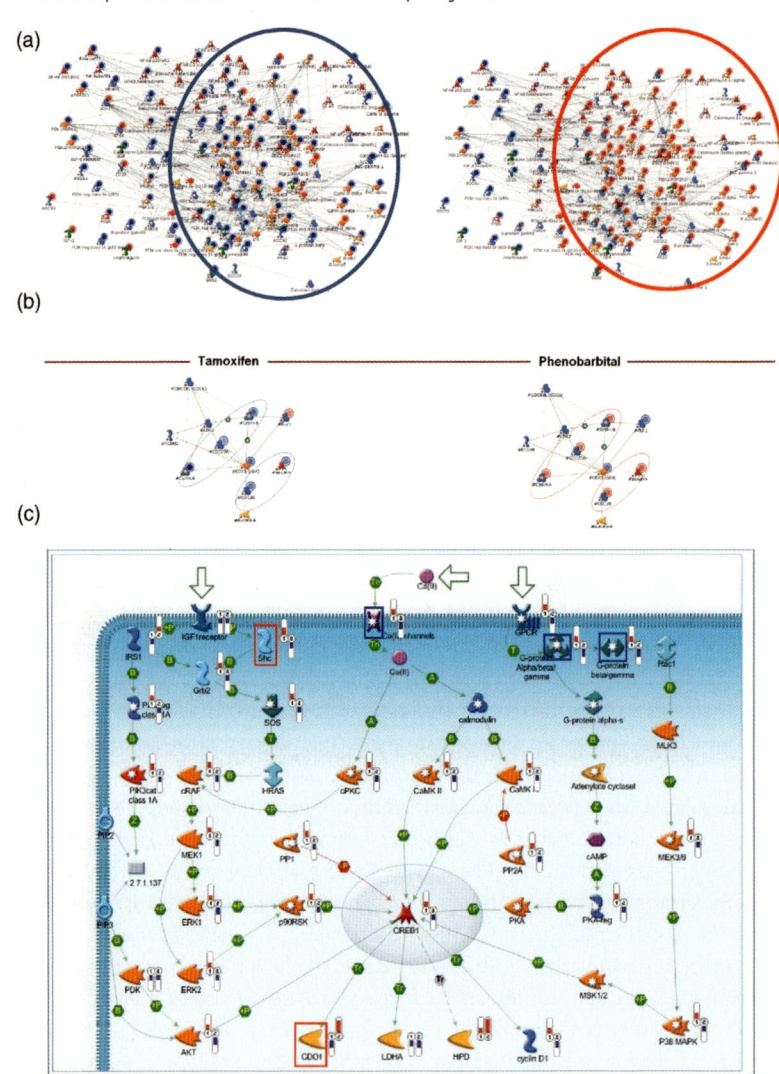

Figure 19.4 (a) Direct interactions network assembled from the genes extracted from the most differentially affected pathways between tamoxifen and phenobarbital profiles. Blue circles = downregulated genes; red circles = upregulated genes. (b) A subgraph from the Direct interactions network connecting genes related to the "mitosis" GO process. (c) The top-scored CREB pathway map for the gene content of network A. "1" = Phenobarbital; "2" = tamoxifen 250 mg/kg. Gene expression was mapped without thresholds. The square blocks mark genes in both signatures expressed at fold change above 1.4 and $p < 0.1$. Red squares = phenobarbital; blue squares = tamoxifen. Note that most genes in both sets show a high consistency of sign of expression, relative level throughput the pathway, in prethreshold levels (fold change below 1.4; $p > 0.1$).

tamoxifen files were mapped and displayed as red (upregulation based on fold change) and blue (downregulation based on fold change) histograms. No threshold was set for fold change and *p*-values. The mapping demonstrates an amazing consistency between data points within individual datasets on the pathways. Almost all phenobarbital genes are upregulated and tamoxifen genes are downregulated. The vast majority of these genes would be excluded from analysis by any combination of statistical procedures currently used for microarray analysis. For instance, a mild restrain threshold of 1.4-fold change and $p < 0.1$ leaves only five out of 32 expressed genes on this map (marked as red and blue squares on Figure 19.4c), which would essentially eliminate the most differentially affected pathways in these two treatments from the analysis. For comparison, we identified 477 "most effected" gene lists by performing conventional analysis of variance and *t*-test analysis (treated versus untreated animals) on the same two datasets. These genes were mapped onto pathway maps in MetaCore and statistically significant maps were selected (data not shown). The highest scored were assorted metabolic maps; they displayed certain generic toxic effects such as downregulation of phase I drug metabolism cytochromes, but no substantial differential expression pattern was observed (data not shown).

19.5
Single Pathway Learning for Identifying Functional Descriptor Pathways

In the next study, we tested the concept of "robustness" of pathway descriptors versus gene feature descriptors by quantitative functional evaluation of gene set descriptors produced by different statistical methods (Figure 19.1). We have also calculated canonical pathways as functional descriptors and directly compared the performance of "pathway descriptors" versus gene set descriptors using confusion matrix statistics [65]. We chose prediction of renal tubule toxicity as the endpoint phenotype and used published expression profiles from the kidneys of male Sprague-Dawley rats treated with 64 nephrotoxic or non-nephrotoxic compounds [66]. To identify differentially expressed probe (DEP) set classifiers, we applied *t*-test, fold-change, *B*-statistics, and RankProd, alone and in combination. The initial probe sets then underwent diagonal linear discriminant analysis (DLDA) modeling to predict the onset of renal tubular toxicity. RankProd generated the signature with the best overall accuracy and specificity. The classifier based on probes selected by *B*-statistics had the best sensitivity and the lowest specificity. Importantly, no probe was common for the top 50 DEPs selected by any of the four methods. This demonstrates that univariate statistics alone is not able to identify the key probes.

In the next step, we tested the proposed hypothesis of superior robustness of pathways compared to gene signatures [28] by mapping the calculated probe sets onto 350 "canonical pathway maps" (GeneGo) and then used the maps in a classifier model. First, we used Hotelling's *T*-square method to select the maps. All the probes with corresponding human EntrezGene IDs were used for the test and a *p*-value was calculated for each map. The top 50 probes generated from each of for univariant methods was performed across 350 maps, and the highest scored (i.e., most "perturbed") maps were used for modeling at the next step. A cross-validation out-of-bag (OOB) method was used to rank and select the "most significant" maps. A selection-with-replacement (SWR) was run on the complete training set (70% of the original data) and validation set (30% of data) to evaluate the prediction performance of each map. The process was repeated 10 times and the average performance was used as the classification performance for each map.

Importantly, although the DEPs identified by the different univariate methods had little overlap, they fell into the same functional categories. Enrichment analysis of the top 50 probes across *GeneGo Pathway Maps* from each method demonstrated that, at the pathway level, two canonical pathway maps – *Regulation of lipid metabolism via PPAR, RXR, and VDR* and *PPAR regulation of lipid metabolism* – were significant by each method ($p < 0.01$). Additionally, three maps – *Nucleocytoplasmic transport of CDK/cyclins, Vitamin B6 metabolism,* and *Unsaturated fatty acid biosynthesis* – were commonly identified among the top five enriched maps in the probe lists generated by at least two methods. In addition to the two commonly affected lipid pathways mentioned above, the enrichment analysis of each of the four lists of DEPs indicated other lipid homeostasis-related processes. This suggests that lipid homeostasis processes can serve as differentiators between the non-nephrotoxic treatments and renal tubular toxicants. As expected, the statistical performance of these "common maps" (accuracy, specificity, sensitivity) was somewhat lower than for the original probe sets.

In the next step, we directly applied a set of 350 maps to the entire set of DEGs, regardless of classifier gene probe set mapping using multivariant analysis (OOB modeling method). The predictive performance of each map was directly estimated using cross-validation or OOB for the training data for each individual map. Canonical pathway maps selected as functional descriptors by this method were more predictive than differentiating and were comparable in performance with gene probe set selection methods. The *Ephrin signaling* map was identified as the best functional descriptor by OOB validation and it had excellent predictive performance on the validation data. As a descriptor, the map had an overall accuracy, specificity, and sensitivity of 82.3, 83.6, and 77.8%, respectively. It was slightly better than the combined approach of *t*-test plus fold-change, although not quite as good the RankProd approach in overall accuracy. Unlike the probe sets with similar performance, the *Ephrin signaling* map made an excellent mechanistic fit with the data. Ephrin receptors and their ligands play a critical role in the development of glomerular microvasculature and in the spatial organization of tubule cells in the adult kidney medulla. The ephrin receptor EphA2 is upregulated in response to renal ischemia-reperfusion injury (IRI) – a common cause of acute kidney injury in both native kidneys and renal allografts. Ephrins, the cell-bound ligands of ephrin receptors, were also strongly expressed in the kidney following IRI. This supports the assumption that interactions between upregulated EphA2 and its ephrin ligands may provide a critical cell contact-dependent, bidirectional signal for cytoskeletal repair in renal IRI.

19.6
Multiple-Path Learning (MPL) Algorithm for Pathway Descriptors

Further development of knowledge-based functional descriptor classifier identification methods incorporates multipathway meta-classification techniques to build classification models based on sets of maps that may represent different aspects of multimechanistic pathology. Multipathway methods are expected to have better performance and biological interpretability of the final classifiers.

The general algorithm for calculation of multipathway classifiers is demonstrated at Figure 19.5. A gene expression profile can be viewed as an $M \times N$ matrix (probes × samples), in which each "cell" x_{ij} is the expression of probe i ($i = 1, \ldots, M$) for sample j ($j = 1, \ldots, N$). Meanwhile, we have K pathways or gene sets annotated according to their biological significance. Each pathway k ($k = 1, \ldots, K$) contains probes (L_k) that overlap between pathway k and the M probes in the array. We split the $M \times N$ matrix into K's submatrices in such a way that each kth submatrix contains exactly the same N samples, but L_k probes, compared with M probes in the original one. In contrast to a traditional classification method that builds a one-layer model on the whole $M \times N$ data matrix or its one subset (probes reduced to one and only one set), in MPL we build a model consisting of two layers, termed as base learning and meta learning layers.

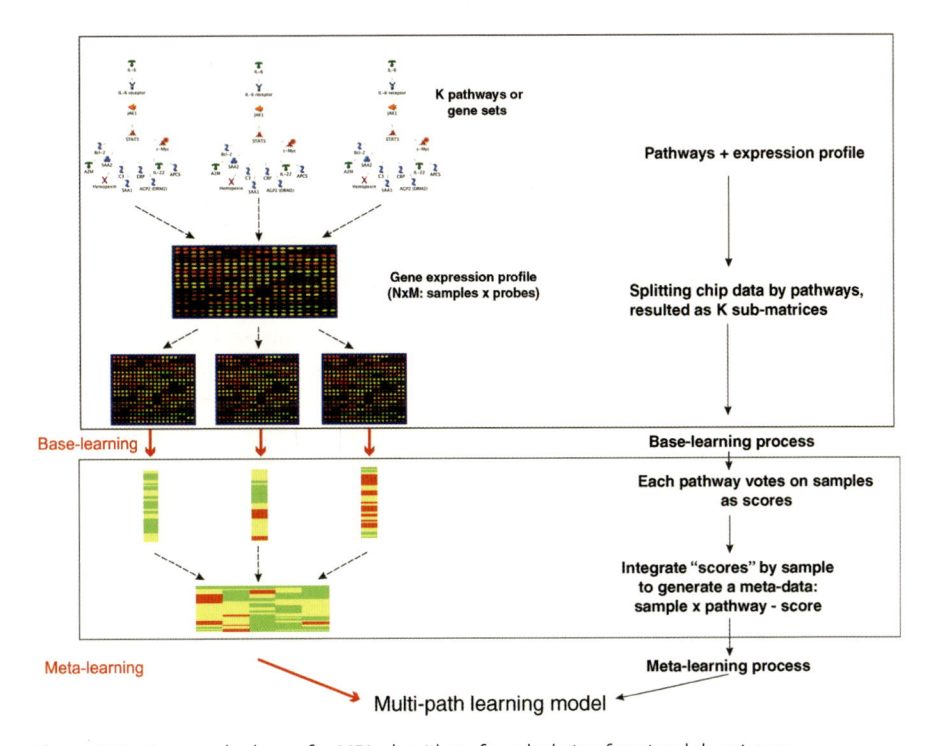

Figure 19.5 A general schema for MPL algorithms for calculating functional descriptors. Comments in text.

For base learning, we have K's $L_k \times N$ submatrices and build a model for each submatrix. That means we end with K models for K pathways. Since the kth model corresponds to the kth pathway's prediction on the same N samples, we can aggregate their predictions by sample and produce another heatmap-like matrix (meta-matrix) as $K \times N$. Compared with the previous gene expression profile $M \times N$, x_{ij} is the expression of a probe i for sample j; in our $K \times N$ meta-matrix, each cell y_{kj} ($k = 1, \ldots, K$, while $j = 1, \ldots, N$) represents the prediction from the kth pathway on the jth sample. Therefore, we term this meta-matrix as a pathway matrix, in which the cell y_{kj} represents the "opinion" or the probability scores that the kth pathway votes on the jth sample. The score, y_{kj}, is named as "pathway activity" since it represents the kth pathway's action for sample j. On such a "pathway activity" matrix ($K \times N$), we then apply another round of the regular modeling process. Since this learning process runs on such meta-data, we name it as a meta-learning process and arrive at a meta-learning model. Such two-layer learning process is done on both training data and test data. For the training data, predictive score generated by cross-validation is used while for the test data, predictive score from training data on validation data is used. We use this model as the final one to classify new samples as well as to identify significant predictors (in this case, multiple pathways), expecting a biological explanation at the level of a set of pathways instead of a set of genes. Since the predictors used here are multiple pathways instead of genes, we named the whole algorithm as the multiple-path learning (MPL) algorithm.

We used the Random Forests statistical algorithm for classification/clustering and nonlinear multiple regression as a learning algorithm. Its useful features in our study are: (i) it uses OOB instead of cross-validation to calculate the training error to estimate the predictive error for new data, (ii) it generates two confusion matrices, one for training data and the other for test data, to report the performance and result in practice, and (iii) it has multiple criteria to rank the importance of features, helping in improvement of model performance and explanation of model and phenomenon. We applied Random Forests in both base-learning and meta-learning steps for its reported high accuracy and its feature-rank measures, called "mean decrease accuracy." Briefly, for each tree classifier, the prediction accuracy on the OOB portion of the data is recorded. Then the same is done after permuting each predictor variable. The difference between the two accuracies is then averaged over all trees and is normalized by their standard deviation as "mean decrease accuracy."

19.7
Applications of MPL-Deduced Pathway Descriptors

19.7.1
Cross-Tissue Prediction of Compound Toxicity Using Functional Descriptors

Recently, we applied an MPL algorithm for identification of genomic biomarkers of drug-induced liver injury from blood and compared the predictive performance of pathway classifiers versus gene content classifiers as part of the MAQC-II project [67].

The blood and liver gene expression data of Lobenhofer et al. [68] was used to identify genes and biological processes in the blood that are predictive of liver necrosis – a particular form of drug-induced liver injury. This work is one of the most comprehensive studies done in this field so far, as expression data of eight compounds causing liver necrosis were used as a training set to formulate predictive models and data from three toxicant exposures (acetaminophen, carbon tetrachloride, and allyl alcohol) were applied as a test to independently validate the derived models. To evaluate cross-tissue predictability, the models generated from blood gene expression training set data were used to make predictions based on the validation set of liver gene expression data and blood gene expression data. Cross-tissue prediction accuracy was relatively high for both gene-based and pathway-based classifiers. Interestingly, however, transferred gene-based classifiers from the blood to the liver test dataset samples gave much better predictions (81.7–88.8%) than accuracy of classifiers within the same tissue (both training and validation sets deriving either from blood or from liver).

The pathway-centered models (GeneGo ontology of canonical pathways) showed that phosphatidylinositol-3,4,5-trisphosphate signaling in B lymphocytes, Toll-like receptor (TLR) signaling pathways leading to a cell proinflammatory response, and regulation of apoptosis by mitochondrial proteins had the best performance of models derived from blood data. Two of the top three pathways identified as highly predictive in the blood also ranked high in cross-tissue prediction, along with the anti-apoptotic tumor necrosis factor/NF-κB/Bcl-2 pathway. Necrosis prediction between blood and liver ranged from 83.6 to 89.3%. These mechanisms are in agreement with the current understanding of drug-related hepatotoxicity. For example, angiogenesis activation may be linked to the formation of new blood vessels during the regeneration of liver to compensate for the loss of hepatocytes. One of the top ranking pathway-based classifiers, the TLR signaling pathway, was also known as a pathway associated with acetaminophen-mediated hepatotoxicity in endotoxin-responsive mice. These results indicate that genomic indicators in blood can serve as biomarkers of drug-induced liver necrosis and possibly be further developed for diagnostic testing of liver damage in humans based on minimally invasive biomaterial sources.

19.7.2
Cross-Platform Reproducibility Using Pathways (MPL Algorithms)

In another MAQC-II study, we cross-tested pathway-based and gene content classifiers derived from the studies on rat liver data on two different microarray platforms, Affymetrix and Agilent [69]. Briefly, eight different hepatotoxicants (1,2- dichlorobenzene, 1,4-dichlorobenzene, bromobenzene, diquat dibromide, galactosamine, monocrotaline, N-nitrosomorpholine, and thioacetamide) were selected based on published literature regarding the differences that exist in the cell types and liver regions that are injured in response to exposure. For each chemical, doses that elicited a subtoxic ("low"), a moderately toxic ("medium"), or a overtly toxic ("high") response 24 h after treatment were selected. Samples were collected for gene expression profiling, clinical chemistry, hematology, and histopathology at 6, 24, and 48 h postexposure. For each compound, four animals were used for each dose (including a vehicle control) and time point group.

Cumulatively, the results suggest that if the signature genes of a classifier were generated using one platform, similar accuracy could be obtained by generating classifiers using the same signature genes within a different microarray platform. Additionally, it was found that the same samples were consistently misclassified in both platforms, which was likely due to the biological outliers as opposed to a shortcoming in the classification approach.

Overall, cross-platform reproducibility of microarray data was greater when the data were examined by pathway classifiers compare to transcript-level data. Several pathways showed higher T-index scores than any of the produced gene signatures.

19.7.3
Predictive Classifiers for Breast Cancer Endpoints

In yet another MAQC-II project, gene-based and pathway-based classifiers were applied for predicting three different endpoints in breast cancer using gene expression data from 230 stage I–III breast cancers [70]. The phenotypes included ER expression (ER status), pathologic complete eradication of cancer after administration of 6 months of preoperative chemotherapy (pCR), and extreme chemotherapy sensitivity (i.e., pCR) among the ER-negative cancers. The three endpoints represented classification problems of varying difficulty, with the latter one the most difficult. The goal of this analysis was to assess how the degree of classification difficulty may affect the accuracy of prediction models and if particular methods perform better on more difficult problems. We applied five different feature selection methods. In total, eight different classifiers were produced, representing a broad range of algorithms including linear discriminant analysis (LDA), DLDA, quadratic discriminant analysis (QDA), logistic regression (LREG), and two versions of support vector machines and K-nearest neighbor methods. In all, 40 different classification models (i.e., gene signature predictors) were developed for each of the three classification problems (five different feature selection × eight different algorithms).

Overall, it was shown that several different feature sets yielded poorly overlapping gene signature predictors with statistically similar performance – a well-documented phenomenon [24, 71] that can be explained by the hypothesis that different probe sets represent different genes from the same complex molecular pathways [28]. Similar to the approach described above [65], we tested this hypothesis by mapping each of the 15 feature sets used in the final validation models to canonical pathways (MetaCore). The different feature sets selected for a particular prediction endpoint had a high level of congruency at the pathway level across all the five different ranking methods. The selected gene sets and pathways were also rather similar to each other for the ER and pCR prediction endpoints. However, the pathways predictive of pCR in ER-negative cancers were very different from the other two informative gene sets. Thirty-six pathways contributed to the prediction of ER and pCR endpoints each, and 17 pathways were included in the pCR prediction for ER-negative cancers in all five-feature selection methods. For the ER endpoint, development, cell adhesion, cytoskeleton remodeling, DNA damage, apoptosis, and ER transcription factor activity were the most significant pathway elements common to all informative feature sets. Most pathways that were involved in pCR prediction (31 out of 36) were the same as those involved in ER status prediction, which is consistent with the known association between pCR rate and ER status. ER-negative cancers have a significantly higher pCR rate than ER-positive cancers. The pathways that were informative to predict pCR in ER-negative cancers were distinct from the pathways that were predictive of pCR (and ER status) in unselected patients and included immune response-related pathways (interleukin-2 and prostaglandin E_2 pathways, T-helper cell activation), opioid receptor signaling, and endothelial cell-related pathways (G-protein-coupled receptors EGD-3 and -5, endothelin receptor B signaling).

19.7.4
Using Pathways as the Measure of Congruency Between Datasets

A different application of pathway descriptors was demonstrated in a comprehensive study on functional analysis of statistically generated predictive gene signatures produced by MAQC-II [72]. Thirty-four data analysis teams generated a set of 262 predictive gene signatures for 13 endpoints on six independent cancer and compound-induced pathology datasets, using over 15 000 statistical models. Signatures of similar performance for the same endpoint varied greatly in gene content. We compared them by similarity (or congruency) using a shared gene content and a pattern of p-value distribution of features of five functional ontologies (disease biomarkers, canonical pathways, GO and GeneGo cellular processes) as the comparative metrics. The κ statistic was used as a statistical measure of inter-rater agreement. The input for κ involves a couple of raters or learners, which classify a set of objects (signature gene content in this case) into categories (gene content of entities in functional ontologies). In order to calculate agreement for gene content, we considered each signature as a learner or rater and each object (probe) in the union of signature probes was categorized by each list of alterations as 1 (selected) or 0 (unselected). Using such a 0/1 matrix (object × learner) as input, we implemented the Cohen κ function {concord} package in [73] to arrive at κs, z-scores and p-values for the congruency. Instead of using Cohen's κ [74] we used Siegel and Castellan's κ, by assuming pooled classification proportions and an adjustment for bias, where the different methods systematically differ in their categorization [75].

Importantly, congruency at the functional level was always higher than that at the gene level for all endpoints. Different statistical methods therefore selected gene signatures with different, sometimes not overlapping, sets of genes. However, genes across different signatures often participated in the same biological processes. Higher congruency at a functional level supports the hypothesis that common underlying biological mechanisms are identified by the different classification algorithms for each endpoint and that the gene signatures do, in fact, comprise genes with biological relevance to the endpoint predicted. We also observed that different feature selection models have a systematic (although unexplainable) bias towards biologically important features of selected genes, such as the relative number of outgoing versus incoming interactions. In other words, some models preferably select "regulator" genes (such as transcription factors and ligands), whereas other models select "effector" genes (encoding metabolic enzymes, proteases, or structural proteins). This result helps to explain the poor biological relevance of individual gene signatures, and confirms the hypothesis that it is, indeed, pathways and processes, not individual genes that determine a particular clinical outcome.

Signature congruency is also in agreement with the observation of "synergy" between signature performances in ontology enrichment analysis. We have shown that multiple gene signatures for a given endpoint make more sense biologically when taken together than when analyzed separately. Lower p-values in functional ontology enrichments were particularly pronounced for the nonredundant gene unions of all signatures for a given endpoint, compared to that of individual signatures. This synergy additionally supports the observation that different statistical models and machine learning algorithms select different subsets of genes from the same pathways and processes.

A strong correlation between signature congruency at both the gene level (gene list stability) and at the functional level was observed with model performance on both training and validation sets. In general, the more congruent the signatures, the better the average prediction performance for an endpoint. Gene signatures with better performance appear to be biologically related and are more reflective of the phenotype. The congruency of different gene signatures for a given endpoint may therefore also be a useful indication of the amenability of the endpoint to accurate prediction using the signature approach.

19.8
Combining Advantages of Pathways and Networks

As demonstrated above, biological pathways and networks have distinct advantages and drawbacks as methods of functional analysis. Specifically, the pathways are experimentally proven as a whole set of interactions, but static. The networks are dynamic, flexible, and unique for the set of "seed" objects, but not validated experimentally. Naturally, we sought to combine the power of the two methods in "combination" tools by developing different "pathway-based" network algorithms, very similarly to the algorithms used in modern GPS systems to select the "shortest time" or "shortest distance" paths among many hundreds of possible driving combinations (Figure 19.6a).

In our quest, we relied on our manually curated "knowledge base" with over 1 000 000 experimentally proven human binary protein interactions and some 150 000 multistep pathways [9]. A typical canonical signaling pathway is linear and, on average, 17 interaction steps long. It starts with a ligand–receptor step, proceeds with

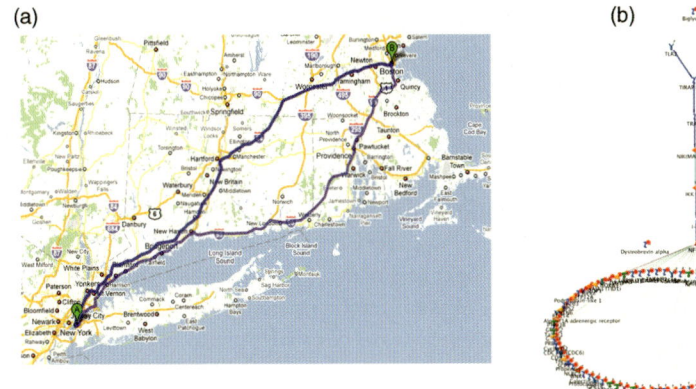

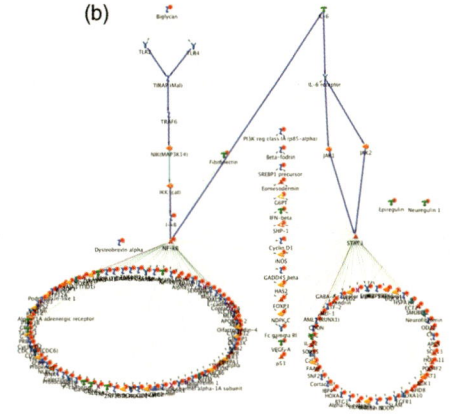

Figure 19.6 Pathway-based network generation algorithm. (a) An analogy with GPS driving directions, which choose "shortest path" or "shortest time" (highlighted) algorithms out of many possible driving combinations. (b) A biological network built for a set of genes genetically altered either in breast or colon cancer by the AN(TF) algorithm (MetaCore). The fragments of canonical signaling pathways are highlighted in blue.

series of signal transduction interactions, and ends up at the transcription factor, which binds to the promoters of target genes.

Biological pathways, annotated at GeneGo, are roughly divided into three types.

- **Normal signaling and metabolic pathways.** which carry out biological functionality in human cells.
- **Pathological pathways.** characteristic for human diseases and toxic categories. Such pathways consist of normal protein interactions, as well as altered "gain" and "loss of function" interactions formed by mutant proteins, such as the truncated chloride transporter CFTR in cystic fibrosis, huntingtin protein elongated by a polyglutamine track in Huntington's disease, and the BCR–ABL fusion kinase in many leukemias.
- **Therapeutic mechanism of action pathways.** Very few drugs directly affect mutant proteins causing a disease, such as Gleevec targeting BCR–ABL kinase or CFTR Δ508 inhibitors. In most cases, drug action is indirect and pleiotropic, as every small molecule hits multiple protein targets involved in complex interconnected networks. Drug action pathways are often the most difficult to reveal, as was recently shown for the 40-year-old "morning sickness" drug thalidomide [76].

It was essential that we had the same set of objects (proteins, protein complexes and groups, compounds, RNA species, etc.) and interactions for both network generation and creation of pathways. The task would be dramatically more difficult using publically available sources of pathways (KEGG, BioCarta) and protein interactions (Human Protein Reference Database, Bind) due to semantic inconsistency between the objects and different definitions of interactions between these sources. It was also very important to have rich meta-data annotated for proteins (over 30 different types) and interactions (direction, effect, 20 mechanisms), as it allowed us to focus pathway-based algorithms on particular key functions, such as transcription factors ("modulators"), receptors, or TR targets ("effectors").

We realized the goal of building "pathway-centered" networks in a series of network algorithms called Analyze Network (AN) (Figure 19.6b). In brief, this algorithm starts by building a supernetwork (which is never visualized), connecting all objects from the input list (seed objects) with all other objects. In the next step this large network is "divided" into smaller fragments of chosen size, from two to 100 nodes. This is done in a cyclical manner (i.e., fragments are created sequentially one by one). Edges used in a fragment are never reused in subsequent fragments. Nodes may be reused, but with different edges leading to them in different fragments. The end result of the AN algorithm is a list of overlapping multiple networks (usually around 30), which can be prioritized based on five parameters: the number of nodes from the input list among all nodes on the network, the number of canonical pathways on the network, and three statistical parameters: p-value, z-score and g-score. Importantly, canonical pathways are considered as "seeds." The degree of the canonical pathway is considered as the number of objects in the pathway ("inner degree"), but not the total number of adjacent edges of pathway participants ("outer degree"). This is achieved by representing a canonical pathway as a dummy node connected to each of the objects participating in the pathway; these canonical pathway nodes are further treated as ordinary nodes. Thus, a pathway with eight nodes is treated as a node with degree of 8 and, therefore, is likely to be selected as a "seed" for the new subnetwork. If the canonical pathway is partly used in earlier iterations, it cannot be used as a "seed" (it is disrupted, as not all of its edges are contained in a subnetwork).

Prioritization within the list of AN networks can be based on different parameters, but follows the same procedure, which we will describe next. A dataset of interest (e.g., the list of all prefiltered nodes) is divided into two random subsets, which overlap in this general case. The size of the intersection between the two sets represents a random variable within the hypergeometric distribution. We apply this fact for numerical scoring and prioritization of the previously discussed node-centered small shortest path networks. Let us consider a general set size of N with R marked objects/events (e.g., the nodes with expression data). The probability of a random subset of size of n, which includes r marked events/objects, is described by the distribution:

$$P(r, n, R, N) = \frac{C_R^r \cdot C_{N-R}^{n-r}}{C_N^n} = \frac{C_n^r \cdot C_{N-n}^{R-r}}{C_N^R}$$

$$= \frac{R! \cdot (N-R)!}{N!} \cdot \frac{n! \cdot (N-n)!}{r! \cdot (R-r)!} \cdot \frac{1}{(n-r)! \cdot (N-R-n+r)!}$$

The mean of this distribution is equal to:

$$\mu = \sum_{r=0}^{n} r \cdot P(r, n, R, N) = \frac{n \cdot R}{N} = n \cdot q$$

where $q = R/N$ defines the ratio of marked objects.

The dispersion of this distribution is described as:

$$\sigma^2 = \sum_{r=0}^{n} r^2 \cdot P(r, n, R, N) - \mu^2 = \frac{n \cdot R \cdot (N-n) \cdot (N-R)}{N^2 \cdot (N-1)}$$

$$= n \cdot q \cdot (1-q) \cdot \left(1 - \frac{n-1}{N-1}\right)$$

It is essential that these equations are invariant in terms of exchange of n for R. This means that the subset and the marked sets are equivalent and symmetrical. Importantly, in the cases of $r > n$, $r > R$ or $r < R + n - N$, $P(r, n, R, N) = 0$.

We will use the following z-scoring for comparison and prioritization of node-specific SP subnetworks:

$$\text{z-score} = \frac{r - n\frac{R}{N}}{\sqrt{n\left(\frac{R}{N}\right)\left(1 - \frac{R}{N}\right)\left(1 - \frac{n-1}{N-1}\right)}} = \frac{r - \mu}{\sigma}$$

where N is the total number of nodes after filtration, R is the number of nodes in the input list or the nodes associated with experimental data, n is the number of the nodes in the network, r is the number of the network's nodes associated with experimental data or included in the input list, and μ and σ are, respectively, the mean and dispersion of the hypergeometric distribution described above.

The AN networks can also be skewed towards two key protein functions in a pathway: transcription factors and receptors. Both algorithms start by creating two lists of objects expanded from the initial list: the list of transcription factors and the list of receptors. Next, the algorithm calculates the shortest paths from the receptors to transcription factors. Then, the shortest paths are prioritized in a similar way. The first algorithm, AN (TFs), connects every transcription factor with the closest receptor by all shortest paths and delivers one specific network per transcription factor in the list. Similarly, the second algorithm, AN(R), delivers a network consisting of all the shortest paths from a receptor in the list to the closest transcription factor; one network per receptor. Since all the edges, and therefore paths, are directional, the resulting networks are not reciprocal.

Every network built by AN algorithms may be optionally enriched with the receptor's ligands and the transcription factor's targets. The networks may be grouped and merged within every group. Namely, if we are building one network for every transcription factor, then all such networks with the same receptors are grouped and merged within each group.

19.9
Key Upstream and Downstream Interactions of Genetically Altered Genes and "Universal Cancer Genes"

In the study discussed above on functional analysis of 260 predictive gene signatures produced in the MAQC-II project, similarity in gene content between signatures for each endpoint was low, which made the signatures virtually incomparable [72]. Cohen's κ-based "pathway congruency" was one method of comparing and classifying the signatures. Our knowledge of network connectivity enabled another approach consisting of calculation of common of sets of genes commonly connected with different signatures via protein interactions. It appeared that signatures for all 13 endpoints featured a disproportionately large fraction of nonoverlapping targets of common transcription factors. Most signature genes were regulated by very few upstream transcription factors in a highly endpoint-specific manner. For instance, nine out of 24 signatures for the chemically induced carcinogenesis endpoint were regulated by NRF2 – a key transcription factor in controlling cellular oxidative stress response. Seventeen of 24 signatures for the breast cancer endpoint included direct targets of HNF3α (epithelial transcription) and 16 of 24 included direct targets of ER1. Downstream signaling from signature genes was also highly channelized. Although very diverse, signature genes regulated a limited number of downstream genes and reactions. Sixteen out of 24 signatures for the chemically induced carcinogenesis endpoint contained drug-metabolizing enzymes; 20 out of 24 signatures for the breast cancer endpoint contained genes regulating tyrosine 3-monooxygenase TY3H; 19 contained genes regulating CGα, and 11 contained genes regulating interleukins. Overall, each endpoint was characterized by a set of upstream "regulators" (triggers and modulators) or downstream "effectors" (Figure 19.7), which are commonly linked to the poorly overlapping sets of DEGs selected as classifiers. The common "regulator" and "effector" genes are not

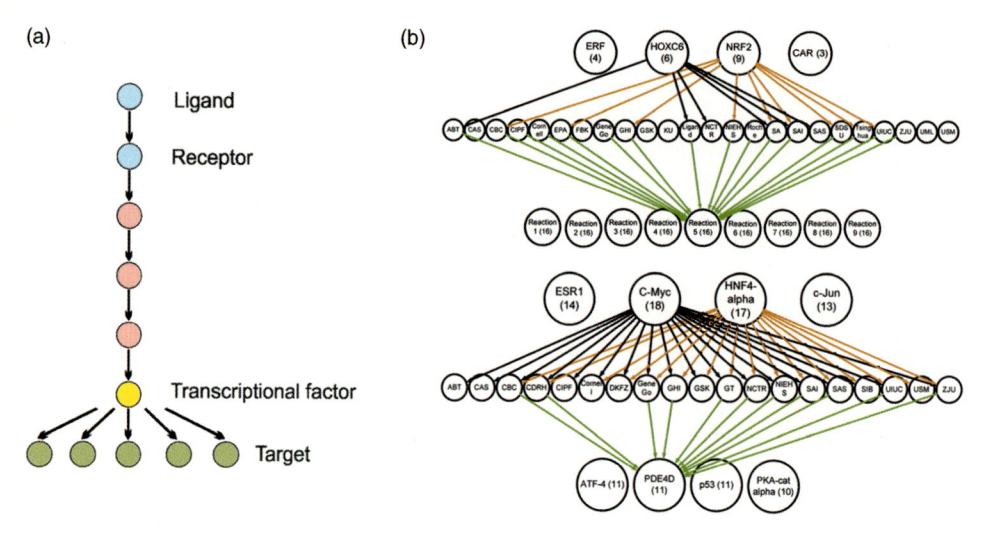

Figure 19.7 Networking poorly overlapping datasets and gene lists by protein interactions. (a) A general schema of a multi-interaction pathway. Proteins of different functions play different roles within the pathways and networks. (b) Gene signatures for toxic drug response (1) and breast cancer (2) endpoints. The gene signatures generated by different methods poorly overlap in gene content but have a limited number of common modulators (upstream transcription factors) and effectors (downstream target genes). From [65].

necessarily overexpressed. Importantly, the lists of common "regulators" are unique for each endpoint, which makes them eligible as potential multivariant classifiers, which are probably more robust than gene-based signatures.

We analyzed the biological sensitivity of subsets of "regulators" and "effectors" in more detail in a study on genetic alterations (somatic mutations, deletions, amplifications) in four human cancers – breast, colon, pancreatic cancer, and glioblastoma – previously revealed by global exon sequencing and SNP array analyses in multiple patients. The 12 gene lists (three alteration types in four cancers) represented pooled data on all genomic alterations detected in primary tumors of 11 patients in both breast and colon cancers, 22 patients in glioblastoma multiforme, and 24 in pancreatic cancer [40, 41, 77]. The gene content of the datasets had very limited overlap both cross-cancer and cross-data type. Thus, only one gene (predictably TP53) was mutated in all four cancer types and not a single gene was commonly amplified or deleted. To check whether these datasets are functionally connected on the level of upstream and downstream objects, we identified the sets of one-step upstream and downstream objects (proteins and complexes) for all 12 alteration datasets and seven nonredundant unions using MetaCore's collection of 300 000 manually curated directional protein–protein interactions. Despite the lack of overlap between altered genes in the sets, we identified 65 common upstream and downstream genes, which we called "universal cancer genes." Sixteen genes were common for two sets, which is expected for highly connected proteins involved in overlapping signaling pathways. Essentially all universal genes formed a tight network module, interconnected by one-step physical protein interactions. The network mostly covers two series of pathways constituting key "decision" points in carcinogenesis: regulation of proliferation and apoptosis, and regulation of angiogenesis. The set of "universal" cancer genes most densely populated key signaling pathways involved in carcinogenesis: BRCA1-regulated transcription in DNA damage, PTEN pathway, transforming growth factor and WNT pathway in cytoskeleton remodeling, G_1/S cell cycle transition, p53 signaling, AKT signaling, and many others.

19.10
Conclusions

Quantitative pathway analysis is a very young field. Using pathways and networks as prognostic and predictive biomarkers seems to be a reasonable future in personalized and translational medicine, and essential for GWAS and upcoming sequence-based studies. It is logical and well-expected that "functional markers" should be more robust and specific than gene content (Figure 19.1). However, to the best of our knowledge no direct comparison between "gene content" and "functional" classifiers of sufficient size has been conducted so far in omics biology. The most comprehensive study done up to date, the MAQC-II project, was still limited in number of samples for all 13 endpoints and the comparative results (roughly similar performance between functional and gene content classifiers) are not conclusive. However, with "pathway biology" quickly becoming a mainstream technology, at least in human genetics, we are very positive about its future in personalized medicine and drug discovery.

A major advantage of using pathways as prognostic and "companion" classifiers is that they provide a meaningful mechanistic insight into the biology of an "endpoint." All three types of pathways (normal, pathological, and drug mechanism) are implicated in complex human phenotypes such as disease or drug response, as we have seen in our 3-year-old MetaMiner program on manual reconstruction of cancers, and central nervous system, metabolic, and respiratory diseases (e.g., cystic fibrosis in [78]). In any disease, the majority of physiological changes are carried out by perturbed normal pathways that are engaged or disengaged in an affected tissue, such as by overexpression of key regulator genes. In addition, a number of pathological pathways are typically involved, often the same pathways in diseases of very different etiology (for instance, tissue necrosis, fibrosis, induced inflammation). The third type, drug mechanism of action pathways, is also relevant for multiple diseases, and provides the basis for drug repositioning and polypharmacology. Each phenotypical endpoint can be characterized by a "barcode" representing a distribution of normal, pathological, and drug mechanism pathways. The distribution pattern itself can be considered as a "classifier" that can be used by quantitative models in multiple diagnostic and therapeutic applications.

References

1 Kanehisa, M. and Goto, S. (2000) KEGG: Kyoto Encyclopedia of Genes and Genomes. *Nucleic Acids Res.*, **28**, 27–30.

2 Ashburner, M., Ball, C.A., Blake, J.A., Botstein, D., Butler, H., Cherry, J.M. et al. (2000) Gene Ontology: tool for the unification of biology. The Gene Ontology Consortium. *Nat. Genet.*, **25**, 25–29.

3 Chang, H.Y., Nuyten, D.S., Sneddon, J.B., Hastie, T., Tibshirani, R., Sørlie, T. et al. (2005) Robustness, scalability, and integration of a wound-response gene expression signature in predicting breast cancer survival. *Proc. Natl. Acad. Sci. USA*, **102**, 3738–3743.

4 van't Veer, L.J., Dai, H., van de Vijver, M.J., He, Y.D., Hart, A.A., Mao, M. et al. (2002) Gene expression profiling predicts clinical outcome of breast cancer. *Nature*, **415**, 530–536.

5 Troester, M.A., Hoadley, K.A., Parker, J.S., and Perou, C.M. (2004) Prediction of toxicant-specific gene expression signatures after chemotherapeutic treatment of breast cell lines. *Environ. Health Perspect.*, **112**, 1607–1613.

6 Kitano, H. (2000) Computational systems biology. *Nature*, **420**, 206–210.

7 Hood, L. and Perlmutter, R.M. (2004) The impact of systems approaches on biological problems in drug discovery. *Nat. Biotechnol.*, **2**, 1218–1219.

8 Barabasi, A.L. and Oltavi, Z.N. (2004) Network biology: understanding the cell's functional organization. *Nat. Genet.*, **5**, 101–113.

9 Nikolsky, Y., Kirillov, E., Zuev, R., Rakhmatulin, E., and Nikolskaya, T. (2009) Functional analysis of OMICs data and small molecule compounds in an integrated "knowledge-based" platform. *Methods Mol. Biol.*, **563**, 177–196.

10 Scheiber, J., Chen, B., Milik, M., Sukuru, S.C., Bender, A., Mikhailov, D. et al. (2009) Gaining insight into off-target mediated effects of drug candidates with a comprehensive systems chemical biology analysis. *J. Chem. Inform. Model.*, **49**, 308–317.

11 Capell, K. (2009) Novartis: radically remaking its drug business. *Business Week*, June 22.

12 Boguski, M.S., Arnaout, R., and Hill, C. (2009) Customized care 2020: how medical sequencing and network biology will enable personalized medicine. *F1000 Biol. Rep.*, **1**, 73.

13 Boguski, M.S., Mandl, K.D., and Sukhatme, V.P. (2009) Repurposing with a difference. *Science*, **324**, 1394–1395.

14 Baranzini, S.E., Galwey, N.W., Wang, J., Khankhanian, P., Lindberg, R., Pelletier, D. et al. (2009) Pathway and network-based analysis of genome-wide association studies in multiple sclerosis. *Hum. Mol. Genet.*, **18**, 2078–2090.

15 Ideker, T. and Sharan, R. (2008) Protein networks in disease. *Genome Res.*, **18**, 644–652.

16 Pedroso, I. (2010) Gaining a pathway insight into genetic association data. *Methods Mol. Biol.*, **628**, 373–382.

17 Perou, C.M., Sørlie, T., Eisen, M.B., van de Rijn, M., Jeffrey, S.S., Rees, C.A. et al. (2000) Molecular portraits of human breast tumours. *Nature*, **406**, 747–752.

18 Sørlie, T., Perou, C.M., Tibshirani, R., Aas, T., Geisler, S., Johnsen, H. et al. (2001) Gene expression patterns of breast carcinomas distinguish tumor subclasses with clinical implications. *Proc. Natl. Acad. Sci. USA*, **98**, 10869–10874.

19 Natsoulis, G., El Ghaoui, L., Lanckriet, G.R., Tolley, A.M., Leroy, F., Dunlea, S. et al. (2005) Classification of a large microarray data set: algorithm comparison and analysis of drug signatures. *Genome Res.*, **15**, 724–736.

20 Huang, Y., Penchala, S., Pham, A.N., and Wang, J. (2008) Genetic variations and gene expression of transporters in drug disposition and response. *Expert Opin. Drug Metab. Toxicol.*, **4**, 237–254.

21 Bonnefoi, H., Potti, A., Delorenzi, M., Mauriac, L., Campone, M., Tubiana-Hulin, M. et al. (2007) Validation of gene signatures that predict the response of breast cancer to neoadjuvant chemotherapy: a substudy of the EORTC 10994/BIG 00-01 clinical trial. *Lancet Oncol.*, **8**, 1071–1078.

22 Eun, J.W., Ryu, S.Y., Noh, J.H., Lee, M.J., Jang, J.J., Ryu, J.C. et al. (2008) Discriminating the molecular basis of hepatotoxicity using the large-scale characteristic molecular signatures of toxicants by expression profiling analysis. *Toxicology*, **249**, 176–183.

23 Buyse, M., Loi, S., van't Veer, L., Viale, G., Delorenzi, M., Glas, A.M. et al. (2006) Validation and clinical utility of a 70-gene prognostic signature for women with node-negative breast cancer. *J. Natl. Cancer Inst.*, **98**, 1183–1192.

24 Buyse, M., Sargent, D.J., Grothey, A., Matheson, A., and de Gramont, A. (2010) Biomarkers and surrogate end points–the challenge of statistical validation. *Nat. Rev. Clin. Oncol.*, **7**, 309–311.

25 Ein-Dor, L., Kela, I., Getz, G., Givol, D., and Domany, E. (2005) Outcome signature genes in breast cancer: is there a unique set? *Bioinformatics*, **21**, 171–178.

26 Ein-Dor, L., Zuk, O., and Domany, E. (2006) Thousands of samples are needed to generate a robust gene list for predicting outcome in cancer. *Proc. Natl. Acad. Sci. USA*, **103**, 5923–5928.

27 Natsoulis, G., Pearson, C.I., Gollub, J.P., Eynon, B., Ferng, J., Nair, R. et al. (2008) The liver pharmacological and xenobiotic gene response repertoire. *Mol. Syst. Biol.*, **4**, 175.

28 Wirapati, P., Sotiriou, C., Kunkel, S., Farmer, P., Pradervand, S., Haibe-Kains, B., Desmedt, C. et al. (2008) Meta-analysis of gene expression profiles in breast cancer: toward a unified understanding of breast cancer subtyping and prognosis signatures. *Breast Cancer Res.*, **10** (4), R65.

29 Hartwell, L.H., Hopfield, J.J., and Leibler, S. (1999) From molecular to modular cell biology. *Nature*, **402** (Suppl.), C47–C50.

30 The MAQC-II Consortium (2010) The MicroArray Quality Control (MAQC)-II Project: a comprehensive study of common practices for the development and validation of microarray-based predictive models. *Nat. Biotechnol.*, **28**, 827–838.

31 Stadler, Z.K., Thom, P., Robson, M.E., Weitzel, J.N., Kauff, N.D., Hurley, K.E. et al. (2010) Genome-wide association studies of cancer. *J. Clin. Oncol.*, **28**, 4255–4267.

32 Seshadri, S., Fitzpatrick, A.L., Ikram, M.A., DeStefano, A., Gudnason, V., Boada, M. *et al.* (2010) Genome-wide analysis of genetic loci associated with Alzheimer disease. *J. Am. Med. Ass.*, **303**, 1832–1840.

33 Cordell, H.J. (2009) Detecting gene–gene interactions that underlie human diseases. *Nat. Rev. Genet.*, **10**, 392–404.

34 Nikolsky, Y. and Bryant, J. (2009) *Protein Networks and Pathway Analysis. Methods in Molecular Biology*, Human Press, Clifton, NJ.

35 Subramanian F A., Tamayo, P., Mootha, V.K., Mukherjee, S., Ebert, B.L., Gillette, M.A. *et al.* (2005) Gene set enrichment analysis: a knowledge-based approach for interpreting genome-wide expression profiles. *Proc. Natl. Acad. Sci. USA*, **102**, 15545–15550.

36 Cusick, M.E., Klitgord, N., Vidal, M., and Hill, D.E. (2005) Interactome: gateway into systems biology. *Hum. Mol. Genet.*, **2**, R171–R181.

37 Vidal, M. (2005) Interactome modeling. *FEBS Lett.*, **579**, 1834–1838.

38 Nikolsky, Y., Ekins, S., Nikolskaya, T., and Bugrim, A. (2005) A novel method for generation of signature networks as biomarkers from complex high-throughput data. *Toxicol. Lett.*, **158**, 20–29.

39 Dezso, Z., Welch, R., Kazandaev, V., Naito, A., Fuscoe, J., Melvin, C. *et al.* (2006) Elucidation of differential response networks from toxicogenomics data. Systems Biology: Integrating Biology, Technology and Computation Conference, Taos, NM.

40 Wood, L.D., Parsons, D.W., Jones, S., Lin, J., Sjöblom, T., Leary, R.J. *et al.* (2007) The genomic landscapes of human breast and colorectal cancers. *Science*, **318**, 1108–1113.

41 Jones, S., Zhang, X., Parsons, D.W., Lin, J.C., Leary, R.J., Angenendt, P. *et al.* (2008) Core signaling pathways in human pancreatic cancers revealed by global genomic analyses. *Science*, **321**, 1801–1806.

42 Parsons, D.W., Jones, S., Zhang, X., Lin, J.C., Leary, R.J., Angenendt, P. *et al.* (2008) An integrated genomic analysis of human glioblastoma multiforme. *Science*, **321**, 1807–1812.

43 Goh, K.I., Cusick, M.E., Valle, D., Childs, B., Vidal, M., and Barabási, A.L. (2007) The human disease network. *Proc. Natl. Acad. Sci. USA*, **104**, 8685–8690.

44 Lee, E., Chuang, H.Y., Kim, J.W., Ideker, T., and Lee, D. (2008) Inferring pathway activity toward precise disease classification. *PLoS Comput. Biol.*, **4**, e1000217.

45 Fischer, H.P. (2005) Towards quantitative biology: integration of biological information to elucidate disease pathways and to guide drug discovery. *Biotechnol. Annu. Rev.*, **11**, 1–68.

46 Kiechle, F.L., Zhang, X., and Holland-Staley, C.A. (2004) The -omics era and its impact. *Arch. Pathol. Lab Med.*, **128**, 1337–1345.

47 Zhang, K., Cui, S., Chang, S., Zhang, L., and Wang, J. (2010) i-GSEA4GWAS: a web server for identification of pathways/gene sets associated with traits by applying an improved gene set enrichment analysis to genome-wide association study. *Nucleic Acids Res.*, **38** (Suppl.), W90–W95.

48 Menashe, I., Maeder, D., Garcia-Closas, M., Figueroa, J., Bhattacharjee, S., Rotunno, M. *et al.* (2010) Pathway analysis of breast cancer genome-wide association study highlights three pathways and one canonical signaling cascade. *Cancer Res.*, **70**, 4453–4459.

49 Luo, L., Peng, G., Zhu, Y., Dong, H., Amos, C.I., and Xiong, M. (2010) Genome-wide gene and pathway analysis. *Eur. J. Hum. Genet.*, **18**, 1045–1053.

50 Zhong, M., Yang, X., Kaplan, L.M., Molony, C., and Schadt, E.E. (2010) Integrating pathway analysis and genetics of gene expression for genome-wide association studies. *Am. J. Hum. Genet.*, **86**, 581–591.

51 Wooten, E.C., Iyer, L.K., Montefusco, M.C., Hedgepeth, A.K., Payne, D.D., Kapur, N.K. *et al.* (2010) Application of gene network analysis techniques identifies AXIN1/PDIA2 and

endoglin haplotypes associated with bicuspid aortic valve. *PLoS ONE*, **5**, e8830.

52 Bureeva, S., Zvereva, S., Romanov, V., and Serebryiskaya, T. (2009) Manual annotation of protein interactions. *Methods Mol. Biol.*, **563**, 75–95.

53 Devidas, S. (2009) Curation of inhibitor-target data: process and impact on pathway analysis. *Methods Mol. Biol.*, **563**, 51–62.

54 Selkov, E., Basmanova, S., Gaasterland, T., Goryanin, I., Gretchkin, Y. et al. (1996) The metabolic pathway collection from EMP: the enzymes and metabolic pathways database. *Nucleic Acids Res.*, **24**, 26–28.

55 Selkov, E. Jr, Grechkin, Y., Mikhailova, N., and Selkov, E. (1998) MPW: the Metabolic Pathways Database. *Nucleic Acids Res.*, **26**, 43–45.

56 Schomburg, I., Chang, A., and Schomburg, D. (2002) BRENDA, enzyme data and metabolic information. *Nucleic Acids Res.*, **30**, 47–49.

57 Orth, J.D., Thiele, I., and Palsson, B.Ø. (2010) What is flux balance analysis? *Nat. Biotechnol.*, **28**, 245–248.

58 Gaasterland, T. and Selkov, E. (1995) Reconstruction of metabolic networks using incomplete information. *Proc. Int. Conf. Intell. Syst. Mol. Biol.*, **3**, 127–135.

59 Papin, J.A. and Palsson, B.O. (2004) The JAK–STAT signaling network in the human B-cell: an extreme signaling pathway analysis. *Biophys. J.*, **87**, 37–46.

60 Hyduke, D.R. and Palsson, B.O. (2010) Towards genome-scale signalling-network reconstructions. *Nat. Rev. Genet.*, **11**, 297–307.

61 Pitkänen, E., Rousu, J., and Ukkonen, E. (2010) Computational methods for metabolic reconstruction. *Curr. Opin. Biotechnol.*, **21**, 70–77.

62 Huang, Y., Tienda-Luna, I.M., and Wang, Y. (2009) A survey of statistical models for reverse engineering gene regulatory networks. *IEEE Signal Process. Mag.*, **26**, 76–97.

63 Hickman, G.J. and Hodgman, T.C. (2009) Inference of gene regulatory networks using boolean-network inference methods. *J. Bioinform. Comput. Biol.*, **7**, 1013–1022.

64 Frades, I. and Matthiesen, R. (2010) Overview on techniques in cluster analysis. *Methods Mol. Biol.*, **593**, 81–107.

65 Shi, W., Bugrim, A., Nikolsky, Y., Nikolskya, T., and Brennan, R.J. (2008) Characteristics of genomic signatures derived using univariate methods and mechanistically-anchored functional descriptors for predicting drug and xenobiotic-induced nephrotoxicity. *Toxicol. Mech. Methods*, **18**, 267–276.

66 Fielden, M.R., Eynon, B.P., Natsoulis, G., Jarnagin, K., Banas, D., and Kolaja, K.L. (2005) A gene expression signature that predicts the future onset of drug-induced renal tubular toxicity. *Toxicol. Pathol.*, **33**, 675–683.

67 Huang, J., Shi, W., Zhang, J., Chou, J.W., Paules, R., Gerrish, K. et al. (2010) Genomic indicators in the blood predict drug-induced liver injury. *Pharmacogenomics J.*, **10** (4), 267–277.

68 Lobenhofer, E.K., Auman, J.T., Blackshear, P.E., Boorman, G.A., Bushel, P.R., Cunningham, M.L. et al. (2008) Gene expression response in target organ and whole blood varies as a function of target organ injury phenotype. *Genome. Biol.*, **9**, R100.

69 Fan, X., Lobenhofer, E.K., Chen, M., Shi, W., Huang, J., Luo, J. et al. (2010) Consistency of predictive signature genes and classifiers generated using different microarray platforms. *Pharmacogenomics J.*, **10**, 247–257.

70 Popovici, V., Hatzis, C., Weijie, C., Shi, W., Gallas, B.G., Nikolsky, Y. et al. (2010) Effect of training-sample size and classification difficulty on the accuracy of genomic predictors. *Breast Cancer Res.*, **12** (1), R5.

71 Fan, C., Oh, D.S., Wessels, L., Weigelt, B., Nuyten, D.S., Nobel, A.B., et al. (2006) Concordance among gene-expression-based predictors for breast cancer. *N. Engl. J. Med.*, **355**, 560–569.

72 Shi, W., Tsyganova, M., Dosymbekov, D., Dezso, Z., Nikolskaya, T., Dudoladova, M. et al. (2010) Functional analysis of multiple genomic signatures demonstrates that classification algorithms choose phenotype-related genes. *Pharmacogenomics J.*, **10** (4), 310–323.

73 Ihaka, R. and Gentleman, R.R. (1996) A language for data analysis and graphics. *J. Comput. Graph. Stat.*, **5**, 299–314.

74 Cohen, J. (1960) A coefficient of agreement for nominal scales. *Educ. Psychol. Meas.*, **20**, 37–46.

75 Siegel, S. and Castellan, N.J. (1988) *Nonparametric Statistics for the Behavioral Sciences*, McGraw-Hill, New York.

76 Ito, T. Ando, H., Suzuki, T., Ogura, T., Hotta, K., Imamura, Y., Yamaguchi, Y. *et al.* (2010) Identification of a primary target of thalidomide teratogenicity. *Science*, **327**, 1345–1349.

77 Leary, R.J., Lin, J.C., Cummins, J., Boca, S., Wood, L.D., Parsons, D.W. *et al.* (2008) Integrated analysis of homozygous deletions, focal amplifications, and sequence alterations in breast and colorectal cancers. *Proc. Natl. Acad. Sci. USA*, **105**, 16224–16229.

78 Wright, J.M., Nikolsky, Y., Serebryiskaya, T., and Wetmore, D.R. (2009) MetaMiner (cystic fibrosis): a disease oriented bioinformatics analysis environment. *Methods Mol. Biol.*, **563**, 353–367.

Index

a
actin 121, 262, 273, 380
active vertices 305, 306, 309, 311
acute lymphoblastic leukemia (ALL) 118
acute myeloid leukemia (AML) 118
adaptive regression process 352, 353
adipocytokine signaling pathway 123
adjacency matrix 308
advanced module identification methods 235
Affymetrix 500K expression data 46
AKT signaling 437
algorithm description approaches
– methodology for identification and 139, 140
– – GP 140–142
– – H_∞ filter 142, 143
Alzheimer's disease 406
AML/ALL dataset 122
– bipartite graph of pathways and genes 125
– network of pathways 123
analysis of variance (ANOVA) 113, 118, 218
analyze network (AN) 434, 435
– algorithms 435
anti-apoptotic tumor necrosis factor/NF-κB/Bcl-2 pathway 429
apoptosis 330, 380, 388, 395, 420, 429, 431, 437
– P53-induced apoptosis effects 420
– regulator BCL2 163
applied semantic knowledgebase (ASK) 262
approximate cliques 234
Arc orientation 84
– assessing methods for 86, 87
– causal Markov condition 85
– faithfulness condition 86
– pseudo-code 86
array technology 359
ask–bid mechanism 106

b
atherosclerotic plaque 260
atrial natriuretic peptide 255

Barabási–Albert graph 312
bar graph 315, 316
basins of attraction 330
Bayesian information criterion (BIC) 93
Bayesian networks 228, 281
– algorithms 76
– applications 95–101
– – computational considerations 101
– – condition-dependent gene relationships, discovery of 96–99
– – elucidation of gene networks 95, 96
– – implementations 91, 92
– – limitations 91
– – MCMC mixture Bayesian network 99–101
– methodology 92–95
– – frequentist-based algorithms 93
– – Gaussian mixture model 93
– – MCMC method 95
– – network learning algorithms 93
– – posterior distribution of 95
B cell lymphoma, prognostic markers 360
Beagle algorithm 46
benchmarking 312–314
– evaluation 312–314
betweenness centrality *(BW)* 205
bicluster analysis, statistical-algorithmic method for 236
binary temporal patterns, types 352
bioinformatics tools 183
biological networks 8, 10, 13, 134, 159, 168, 270, 407, 417, 433
– community detection
– – centrality measures 300–302
– – complex systems study 302

Applied Statistics for Network Biology: Methods in Systems Biology, First Edition.
Edited by M. Dehmer, F. Emmert-Streib, A. Graber, and A. Salvador.
© 2011 Wiley-VCH Verlag GmbH & Co. KGaA. Published 2011 by Wiley-VCH Verlag GmbH & Co. KGaA.

– – conclusions 324
– – experiments 312–323
– – further improvements 323, 324
– – introduction 299, 300
– – overview 302–304
– – proposed algorithm 304–312
– function of 5
– local communities in 302
– majority 231
biological pathways 13, 40, 111, 117, 120, 162, 233, 251, 252, 379, 418–420, 432, 433
– advantages of 432–435
– applications of 420–425
– graphical representations of 418
– MPL-deduced pathway, applications of 428–432
– multiple-path learning (MPL) algorithm for 427, 428
– robustness of 425
biological phase transitions 383
biological phenomena 282, 357
biological processes 352
– classes 108
biological systems models 184
– paradigm for 106
biomarkers 237
Boolean analysis 349, 350
– gene regulation 349
Boolean functions 329, 331
Boolean implication network 351, 357, 359, 366, 370
– application of 349
– comparison 365
– extraction process 369
– generating 367
– for human, mouse, and fruit fly 359
– properties 364, 366, 367
– properties of 364, 365, 366
– scale-free 365–367
– vs. correlation-based network 371
BooleanNet method 351, 355–371
– algorithm 367–371
– – Boolean relationships, discovery 368–370
– – connected component analysis 371
– – conserved boolean relationships, discovery 371
– – data collection and preprocessing 368
– – FDR computation 371
– – human CD genes, correlation network 371
– Boolean implications
– – biological interpretations 360, 361
– – networks are not scale-free 365–367

– comparison against correlation network 364, 365
– computational efficiency 367
– conserved boolean implications 361–364
– gene expression microarray data, boolean implications in 357–360
Boolean networks (BNs) 69, 281, 329, 331, 332
– matrices 344, 345
– size 359
– truth table 337
Boolean operations
– various representations 350
breast cancer endpoints
– predictive classifiers for 431

c
Caenorhabditis elegans 107
cancer
– anomalous attractor 386, 387
– as anomalous attractor 386, 387
– – phenotypes 386
– cell shape transitions 392
Cancer Genome Atlas 237
candidate gene approach 405
canonical pathway 420, 426
– for DNA damage 419
– GeneGo ontology of 429
– maps 425
– vs. networks 419
carbon compounds, significant network 291
causal graph 283, 284
ccytoskeleton organization 386
CDK/cyclins
– nucleocytoplasmic transport of 426
cell–cell communication 293, 294
cell–cell signaling 107
cell cycle 362, 363
cell morphogenesis
cell shape 381, 382, 391
– control 382
– distortion 382
– DNA, role of 382
– genetic program 382
– modifications, precede tumor metabolome reversion 391–395
– – β-catenin distribution 393
– – cell shape transitions after exposure to 392
– – E-cadherin release 394
– – EMT-treated cells 392
– – Prigogine–Wiame theory 395
– – principal component analysis (PCA) 394
– molecular interactions and physical forces 382

– molecular mechanism 380
– morphologic and phenotypic traits 382
– patterns of change and 382
– shapes as system descriptors 388, 389
cell signaling pathways 14
cellular metabolic network 202
centrality measure 120, 205, 206, 300, 301
chemical master equation (CME) 106
chromatin immunoprecipitation (ChIP)
  networks 203, 251, 282, 372
– network for anaerobic respiration 292
– network screening 282
– – on iPSCs 292–294
– use 282
clique percolation method 235
closeness centrality concept 205, 301
CLR algorithm 274
cluster analysis of tumor transcriptomes 386
clustering techniques 229
c-Myc activation dataset (MYC) 117
c-Myc regulation cascade, 116
c-Myc transregulation cascade 108
coefficients of variation (CVs) 256
– analytical platforms 257
– properties 257
complementary DNA 217
complementary RNA 217
complexity in living systems, emergence
  of 385
complex measurement process 330
complex networks 105, 106
complex phenotypes, determined by molecular
  network 4
complex systems 105, 232
– study 302
computational cost analysis 338, 341
computed networks, characterization of 171
– specific protein–protein interactions
– – application of 175–177
– – evaluation of 171–175
computer-generated dynamic algorithm 303
computer-generated experiments 314–317
computer-generated networks, benchmarking
  on 302
conditional mutual information 69, 71
– characteristics 71
– inference based on 76
– – approximated conditional mutual
    information 78
– – constraint-based methods 77
– – variable selection algorithms 78–80
confusion matrix 313
conjugate gradient (CG) method 330, 334,
  335, 345, 346
– disadvantages 345
connectivity pattern 299, 300
constrained two-way model (CTWM) 50
– applying CTWM to HapMap data 52
– – characterizing putative eQTL identified
    by 52
– – justification of model assumptions 53
– modeling SNP–GE association in pooled data
    by 50, 51
continuous stochastic modeling 26
– deterministic models, with threshold
  values 29, 30
– external noise, stochastic models for 28, 29
– for λ phage network 26–28
– stochastic switching 30, 31
continuum approaches
– Langevin and Fokker–Planck
  formalism 106
contrapositive law 350
correlation coefficients 115, 365
– distribution 258
– histogram of 114–116
– p-value vs. r-value plot 259
correlation network analysis 253–259, 364
– correlation statistics, distributions 258, 259
– example 254
– introduction 251, 252
– multimodal experimental data 268
– selecting nodes and edges for
  networks 255–257
– semantic web approaches 252, 253
– systems biology data quandaries 252
– – better tools for stratifying key
    observations 274, 275
– – improved background corrections 274
– – knowledge-based applications, new
    classes 277, 278
– – public datasets, expanded sharing and
    integration 275, 276
– – specialized content integration, chemical
    structure and images 275
– – text and structured data, improved
    integration 276
correlation network analysis on time-series
  data 112–117
– dataset (MYC dataset) 112, 113
– – transition from unimodal to bimodal
    behavior for 117
– *D. melanogaster* dataset 113
– human aging dataset 113
– tamoxifen-induced dataset 114
covariance matrix 258
CREB pathway 423, 424
cross-omics expression 261

CTWM-GS method 53, 54
– applying CTWM-GS to HapMap data 56, 57
–– applying GS to population studies 57–60
–– Heatmap of hierarchical clustering 59
– estimators of BD and GS 55, 56
– solving normal equations in 54, 55
– testing BD and GS 56
cytochalasins 380
cytokine–cytokine receptor interaction 121
cytoskeleton organization 387

d

data integration methods 230
data interpretation 415
DAVID functional annotation tools 363
degree centrality 300
degrees of freedom 354, 355
dendrogram plot 315
diagonal linear discriminant analysis (DLDA) 425
differentially expressed genes (DEGs) 417
differentially expressed probe (DEP) 425
directed acyclic graph (DAG) 111, 283. see also causal graph
– biological process 111
– cellular component 111
– molecular function 111
directed network 5
discrete probabilistic approach 106
discrete stochastic modeling method 14, 20–22
– stochastic simulation algorithm (SSA) 15
–– accelerating τ-leap methods 16
–– binomial τ-leap method 18
–– direct method 15, 16
–– Langevin approach 19, 20
–– Poisson τ-leap method 17, 18
– Toggle switch
–– other models for 24–26
–– with SOS pathway 22, 23
disease-causing proteins 405
disease gene identification
– protein interaction network analysis-based methods 405
DNA 349
– damage 437
– information in 216
– microarray data (see microarray data)
DNA microarray data analysis
– gene coexpression networks 215
–– analysis 230–237
–– background 216–218
–– for cancer study 237

–– construction 218–224
–– integration with other data 224–230
–– introduction 215, 216
Drosophila Melanogaster 107, 113
– change point analysis 114
drug-metabolizing enzymes 436

e

eccentricity *(EC)* 205
eigenvector centrality 206, 300
empirical Bayes method 118
Ephrin signaling map 426
epigenetic landscape 3
ER-positive cancers 431
error rate 370
Escherichia coli
– networks under anaerobic conditions, network screening for 290–292
– SOS network, evaluation 289
estrogen receptor (ER) 420
Euclidean distance 220, 422
EU Gene database 371
European Bioinformatics Institute (EBI) 275
experimental techniques 230
expression quantitative trait loci (eQTL) studies 39
– current 40
–– effects of allele frequency and population-level expression 45
–– GE data, on HapMap cell lines 42
–– in multiple human populations 43, 44
–– in single human population 40, 43
– data structure in 39, 40
–– microarray eQTL mapping pipeline 41
– transcript *C8orf13* 61
expression vectors 220
external (extrinsic) noise 13
extracellular matrix (ECM) 379

f

false discovery rate (FDR) 223, 259, 290, 351, 355, 359
– computation 371
fatty acid metabolism 123
Fisher exact test 119
Fisher's Z-Transformation 222
formalization 71, 72
fractal analysis 389, 390
– and cancer 390–391
– dimension 389
– of living organisms 389–390
–– Bendixon–Poincaré theorem 389
–– self-organization 390
–– stable steady state 390

frequentist-based algorithms   93–95
F-statistic   353
functional annotation-based method   410
functional classifiers   416–418, 420–425
functional multimodal correlation   271

g

GABRB1 receptor   363
Gaussian distribution   115
Gaussian network (GN)   284
Gaussian white noise   144
gene activity profile (GAP)   331
Gene Atlas Project   275
gene coexpression networks (GCNs)   216, 230
– analysis   230–237
– for cancer study   237
– construction   218–224, 220
– integration with other data   224–230
gene content classifiers   416–418, 428
gene control policy   330
gene-dependent disease   387
gene expression (GE)   39, 379, 421, 424
– Boolean implications in microarray data   357
– BooleanNet zis   351
– calculation of distances between   421
– changes in   22, 97
– datasets from microarrays   212
– datasets of time-course gene expression arrays   112
– degree of temporal synchronization or polarization of   109
– enhanced by use of gene categorization and   126
– Gene Expression Omnibus (GEO) database   40, 225, 275, 368
– log-ratios of   423
– measurements   93, 96
– – complex networks   105, 106
– – conclusions   127, 128
– – correlation network analysis on time-series data, examples and methods   112–117
– – discussion   126, 127
– – gene expression data by priori biological knowledge, network reconstruction   110–112
– – gene interaction networks from gene expression measurements   107, 108
– – multiscale network reconstruction from   105
– – pathway network analysis, examples and methods   117–125
– – perturbation method   108, 109
– – time-series gene expression data, network reconstruction by correlation method   109, 110
– microarray data   356
– – Boolean implications in   358
– network reconstruction from   110
– noise effects of   135
– patterns through alteration in ECM   380
– profiles   161, 173, 226, 427, 428
– – associated with cell fates   384
– properties in   134
– statistical dependence between   70
– stochastic models for microarray data   33–34
gene expression datasets   421
– Boolean analysis   349, 350
– – BooleanNet   355–367
– – BooleanNet algorithm   367–371
– – conclusions   371, 372
– – introduction   349
– – main organization   351
– – StepMiner algorithm   352–355
Gene Expression Omnibus (GEO)   40, 225, 275, 368
gene function   3
gene–gene interactions   116
GeneGo processes   423
– pathway maps   426
gene interaction networks, from gene expression measurements   107, 108
gene modules   231, 232
– annotation   236
– identification   233
– – expression vector clustering methods   233
– – network-based methods   233–236
gene network analysis   3
gene ontology (GO)   111, 236, 415
– annotations   351
– database   111
generalized extreme value (GEV) distribution   284, 285
generalized mass action (GMA) formulation   185, 186
– format   185
gene regulations   372
– modeling and simulation of   329
– noise   147
– post-transcriptional   162
gene regulatory networks inference   133
generic S-system format   186
gene set enrichment analysis (GSEA)   111, 218
gene's expression vector   219

gene signatures 416, 417, 425, 436
– identification, statistical methods of 417
genetic programming (GP) 134
GeneTrace 113
gene transcription 216, 217
– regulation, roles in 232
gene transcripts, PPARα/RXR activation in 263
gene vector 223
genome–proteome–metabolome network 384
genome, regulatory code 217
genome-wide association studies (GWAS) 415
– for network analysis 8
genomics, advances in 299
gonadotropin-releasing hormone (GnRH) signaling pathway 121, 126
goodness-of-fit model 284
G-protein-coupled synaptic signaling 320
graph consistency probability (GCP) 282
graphical Gaussian models (GGMs) 110, 281
graph-theoretic methods 218
green fluorescent protein (GFP) 289
"guilt by association" principle 226, 236

h

hematopoietic stem cell marker 361
heterogeneity 252
heuristic algorithms 330
– algorithm 15.1 339
– algorithm 15.2 339, 340
high-density lipoprotein (HDL) levels 254
highly connected subgraphs (HCS) algorithm 235
high probabilistic coefficient 308
high-risk plaque (HRP) 260
high-throughput gene expression analysis 110
high-throughput techniques 109, 224, 238
– development 215, 227
HITS algorithm 408
homeostasis 380
hub genes 125, 126
human cancers
– breast, colon, pancreatic cancer, and glioblastoma 437
– ER-negative 430, 431
human ChIP network 295
human diseases 405–409
– characteristic for 433
– molecular basis of 405, 406

Human Metabolome Database (HMDB) Compound 263

i

Illumina's commercial whole-genome expression array 46
independent group model (IGM) method 47
– to HapMap data 48, 49
–– characterizing putative eQTL identified by IGM 49, 50
– integrating hypotheses to identify common eQTL 48
– modeling SNP–GE association in a single population 47, 48
induced pluripotent stem cells (iPSCs) research 282
– activate networks in 294
– challenges for 293
– surface glycan composition in 294
inference network approach 282
inflammatory response subnetwork 273
integral informatics approach 263–274
– annotating biological networks with public knowledge resources 269–271
– biological networks in semantic web context 265–269
– histidine pathway focus 272–274
integrins 379, 383
internal (intrinsic) noise 13
intracellular signaling pathways 380
inverse problem 110, 187, 330, 333, 334, 338
ischemia-reperfusion injury (IRI) 426
iterative agglomeration technique 324

j

Jak–STAT signaling pathway 121

k

Kalman filter and H_∞ filter, comparison between 148–150
k-clique 234
knowledge annotation 265
– for networks 259–274
knowledge annotation for networks 259–274
– annotation with public sources and ontologies 261, 262
– HRP and paired-plaque study 260, 261
– results and benefits of the approach 262–274
knowledge-based functional groups 417
knowledge integration
– correlation network analysis 253–259

-- correlation statistics, distributions 258, 259
-- selecting nodes and edges for networks 255–257
- future developments 274–278
-- better tools for stratifying key observations 274, 275
-- improved background corrections 274
-- knowledge-based applications, new classes 277, 278
-- public datasets, expanded sharing and integration 275, 276
-- specialized content integration, chemical structure and images 275
-- text and structured data, improved integration 276
- introduction 251, 252
- semantic web approaches 252, 253
- systems biology data quandaries 252
knowledge network 271
K-StepMarkov approach 409
Kyoto Encyclopedia of Genes and Genomes (KEGG) 8, 111, 127, 236, 263, 415
- database 96
- KEGG database via R software (KEGGSOAP package) 120

l

learning algorithm 428
linear discriminant analysis (LDA) 430
linear Markov models 107
- and correlation approaches for network reconstruction 107
linear programming optimization 10
Linked Open Drug Discovery (LODD) data 275
Lin Log energy model 8
liver toxicity biomarker study (LTBS) 255
local community structure concept 303
local ontology 269
- example 266
local optimization 308
- community ranking 310, 311
- vertex ranking 308–310
logistic regression (LREG) 430
log-likelihood, estimated generalized extreme distribution 287
log-likelihood score (LLS) 229
log–log plot 366

m

machine learning algorithms 208–210
- applications of 210

-- training with data from one organism to predict essential genes for 211
-- validating an experimental knock-out screen 210
mammalian target of rapamycin (mTOR) signaling pathway 123
MAP kinases signaling pathway 126
Markov chain process 333
- randomwalk 306
- sparse transition probability matrix 345
- transition probability matrix 333
Markov clustering algorithm 8
Massachusetts Institute of Technology (MIT) 304
MATLAB Bioinformatics tool 314
matrix approximation method 330
matrix metalloproteinase-2 255
matrix-vector multiplication 335
maximal cliques 234
mean field theory 105, 106
- drawback 105
means of the gene set database 293
mestranol 423
mestranol–phenobarbital distances 422
meta-analysis 224, 226
metabolic networks 202
- representations of 203
- *in silico* evolution 299
metabolic pathways 419
metabolite–metabolite correlations 258
MetaCore/MetaDrug data analysis suite 415, 418
N-methyl-D-aspartate receptor complex/ MAGUK associated signaling complex (NRC/MASC) receptor complexes 304, 321
Michaelis–Menten rate 184
microarray analysis 221, 237, 289, 425
microarray data 33, 70, 80, 145, 217, 220, 225, 229, 230, 236, 352, 372, 430
- integration, applications 229
- matrix 232
microarray experiment 217
- goal 232
microarray expression 417
minimum redundancy networks (MRNET) 83, 84
mining overlapping dense subgraphs (MODES) 235
mitogen-activated protein (MAP) kinase signaling pathway 121
mobile users, macro mobility 323
model design 184
- BST models 186, 187

– – advantages   186
– – disadvantages   186, 187
– – generic S-system format   186
– – GMA formulation   186
– challenges   185
– selection   184
molecular complex detection (MCODE) algorithm   8, 235
morphogenesis   379
morphological plasticity   381
mRNA profiling   10
multimodal subnetwork   268
multiple expression datasets, integration   225–227
– integrating data across species   226, 227
– integrating data within species   226
multiple higher-confidence datasets   224
multiple-path learning (MPL) algorithm   427, 428
– schema for   427
mutual exchange criteria   311, 312
– dynamic optimization   311, 312
mutual information base network   281
– in network inference   69, 70
– – characteristics of   70
– pairwise, inference based on   80
– – ARACNE networks, algorithm for   82
– – Chow–Liu tree   81, 82
– – context likelihood of relatedness (CLR)   81
– – minimum redundancy networks (MRNET)   83, 84
– – relevance network (RELNET)   80, 81
MYC dataset   121
– bipartite graph of pathways and genes   124
– network of pathways   122
MYC-induced tumorigenesis   363

n

National Cancer Institute's Meta Thesaurus   275
National Center for Biomedical Ontology (NCBO)   264
National Center for Biotechnology Information (NCBI)   263
National Center for Biotechnology Information's Gene Expression Omnibus (GEO)   290–292, 368
natural language processing (NLP)   276
neoplastic transformation   380
network aggregation methods   227
network algorithms   434
network analysis   3
– biological functions, understanding of   8

– extracting pathway-linked regulators and effectors   10
– genes, identification of   5
– – betweenness   6
– – degrees   5
– – hierarchical structure   7
– – network motifs   6, 7
– inferring functional relationships and functional genes through   8, 9
– methods of   416
network-based methods   128, 218
– data integration approaches, overview   225
NetworkBlast algorithm   235
network-building algorithm   423
network cliques   234
network construction process, overview   216
network descriptors   203
– topological features for bipartite graph of metabolic networks   206
– – chokepoints   207
– – deviations   207, 208
– – global networks, damage in   208
– – load scores   207
– – stoichiometric properties   206
– topological features, for undirected graphs   204
– – centrality measures   205, 206
– – clustering coefficient   205
– – connectivity   205
network hubs   230–232
network learning algorithms   93–95
network mining techniques, overview   231
network motifs   230, 232
network reconstruction from gene expression data
– by *priori* biological knowledge   110–112
network screening   282, 287–289, 287–294
network theory   108, 302
neurotransmitter receptor   319–321
– communities generated for   321
– network representation of   318
128-node network
– performance analysis of algorithm   313
512-node rewired network
– performance analysis of algorithm   314
noise in gene expression   13, 134–136
– modeling of gene regulatory networks with noise   136
– – Bayesian networks model   136, 137
– – Boolean networks model   136
– – linear additive regulation model   137
– – neural networks model   137, 138
– proposed nonlinear ODE model   138, 139

nonparametric method 287
normalized mutual information (NMI) 313
null hypothesis 119
null log-likelihood distribution
– in randomized graphs 288

o
omics data
– statistical analysis of 415
omics profiles
– analyzing omics data on network level 157–159
– selecting relevant features from 156, 157
Oncotype DX 416
one gene–one phenotype 3
online mendelian inheritance in man (OMIM) database 8, 263, 406
open source bioinformatics network visualization platform 260
ordinary differential equations (ODEs) 14, 134
out-of-bag (OOB) method 426

p
PageRank 409
paradigm shift 415
parameter estimation 187–190
– typical problems 190
– – acceptability, of fit 192–195
– – better fit, need for 195–197
– – data fit, unacceptable 190, 191
– – extension of concept of generic approximations 197
– – MCMC strategy 195
– – models, difficult to compare 191, 192
– – overall SSE 197
– – parameterization 196
partial correlation coefficients 110
particular regulatory program 226
pathway-centered models 429
pathway classifiers 417
pathway network analysis 117, 118–120
– for AML/ALL dataset 122, 123, 125
– gene selection and pathway grouping 118
– KEGG database 118, 120
– for MYC dataset 121, 122, 124
– pathway significance 118–120
– 2×2 contingency table 119
pathway ontology 270
patient-centric predictive screening 278
Pearson correlation 223, 253
Pearson distance 422
Pearson's correlation coefficient 219, 220, 221

permutation-based thresholds 223
permutation testing 110
permuted data matrix 223
peroxisome proliferator-activated receptor-α (PPARα) 263
perturbation method, analysis of functional genomics data 108, 109
phase transitions 109, 382–386
– gas/liquid/solid 389
phenobarbital genes 425
phenobarbital profiles 424
phylogenetic tree 111
plaque biology study 260, 261, 264, 265
– key finding 262
POINTILLIST method 229
power graph analysis 275
predictor function. see Boolean functions
preoperative chemotherapy (pCR) 430
probabilistic Boolean networks (PBNs) 332, 333
– dynamics (transitions) 330
– generation, inverse problems in 329
– – BNs 331, 332
– – conclusions 345, 346
– – construction from stationary distribution 333–338
– – construction from transition probability matrix 338–345
– – introduction 329, 330
probabilistic methods 224, 228, 229
probability distribution, entropy 342
proliferator-activated receptor (PPAR) signaling pathway 123
protein arginine methyltransferases (PRMT1) 125
protein–disease relationships 406
protein interaction datasets 405
protein interaction networks 159, 405–409, 410, 436
– analysis-based methods 405
– disease-causing proteins 406
– network categories 159
– – metabolic networks 159, 160
– – paralog networks 160
– – physical interaction networks 160, 161
– prioritization methods 407–409
– for prioritizing disease genes 406
– protein annotation, parameters for 161
– – data preparation 163–166
– – deriving models 166–169
– – gene annotation 161, 162
– – gene expression profiles 161
– – microRNAs (miRNAs) 162
– – pathways 162, 163

-- subcellular location   161
-- transcription factors   162
-- validation procedures   169, 170
- ranking algorithms for   406, 407
- ToppGene suite of applications   409, 410
- ToppNet   405–410
protein network analysis-based method   410
protein-protein interaction (PPIs)   4, 107, 203, 251, 299, 372, 409
- evaluation of   409
proteins
- PPARα/RXR activation in   265
- RNA interactions   418
- three-dimensional structure   300
pseudo-code
- to compute possible exchange of vertices   310
- to find local community structure   309
PTEN pathway   437
public-domain gene expression data   356
"pull-down" experiments   251
p-value   354

q

quadratic discriminant analysis (QDA)   430
quantitative functional analysis   417
quantitative mathematical models   14
quantitative pathway   416
quantitative trait loci (QTL) analysis   230
- mapping   47

r

random graphs   286
random walk algorithm   305–307, 306
- communities generation   307
rank of community   311
rank-transformed correlation   254
real networks, application to   317
- neurotransmitter receptor complexes   319–321
- zachary karate club   318, 319
reference sequence (RefSeq) transcripts   46
regulatory impact factor (RIF) algorithm   9
regulatory networks   202
resilience of biological systems   384
resource description framework (RDF)   253
- triplestore databases
-- Franz's Allegrograph   253
-- Garlik's 4Store   253
-- Open Link's Virtuoso   253
-- Oracle's Spatial 11G   253
response net   10

retinoid X receptor-α (RXRα) signaling   263
RMA algorithm   368
RNA sequencing technology   10
ruptured/stable plaque comparison
- pathways/processes, enrichment of   267

s

Saccharomyces cerevisiae   226
selection-with-replacement (SWR)   426
self-organizing maps (SOMs)   110
self organizing structures   384
semantic data integration methods   273
semantic network graph, example   276
semantic web approaches   252, 253
semantic web technology   264–269
Sentrix Human-6 Expression BeadChip   46
shell-spreading method   303
signaling pathway   112
signaling subnetwork   124
signal transduction networks   69, 379
simulation evaluation   144
- microarray data   145, 146
-- experimental learning curves of E-cell model, by H_∞ filtering   146
-- interactions among CYB2, HAP1, and CYC7   147
-- model obtained by proposed algorithm   146
-- time series for E-cell simulation, by Kalman and H_∞ filtering   145
-- trajectories for CYB2, HAP1, and CYC7   147
- synthetic data   144, 145
-- GP plus Kalman filter, performance   144, 145
-- Runge–Kutta method   144
single-cell internal signaling   128
single-gene analysis   117
single nucleotide polymorphisms (SNPs)   39, 107, 156, 416
- array analyses   437
- SNP–GE associations   46
Smyth's methods   407, 408
Smyth's pagerank algorithm   408
social network analysis algorithms   406
SOS DNA repair system   289
SPARQL queries   271, 277
- arrays   271
- graphical query generation   272
- represented network graphs   271
Spearman's method   255
Spearman's rank correlation   219, 220, 221
- coefficient   221, 222

Stanford Microarray Database   225
Statistical methods for evaluation
– GEV distribution   284, 285
– network screening   287–289
– nonparametric statistic   285, 286
– simulation study   286, 287
StepMiner algorithm   368, 369
– best step function, selection   353
– binary temporal patterns, types of   352
– degrees of freedom   353–355
– FDR   355
– fitting one/two-step functions   352, 353
StepMiner method   352
– false discovery rate (FDR)   351, 355
– time-course microarray data, analysis of   349
StepMiner tool   372
stochastic differential equations (SDEs)   14
stochastic models
– for internal and external noise   31–33
– – chemical Langevin equations   32
– – for microarray gene expression data   33, 34
– – Rosetta error model   33, 34
– – simulations and statistical properties   31
– – stationary fluctuations   32
– – Stratonovich interpretation   33
– for range of biological systems   13
– stochastic Boolean models   14
stochastic Petri nets   14
structural equation model (SEM)   295
sum of squares (SSR)   353, 354
system descriptors   388, 389
systems biology   107
– data quandaries   252
– framework   109
– project   252
systems biology approaches   262, 277

t

tamoxifen   423, 424
– induced gene expression datasets
– – principal network parameters   114
– specific c-Myc–estrogen receptor fusion protein   112
TCA cycle   419
T-helper cell activation   431
thresholding method   222
time-course gene expression arrays   112
time-course microarray data   349
time-dependent interaction network   230
time-series gene expression data
– network reconstruction by correlation method from   109
– – graphical Gaussian models (GGMs)   110

tissue morphogenesis   379
toll-like receptor (TLR) signaling pathways   429
ToppGene Suite   409. *see also* functional annotation-based method
– homepage   410
– web site   410
ToppNet. *see* protein network analysis-based method
toy model, for simulation study   286
training error   428
transcriptional networks
– regulatory networks   69, 232
– unraveling transcriptional regulations from   9, 10
transcription factors   217, 220, 226
transition probability matrix   330, 332, 333, 340, 343
truth table   337, 343
tumor protein p53 (*TP53*)   125, 126
tyrosine protein kinase   320

u

ultrabithorax (UBX)   361
undirected network inference, performance measures in   72
– causal subset selection   74
– – causality implies stochastic dependency   75
– – causal sufficiency condition   75
– – causal transitivity   76
– – common-cause effect   75
– – dependency implying causality   75
– confusion matrix   72
– *F*-scores   74
– precision–recall (PR) curves   73
union-find algorithm   371
universal cancer genes   435–437
universal resource identifier (URI)   253
unsaturated fatty acid biosynthesis   426

v

vascular endothelial growth factor (VEGF) signaling pathway   121, 126
Venn diagram   350
vertices   105
– adjacency matrix   301
– mutual exchange   311
– neighborhood   305
vitamin B6 metabolism   426

w

web network analysis algorithms   406
web ontology language (OWL)   253

Welch's *t*-test   52
Wet-lab network validation   420
White's methods   407, 408
whole genome methylation   107
Wilcoxon signed-rank test   268
wireless access points   322
wireless mobile users study
   321–323
WNT signaling pathway   419, 437
World Wide Web Consortium   263

y
Yamanaka factor networks   289, 293
yeast cell cycle   96

z
Zachary Karate Club   303, 304, 307,
   317–320, 324
– actual breakdown   319
– convergence graph   320
– local factions in   319